ALTERNATIVE FUEL

Edited by **Maximino Manzanera**

Alternative Fuel
Edited by Maximino Manzanera

Published by InTech
Janeza Trdine 9, 51000 Rijeka, Croatia

Publishing Process Manager Mia Devic

Technical Editor Teodora Smiljanic

Cover Designer InTech Design Team

Additional hard copies can be obtained from orders@intechopen.com

Alternative Fuel, Edited by Maximino Manzanera
p. cm.
ISBN 978-953-307-372-9

For EU product safety concerns:
IN TECH d.o.o., Prolaz Marije Krucifikse Kozulić 3, 51000 Rijeka, Croatia,
info@intechopen.com or visit our website at intechopen.com.

Meet the editor

Dr. Maximino Manzanera was born in Barcelona, Spain, in 1972. He received a degree in Biological Sciences from the University of Granada in 1994. After his degree he began his doctoral studies at the Agricultural Sciences Institute (Estación Experimental del Zaidín) as part of the Spanish High Research Council (Consejo Superior de Investigaciones Científicas, CSIC) where he received his Ph.D. in 2000. He then joined the University of Cambridge as Research Associate working in different centres of the University, including the Institute of Biotechnology and Departments of Genetics, Chemistry and CIMR. In mid-2006 he was awarded with the Ramón y Cajal Fellowship with which he joined the Institute for Water Research at the University of Granada as a Lead Researcher. Maximino is participating actively in three main lines of research: Anhydrobiotic Engineering, Rhizoremediation and Biofuels of microbial origin.

Contents

Preface XI

Part 1 Biodiesel 1

Chapter 1 **Biodiesel Production by Using Heterogeneous Catalysts 3**
Rubi Romero, Sandra Luz Martínez and Reyna Natividad

Chapter 2 **An Overview of Enzyme-Catalyzed Reactions and Alternative Feedstock for Biodiesel Production 21**
Maria Manuela Camino Feltes, Débora de Oliveira, Jorge Luiz Ninow and José Vladimir de Oliveira

Chapter 3 **Biodiesel Fuel Production by Enzymatic Transesterification of Oils: Recent Trends, Challenges and Future Perspectives 47**
Nevena Luković, Zorica Knežević-Jugović and Dejan Bezbradica

Chapter 4 **Engine Test of Bio-Diesel Manufactured from Waste Cooking Oil and Reward Preferential Benefit Analyses for Its Promotion 73**
Jai-Houng Leu

Chapter 5 **Waste Animal Fats with High FFA as a Renewable Energy Source for Biodiesel Production - Concept, Experimental Production and Impact Evaluation on Air Quality 93**
Francisc Popescu and Ioana Ionel

Part 2 Bioethanol 109

Chapter 6 **Delignification Process of Agro-Industrial Wastes an Alternative to Obtain Fermentable Carbohydrates for Producing Fuel 111**
Oscar Sánchez, Roció Sierra and Carlos J. Alméciga-Díaz

Chapter 7 **Performance and Emission Characteristics of Spark Ignition Engine Fuelled with Ethanol and Methanol Gasoline Blended Fuels 155**
Ioannis Gravalos, Dimitrios Moshou, Theodoros Gialamas, Panagiotis Xyradakis, Dimitrios Kateris and Zisis Tsiropoulos

Part 3 Biomass 175

Chapter 8 **Use of Charcoal and Partially Pirolysed Biomaterial in Fly Ash to Produce Briquettes: Sugarcane Bagasse 177**
Silvio Rainho Teixeira, Agda Eunice de Souza, Angel Fidel Vilche Peña, Regiane Godoy de Lima and Álvaro Gil Miguel

Chapter 9 **Biomass Waste as a Renewable Source of Biogas Production - Experiments 201**
Adrian Eugen Cioablă and Ioana Ionel

Chapter 10 **Influential Parameters on Biomethane Generation in Anaerobic Wastewater Treatment Plants 227**
Meisam Tabatabaei, Alawi Sulaiman, Ali M. Nikbakht, Norjan Yusof and Ghasem Najafpour

Chapter 11 **Alternative Fuels in Cement Manufacturing 263**
Moses P.M. Chinyama

Chapter 12 **Waste Biomass as Alternative Bio-Fuel - Co-Firing versus Direct Combustion 285**
Gavrilă Trif-Tordai and Ioana Ionel

Part 4 Hydrogen 307

Chapter 13 **Nano-Porous Silicon-Based Mini Hydrogen Fuel Cells 309**
Tayyar Dzhafarov and Sureyya Aydin Yuksel

Chapter 14 **Hydrogen - The Ecologically Ideal Energy Vector 335**
Mircea Cârdu and Malvina Baica

Preface

Renewable energy sources such as biodiesel, bioethanol, biomethane, biomass from wastes or hydrogen are subjects of great interest in the current energy scene. These fuels contribute to the reduction of prices and dependence on fossil fuels. In addition, energy sources such as these could partially replace the use of what is considered as the major factor responsible for global warming and the main source of local environmental pollution. For these reasons they are known as "alternative fuels". There is an urgent need to find and optimise the use of alternative fuels to provide a net energy gain, to be economically competitive and to be producible in large quantities without compromising food resources.

This book presents an overview of biodiesel, bioethanol, biomethane, biomass and hydrogen for the implementation of systems of production for these alternative fuels. The first section focuses on the optimization of biodiesel production addressing some drawbacks derived from the chemically catalyzed reactions. Beginning with Chapter 1 the use of heterogeneous catalysts for biodiesel production are reviewed as well as the effects of the different parameters on the reaction yield, and the common feedstock used for biodiesel production using heterogenous catalysts. There are another two chapters on the description of enzymes as alternative catalysts. In the first one (Chapter 2) the authors guide the reader through the advances in biodiesel-making, from a global perspective to the specific issues of a real case, for example the particularities found in Brazil, paying special attention to immobilization of lipase enzymes. The second of these two chapters (Chapter 3) is dedicated to reviewing the biotechnological uses of lipases for biodiesel production with special interest in achieving enzymatic transesterification that would be competitive on an industrial scale. Solvent engineering, lipases immobilization, selection of acyl acceptor, and selection of the reactor system are reviewed in this chapter in order to offer solutions to the current problems and to look for new perspectives on the industrial application of enzymatic biodiesel synthesis. The section dedicated to biodiesel is complemented with an applied case where biodiesel manufactured from waste cooking oil is tested on a real engine, testing the brake horsepower, the emissions of pollutants, and includes an economical analysis for production cost on a real case such as Taiwan (Chapter 4). The last chapter (Chapter 5) is dedicated to the production of biodiesel from waste animal fats and its effect on air quality.

The second section details the use of bioethanol as an alternative fuel including a review (Chapter 6) of delignification by chemical and biological pretreatments and the use of lignocellulosic wastes with references to the compositional changes of these substrates necessary for their use in the biofuels production industry. This second section also includes a chapter (Chapter 7) on the production of bioalcohol for mixtures of alcohol and gasoline with a study on the performance of these mixtures in comparison with ethanol and methanol and their effect on the brake power and emissions, using an engine without major modifications.

The third section focuses on residues as feedstock for alternative fuel production, with biomass as the main subject of the study. This section starts with the use of sugarcane bagasse waste from the bioethanol industry (Chapter 8) and focuses on the use of charcoal and partially pirolysed biomaterial for the production of energy as an important element for a strategy to achieve renewable energy goals and to reduce waste disposal and environmental pollution in the context of Brazil as a leading country in the bioethanol production. Chapter 9 deals with biomass waste as a renewable source for biogas production and Chapter 10 presents a very interesting study on the parameters that influence biomethane generation in anaerobic wastewater treatment plants with special remarks on the main factors that affect the generation of biomethane, the efficient treatment of different types of wastewaters and possible applications of computer modelling techniques to provide support for an optimal operation of the plant and the correct decision-making. Also in this section (Chapter 11) the author highlights the potential for alternative fuels to be used in the cement industry to avoid negative effects on the final product.

The final section focuses on hydrogen as an alternative fuel, with two different chapters. Chapter 13 reviews the main characteristics of hydrogen as an energy vector focusing on its production, storage and transportation. The authors evaluate the main advantages and drawbacks of hydrogen from an ecological perspective. The last chapter (Chapter 14) reviews the use of proton exchange membrane (PEM)-type fuel cells for the production of hydrogen gas as an alternative energy source, based on the production and characterization of nano-porous silicon-based direct-fuelled hydrogen fuel cells.

Dr. Maximino Manzanera
Institute for Water Research
University of Granada
CL. Ramon y Cajal No. 4
Granada 18071 SPAIN

Part 1

Biodiesel

1

Biodiesel Production by Using Heterogeneous Catalysts

Rubi Romero, Sandra Luz Martínez and Reyna Natividad
Centro Conjunto de Investigación en Química Sustentable UAEM-UNAM
Carretera Toluca-Atlacomulco Km 14.5
Toluca Estado de México
México

1. Introduction

In recent years, biodiesel has gained international attention as a source of alternative fuel due to characteristics like high degradability, no toxicity, low emission of carbon monoxide, particulate matter and unburned hydrocarbons (Al Zuhair, 2007; Vicente et al., 1998).

Biodiesel is a mixture of alkyl esters and it can be used in conventional compression ignitions engines, which need almost no modification. As well, biodiesel can be used as heating oil and as fuel (Mushrush et al., 2001; Wardle, 2003). So far, this alternative fuel has been successfully produced by transesterification of vegetable oils and animal fats using homogeneous basic catalysts (mainly sodium or potassium hydroxide dissolved in methanol). Traditional homogeneous catalysts (basic or acid) possess advantages including high activity (complete conversion within 1 h) and mild reaction conditions (from 40 to 65 °C and atmospheric pressure). However, the use of homogeneous catalysts leads to soap production. Besides, in the homogeneous process the catalyst is consumed thus reducing the catalytic efficiency. This causes an increase in viscosity and the formation of gels. In addition, the method for the removal of the catalyst after reaction is technically difficult and a large amount of wastewater is produced in order to separate and clean the products, which increases the overall cost of the process. Thus, the total cost of the biodiesel production based on homogeneous catalysis, is not yet sufficiently competitive as compared to the cost of diesel production from petroleum. An alternative is the development of heterogeneous catalysts that could eliminate the additional running costs associated with the aforementioned stages of separation and purification. In addition, the use of heterogeneous catalysts does not produce soap through free fatty acid neutralization and triglyceride saponification. Therefore, development of efficient heterogeneous catalysts is important since opens up the possibility of another pathway for biodiesel production. The efficiency of the heterogeneous process depends, however, on several variables such as type of oil, molar ratio alcohol to oil, temperature and catalyst type. Thus, the objective of this chapter is to present a review of the effect of the aforesaid variables on important characteristics of biodiesel such as methyl esters content. Some characterization techniques for both, biodiesel and heterogeneous catalysts will also be addressed.

2. Transesterification reaction

Nowadays, there are four known methods to reduce the high viscosity of vegetable oils to enable their use in conventional compression ignitions engines: blending with diesel, pyrolysis, emulsification and transesterification. The pyrolysis and the emulsification, however, produce heavy carbon deposits, incomplete combustion, an increase of lubricating oil viscosity and undesirable side products such as alkanes, alkenes, alkadienes, aromatic compounds and carboxylic acids. Regarding the direct use of vegetables oils as fuel for combustion engines, this requires the engines to be modified (Demirbas, 2005; Ma & Hanna, 1999). Also, the direct use of vegetables oils is not feasible due to their high viscosity and low volatility which affect the atomization and spray pattern of fuel, leading to incomplete combustion, severe carbon deposits, injector choking and piston ring sticking (Ryan et al., 1984; Xie & Li, 2006). Thus, the most common way to produce biodiesel is by transesterification of triglycerides of refined/edible types of oils using alcohol, in presence of an acid or a basic catalyst (López et al., 2005). The alcohol used for transesterification is usually methanol. Producing biodiesel is a bulk process; the general scheme of the trasesterification reaction is presented in Figure 1, where R is a mixture of various fatty acid chains.

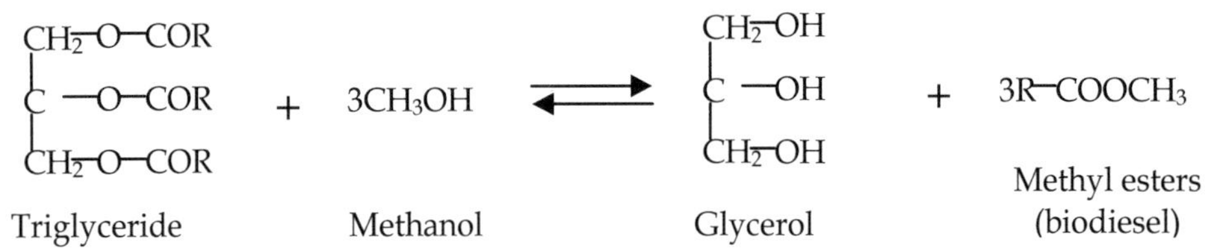

Fig. 1. Reaction for oil transesterification.

In principle, transesterification is a reversible reaction, although in the production of biodiesel, the back reaction does not occur or is negligible because the glycerol formed is not miscible with the product, leading to a two-phase system. Nevertheless, an excess of alcohol is usually employed to force reaction towards the right side. The stoichiometry of reaction is a 3:1 molar ratio of alcohol to oil, to produce 3 mol of biodiesel and 1 mol of glycerol. Though, in practice it is usually increased from 6:1 to 1000:1 to favor the formation of products and increase its performance. In this context, the amount of alcohol used can be reduced by conducting the transesterification in steps: part of the alcohol and catalyst are added at the start of each step, and the glycerol is removed at the end of each step (Encinar et al., 2005; Gerpen, 2005; Harvey et al., 2003; Verziu et al., 2008). Complete conversion of the triglyceride involves three consecutive reactions with monoglyceride and diglyceride intermediates which are reversible reactions as shown in Figure 2 (Harvey et al., 2003; Suppes et al., 2004).

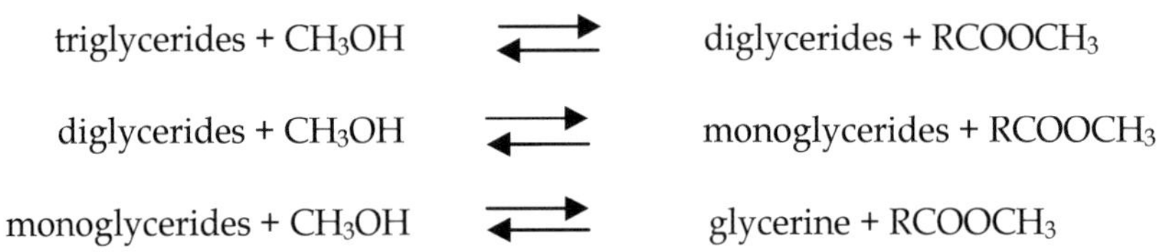

Fig. 2. Reaction scheme for oil transesterification

While transesterification is an equilibrium reaction between esters and alcohols, the reaction may be under kinetic control before thermodynamic equilibrium is achieved, and this would favor the formation of monoalkyl esters (Meneghetti et al., 2006).

The transesterification reaction produces two liquid phases: alkyl esters and crude glycerol (the heavier liquid). In a typical stirred tank reactor, glycerol is collected at the bottom after some time of settling. Phase separation can be observed within short time (approximate 10 minutes) and can be complete within 2 to 20 h, when the reaction is carried out at laboratory scale (Demirbas, 2005).

In the case of alcohols, these can be primary or secondary monohydric aliphatic alcohols having from 1 to 8 carbon atoms. Among the alcohols that have been used to produce biodiesel, either homogeneously or heterogeneously, are methanol, ethanol, propanol, isopropanol, butanol, pentanol and amyl alcohol (Demirbas, 2005; Fukuda et al., 2001; Meneghetti et al., 2006). The use of methanol is advantageous as it can quickly react with triglycerides (polar and shortest chain alcohol) and is a relatively inexpensive alcohol, while the same reaction using ethanol has as drawback that the produced ethyl esters are less stable and a carbon residue is observed after reaction. The use of ethanol as solvent, however, is becoming more popular since this alcohol is a renewable resource and does not raise the same toxicity concerns than methanol (Demirbas, 2005; Geise, 2002; Meneghetti et al., 2006). Similar yields of biodiesel can be obtained using either methanol or ethanol. With the former, however, mild reaction temperature (approximately 60 °C) can be employed, whereas for the latter and other alcohols (butanol) at similar molar ratios higher temperatures (75 and 114 °C, respectively) are required for optimum conversion (Geise, 2002). Also, the reaction time is shorter in the methanolysis because of the physical and chemical properties of methanol: polar character and the short chain alcohol. For instance, Meneghetti et al. (2006) reported that the production of biodiesel from castor oil was faster with methanol compared with ethanol. In such a study, maximum yields of esters were obtained after 1h of reaction time with methanol or 5 h with ethanol.

Biodiesel is usually prepared in the presence of homogeneous base or acid catalysts. With homogenous base catalysts (sodium and potassium hydroxides, carbonates, sodium and potassium alkoxides, principally) the reaction is faster than with acid catalysts (sulfuric acid, phosphoric acid, hydrochloric and sulfonic acid principally) (Fukuda et al., 2001; Ma & Hanna, 1999). However, the main disadvantage of the aforementioned homogeneous catalysts is the undesirable production of both, soap and glycerol. This fact increases the production costs. On the other hand, heterogeneous catalysts could improve the synthesis methods by eliminating the neutralization salts in the glycerol and therefore the number of separation steps can be reduced (MacLeod et al., 2008). Also, heterogeneous catalysts exhibit a less corrosive character and can be used in a fixed-bed reactor, leading to safer, cheaper and more environment-friendly operation (Dossin et al., 2006b). In addition to the type of catalyst, important parameters of the transesterification reaction are the molar ratio of alcohol, type of alcohol, temperature, reaction time and degree of refinement of the vegetable oil (Fukuda et al., 2001; Geise, 2002; Ma & Hanna, 1999; Marchetti et al., 2007; Vicente et al., 2004). Also, stirring is a critical point in the efficiency of the process, higher stirring is recommended to create a homogeneous phase. We must remember that the insolubility of fat materials in short chain alcohols reduces the rate of transesterification. That means that transesterification does not proceed properly, unless the reaction mixture is homogenized in some way, at least during the initial stages of the process (Mittelbach & Remschmidt, 2004). In consequence, it has been shown that the use of a cosolvent greatly accelerates the reaction so that it reaches substantial completion within a few minutes. The primary concerns with this method are the additional complexity of recovering and recycling the cosolvent. Although this can be simplified by choosing a compound with a boiling point near to that of the alcohol in use. The most commonly used cosolvents are tetrahydrofuran and methyl tertiary butyl ether. Nevertheless, hexane has been successfully employed as co-solvent to obtain a 95.5 % of methyl ester content (Gerpen, 2005; Mittelbach & Remschmidt, 2004; Peña et al., 2009) when assessing three alkaline catalysts (CH_3ONa, NaOH, and KOH).

Usually, the choice of feedstock depends largely on the resources availability, and depending on the origin and quality of the feedstock, changes to the production process may be necessary. The use of non-edible oils or spent oils as well as heterogeneous systems is preferred because they are more environmentally friendly.

3. Oil type and catalysts

3.1 Oils

For the production of biodiesel there are not technical restrictions regarding the use of vegetable oils or animal fats. Nevertheless, there are preferred vegetable oils with high fatty acid content and whose wide world production is significant.

Constituent fatty acids of vegetable oils are mostly unsaturated. Oils, therefore, are liquid at room temperature, so that their use as diesel fuel depends mainly on their viscosity. Moreover, animal fats, because of their higher content of saturated fatty acids are solid at room temperature, and cannot be used in diesel engines in its original form. Although it is not common to use mixtures of vegetable oils with diesel in different proportions, depending on the viscosity of oil, these blends can be used in diesel engines. Nowadays, the major production of canola, followed by sunflower, soybean and palm oils is an important factor to select them to produce biodiesel (Körbitz, 1998). Though, the most assessed vegetable oils in the transesterification reaction are the castor, corn, cottonseed, crambe, peanut, soybean, palm, rapeseed and sunflower oils, mainly due to their content of glycerides (Demirbas, 2005). Animal fats have not been studied to the same extent as vegetables oils, however there are some works about poultry fat used to produce biodiesel, for example. Oil from algae, bacteria and fungi also has been investigated (Hernando et al., 2007; Leadbeater & Stencel, 2006; Ma & Hanna, 1999; Reddy et al., 2006). In addition to vegetable oils and animal fats, other materials such as spent frying oils have been used for biodiesel production; however, some changes in the reaction procedure frequently have to be made due to the presence of water or free fatty acids in the biodiesel (Bockisch, 1998). In consequence, the main raw material is vegetable oil. Some natural glycerides (vegetable oils) contain higher levels of unsaturated fatty acids (Table 1) (Ma & Hanna, 1999). These oils are liquids at room temperature and their direct use as fuel is precluded by high viscosities and requirement of engines modification. Therefore, it is convenient that vegetable oils are converted into their alkyl esters (biodiesel) by transesterification.

Relevant characteristics of oils typically used for biodiesel production are given in table 2.

3.2 Catalysts

At the moment almost all commercial biodiesel production plants are using homogenous alkaline catalysts. However, the major disadvantage of homogeneous catalysts is the fact that these cannot be reused. Besides, as above explained, the homogeneous process implies further stages of washing, which involves an increase in production costs. Recently, the biodiesel production using heterogeneous catalysts has been carried out at industrial level and in such a process the employed catalyst has been reported to be a mixed oxide of zinc and aluminium (Bournay et al., 2005). Indeed, the development of solid acid or basic catalysts for the transesterification reaction has been an important issue addressed by the scientific community. As a result, various types of catalysts have been assessed such as alkali earth oxides, alkali oxides, not metal oxides, metal oxides, cation exchange resins, metal phosphates and acid supported on different materials. Despite all the efforts, heterogeneous catalysts for biodiesel production have not been widely exploited at industrial level, yet.

Vegetable oil	Fatty acid composition (% weight)								
	16:1	18:0	20:0	22:0	24:0	18:1	22:1	18:2	18:3
Corn	11.67	1.85	0.24	0.00	0.00	25.16	0.00	60.60	0.48
Cottonseed	28.33	0.89	0.00	0.00	0.00	13.27	0.00	57.51	0.00
Crambe	20.7	0.70	2.09	0.80	1.12	18.86	58.51	9.00	6.85
Peanut	11.38	2.39	1.32	2.52	1.23	48.28	0.00	31.95	0.93
Rapeseed	3.49	0.85	0.00	0.00	0.00	64.4	0.00	22.30	8.23
Soybean	11.75	3.15	0.00	0.00	0.00	23.26	0.00	55.53	6.31
Sunflower	6.08	3.26	0.00	0.00	0.00	16.93	0.00	73.73	0.00
Castor	0.00	0.00	0.00	0.00	0.00	87.0	0.00	0.00	11.2
Palm	10.2	3.7	0.00	0.00	0.00	22.8	0.00	53.7	8.6

Table 1. Fatty acid composition

Vegetable oil	Kinematics viscosity (mm^2/s)	Flash point (°C)	Density (kg/l)
Rapseed	37	246	0.911
Castor	227	230	0.961
Palm	36	164	0.880
Peanut	39.6	271	0.902
Soybean	32.6	254	0.914
Sunflower	33.9	174	0.916

Table 2. Properties of typical vegetable oils employed to produce biodiesel from transesterification reaction (Demirbas, 2005).

Heterogeneous acid and basic catalysts could be classified as Brönsted or Lewis catalysts. This catalyst character determines the transesterification reaction rate. It has been concluded that the stronger basicity and therefore the presence of more active sites improves the performance of catalysts in the transesterification reaction. Hence, biodiesel is usually produced in the presence of an alkaline catalyst. Different studies, however, have been carried out using acid catalysts (Di Serio et al., 2005; Lotero et al., 2005; MacLeod et al., 2008; Marchetti et al., 2007; Zhu et al., 2006).

One should bear on mind that the benefit with solid catalysts, acid or basic, is the lesser consumption of catalyst in the reaction. For example, to produce 8000 tons of biodiesel, 88 tones of sodium hydroxide may be required, while only 5.6 tons of supported MgO are sufficient for the production of 100,000 tones of biodiesel. Besides, heterogeneous catalysts exhibit a less corrosive character and can be used in a fixed-bed reactor, leading to safer, cheaper and more environment-friendly operations and the number of separation steps is less than when using homogeneous catalysts. The heterogeneous catalysts do not leave neutralization salts in the glycerol, and are plausible to be retained in the reactor by filtration (Di Serio et al., 2008; Kaita et al., 2002; Kawashima et al., 2009; Kovacheva et al., 2001; MacLeod et al., 2008; Madje et al., 2004; Park et al., 2010; Suppes et al., 2004; Waghoo et al., 1999; Xie & Li, 2006; Yan et al., 2008).

The most commonly studied heterogeneous basic catalysts are alkaline metals carbonates (Na_2CO_3, K_2CO_3), alkaline earth metal carbonates ($CaCO_3$), alkaline earth metal oxides (CaO, MgO, SrO, BaO) and other oxides as ZnO (Dossin et al., 2006b; Encinar et al., 2005; Fukuda et al., 2001; Liu et al., 2007; López et al., 2005; Ma & Hanna, 1999; MacLeod et al., 2008; Verziu et al., 2008). A comparison of attained yields with the aforementioned catalysts would lack of objectivity since all the related studies have been performed under significant different operating conditions such as temperature, raw material and molar oil:alcohol ratio. Therefore, in the following paragraphs a summary of the most relevant results will be presented rather than a comparison. Moreover, the effect of important catalyst characteristics such as active phase, calcination temperature, catalytic support and catalyst concentration will be addressed.

For instance, the transesterification of rapeseed oil catalyzed by MgO, CaO, SrO and BaO at the same reaction conditions (64.5°C, reaction time 3.5 h, 18:1 methanol/oil molar ratio and 10 % catalyst dosage) resulted in <5 %, 58 %, 60 % and 86 % of methyl esters content, respectively. Actually, the catalytic activities decrease in the order of BaO>SrO>CaO>MgO, which suggests that the catalytic activities of alkaline earth metal oxides are associated with their alkalinity. However, the use of BaO is not practical enough since it is soluble in methanol and also forms highly toxic compounds. Regarding SrO, this oxide possesses a strong tendency to react with CO_2 and water present in air to form strontium hydroxide and strontium carbonate, thus losing its catalytic ability (Yan et al., 2008). Among the above-mentioned oxides, the CaO and MgO have been extensively studied in the transesterification reaction. About the former, its catalytic activity has been compared with other calcium compounds (calcium hydroxide and calcium alkoxides) at the same reaction conditions. The reported reactivity order is $Ca(OH)_2 < CaO < Ca(CH_3O)$. This is in agreement with Lewis theory: the methoxides of alkaline-earth metals are more basic than their oxides and these are more basic than their hydroxides (Gryglewicz, 1999; Kawashima et al., 2009).

Veljkovick et al. (2009) studied CaO catalysts for the transesterification of sunflower oil at 6:1 molar ratio of methanol to oil, 1 % wt catalyst based on oil weight and the CaO calcination temperature was 550 °C. A 98 % conversion was achieved at 60 °C and 2 hr reaction time. Yoosuk et al. (2010), attempted to increase the CaO catalytic activity by producing it from $CaCO_3$ calcination, using a calcination temperature of 800 °C for 3 h. This treatment is likely associated with crystallites fracture and the generation of more porosity and basic sites. This study was carried out using palm oil with methanol at 15:1 molar ratio of methanol to oil, 7 % wt catalyst, 60 °C and 1 hr reaction time. The catalysts tested were the modified calcium oxide and commercial calcium oxide. A 94 % of methyl esters content was obtained with the former while the latter led to attain only 75.5 % wt. Thus, it was concluded that the hydration treatment and subsequent calcination favors the formation of stronger basic sites and possesses a strong effect over crystallinity and crystallite size. López et al., (2007) studied the catalytic activity of calcium oxide for biodiesel production from sunflower oil (at 60 °C, 13:1 methanol to oil molar ratio, 3 % wt catalyst). The reaction was complete in 100 min giving 94 % conversion. In this study, the poisoning of active surface sites of calcium oxide by the atmospheric H_2O and CO_2 was observed. In the same study, to improve the catalytic activity of CaO, this was subjected to an activation treatment at high temperature (> 700 °C) prior reaction and as a result the carbonate group (main poisoning specie) was removed from the catalytic surface. Regarding MgO, Dossin et al. (2006a) studied the ethyl acetate transesterification reaction with methanol catalyzed with MgO at 283-323 K and 0.1-10 methanol/ethyl acetate molar ratio, with 1.6-2.2x10^{-4} kg of catalyst mass. A biodiesel with 99 % purity was obtained after 2 h. This confirms the MgO as a viable heterogeneous solid base catalyst for the transesterification of ethyl acetate, not good enough for other oils as rapeseed or soybean oil though. This has been evidenced by Di Serio et al. (2006), who tested MgO in the industrial production of biodiesel from soybean oil transesterification at 180 °C and 12:1methanol to oil molar ratio. Only a 72 % conversion was achieved.

On another study (Yoo et al., 2010), using CaO under supercritical state (at 200°C to 220°C, 40:1 methanol/oil molar ratio and 1.0 % wt of catalyst), the achieved conversion was near to 96 %. At the same conditions, the ZnO was the optimum catalyst for the transesterification of rapeseed oil with more than 96 % conversion towards biodiesel.

Broadly speaking, a strategy to increase catalysts activity has been to make them nanocrystalline. This type of materials has recently received much attention because of the effect of particle size and shape in their physical and chemical properties. The unusual physical and chemical properties are associated with crystal faces, edges, corners, and defects that lead to enhanced surface reactivity; and the percent dispersion as a measure of the available fraction of atoms on the surface for incoming reactants. Thus, the successful nanocatalytic process is due to an increased surface area and greater concentrations of highly reactive sites. Although nanotechnology is rapidly developing into a stand-alone industry that covers a host of applications and industries, there are only few works related with nanocatalysts applied in the transesterification reaction reaction (Edelstein & Cammarata, 1996; Klabunde, 2001).

As example of this we can mention the nanocrystalline calcium and magnesium oxides. CaO has been employed as heterogeneous catalyst for the transesterification of soybean oil. In such study the achieved conversion was 99 % wt at a reaction time of 12 h, room temperature and 1:27 oil/methanol molar ratio (Reddy et al., 2006). In particular, the reactivity of nanosized oxides is associated whit their increased surface area, greater concentrations of highly reactive edge and corner defect sites, and stabilized lattice planes. Improvements in the results with nanocrystalline calcium oxide are not surprising since it is well documented that nanoparticles and nanocrystals possess unusual physical and chemical properties. Verziu et al. (2008), studied the catalytic activity of nanocrystalline magnesium oxide, with three different morphologies, in the transesterification of sunflower

oil. The obtained yield was 90 % at 70 °C and 4:1 methanol to oil molar ratio. However, there was the detachment of magnesium caused by saponification.
As for numerous heterogeneous catalytic systems, another important variable affecting the catalytic activity in the transesterification reaction is the support of the catalyst, which also affects surface area, basicity, mechanical strength and cost. Some studies have reported the use of CaO or MgO supported on mesoporous silicas, aluminas or zeolites, to provide greater surface area and therefore basicity (Di Serio et al., 2006; MacLeod et al., 2008; Martínez et al., 2011; Suppes et al., 2004; Xie et al., 2006; Di Serio et al. 2006, Yan et al., 2008). For instance, samples of CaO/MgO, CaO/SiO_2, CaO/Al_2O_3 and CaO/Zeolite HY were employed to catalyze the transesterification of rapeseed oil at 64.5 °C during 3.5 h and with a 18:1 methanol:rapeseed oil molar ratio. It was shown that with a 10 % catalyst content a 92 %, 60 %, 36 % and 23 % of biodiesel was obtained, respectively (Yan et al., 2008). In this context, it is worth mentioning that Di Serio et al. (2006) studied the transestification of soybean oil at 180 °C and 12:1 methanol to oil molar ratio and the obtained conversion was 92 %. In another work (Samart et al., 2009), a 90 % soybean oil conversion was reported when using mesoporous silica loaded with 15 % wt of KI at 70 °C using 1 mol of vegetable oil, 16 mol of methanol, 5 % wt of catalyst and after 8 h of reaction. NaX zeolite loaded with 16 % nanoparticles CaO was reported as a base catalyst in sunflower oil transesterification at 60 °C, 6:1 molar ratio of methanol to oil and 10 % wt catalyst, 93.5 % biodiesel was achieved within 6 h (Martínez et al., 2011). All the aforesaid show the relevance of the catalytic support.
On the other hand, regarding acid catalysts, these possess the advantage of being active for both, esterification and transesterification reactions. Lopez et al. (2005) tested the activity of several heterogeneous acid catalysts and reported the following reactivity order: Amberlyst 15 (79 %)> sulfated zirconia (57 %)> Nafion NR50 (33 %)> tungstated zirconia > others, in the transesterification of triacetin with methanol at 60 °C, and 6:1 alcohol to oil molar ratio. Amberlyst-15, cation-exchange resin, and Amberlyst A27, anion-exchange resin, have also been studied for transesterification reactions. A relatively low reaction temperature (60 °C) resulted in poor sunflower oil, conversion: 0.7 and 0.4 % respectively, when carrying out the reaction at atmospheric pressure and using a 6:1 methanol-to-oil initial molar ratio (Vicente et al., 1998). A further possibility for accelerating the transesterification reaction with acid catalysts is the use of heterogeneous catalysts such as WO_3/ZrO_2, SO_4^{2-}/ZrO_2 and Amberlyst-15 catalysts. These systems were tested in the esterification of 4 % wt oleic acid in soybean oil. The reactions were carried out using 9:1 molar ratio of methanol to oil, 0.29 g of catalyst per mol of oil at 75 °C. At all cases, a conversion of about 93 % was obtained (Park et al., 2010). Even when the results regarding conversion may seem promising, one should not forget that acid catalysts are more corrosive than basic ones. This being a main drawback of such catalytic systems.
Sodium molybdate (Na_2MoO_4) was used for the transesterification of different types of lipids derived from soybean oil with methanol. In this case, a high methanol-to-oil molar ratio (54:1) was required in order to be effective, at 65 °C, with 5 % wt catalyst contents in 3 h. The achieved conversion was higher than 92.5 % (Nakagaki et al., 2008). Besides, Carma et al. studied the Al-MCM-41 mesoporous molecular sieves with Si/Al for esterification of palmitic acid with some alcohols (methanol, ethanol and isopropanol), at 130 °C, 0.6 % wt catalysts and 60:1 alcohol to oil molar ratio. The found conversions were 79, 67 and 59 %, respectively to each alcohol (Carmo Jr. et al., 2009).

The application of beta zeolite modified with La in the methanolysis of soybean oil was reported by Shu et al. (2007). In this system, a 49 % triglyceride conversion was obtained. Besides, Karmee and Chadha (2005) have been studied Hβ-zeolite, montmorillonite K-10 and ZnO catalysts in the transesterification of non-edible oil of Pongammia Pinnata crude at 120 °C with 10:1 ratio of methanol to oil, and 0.115 % wt catalyst, and attained 59, 47 and 83 % conversion, respectively (Karmee & Chadha, 2005).

The NKC-9 cation-exchange resin has been used in a fixed-bed reactor to carry out the esterification of oleic acid. The conversion was over 98 % at 65 °C, 2.8:1 molar ratio of methanol to oil and 500 h of continuous running, showing an excellent operational stability (Feng et al., 2011).

The use of seashell and eggshells heterogeneous catalyst of the methanolysis of vegetables oils, has provided promising results (Sarin et al., 2009). They use various vegetables oils such as jatropha, castor, sunflower, soybean, rapeseed, cotton, corn, coconut, and observed 98 % conversion. The operating conditions were 2 hr reaction time, 6:1 molar ratio of methanol to oil and 4 % wt of catalyst.

According to the aforementioned works, the advantages of working with heterogeneous alkaline catalysts are evident, and therefore any variable increasing the basicity of the catalytic system will positively impact its performance.

3.2.1 Heterogeneous catalysts characterization

Though there are a wide range of analytical techniques for heterogeneous catalysts characterization, the selection depends on the information provided by the techniques. The typical characterization techniques are: pore size distribution and surface area (BET), which are established by adsorption and desorption of nitrogen. In this method, the solid samples are evacuated under vacuum and temperature in such a way that the catalytic surface be free from water and other impurities and thus available for being occupied with nitrogen molecules. Specific total surface area is calculated using the BET equation, whereas specific total pore volume is evaluated from N_2 uptake at a relative pressure (P/Po) of N_2 equal to 0.99.

X-Ray Diffraction (XRD) is the most widely employed technique for general crystalline material characterization. This technique allows the identification of crystalline phases in bulk materials and the determination of crystallite size and shape from diffraction peak characteristics. Figure 3 shows an example of XRD patterns for samples of NaX and CaO nano-16%/NaX used in the transesterification of sunflower oil.

Total basic site density and acid strength distribution of the heterogeneous catalysts can be measured by temperature-programmed desorption of CO_2 (CO_2-TPD). The temperature of desorption and the maximum desorbed CO_2 are illustrative of the strength and amount of basic sites, respectively. The basic site density is obtained by integration of area under the curve. Figure 4 shows the CO_2-TPD profiles for 3NaX and ion–exchanged (NaX-ie) samples. Both samples exhibit a desorption peak at approximately 150 °C. This peak can be ascribed to the interaction of CO_2 with sites of weak basic strength. A second desorption peak (3NaX sample) appeared at higher temperature (700 °C), which may be attributed to the interaction of CO_2 with occluded sodium oxide species that possess stronger basic sites. These results demonstrated that the stronger basic sites (3NaX sample) were responsible of the high activity in the transesterification of sunflower oil (Ramos et al., 2008).

Scanning Electron Microscopy (SEM) provides a high resolution image of the surface of a catalyst (topographical information) over a length scale from nanometers up to several hundred micrometers. It can operate at magnifications that are adjusted from about 20X-5(10^5). It provides of information concerning catalytic particle morphology, active phase homogeneity and composition near surface regions of the material since X rays can be detected from all the elements in the periodic table (Energy Dispersive X-Ray Spectroscopy, EDS). Figure 5 depicts the scanning electron micrograph of supported 16 % of CaO nanoparticles on NaX zeolite. Typical quasi-spherical morphology of NaX zeolite can be observed.

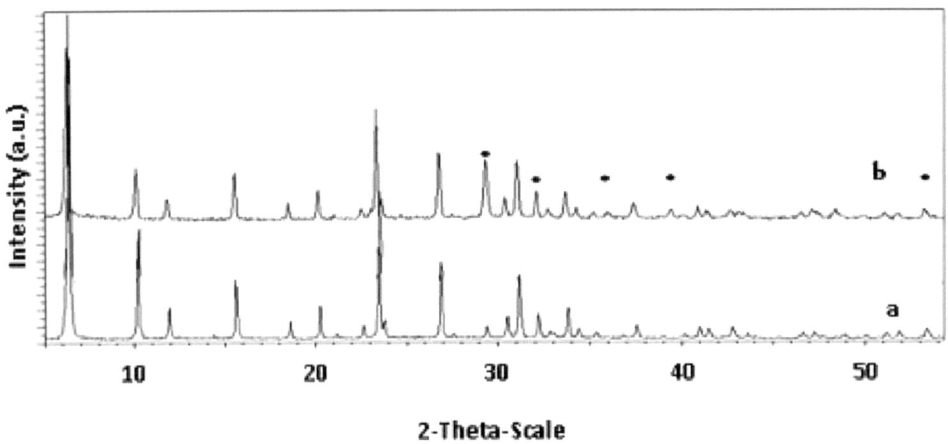

Fig. 3. XRD patterns for samples (a)NaX and (b)CaO nano-16%/NaX. (Martínez et al., 2011).

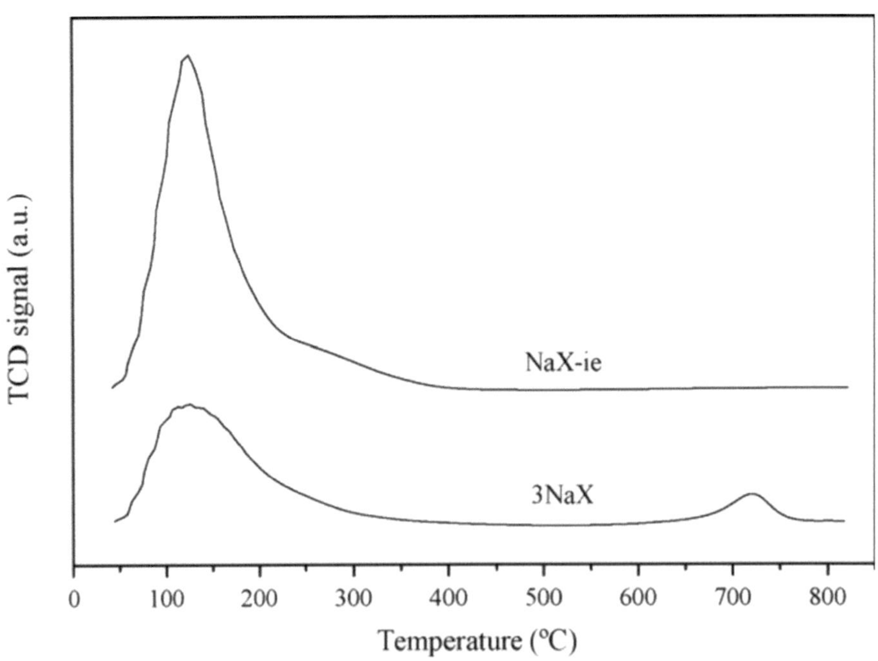

Fig. 4. CO_2-TPD profiles for impregnated (3NaX) and ion-exchanged (NaX-ie) samples. (Ramos et al., 2008).

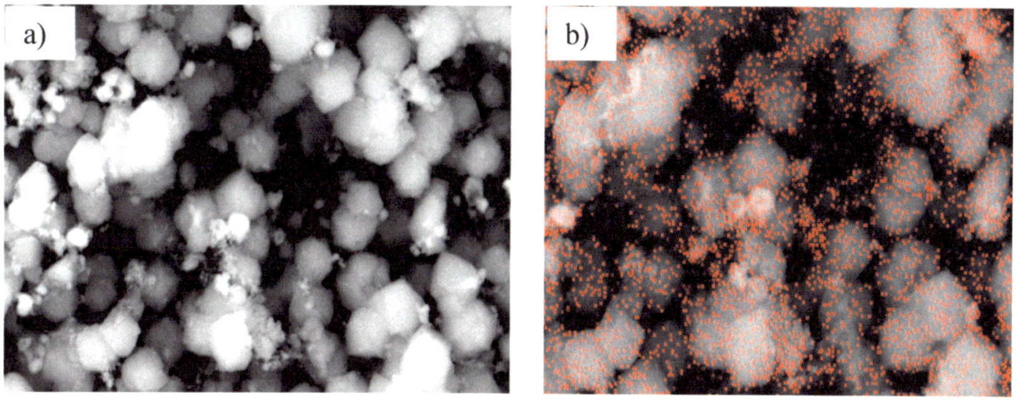

Fig. 5. (a) SEM image of the CaO nano-16%/NaX Catalyst; (b) EDS chemical mapping for Ca (Matínez et al., 2011).

X Ray Photoelectron Spectrometry is an analytical technique that utilizes photo-ionization and analysis of the kinetic energy distribution of emitted photoelectrons to estimate the elemental composition and chemical state of the elements on the surface of a material by projecting soft X-ray onto the surface and detecting the energy of photoelectrons emitted from areas a few nm (up to 5) from the sample surface. Figure 6 shows the depth profile analysis with Ar^+ gun of the 16% nano-CaO/NaX material. In this example the results suggest a homogeneous distribution of CaO nanoparticles on the NaX surface.

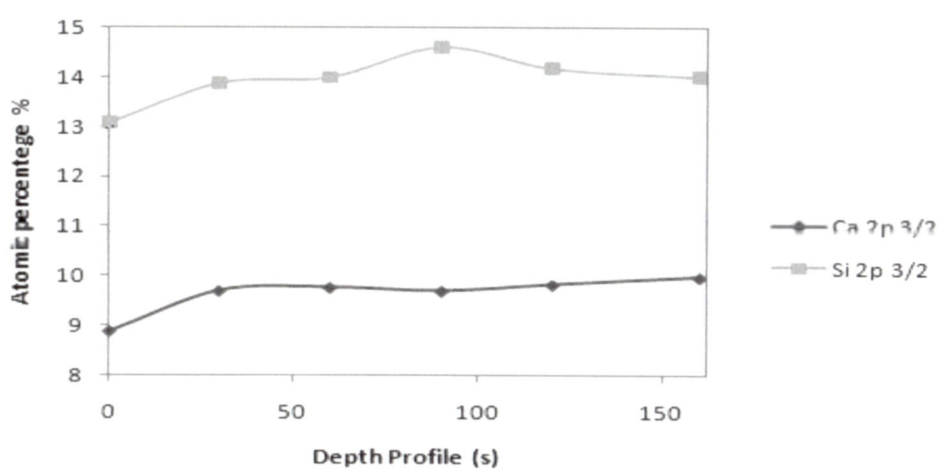

Fig. 6. XPS depth profile for CaO nano-16%/NaX. (Martínez et al., 2011).

4. Biodiesel characterization

The European standard EN 14214, which went into effect in 2003, supersedes the biodiesel standards in European countries that are members of the European Committee for Standardization (CEN). Therefore, no standards from individual European countries are given. The European standard EN 590 for conventional diesel fuel contains a provision that conventional diesel fuel can contain up to 5 % FAME meeting the standard EN 14214.

In addition to the biodiesel standards, analytical standards have been developed in the United States and Europe for the purpose of including them as prescribed methods in biodiesel standards. Table 3 lists such relevant analytical standards.
Some more important parameters of biodiesel are methyl ester content, kinematic viscosity, density, methanol, acid value, flash point and cetane number.

Property	Test method	Limits		Unit
		Min	Max	
Ester content	EN 14103	96.5		% (m/m)
Density at 15°C	EN ISO 3675 EN ISO 12185	860	900	Kg/m^3
Kinematics viscosity at 40°C	EN ISO 3104	3.5	5.0	mm^2/s
Flash point	EN ISO 3679	120		°C
Sulfur content	UNE-EN ISO 20846 UNE-EN ISO 20884		10.0	mg/kg
Carbon residue (10% dist. residue)	EN ISO 10370		0.30	% (m/m)
Sulfated ash	ISO 3987		0.02	% (m/m)
Water content	EN ISO 12937		500	mg/kg
Total contamination	EN 12662		24	mg/kg
Oxidative stability at 110°C	EN 14112	6.0		H
Acid value	EN 14104		0.5	mg KOH/g
Iodine value	EN 14111		120	g iodine/100 g
Monoglyceride content	EN 14105		0.8	% (m/m)
Diglyceride content	EN 14105		0.20	% (m/m)
Triglyceride content	EN 14105		0.20	% (m/m)
Linoleic acid content	EN 14103		12	% (m/m)
Free glycerine	EN 14105 EN 14106		0.02	% (m/m)
Total glycerine	EN 14105		0.25	% (m/m)
Methanol content	EN 14110		0.2	% (m/m)
Cold-filter plugging point	EN 116			
Pour point	ISO 3016		0	°C
Cetane number	EN ISO 5165		51	
Alkali metals (Na+K)	EN 14108 EN 14109		5.0	mg/kg
Earth metals (Ca+Mg)	pr EN 14538		5.0	mg/kg
Phosphorous content	EN 14107		10.0	mg/kg
Copper strip (3h, 50°C)	EN ISO 2160		1	
Heating value	DIN 51900-1	35		MJ/kg

Table 3. European Standard EN 14214 for Biodiesel

The methyl esters are quantified according to the European norm UNE-EN 14103. This indicates a minimum of 96.5 % of fatty acid methyl esters for biodiesel fuels. The standard analytical procedure for determination of methyl esters content involves capillary gas chromatography on polar stationary phases, applying internal calibration with methyl heptadecanoate and detection via flame ionization.

Viscosity is an important property of biodiesel since it affects the operation of fuel injection equipment (at low temperatures affects the fluidity of biodiesel) (Demirbas, 2005). Higher viscosity leads to a higher drag in the injection pump and thus causes higher pressures and injection volumes, especially at low engine operating temperatures (Wörgetter et al., 1998). Viscosity is closely related to the fatty acid composition of a given biodiesel sample. It increases with increasing length of both the fatty acid chain and the alcohol group, so that ethyl esters show slightly higher values than the respective methyl compounds. Kinematic viscosity at 40 °C is measured according to EN ISO 3104. The standard analytical procedure for the kinematic viscosity indicates the measure of the time that a given volume of liquid flows by gravity through a capillary viscometer at a known and well controlled temperature. Kinematic viscosity is the product of time of flow and the viscometer calibration constant.

Density and therefore specific gravity is another important parameter of biodiesel quality. Fuel injection equipment operates on a volume metering system, hence a higher density for biodiesel results in the delivery of a slightly greater mass of fuel (Demirbas, 2005). The specific gravity of biodiesel ranges between 860-900 kg/m^3 at 15 °C. The standard analytical procedure for the determination of density of biodiesel fuel involves the use of a standardized glass hydrometer or an oscillating U-tube at the prescribed temperature.

Regarding methanol content, this must not exceed 0.20 % in biodiesel fuels (EN 14214). This parameter is important for fuel quality since high methanol content poses safety risks in biodiesel transport and storage due to correspondingly low flash points.

Acid value or neutralization number is a measure of mineral acids and free fatty acids contained in a fuel sample. It is expressed in milligrams of KOH required to neutralize 1 g of fatty acid methyl esters and it sets to a maximum value of 0.5 mg KOH/g in the European norm (EN-14214).

Flash point is a measure of the flammability of fuels and thus, an important parameter for assessing hazards during fuel transport and storage. The flash point is measured following the European norm EN ISO 3679 (>120 °C). The technique indicates that a defined volume of the sample is introduced into a closed cup flash point test, which holds at the flash point temperature estimated of the material to be analyzed. After a period of time, a flame is applied and presence or absence of inflammation is observed. Additional tests are performed at different temperatures with new samples until it determining the flash point with the specified sensitivity.

Cetane number is a dimensionless indicator of the ignition quality of a biodiesel, if the cetane number is high, combustion can occur before the fuel and air are properly mixed, resulting in incomplete combustion and smoke, and if is low, engine roughness, misfiring, higher air temperatures, slower engine warm-up, and also incomplete combustion occur. The European norm EN 14214 indicates a minimum cetane number of 51.

All the aforementioned parameters determine the quality of biodiesel produced either homogenously or heterogeneously. Methyl esters content, however, can be considered as the most important property since its value dictates the value of the other parameters. This means that if the standard regarding methyl esters content is fulfilled is very likely that the standards of the other parameters are also satisfied. This may be the reason for the literature not being so keen on reporting a full biodiesel characterization and then methyl esters content has become the usual reported indicator of biodiesel quality. An exemption would be the castor oil whose initial viscosity is already high and hence producing biodiesel with high viscosity as well. Viscosity possesses an inverse correlation with methyl esters content. As above explained, viscosity is also a property of paramount importance that affects the fuel injection equipment, mainly at low temperature when an increase in fuel viscosity affects its flow.

In the section regarding catalysts (3.2), it was mentioned that a methyl esters content between 79 and 99 % can be achieved via heterogeneous catalysis. It is worth mentioning that the so produced biodiesel (with a methyl esters content near to the norm) can be employed in blends consisting of 20 % biodiesel and 80 % petroleum diesel (Demirbas, 2005). This simple action would lead to a large reduction on carbon dioxide emissions. One should remember that carbon dioxide has been identified as a compound that importantly affects the climate global warming.

Finally, it can be said that the heterogeneous catalysts that nowadays are more promising for biodiesel production are calcium and magnesium based since high conversions and methyl esters content are attained with them even at mild reaction conditions (see section 3.2).

5. Conclusion

The development of efficient heterogeneous catalysts is important since opens up the possibility of another pathway for biodiesel production. The efficiency of the heterogeneous process depends, however, on several variables such as type of oil, molar ratio alcohol to oil, temperature, catalyst type, even type of reactor. According to the aforementioned works, the advantages of working with heterogeneous alkaline catalysts are evident and encourage to further and deeper research with them, mainly in the area of reaction engineering (kinetics and reactor design).

6. Acknowledgment

The authors are grateful to Universidad Autónoma del Estado de México (2886/2010 U).

7. References

Al Zuhair, S. (2007). Production of biodiesel: possibilities and challenges. *Biofuels Bioproducts and Biorefining,* Vol.1, No.1, (September 2007) p.p. 57-66, ISSN 1932-1031

Bockisch, M. (1998). *Fats and Oils Handbook.* AOCS Press. ISBN 0-935315-82-9. Champaign Illinois. U.S.A.

Bournay, L., Casanave, D., Delfort, B., Hillion, G. & Chodorge, J.A. (2005). New heterogeneous process for biodiesel production: A way to improve the quality and the value of the crude glycerin produced by biodiesel plants. *Catalysis Today,* Vol. 106, No. 1-4, (October 2005), p.p. 190-192, ISSN 0920-5861

Carmo Jr. A.C., De Souza, L.K.C., Da Costa, C.E.F., Longo, E., Zamian, J.R. & Da Rocha Filho, G.N. (2009). Production of biodiesel by esterification of palmitic acido over mesoporous aluminosilicate Al-MCM-41. *Fuel,* Vol. 88, No. 3, (March 2009), p.p. 461-468, ISSN: 00162361

Demirbas, A. (2005). Biodiesel production from vegetable oils via catalytic and non-catalytic supercritical methanol transesterification methods. *Progress in Energy and Combustion Science*, Vol. 31, No. 5-6, (December 2005), p.p. 466-487, ISSN 0360-1285

Di Serio, M., Ledda, M., Cozzolino, M., Minutillo, G., Tesser, R. & Santacesaria, E. (2006). Transesterification of soybean oil to biodiesel by using heterogeneous basic catalysts. *Industrial and Engineering Chemistry Research*, VOL. 45, No. 9, (April, 2006), p.p. 3009-3014, 0888-5885

Di Serio, M., Tesser, R., Dimiccoli, M., Cammarota, F., Nastasi, M. & Santacesaria, E. (2005) Synthesis of biodiesel via homogeneous Lewis acid catalyst. *Journal of Molecular Catalysis A: Chemical* Vol. 239, No. 1-2, (September 2005), p.p. 111–115 ISSN 1381-1169

Di Serio, M., Tesser, R., Pengmei, L. & Santacesaria, E. (2008). Heterogeneous Catalysts for Biodiesel Production. *Energy and Fuels*, Vol. 22, No. 1, (January 2008), p.p. 207-217, ISSN 0887-0624

Dossin, T.F., Reyniers, M.F. & Marin, F.B. (2006a). Kinetics of Heterogeneously MgO-Catalyzed Transesterification. *Applied Catalysis B: Environmental*, VOL. 62, No. 1-2, (January 2006), p.p. 35-45, ISSN 0926-3373

Dossin, T. F., Reyniers, M. F., Berger, R. J. & Marin, G. B. (2006b). Simulation of heterogeneously MgO-catalyzed transesterification for fine-chemical and biodiesel industrial production. Applied *Catalysis B: Environmental*, Vol. 67, No. 1-2, (September 2006), p.p. 136-148, ISSN: 0926-3373

Edelstein, A.S. & Cammarata, R.C. (Eds.) (1996). *Nanomaterials: Synthesis, Properties and Applications*. Taylor and Francis. ISBN 0750303581. Philadelphia, P.A. United States.

Encinar, J. M., González, J. F. & Rodríguez-Reinares, A. (2005). Biodiesel from Used Frying Oil.Variables Affecting the Yields and Characteristics of the Biodiesel. *Industrial and Engineering Chemistry Research*, Vol. 44, No. 15, (June 2005), p.p. 5491- 5499, ISSN 0888-5885

Feng, Y., Zhang, A., Li, J. & He, B. (2011). A continuous process for biodiesel production in a fixed bed reactor packed with cation-exchange resin as heterogeneous catalyst. 2010. *Bioresource Technology*, Vol. 102, No. 3, (February 2011), pp. 3607-3609, ISSN 0960-8524

Fukuda, H., Kondo, A. & Noda, H. (2001). Biodiesel Fuel Production by Transesterification of Oils. *Journal of Bioscience and Bioengineering*, Vol. 92 No. 5, (January 2001), p.p. 405-416, ISSN 1389-1723

Geise, R. (2002). Biodiesel's Bright Future Deserves Equality. *Render Magazine*, Vol. 31, p.p. 16–17, ISSN 0090-8932

Gerpen, J.V. (2005). Biodiesel processing and production. *Fuel Processing Technology*, Vol. 86, No. 10, (June 2005), p.p. 1097-1107, ISSN 0378-3820

Gryglewicz, S. (1999). Rapeseed oil methyl esters preparation using heterogeneous catalysts. *Bioresource Technology*, VOL. 70, NO. 3,, (December 1999) p.p. 249-253, ISSN 0960-8524

Harvey, A.P., Mackley, M.R., & Seliger, T. (2003). Process Intensification of Biodiesel Production Using a Continuous Oscillatory Flow Reactor. *Journal of Chemical Technology & Biotechnology*, Vol. 78, No. 2-3, (February - March 2003) p.p. 338–341. ISSN 0268-2575

Hernando, J., Leton, P., Matia, M.P., Novella, J.L. & Alvarez Builla, J. (2007). Biodiesel and FAME synthesis assisted by microwaves: homogeneous batch and flow processes. *Fuel*. Vol. 86 No.10-11, (July-August 2007), p.p. 1641-1644, ISSN 00162361

Kaita, J., Mimura, T., Fukuoda, N., & Hattori, Y. (2002). *Catalysts for transesterification*. U.S. Patent 6407269, June 18, 2002.

Karmee, S.K. & Chadha, A. (2005). Preparation of biodiesel from crude oil of Pongamia pinnata. *Bioresource Technology*, Vol.96, No.13, (September 2005), p.p. 1425-1429, ISSN 0960-8524

Kawashima, A., Matsubara, K. & Honda, K. (2009). Acceleration of catalytic activity of calcium oxide for biodiesel production. *Bioresource Technology*, Vol. 100, No. 2 (January 2009), p.p. 696-700, ISSN 0960-8524

Klabunde, K.J. (Ed.) (2001). *Nanoscale Materials in Chemistry*, Jonh Wiley & Sons, Inc. ISBN 0-471-38395-3, New York, United States.

Körbitz, W. (1998). Multi-Feed-Stock-Biodiesel:The Modern & Profitable FAME. Production Plant, *Proceeding of International Liquid Biofuels Congress*. Curitaba-Paraná, Brazil, July 19-22, 1998

Kovacheva, P., Arishtirova, K. & Vassilev, S. (2001). MgO/NaX zeolite as basic catalyst for oxidative methylation of toluene with methane. *Applied Catalysis A: General*. Vol. 210, No. 1-2, (March 2001), p.p. 391-395, ISSN 0926-860X

Leadbeater, N.E. & Stencel, L.M. (2006). Fast, easy preparation of biodiesel using microwave heating. *Energy and Fuels*, Vol.20, No. 5, (September 2006) p.p. 2281-2283, ISSN 08870624

Liu, X., He, H., Wang, Y. & Zhu, S. (2007). Transesterification of soybean oil to biodiesel using SrO as a solid base catalyst, *Catalysis Communications*, Vol. 8, No. 7, (July 2007), p.p. 1107-1111, ISSN 1566-7367

López Granados, M., Zafra Poves, M.D., Martín Alonso, D., Mariscal, R., Cabello Galisteo, F., Moreno-Tost, R., Santamaría, J. & Fierro, J.L.G. (2007). Biodiesel from sunflower oil by using activated calcium oxide. *Applied Catalysis B: Environmental*, Vol. 73, No. 3-4, (May 2007), p.p. 317-326, ISSN: 0926-3373

López, D.E., Goodwin Jr., J.G., Bruce, D.A. & Lotero, E. (2005). Transesterification of triacetin with methanol on solid acid and base catalysts. *Applied Catalysis A: General*, Vol. 295, No. 2, (November 2005), p.p. 97-105, ISSN 0926-860X

Lotero, E., Liu, Y., Lopez, D.E., Suwannakarn, K., Bruce, D.A., & Goodwin, J.G. (2005). Synthesis of Biodiesel via Acid Catalysis. *Industrial and Engineering Chemistry Research*, Vol. 44 No. 14, (July 2005), p.p. 5353 ISSN 0888-5885

Ma, F. & Hanna, M. A. (1999). Biodiesel production: a review. *Bioresource Technology*, Vol. 70, No. 1, (October 1999), p.p. 1-15, ISSN 09608524

MacLeod, C. S., Harvey, A. P., Lee, A. F., & Wilson, K. (2008). Evaluation of the activity and stability of alkali-doped metal oxide catalysts for application to an intensified method of biodiesel production. *Chemical Engineering Journal*, Vol. 135, No. 1-2, (January 2008), p.p. 63-70, ISSN 1385-8947

Madje, B. R., Patil, P. T., Shindalkar, S. S., Benjamin, S.B., Shingare, M. S. & Dongare, M. K. (2004). Facile transesterification of beta-ketoesters under solvent-free condition using borate zirconia solid acid catalyst. *Catalysis Communications*, Vol. 5, No. 7, (July 2004), p.p. 353-357, ISSN 1566-7367

Marchetti, J.M., Miguel, V.U. & Errazu, A.F. (2007). Possible methods for biodiesel production. *Renewable and Sustainable Energy Reviews*, Vol. 11, No. 6 (August 2007) p.p. 300-1311, ISSN: 1364-0321

Martínez, S.L., Romero, R., López J.C., Romero,A., Sánchez, V. & Natividad, R. (2011). Preparation and Characterization of CaO Nanoparticles/NaX Zeolite Caralysts for the Transesterification of Sunflower Oil. *Industrial and Engineering Chemistry Research*, Vol. 50, No. 5, (March 2011), p.p. 2665-2670, ISSN 0888-5885

Meneghetti, S.M.P., Meneghetti, M.R, Wolf, C.R, Silva, E.C, Lima, G.E.S, De Lira Silva, L., Serra, T.M., Cauduro, F., & De Oliveira, L.G. (2006). Biodiesel from Castor Oil: A Comparison of Ethanolysis versus Methanolysis. *Energy and Fuels*, Vol. 20, No. 5, (September 2006), p.p. 2262-2265, ISSN 08870624.

Mittelbach, M. & Remschmidt, C. (Eds.) (2004). *Biodiesel: The Comprehensive Handbook*, Boersedruck Ges.m.b.H, ISBN 3-200-00249-2, Vienna, Austria

Mushrush, G., Beal, E.J., Spencer, G., Wynne, J.H., Lloyd, C.L., Hughes, J.M., Walls, C.L., & Hardy, D.R. (2001). An Environmentally Benign Soybean Derived Fuel as a Blending Stock or Replacement for Home Heating Oil. *Journal of Environmental Science and Health, Part A: Toxic/Hazardous Substances and Environmental Engineering*, Vol.36, No5., (May 2001) p.p. 613–622, ISSN 1532-4117.

Nakagaki, S., Bail, A., DosSantos, V.C., DeSouza, V.H.R., Vrubel, H., Nune, F.S., & Ramos, L.P. (2008). Use of anhydrous sodium molybdate as an efficient heterogeneous catalyst for soybean oil methanolysis. *Applied Catalysis A: General*, Vol. 351, No. 2, (December 2008), p.p. 267-274, ISSN 0926-860X

Park, Y-M., Lee, J.Y., Chung, S-H., Park, I.S., Lee, S-Y. Kim, D-K., Lee, J-S. & Lee, K-Y. (2010). Esterification of used vegetable oils using the heterogeneous WO_3/ZrO_2 catalyst for production of biodiesel. *Bioresource Technology*, Vol. 101, No.1, (January 2010), p.p. S59-S61, ISSN 0960-8524

Peña, R., Romero, R., Martínez S.L., Ramos, M.J., Martínez, A. & Natividad, R. (2009). Transesterification of Castor Oil: Effect of Catalyst and Co-Solvent. *Industrial and Engineering Chemistry Research*, Vol. 48, No. 3, (February 2009), p.p.1186–1189, ISSN 0888-5885

Ramos, M.J.; Casas, A.; Rodriguez, L.; Romero, R, & Perez, A. (2008). Transesterification of sunflower oil over zeolites using different metal loading: A case of leaching and agglomeration studies. *Applied Catalysis A. General*, Vol. 346, No. 1-2, (August 2008), p.p. 79-85, ISSN 0926-860X

Reddy, Ch., Reddy, V., Oshel, Reed. & Verkade, J.G. (2006). Room-temperature of soybean oil and poultry fat to biodiesel catalyzed by nanocrystalline calcium oxides. *Energy and Fuels*, Vol. 20, No.3, (May 2006) p.p. 1310-1314, ISSN 08870624

Ryan, T.W., Dogne, L.G. & Callahan, T.J. (1984). The effects of vetable oil properties on injection and combustion in two different diesel engines. *Journal of the American Oil Chemists' Society, Vol.* 61, No. 10, (October 1984), p.p. 1610-1619, ISSN 0003-021X

Samart, C. Sreetongkittikal, P. & Sookman, C. (2009). Heterogeneous catalysis of transesterification of soybean oil using KI/mesoporous silica. *Fuel Processing Technology*, Vol. 90, No.7-8, (July-August 2009), p.p.922-925, ISSN 0378-3820

Sarin, R., Arora, A.K., Puri, S.K., Prakash, S., Ranjan, R., Christopher, J., Tuli, D.K., Malhotra, R.K. & Kumar, A. (2009). *Novel catalyst composition for biodiesel production and a process for producing biodiesel and product thereof.* Patent applied (Appl. No. 1348IN033).

Shu, Q., Yang, B., Yuan, H., Qing, S. & Zhu, G. (2007). Synthesis of biodiesel from soybean oil and metanol catalyzed by zeolite beta modified with La^{3+}. *Catalalysis Communications*, Vol. 8, No. 12, (December 2007), p.p. 2159-2165, ISSN 1566-7367

Suppes, G.J., Dasari, M.A., Doskocil, E.J., Mankidy, P.J., & Goff, M.J. (2004). Transesterification of soybean oil with zeolite and metal catalysts. *Applied Catalysis A: General*, Vol. 257, No. 2, (January 2004), p.p. 213-223, ISSN: 0926-860X

Veljkovic, V.B., Stamenkovic, O.S., Todorovic, Z.B., Lazic, M.L. & Skala, D.V. (2009). Kinetics of sunflower oil methanolysis catalyzed by calcium oxide. *Fuel*, Vol. 88, No. 9, (September 2009), p.p.1554-1562, ISSN 00162361

Verziu M., Cojjocaru B., Hu J., Richards R., Ciuculescu C., & Filip P. (2008). Sunflower and rapeseed oil transesterification to biodiesel over different nanocrystalline MgO catalysts. *Green Chemistry*, Vol. 10, No. 4, p.p.373-381, ISSN 1463-9262

Vicente, G., Coteron, A., Martinez, M. & Aracil, J. (1998). Application of the factorial design of experiments and response surface methodology to optimize biodiesel production. *Industrial Crops and Products*, Vol.8, No.1, (March 1998) p.p. 29-35, ISSN 0926-6690

Vicente, G., Martinez, M. & Aracil, J. (2004). Integrated biodiesel production: a comparison of different homogeneous catalysts systems. *Bioresource Technology*, Vol. 92, No. 3, (May 2004), p.p. 297-305, ISSN 0960-8524

Waghoo, G., Jayaram, R. V. & Joshi, M. V. (1999). Heterogeneous, catalytic conversions with hydrous SnO2. *Synthetic Communications*, Vol. 29, No. 3, p.p. 513-520, ISSN 0039-7911

Wardle, D.A. (2003). Global Sale of Green Air Travel Supported Using Biodiesel. *Renewable and Sustainable Energy Reviews*, Vol. 7, No. 1, (February 2003), pp. 1–64, ISSN 1364-0321

Wörgetter, M., Prankl H. & Rathbauer, J. (1998). Eigenschaften von Biodiesel. *Landbauforschung Völkenrode*. Sonderheft 190 (Biodiesel-Optimierungspotentiale und Umwelteffkte), p.p. 31-43, ISSN: 04586859

Xie, W. & Li, H. (2006). Alumni-supported potassium iodide as a heterogeneous catalyst for biodiesel production from soybean oil. *Journal of Molecular Catalysis A: Chemical*, Vol. 255, No. 1-2, (August 2006), p.p. 1-9, ISSN 1381-1169

Xie, W., Peng, H. & Chen, L. (2006). Transesterification of soybean oil catalyzed by potassium loaded on alumina as a solid-base catalyst. *Applied Catalysis A: General*, Vol. 300, No. 1, (January 2006), p.p. 67-74, ISSN 0926-860X

Yan, S., Lu, H. & Liang, B. (2008). Supported CaO Catalysts Used in the Transesterification of Rapeseed Oil for the Purpose of Biodiesel Production. *Energy and Fuels, Vol.* 22. No. 1, (January 2008), p.p. 646-651, ISSN 08870624

Yoo, S.J., Lee, H., Veriansyah, B., Kim, J., Kim, J-D., & Lee, Y-W. (2010). Synthesis of biodiesel from rapeseed oil using supercritical methanol with metal oxide catalysts. *Bioresource Technology*, Vol. 101, No. 22, (November 2010), p.p. 8686-8689, ISSN 0960-8524

Yoosuk, B.; Udomsap, P.; Pattasawat, B. & Krasae, P. (2010). Improving transesterification activity of CaO with hydration technique. *Bioresource Technology*. Vol. 101, No. 10, (May 2010), p.p. 3784-3786, ISSN 0960-8524

Zhu, H., Wu, Z., Chen, Y., Zhang, P., Duan, S., Liu, X. & Mao, Z. (2006). Preparation of Biodiesel Catalyzed by Solid Super Base of Calcium Oxide and Its Refining Process. *Chinese Journal of Catalysis*, Vol. 27, No. 5, (May 2006), p.p. 391-396, ISSN: 1872-2067

2

An Overview of Enzyme-Catalyzed Reactions and Alternative Feedstock for Biodiesel Production

Maria Manuela Camino Feltes, Débora de Oliveira,
Jorge Luiz Ninow and José Vladimir de Oliveira
Federal University of Santa Catarina
Brazil

1. Introduction

There is an increasing interest in biodiesel production because of the environmental and economic advantages of this biofuel over traditional fuels derived from non-renewable resources (Akoh et al., 2007). Biodiesel is a renewable, non-toxic and biodegradable fuel (Van Gerpen et al., 2004) defined as the mono alkyl esters of long chain fatty acids derived from vegetable or animal fats, for use in compression-ignition (diesel) engines (American Society for Testing and Material [ASTM], 1989). This fuel can be used pure or in a blend with petroleum-based diesel (Van Gerpen et al., 2004). Currently, large-scale biodiesel production has been done by means of alcoholysis reactions of fats and oils (Reaney et al., 2005). The choice of the suitable alcohol for this process will depend on factors like its cost, availability, toxicity, easy of recovering and recycling, and the amount of alcohol needed for the reaction. Commercially, biodiesel has been produced using methanol as alkyl donor due to advantages like easy recovering and low cost (Van Gerpen et al., 2004). Anhydrous ethanol has also a great potential for biodiesel production in Brazil, one of the largest world ethanol producers (Schuchardt et al., 1998).

Renewable sources from both vegetable (Reaney et al., 2005) and animal (Lee et al., 2002) origin can be used as raw materials for biodiesel production. Microalgal oils have also received much attention as substrate for esters formation (Miao & Wu, 2006). Many agro-industrial processing operations generate a significant amount of lipids as by-products, such as chicken fat, beef tallow and fish oil. This feedstock, usually applied in livestock feeding, is a viable resource to produce biodiesel, and this use is a way for adding value to this material (Feltes et al. 2009). The conversion of used cooking oils and restaurant grease into alkyl esters is another proposal of particular interest in the field of alternative fuels (Lee et al., 2002). The use of biomass as feedstock is an attractive alternative for biodiesel production in tropical countries such as Brazil. The agricultural and farming industries are activities of great importance in many parts of this country. Also, government policies and legislative action may rapidly increase biodiesel production in such region (Ferrari et al., 2005; Meneghetti et al., 2006).

Chemical catalysis is a well established process for biodiesel production (Bournay et al., 2005). The homogeneous alkali-catalyzed transesterification process has been extensively applied to the large-scale synthesis of alkyl esters, especially due to the low cost of base catalysts and their efficiency even at low concentrations. The chemical reaction, however, has some disadvantages as it is energy-intensive, requires several separation/purification steps and generates significant amounts of wastewater to be treated. Additionally, high temperature alkali catalysis accelerates the oil oxidation, which is a serious drawback when dealing with a feedstock rich in polyunsaturated fatty acids (Damstrup et al., 2006). Also, ester phase processing and waste disposal are great problems when dealing with alkali-catalyzed reactions (Hájek & Skopal, 2010, Van Gerpen et al., 2004).

Many investigations have been conducted for the synthesis of biodiesel mediated by lipases. The enzymatic approach for the modification of oils and fats takes advantage of the specificity of some lipases, leading to high purity products, therefore reducing post processing operations and costs. Due to the mild conditions under which the enzymatic reaction occurs, oil oxidation becomes a negligible problem and less energy is required. Moreover, wastes are minimized and by-products are more easily purified (Gunstone, 1999). It is worth mentioning that the alcohol used in the reaction can be inhibitory to some enzymes (Van Gerpen et al., 2004). Moreover, compared to classical chemical catalysts, the relatively low stability of enzymes in their native state as well as the high cost of commercial lipases are important drawbacks to the use of these biocatalysts in industrial processes (Villeneuve et al., 2000). For these reasons, much effort has been directed for reducing lipases costs therefore allowing the development of competitive enzymatic processes with potential for industrial application. The production of lipases from new sources (Rigo et al., 2010; Wolski et al., 2009), the development of techniques for lipases immobilization (Villeneuve et al., 2000) as well as to perform enzyme-catalyzed reactions in compressed or supercritical fluids, such as propane, *n*-butane or carbon dioxide have appeared as attractive alternatives for reaching these goals (Dalla Rosa et al., 2009; Oliveira et al., 2006).

In light of the increasing interest in the development of alternative energy sources, the aim of this section is to make a review of an enzymatic approach for biodiesel production, focusing on a sustainable process. A section of this chapter will be dedicated to the raw materials that have been used as feedstock for biodiesel production, especially in Brazil, outlining the use of agro-industrial residues. We will survey recent researches for the production of lipases with great potential for biodiesel synthesis. The enzyme-catalyzed production of this biofuel in compressed or supercritical fluids will also be reviewed.

2. Biodiesel as fuel

Recently, there are several concerns about the dependence on foreign-controlled fuels sources of finite supply, reserves of petroleum shortages and rising prices of petroleum-based fuels. These problems have encouraged the development of alternative fuels, obtained especially from renewable sources (Reaney et al., 2005) like fats and oils. Oils can be used as liquid fuel for combustion in compression-ignition (diesel) engines directly or in a blend with other fuels. This use, however, has many problems (Rathore & Madras, 2007), that can be overcome by means of pyrolysis, micro-emulsification or transesterification in order to modify raw fats or oils. Transesterification is by far the most used process for oil modification, yielding fatty acid alkyl esters, the so-called biodiesel (Cerveró et al., 2008). This biofuel has a viscosity similar to that of the petrodiesel, as the oil viscosity decreases as transesterification reaction proceeds, as verified for alkyl esters of refined soybean oil (Costa Neto et al., 2004).

Biodiesel possesses several technical and environmental advantages over the conventional diesel. Besides being renewable and of domestic origin, advantages of biodiesel compared to petrodiesel include biodegradability, higher flash point, reduction of most regulated exhaust emissions, miscibility in all ratios with petrodiesel, compatibility with the existing fuel distribution infrastructure, and inherent lubricity (Knothe, 2008; Moser, 2009). Ferrari et al. (2005) produced ethyl esters from neutral soybean oil that were further mixed with petroleum-based diesel and tested as fuel in an energy generator. The diesel oil consumption ($0.6981L.h^{-1}$) was reduced when this fuel was mixed with up to 20% of biodiesel, when the fuel consumption was $0.6875L.h^{-1}$. Technical problems with biodiesel include oxidative stability, cold flow, and increased NOx exhaust emissions (Knothe, 2008).
Solutions to some of the aforementioned problems have included the use of additives, the modification of the fatty acid composition of the feedstock, or mixtures of components with advantageous properties (Knothe, 2008).
In Brazil, the Law 11.097/2005 introduced biodiesel in the energy sources to be used in this country (Agência Nacional do Petróleo, Gás Natural e Biocombustíveis, 2005). Since January 2010, with the National Biodiesel Program, the blend of 5% of biodiesel has been mandatory, according to the Governmental regulation 6/2009 (Conselho Nacional de Política Energética, 2009). The biodiesel produced in Brazil in 2010 was about 2.4 million m^3 in 67 facilities whose total capability was $16{,}414.51m^3.day^{-1}$ (Agência Nacional do Petróleo, Gás Natural e Biocombustíveis, 2011).

3. Interesterification reactions

Despite of some divergences between authors, acidolysis, alcoholysis (including glycerolysis), and interesterification, all fall under the heading transesterification. Interesterification or ester-ester interchange is the exchange of acyl groups between two esters, namely, two triacylglycerols. Acidolysis is the transfer of an acyl group between an acid and an ester. Alcoholysis is the process of exchanging acyl groups between an ester and an alcohol (Akoh et al., 2007). The most common method now in use for biodiesel production is reversible transesterification reactions in which the oil or fat is reacted with a monohydric alcohol in the presence of a catalyst (Figure 1), where the raw glycerol (glycerin) is obtained as co-product (Knothe & Dunn, 2005).

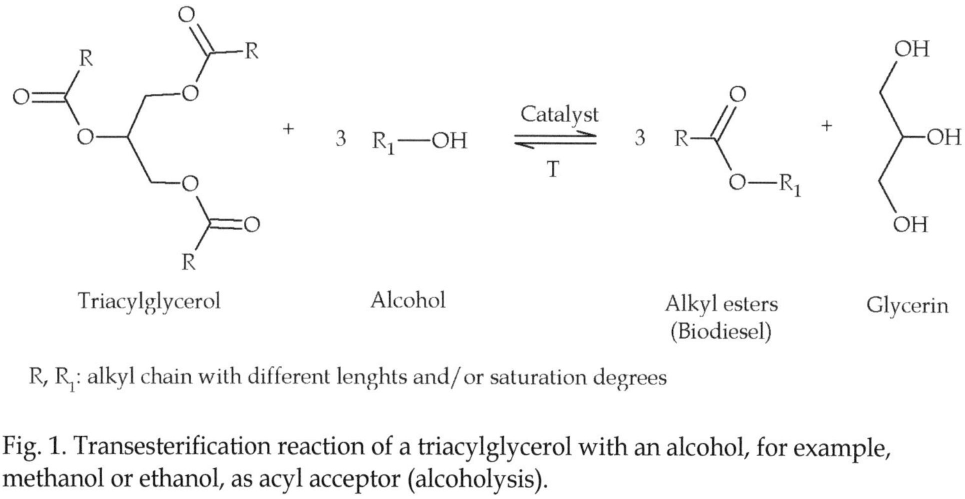

Fig. 1. Transesterification reaction of a triacylglycerol with an alcohol, for example, methanol or ethanol, as acyl acceptor (alcoholysis).

The glycerin obtained at large amounts and with a low cost in biodiesel production (Yazdani & Gonzalez, 2007) can be used as substrate for the synthesis of monoacylglycerols for application in the food industry, as suggested by Freitas et al. (2009). One has to consider, however, additional steps for products purification should be done when glycerin is obtained from the chemical-catalyzed biodiesel process, in order to assure a high-quality ingredient to human consumption (Hájek & Skopal, 2010).
Alternatively, as shown in Figure 2, methyl acetate can be used as acyl acceptor, where triacetyn, a valuable biodiesel additive, is produced as side product rather than glycerin, which has lower commercial value (Tan et al., 2010). Additionally, alkyl esters can be produced by means of direct esterification of fatty acids with alcohols (Figure 3). In this reaction, water must be removed to increase the yield, as the reverse reaction is the hydrolysis. Direct esterification is possible but not common for biodiesel preparation (Akoh et al., 2007).

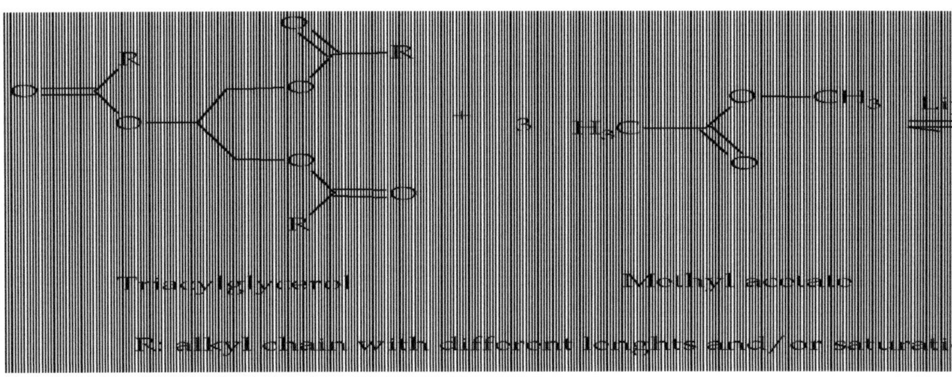

Fig. 2. Transesterification reaction of a triacylglycerol with methyl acetate as acyl acceptor (interesterification or ester-ester interchange) (Adapted from Du et al., 2004).

Since reactions are reversible, it is advantageous to try to shift the equilibrium toward the synthesis, by using an excess of one of the substrates or continuously removing water eventually formed during the reaction (Villeneuve, 2007).

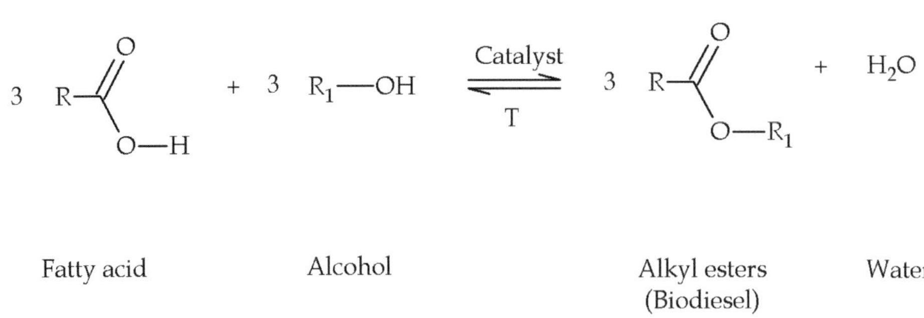

R, R1: alkyl chain with different lenghts and/or saturation degrees

Fig. 3. Transesterification reaction of a fatty acid with an alcohol as acyl acceptor (direct esterification). The reverse reaction is the hydrolysis.

3.1 Chemical versus lipase-catalyzed reactions

Currently, alcoholysis reaction by means of chemical catalysis is the most used method for the industrial production of biodiesel, especially because of the low cost of the catalysts, and high conversions reached in short time (Van Gerpen et al., 2004; Vicente et al., 1998). Many processes used in the oleochemical industry, however, are energy-intensive, requiring high temperatures (often above 200°C) and pressures. The resulting products generally need to be redistilled to remove impurities formed via thermal degradation. Moreover, highly unsaturated oils are thermally labile (Sheldon, 1996). Additionally, free fatty acids and water from the alcohol or the oil reduce ester yields, as they lead to the formation of soap and further free fatty acids, respectively (Fukuda et al., 2001). In the methanolysis of refined sunflower oil catalyzed by sodium hydroxide, for example, high temperatures (above 60°C) and catalyst concentrations (above 1.5%) led to the production of large amounts of soap (Vicente et al., 1998).

The development of green processes for biodiesel production has received much attention, involving the use of heterogenous catalysts, either chemical (Bournay et al., 2005; Boey et al., 2011) or enzymatic (Adamczak et al., 2009). In this sense, enzymatic processes can provide significant advantages over chemical process for biodiesel production. The mild conditions (pH, temperature and pressure), usually applied in the biocatalyzed reaction, allow energy saving and product quality improvement, because of the minimal thermal degradation of the substrates. Free fatty acids contained in waste oils and fats can be completely converted to alkyl esters, thus making their previous removal unnecessary (Van Gerpen et al., 2004).

In lipase-catalyzed reactions, glycerin can easily be recovered with simple separation processes (Cerveró et al., 2008). Moreover, the purification of fatty acid alkyl esters is simple to accomplish (Fukuda et al., 2001). These characteristics allow the reduction of residues treatment costs and of environmental concerns. Some disadvantages related to lipase-catalyzed reactions for biodiesel production include high cost (Rousseau & Marangoni, 2002), inactivation by acyl acceptors such as methanol depending on the enzyme, inactivation by minor components in the crude oil and waste oils, desorption from immobilization carrier, and fouling in packed bed bioreactors, as discussed by Akoh et al. (2007).

It should be emphasized that enzyme-catalyzed transesterification is a very good option to all chemical catalyzed reactions; however, it must be developed for its industrial application for esters formation (Kulkarni & Dalai, 2006). In order to use lipases more economically and efficiently in aqueous and non-aqueous solvents, techniques for their modification, to increase their activity, selectivity or stability and to improve their solubility in organic solvents, have been the subject of increased interest (Villeneuve et al., 2007).

4. Lipases as catalysts

Lipases (triacylglycerol lipases, E.C. 3.1.1.3) are carboxylic ester hydrolases with catalytic activity not only in an aqueous solution, but also in nonaqueous solvents (Rousseau and Marangoni 2002). The amount of water present in the reaction mixture influences biocatalysis in several ways. Often optimal water content can be found; further water addition will cause decreased enzyme activity. Water probably activates the enzyme by increasing the internal flexibility of the catalyst molecule. However, water can also act as a substrate in the enzymatic reaction, especially in the reactions of hydrolytic enzymes. This results in side-reactions and lower product yields (Adlercreutz, 1996). The best way to characterize the degree of hydration of a biocatalytic system in organic media is to use the thermodynamic water activity (a_w) as the parameter (Adlercreutz, 1996; Halling, 2002).

The optimum water content to the highest enzyme activity depends on the catalyst. The best results for the majority of lipases preparations are with a_w values ranging between 0.25 and 0.45, which corresponds to an water content from 0.5% to 1%. The optimal water content, however, can vary up to values as high as 11% (w/v) (Villeneuve, 2007).

4.1 New sources of lipases

Enzymes are produced by microorganisms, plants and animals. Purified enzymes, though expensive, are easy to apply for biotransformation in organic media and their use is increasing. Barriers to the production of industrial enzymes include economic factor, the availability of optimal enzymes and safety issues. The production and purification processes differ for extracellular and intracellular enzymes, liquid and solid culture in the case of fermentation, and enzyme application. Liquid fermentation is useful for the production of both intracellular and extracellular enzymes, especially from bacteria and yeast. It is good for scale-up and reproduction. The use of *Escherichia coli* genetically modified is very popular for the production of the enzyme inside the cell. Solid fermentation is also used to produce many kinds of enzymes mainly from fungal species (Hirose, 2002).

Solid state fermentation could be a perfect technology for adding value for low-cost agro-industrial residues aiming the production of enzymes, and for helping reducing pollution problems (Soccol & Vandenberghe, 2003). Salum et al. (2010) produced a lipase from *Burkholderia cepacia* LTEB11 (234 units of hydrolyzing activity U against *p*-nitrophenyl palmitate per gram of dry solids) by solid-state fermentation of a low-cost agro-industrial substrate (sugarcane bagasse and sunflower seed meal). Ramani et al. (2010) used beef tallow, a slaughter house waste, as substrate for lipase production by *Pseudomonas gessardii*. This enzyme showed high hydrolytic activity towards oils from different origin. Nelofer et al. (2010) reported the optimization of the production of a thermostable and organic solvent tolerant lipase by submerged culture of a recombinant *Escherichia coli* BL21. The lipase activity was 50.2 $U.mL^{-1}$ against an olive oil emulsion.

Griebeler et al. (2009) carried out the solid state fermentation with soybean bran of different microorganisms isolated from agro-industrial sources. They reported the most promising fungi with hydrolytic activity against tributyrin belonged to *Penicillium* and *Aspergillus* genera. It is worth mentioning Zanotto et al. (2009) pointed out triacylglycerols such as triolein or tributyrin are suitable substrates for the determination of lipase and esterase activity, respectively. Wolski et al. (2009) reported the partial characterization of enzymatic extracts produced by a newly isolated *Penicillium* sp. in submerged and solid state fermentation with, respectively, 15.17$U.mL^{-1}$ and 11.28$U.mL^{-1}$, by using an olive oil emulsion. In the solid state fermentation, soybean bran was used as a low-cost substrate. Zanotto et al. (2009) were able to isolate three fungi from plants of the Amazon region with a great potential as providers of mycelium-bound lipases for use in biotransformations (good hydrolysis activity, and good esterification and transesterification capacity in organic media). Chen et al. (2009) attained 91.08% of fatty acid methyl esters in the alcoholysis of an waste cooking oil catalyzed by the lipase derived from *Candida* sp. 99-125, previously identified from sewage water in north China. The enzyme was immobilized on a carrier textile cloth by using co-fixing agents (polyethylene glycerol, tween and span, gelatine and lecithin).

Vargas et al. (2008) studied the lipase production by *Penicillium simplicissimum* isolated from the babassu oil industry by using soybean meal as substrate. The enzyme activity was

30U.g1 dry substrate in the reaction with an olive oil emulsion. Volpato et al. (2008) reported the use of the low-cost glycerin obtained as co-product in biodiesel production as an alternative feedstock for the production of an organic solvent tolerant lipase by submerged culture of *Staphylococcus caseolyticus* EX17 (activity of 127.3U.L^{-1} against *p*-nitrophenyl palmitate). Castilho et al. (2000) performed an economic analysis of the production of *Penicillium restrictum* lipase in submerged fermentation as well as by solid-state fermentation. The latter process turned out to be very attractive from an economic point of view, especially because of the extremely cheap raw material used as main substrate (babassu cake and olive oil).

The use of intracellular lipase as a whole cell biocatalyst immobilized within biomass carrier particles is significantly advantageous since immobilization can be achieved spontaneously during batch cultivation, and in addition, no purification is necessary (Fukuda et al., 2001). For these reasons, Robles-Medina et al. (2009) argued that the use of the whole cell biocatalyst can reduce both the cost of the biocatalyst and dependence on lipase manufacturers. For more details about microbial lipase production, the reader is referred to a recent review on this subject (Treichel et al., 2010).

According to Foglia & Villeneuve (1997), the high cost often associated with an enzymatic process can be overcome by the utilization of plant enzymes, which take the advantage of the lower cost, when compared to microbial lipases, and ready availability. These authors described high activity in synthesis reactions with *Carica papaya* latex lipase. Cambon et al. (2009) verified the lipase from the unripe fruit of the babaco plant (*Vasconcellea heilbornii*; ex. *Carica pentagona*, activity of 280IU.g^{-1} against olive oil), was able to catalyse the alcoholysis reaction of sunflower oil with alcohols varying in chains length.

5. Substrates and reaction media

The following sections are dedicated to an overview of sources of alcohol, fats and oils that have been used for biodiesel production, focusing on lipase-catalyzed reactions.

5.1 Sources of alcohol

For economic reasons, methanol is most commonly used in transesterification reactions to the commercial production of biodiesel, although other alcohols can also be applied as alkyl donors (Moser, 2009), such as ethanol, propanol, and butanol (Knothe & Dunn, 2005). In general, improved conversion is achieved by adding extra alcohol (Reaney et al., 2005). It is known that largest conversions to methyl esters, for example, are obtained for a methanol to triacylglycerols molar ratio of 6:1 (Vicente et al., 1998). Methanol in excess, however, may lead to lipase inactivation. According to Soumanou & Bornscheuer (2003), particularly with water-miscible alcohol substrates such as methanol, water required to maintain lipase structure can be stripped leading to lower activity and eventual biocatalyst inactivation. This problem can be overcome, for example, by using methyl acetate as acyl acceptor, as already verified in the transformation of soybean oil for biodiesel production catalyzed by the commercial lipase Novozym 435 from *Candida antarctica* (Du et al., 2004). Alternatively, a method of adding methanol stepwise was successful for avoiding the enzyme inactivation in the methanolysis of degummed soybean oil catalyzed by the immobilized *Candida antarctica* lipase (Watanabe et al., 2002). Another possibility to prevent methanol inactivation during lipase-catalyzed reactions is the addition of *tert*-butanol (Akoh et al., 2007).

Rodrigues et al. (2008b) presented the alcoholysis of different vegetable oils (soybean, sunflower and rice bran) catalyzed by three immobilized commercial lipases. Novozym 435 from *Candida antarctica*, Lipozyme TL-IM from *Thermomyces lanuginosus* and Lipozyme RM-IM from *Rhizomucor miehei* presented higher activity in methanolysis (with an alcohol:oil molar ratio of 5:1), ethanolysis (7:1) and butanolysis (9:1), respectively, in the range of 30–35°C. The assessment of enzyme stability over repeated batches was carried out by washing the immobilized enzymes with different solvents. When washing with *n*-hexane, approximately 90% of the enzyme activity remained after seven synthesis cycles.

Higher molecular weight alcohols have lower polarity and therefore their separation from alkyl esters becomes more difficult after reaction completion, which can be overcome by mixing higher alcohols with methanol, glycerol or water (Reaney et al., 2005). Ethanol has a great potential for use in biodiesel production in countries like Brazil, where it is a cheap and abundant commodity produced from the fermentation of sucrose from sugarcane (Rodrigues et al., 2008b). In 2009, Brazil produced 26.10 million m^3 of ethanol, whose major producer was by far São Paulo State (58% of the total production) (Agência Nacional do Petróleo, Gás Natural e Biocombustíveis, 2009). In spite of its renewable origin, the ethanolysis reaction has some drawbacks to the industrial production of biodiesel. The yield of fatty acid ethyl esters is less compared to methyl esters and separation of glycerol is the main constrain in the process of ethanolysis (Meher et al., 2006).

Propanol or butanol can also be used in alcoholysis reactions, especially because these two alcohols promote a better miscibility between the alcohol and the oil phases (Iso et al., 2001). Soumanou & Bornscheuer (2003) were able to produce fatty acid alkyl esters from various vegetable oils (cottonseed, peanut, sunflower, palm olein, coconut and palm kernel) by transesterification with different alcohols (methanol, ethanol, 1-propanol, 1-butanol and isobutanol) using commercial immobilized lipases (AK lipase from *Pseudomonas fluorescens*, RM lipase from *Rhizomucor miehei*, and lipase from *Thermomyces lanuginosa*). Using *n*-hexane as organic solvent, all immobilized lipases tested were found to be active during methanolysis. Highest conversion (97%) was observed with *Thermomyces lanuginosa* lipase after 24h.

5.2 Sources of fats and oils

Along with vegetable oils, animal fats and oils also have a great potential to be used as substrate for biodiesel production. The choice of the source depends largely on the country and its climate (Knothe & Dunn, 2005), and also on economics (Rodrigues et al., 2008b). A wide variety of feedstock is expected to be used in Brazil for biodiesel production. Currently, the main resource for biodiesel production in this country is soybean oil (*Glycine max*), comprising about 80% of total feedstock, followed by tallow, cottonseed oil and other sources such as chicken fat, pork fat, waste cooking oil, oilseed radish (*Raphanus sativus* L.) (Agência Nacional do Petróleo, Gás Natural e Biocombustíveis, 2011).

The use of biodiesel has not expanded in developing countries mostly due to the high production cost which is associated with the expensive high-quality virgin oil feedstocks (Hoque et al., 2010). The cost of the raw materials currently represents about 70% of the total production costs (Robles-Medina et al., 2009).

Fats and oils of differing origin have different fatty acid compositions (Table 1). The properties of fats and oils, and hence their commercial value, depend on the fatty acid composition and of the triacylglycerol structure (Sheldon, 1996). For example, a high content of saturated fatty acids in the raw feedstock (i.e. myristic, palmitic, stearic acids) rises turbidity and the cetane number, decreases NOx emissions, and shows better stability. Furthermore, a high ratio of polyunsaturated fatty acids (i.e. linoleic or linolenic acids) has the opposite effect (Cerveró et al., 2008). Knothe & Dunn (2005) discussed in detail the influence of fats and oils compositions on biodiesel properties.

One of the main criteria for the quality of biodiesel is the storage stability which, in turn, depends on the fatty acid composition of the feedstock. Vegetable oil derivatives especially tend to deteriorate owing to hydrolytic and oxidative reactions. Their degree of unsaturation makes them susceptible to thermal and/or oxidative polymerization, which may lead to the formation of insoluble products that cause problems within the fuel system, especially in the injection pump (Meher et al., 2006). Santos et al. (2010) outlined the high amount of unsaturated fatty acids found in fish oil may result in higher oxidation problems. It is also a critical parameter when dealing with phytoplankters and seaweeds rich in long-chain polyunsaturated fatty acids. Autooxidation of unsaturated oils can be limited by addition of suitable antioxidants (Lands, 2005).

Fatty acid	Soybean oil[a]	Cotton seed oil[b]	Beef tallow[b]	Chicken fat[b]
C12:0, lauric			1	
C14:0, myristic		1	4	1
C16:0, palmitic	11	23	26	25
C16:1, palmitoleic		1	4	8
C18:0, stearic	4	2	20	6
C18:1, oleic	25	17	28	41
C18:2, linoleic	53	56	3	18
C18:3, linolenic	7			1
Other			14	

[a] From Liu (2004); [b] From Gunstone & Harwood (2007).

Table 1. Typical fatty acid composition (wt%) found in vegetable and animal oils and fats that may be used for biodiesel production.

Knothe & Dunn (2005) outlined biodiesel is susceptible to start-up and operability problems when subjected to cold temperatures. Biodiesel made from feedstock containing larger amounts of high-melting point long-chain saturated fatty acids tends to have relatively poor cold flow properties. This drawback can be overcome by the transesterification of triacylglycerols with medium chain-length (C3-C8) or branched-chain alkyl alcohols.

Despite of the economic, environmental and social advantages of biodiesel as an energy source, there is concern about the utilization of edible oils as substrates for its production. Taking it into consideration, non-edible and waste oils have been investigated as starting material (Robles-Medina et al., 2009).

5.2.1 Vegetable fats and oils

Traditional oilseed feedstocks for biodiesel production predominantly include soybean, rapeseed/canola, palm, corn, sunflower, cottonseed, peanut, and coconut oils (Knothe & Dunn, 2005; Moser, 2009).

According to Liu (2004), the range oil content in soybean is 12-30% (dry weight; typical oil content 20%), whose major components are triacylglycerols. There is a large genetic variation in fatty acid composition of soybean oil, mainly resulting from plant breeding. The range of fatty composition (% relative to total oil) among soybean germplasm is palmitic acid, 4-23%; stearic acid, 3-30%; oleic acid, 25-86%; linoleic acid, 25-60%; linolenic acid, 1-15%. This oil contains minor components (phospholipids, unsaponifiable material, free fatty acids, and trace metals) that are reduced after oil refinement. These substances affect lipase activity, as verified in the methanolysis of unrefined soybean oil, containing phospholipids, catalyzed by the immobilized *Candida antarctica* lipase (Watanabe et al., 2002).

Palm-base alkyl esters are potential substitutes for diesel fuel, as already investigated by Dors (2011). Palm trees are an attractive alternative feedstock, especially because of their great productivity (610 gallons of oil per acre) (Kurki et al., 2010). These plants produce two types of oils, palm oil from the fibrous mesocarp and lauric oil from the palm kernel. The former contains mainly palmitic (44.0%) and oleic acids (39.2%), while the latter is rich in lauric (48.3%), myristic (15.6%) and oleic acids (15.1%) (Teoh, 2002). The world´s two largest producing nations are Indonesia and Malaysia (United States Department of Agriculture, [USDA], 2011). Batistella et al. (2002) proposed an approach to recover carotenoids, valuable pigments for the food industry, and biodiesel from palm oil by means of suitable techniques.

Jatropha curcas oil, a non-edible vegetable oil, has been considered as a potential alternative feedstock for biodiesel production. *Jatropha curcas* is a large shrub or tree native to the American tropics but commonly found and utilized throughout most of the tropical and subtropical regions of the world. It is a hardness plant with rapid growth, easy propagation and wide ranging usefulness. The oil content of jatropha seed ranges from 30 to 50% by weight and the kernel itself ranges from 45 to 60%. The major fatty acids found in jatropha oil are the oleic and linoleic acids. The seeds and oil are toxic due to the presence of cursive and curcasive (Pramanik, 2003). The properties of this oil make it suitable for alkyl esters production, as they have lower viscosity than the starting material. Berchmans & Hirata (2008) produced biodiesel from crude *Jatropha curcas* L. seed with a high content of free fatty acids (15%) by means of a two-step chemical reaction.

Castor oil is a unique commercial resource containing ricinoleic acid (12-hydroxy-9-octadecene acid; C18:1, OH) with a major share of up to about 90%. Castor oil is obtained from the seeds of *Ricinus communis*, with a yield of about 50%. It is non-edible, virtually insoluble in water, chemically stable and fully biodegradable. It is widely used in oleochemical applications, mainly in special productions. The countries of production are primarily India and Brazil, where the plant grows native with high hectare yields. Cvengros et al. (2006) suggested a dilution of castor oil-based methyl and ethyl esters with esters based on oils/fats without -OH group, or a blending with conventional diesel fuel and addition of a cetane enhancer, as alternatives for reaching EN 14214 standards. Oliveira et al. (2004a, 2004b) investigated the lipase-catalyzed transesterification of castor oil to produce biodiesel.

An alternative feedstock for biodiesel production is the non-edible oil extracted from babassu (*Orbinya martiana*), a palm tree available in great amount in tropical forests. Their coconuts contain in average 7 wt% of almonds with 62 wt% of oil rich in lauric acid. The possibility of the whole usage of the coconut for other purposes makes babassu a potential source for biodiesel production (Freitas et al., 2009). Rubber tree (*Hevea brasiliensis*) is another potential feedstock for biodiesel production (Ramadhas et al., 2005). Rubber seed kernels (50–60% of seed) contain 40–50% of brown color oil, consisting of 18.9% saturated (palmitic and stearic acids) and 80.5% unsaturated fatty acids (oleic, linoleic and linolenic acids). It may contain up to 17% free fatty acids.

5.2.2 Animal fats and oils

Although animal fat such as beef tallow is considered an industrial residue, and not a renewable source of energy, if compared to oleaginous seeds, there is an increasing interest in its use as feedstock for biodiesel production. Additionally, the use of animal fat from residues as starting material for alkyl esters production plays an important role in the pollution reduction (Krause, 2008). According to Lee et al. (2002), lard is an inexpensive co-product of the meat-packing industry that could be further converted to value-added products such as biodiesel. However, the relatively high contents of saturated fatty acids in lard result in the poor cold temperature properties of its alkyl esters. To improve cold temperature properties, the content of saturated fatty acid in lard could be reduced through a fractionation process before esterification. As a consequence of this procedure, biodiesel from fractionated lard resists forming crystals at low temperatures, thus improving cold-temperature properties.

Chicken fat obtained from processing by-products is a promising alternative feedstock for biodiesel production because of its great availability and low cost (Feltes et al., 2009). Brazil is the third world´s largest producer of chicken meat and the first largest exporter of chicken meat, whose total production was 10.98 million tonnes in 2010 (Associação Brasileira dos Exportadores de Frango, 2011). Boye et al. (2011) studied the methanolysis of low free fatty acids chicken fat by using shells as catalysts. Krause (2008) described the economic feasibility of biodiesel production from animal fats (beef tallow and chicken fat), by using the alkaline methanolic route. The same research group reported the alkali-catalyzed methanolysis of biodiesel from beef tallow in a pilot plant (Cunha et al., 2009). Hoque et al. (2010) demonstrated a high potential of producing economically viable biodiesel from low cost feedstock (beef fat, chicken fat and used cooking oil) with proper optimization of the process parameters.

According to Food and Agriculture Organization of the United Nations [FAO] (2010), capture fisheries and aquaculture supplied the world with about 142 million tonnes of fish in 2008, excluding aquatic plants. Of this, 27.2 million tonnes were used for non-food purposes. The dominant species in marine fishery catches were anchoveta (*Engraulis ringens*), Alaska Pollock (*Theragra chalcogramma*) and Atlantic herring (*Clupea harengus*). China remained by far the largest fish-producing country, with production of 47.5 million tonnes in 2008. Brazil was one of the major producer countries of inland capture fisheries, accounting for an estimated amount of 243,000 tonnes in 2008.

Total fish oil production by the five main exporting countries (Peru, Chile, Iceland, Norway and Denmark) in 2009 was 530,000 tonnes, a decline of 100,000 tonnes compared with 2008. Fish-oil prices reached US$950/tonne in March 2010, which was 50 percent higher than an year earlier. For fish oil, almost 85 percent of production was used as an ingredient in fish and shrimp feeds (FAO, 2010). This oil could be applied as substrate for the production of biodiesel (Feltes et al., 2010).

In 2009, the total fisheries production in Brazil was more than 1.24 million tonnes. The Northeast region was the major national producer (34%), but the state of Santa Catarina, at the South, was the largest producer (more than 199,000 tonnes). The sardine (*Sardinella brasiliensis*) was the main fish resource caught in this country (more than 83,000 tonnes) in the aforementioned year (Ministério da Pesca e Agricultura, 2010). According to Feltes (2006), the oil extracted from sardine (*Sardinella brasiliensis*) viscera and heads by the Bligh & Dyer method, contains 5.43% of lipids (wet weight), with 12.1% of EPA and 16.5% of DHA. These residues are therefore an alternative substrate for the production of biodiesel in this region.

Santos et al. (2010) investigated the chemical production of biodiesel from oil extracted from the viscera of *Oreochromis niloticus*, containing 15.9%, 12.1%, 18.1% and 15.8% of palmitic, oleic, eicosapentaenoic and docosahexaenoic acids, respectively. These authors converted the Nile tilapia oil into free fatty acids, further esterified with methanol by using sulphuric acid as catalyst in ultrasonic system. Chiou et al. (2008) showed that despite the differences in the fatty acid profile, waste Alaska pink salmon (*Oncorhynchus gorbuscha*) oil and corn oil methyl esters had comparable physical properties (viscosity, volatility, low temperature properties, oxidative stability, heating value, acid value, and specific gravity), and could be used as automotive fuel.

5.2.3 Waste fats and oils

Low-cost sources such as restaurant greases (Lee et al., 2002) and soapstock can be used for biodiesel production. These sources, however, are of lower quality than refined vegetable oils, especially because of the high content of free fatty acids, what is a major problem when dealing with transesterification reactions catalyzed by alkali (Knothe & Dunn, 2005). Along with free fatty acids, restaurant grease contains other substances formed during cooking, especially deep-frying (Akoh & Min, 2008), that must be removed prior to alcoholysis in order to a high yield in alkaline-catalyzed reactions (Lee et al., 2002).

The use of vegetable oils as frying oils produces significant amounts of used oils which may present a disposal problem. Their use for biodiesel production has the advantage of their low price. Free fatty acids present in this feedstock can be a drawback when dealing with alkaline catalyzed-processes, but acid or enzymatic reactions can be a practical solution to the production of this biofuel (Knothe & Dunn, 2005). Besides free fatty acids, this starting material contains other degradation products as polymers, as revised by Kulkarni & Dalai (2006). These compounds can affect the catalyst activity, but they can be removed by means of simple filtration. Waste oils have higher amounts of saturated fatty acids, leading to less favourable cold flow properties (Knothe & Dunn, 2005).

Charoenchaitrakool & Thienmethangkoon (2011) investigated the optimum conditions in biodiesel production from waste frying oil using two-step catalyzed process (acid catalysis followed by alkali-catalysis). Although liquid mineral acids can be used to catalyze the modification of an oil with a high free fatty acids content, Silva et al. (2011) pointed out serious drawbacks, such as high corrosiveness, large effluent generation, the impossibility of catalyst reuse, compromise their application. Contrarily, solid acid catalysts appear to be an attractive option; however, the water present or generated during free fatty acids esterification causes the leaching and deactivation of these catalysts.

Coconut and palm kernel distillates containing a high content of free fatty acids (71.3 and 61.2%, w/w as lauric acid), rich in saturated fatty acids, especially lauric acid, a low content of unsaponifiable matter and neutral acylglycerols (Nandi et al., 2005) could be used as starting material for biodiesel production.

5.2.4 Aquatic plants

According to FAO (2010), procedures for the industrial preparation of biofuel not only from fish waste but also from seaweeds are being developed. Aquaculture produced 15.8 million tonnes (live weight equivalent) of aquatic plants in 2008.

Chile was the most important seaweed culturing country outside Asia, producing 21,700 tonnes in 2008, followed by Africa (14,700 tonnes). In 2008, the highest production of cultured seaweed was of Japanese kelp (*Laminaria japonica*, 4.8 million tonnes), followed by

Eucheuma seaweeds (*Kappaphycus alvarezii* and *Eucheuma* spp., 3.8 million tonnes), Wakame (*Undaria pinnatifida*, 1.8 million tonnes), *Gracilaria* spp. (1.4 million tonnes) and Nori (*Porphyra* spp., 1.4 million tonnes) (FAO, 2010).

The production of algae culture in freshwater was 68,400 tonnes in 2008, especially of *Spirulina* from China (62,300 tonnes) and Chile (6,000 tonnes) (FAO, 2010). Some microalgae produce substantial amounts of lipids (up to 70% w/w of biomass) with widely varying fatty acid composition in their biomass. The microalgae *Phaeodactylum triconutum*, for example, can produce 20% of oil (w/w) in biomass (5g lipids.m^{-2}.day^{-1}) (Robles-Medina et al., 2009). *Chlorella vulgaris* can produce 39% (w/w) of biomass rich in oleic acid (58% w/w of total) (Weber & Mukherjee, 2008). The latest development in freshwater algae culture is the development of lipid-rich species for biofuel production (FAO, 2010). The production cost of high grade algae oils, however, is still high, despite of the recent advances in photobioreactors, hindering the use of this feedstock in the short term, as outlined by Robles-Medina et al. (2009).

5.3 Reaction media

Several approaches have been made in order to optimize transesterification reactions for biodiesel production in batch systems (Table 2). Many studies are carried out in order to reduce enzymatic processes costs. In an attempt to produce biodiesel from used palm oil and ethanol using immobilized lipases in a solvent-free system, Tongboriboon et al. (2010) verified higher conversions could be achieved using the combination of Lipase AK (from *Pseudomonas fluorescens*) and Lipase AY (from *Candida rugosa*).

As can be seen in Table 2, several enzymatic reactions are performed in organic media. An organic solvent is used in order to dissolve the reactants, to enable interesterification rather than simple hydrolysis to occur, because excess water would cause the enzyme to desorb from its support (Cheetham, 1995), and also to improve the mass transfer during the reaction, increasing its rates. In all the solvent-containing biocatalytic systems, however, the nature of the solvent influences the activity and the stability of the enzyme to a large extent. The tendency of solvents to inactivate enzymes does not depend only on their hydrophobicity, expressed as their log *P* value (partition coefficient of the solvent in an octanol/water biphasic system), but also on other physicochemical characteristics of solvents, such as the solvating ability and molecular geometry (Adlercreutz, 1996).

Hydrophilic solvents with log P values < 2 such as *tert*-butanol often lead to enzyme deactivation if present in high concentrations; in contrast, apolar solvents with log $P \geq 4$ (*n*-heptane) are compatible with enzymes, leaving the essential layer of water molecules on the polar surface regions unperturbed. The results using solvents with intermediate values of log *P* (2-4) such as *n*-hexane are unpredictable and depend very much on the individual case (Kula, 2002). Laane et al. (1987) indicated lipases show a good stability and activity in hydrophobic solvents with $2 <$ log $P < 4$. Enzymes generally remain active and will not be rapidly deactivated when using a solvent with log $P > 3$ (Villeneuve, 2007). The choice of the suitable solvent for an enzymatic reaction must to be made taking into consideration both enzyme stability and substrate solubility (Adlercreutz, 1996).

When using Novozym 435, a commercial immobilized lipase, as catalyst, *tert*-butanol proved to be a suitable solvent for assure substrates homogeneity in the continuous biodiesel production from palm oil. In the ethanolysis of palm oil in the same system, however, the lipase AK immobilized on polysiloxane-polyvinyl alcohol hybrid support was able to produce ethyl esters without the use of this tertiary alcohol. These finding is extremely interesting for process costs (Dors, 2011). For industrial-scale production, despite of the benefits, the presence of organic solvents introduces problems like reduction of capacity (as the solvent takes up volume), environmental issues (toxicity, emissions) and costs (recovery and losses), as discussed by Nielsen et al. (2008).

Lipase	Oil or fat source	Acyl acceptor	Result	Conditions and additional information	Ref.
Lipozyme IM (Novo Nordisk) from *Rhizomucor miehei*, IM	Castor oil	Ethanol	99.5% yield	*n*-hexane; 6.5:1 ethanol to oil molar ratio; 6h; 200rpm; 50°C; 12.5% lipase (w/w based on oil and alcohol)	Oliveira et al. (2004a)
Lipozyme IM-77 (Novo Nordisk) from *Rhizomucor miehei*, IM	Soybean oil	Methanol	92.2% yield	*n*-hexane; 3.4:1(methanol to oil molar ratio); 6.3h; 200rpm; 36.5°C; 0.9BAUN* of lipase; added water 5.8wt% of oil	Shieh et al. (2003)
Lipozyme TL IM (Novozymes) from *Thermomyces lanuginosus*, IM	Soybean oil	Ethanol	96% yield	7.5:1 ethanol to oil molar ratio; 31.5°C; 5h; enzyme content, 15wt% (based on oil); added water, 4wt% (based on oil)	Rodrigues et al. (2008a)
Novozym 435 (Novozymes) from *Candida antarctica*, IM	Refined soybean oil	Methanol	95.9% yield	3:1 alcohol to oil molar ratio; methanol added in three steps; 48h; 130rpm; 30°C; 4wt.% lipase	Watanabe et al. (2002)
Novozym 435	Degummed soybean oil	Methanol	93.8% yield	3:1 alcohol to oil molar ratio; methanol added in three steps; 48h; 130rpm; 30°C; 4wt.% lipase; 25 cycles of reuse with no loss of activity	Watanabe et al. (2002)
Novozym 435	Crude and refined soybean oil	Methanol	-	Lipase inactivation with a molar ratio of methanol to oil > 1:1	Du et al. (2004)

Table 2. (Continued). Lipase-catalyzed reactions for biodiesel production in batch systems.

Novozym 435	Crude and refined soybean oil	Methyl acetate	92% yield	12:1 methyl acetate to oil molar ratio; 14h; 150rpm; 40°C; 30wt% enzyme (based on oil); 100 cycles of reuse with no loss of activity	Du et al. (2004)
Novozym 435	Soybean oil	Ethanol	100% conversion	6:1 ethanol to oil molar ratio; 6h; 1000rpm; 65°C; 50bar; 5wt% lipase (based on oil and ethanol); compressed propane (solvent to substrates weight ratio of 2:1)	Dalla Rosa et al. (2008)

* Abbreviations: IM: immobilized. BAUN: Batch Acidolysis Units Novozymes

Table 2. Lipase-catalyzed reactions for biodiesel production in batch systems.

The key issue in enzymatic methods is to reduce the cost of the catalyst or to improve its operational life (Du et al., 2004). The reuse of the catalyst is crucial to make the process economically feasible to a large scale application (Villeneuve et al., 2000). Furthermore, different reaction media or technologies have been applied for the development of a cost-effective enzymatic system for biodiesel production, as for example the use of low-frequency (10–60 kHz) ultrasound (Rokhina et al., 2009; Kumar et al., 2011), ionic liquids (Gamba et al., 2008). Supercritical fluids have also received much attention, as it will be discussed in the next section.

0. Enzymatic reactions in compressed and supercritical fluids

When a compound is compressed and heat above its critical point, it enters a phase that is referred to as its supercritical phase. The compound in the supercritical region is called as supercritical fluid (Dunford, 2004). An experimental setup used for alkyl esters production with immobilized lipase in pressurized fluids is schematically shown in Figure 4. Potential advantages with supercritical media are the high diffusivity and low surface tension which lead to low mass-transfer resistance (Adlercreutz, 1996).

High temperatures and pressures are needed during the non-catalytic transesterification process (Kulkarni & Dalai, 2006; Silva et al., 2007; Tan et al., 2010). For energy saving, transesterification reactions in supercritical fluids can be enzyme-catalyzed (Adlercreutz, 1996). Moisture content, particle size, and solvent flow rate are important parameters that must be optimized when using supercritical fluid technology for oil and oilseed processing (Dunford, 2004). Supercritical and near-supercritical fluids have been used for enzymatic reactions in a large number of studies. The most widely used medium in this group is supercritical carbon dioxide, whose main advantage is the easy removal after the reaction by just decreasing the pressure. Furthermore, supercritical carbon dioxide has low toxicity compared to organic solvents. As solvent, it resembles hexane and accordingly, hydrophobic compounds can be dissolved, but the solubility of more polar compounds is low, but this can be overcome by addition of organic co-solvents, when necessary. Enzymes are normally not soluble in the supercritical fluids, and therefore heterogeneous systems with solid enzyme preparations are normally used (Adlercreutz, 1996)

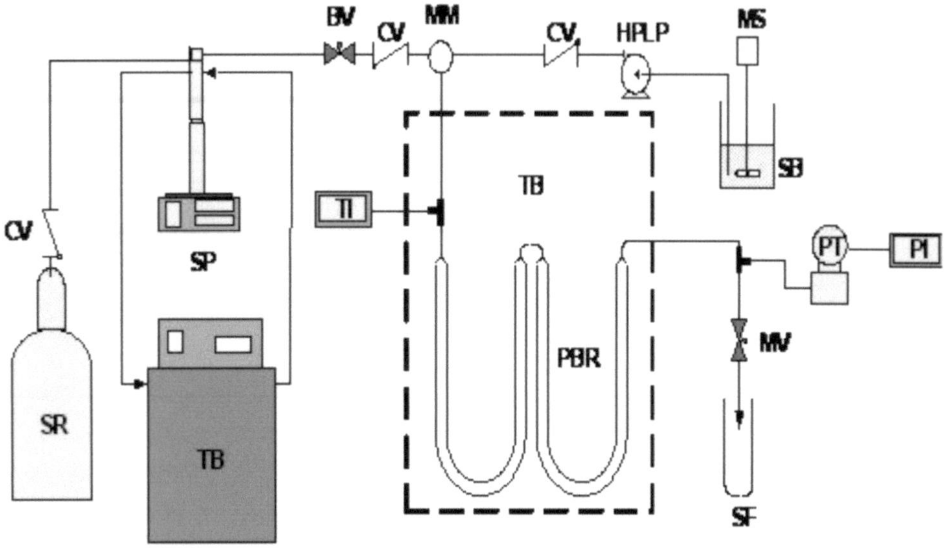

Fig. 4. Schematic diagram of the packed-bed experimental apparatus: SR- solvent reservoir; TB- thermostatic bath; SP- syringe pump; CV- check valve; MM- micro mixing; TI- temperature indicator; PI- pressure indicator; PBR- packed-bed reactor; HPLP- high-pressure liquid pump; SB- substrates; MS- mechanical stirrer; MV- micrometering valve; SF- sample flask collector. From Dalla Rosa et al. (2009).

Dalla Rosa et al. (2009) indicated the continuous lipase-catalyzed ethanolysis of soybean oil in a packed-bed reactor using compressed propane as solvent as a potential alternative to conventional process for biodiesel production. Good reaction conversions were obtained at mild temperature (70°C) and pressure (60bar), opening possibilities to manipulate process variables as well as reactor configurations towards reaching acceptable yields. Rathore & Madras (2007) investigated the synthesis of biodiesel from edible (palm oil, groundnut oil) and non-edible oils (*Pongamia pinnata* and *Jatropha curcas*) in supercritical methanol and ethanol without using any catalyst from 200 to 400°C at 200bar. Very high conversions (>80%) were obtained within 10min and nearly complete conversions were obtained after 40min. Biodiesel was also produced by the lipase Novozym 435 in presence of supercritical carbon dioxide. Conversions of only 60-70% were obtained in the enzymatic process even after 8h. One should consider, however, temperature reaction for the lipase-catalyzed method (45°C) was strongly lower than that used for the non-catalytic reaction. Moreover, the authors discussed stepwise methanol feeding and glycerol removal by dialysis could be adopted to avoid strong substrate and product inhibition, respectively, therefore increasing reaction conversion.

A preliminary study comparing conventional and supercritical media for the ethanolysis of palm kernel oil was carried out using two commercial immobilized lipases (Lipozyme IM from *Mucor miehei* and Novozym 435) (Oliveira & Oliveira, 2001). Higher conversions (up to

77.5%) were obtained in the presence of *n*-hexane for the reaction catalyzed by Lipozyme IM. Despite of enzyme activity losses and lower conversions (up to 63.2% with Novozym 435), supercritical carbon dioxide showed to be an appropriate solvent for enzymatic alcoholysis. Other studies involving lipase-catalyzed in pressurized fluids indicated this process as a potential technology for biodiesel production (Hildebrand et al., 2009).

According to Oliveira et al. (2006), the enzyme behaviour in compressed fluids is of primary importance as the loss of enzyme activity may lead to undesirable poor reaction rates and low yields of target products. Enzyme stability and activity may depend on the enzyme species, characteristics of compressed fluid, water content of the enzyme/carrier/reaction mixture and process variables manipulated. These authors investigated the behaviour of two commercial immobilized lipases submitted to compressed carbon dioxide, propane and *n*-butane under different reaction conditions (temperature from 35 to 75°C, pressure range of 10-280bar, exposure times from 1 to 6h, with distinct decompression rates). The lipase Lipozyme IM from *Mucor miehei* showed activity losses in all compressed solvents, markedly in carbon dioxide. For Novozym 435, treatment in carbon dioxide also led to activity losses, while the use of propane and *n*-butane promoted enhancements of the enzyme activity. Thermogravimetric analyses showed that the thermal profiles of Novozym 435 treated in *n*-butane and in carbon dioxide were similar to that of the untreated enzyme. Scanning electron micrographs (SEM) of Novozym 435 indicated that the material submitted to carbon dioxide presented morphological alterations when compared to the untreated enzyme (Figure 5).

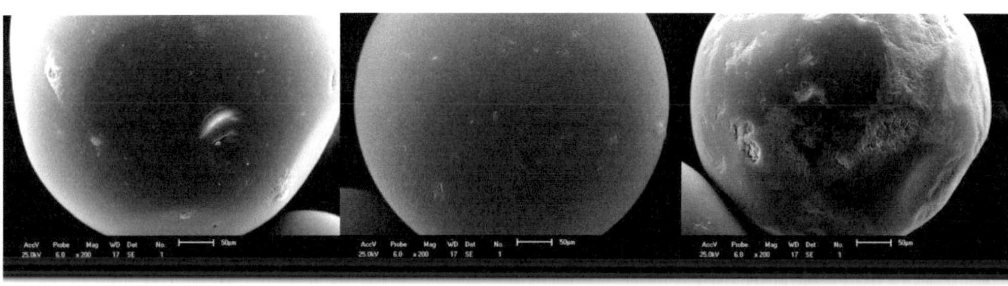

(a)

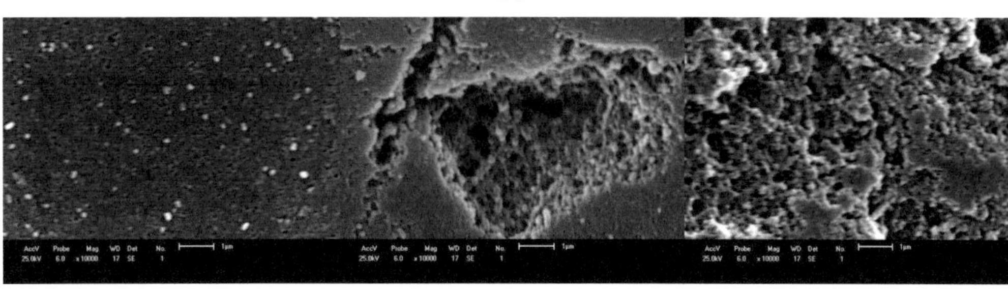

(b)

Fig. 5. SEM micrographs of Novozym 435 from *Candida antarctica* samples: (a) amplification of 200 times, (b) amplification 10,000 times. From left to right side, SEM micrographs refer to untreated enzyme, treated in *n*-butane (75°C, 250bar, 6h, decompression rate 5050bar.min^{-1}), and in carbon dioxide (75°C, 6h, decompression rate 200kg.m^{-3}.min^{-1}, reduced density 1.6, and related pressure 276.4bar), respectively. From Oliveira et al. (2006).

Lanza et al. (2004) observed the commercial immobilized lipase from *Candida antarctica* (Novozym 435) was stable in supercritical carbon dioxide under the investigated conditions (30 to 70°C, pressure range 70-250bar, 60-360min, decompression rates 10-200kg.m^{-3}.min^{-1}). These results therefore indicate the tested lipase can be used with low activity loss at typical conditions of temperature and pressure employed in many biotransformation reactions of raw materials in this supercritical fluid.

The commercial exploitation of supercritical fluids technology has been limited to a large-scale biodiesel production. One factor that may contribute to the development of this technology is the study of the phase behaviour of the high-pressure multicomponent systems found in transesterification reactions, as it is essential for the design of any supercritical fluid reaction system. Furthermore, these experimental data are of primary importance for selecting appropriate operational conditions to accomplish enzyme-catalyzed transesterification reactions in supercritical media. The phase behaviour of castor and soybean oils and their fatty acid ethyl esters in carbon dioxide at high pressures was already investigated, for example, by Ndiaye et al. (2006).

7. Conclusion

Several approaches have been made for a greener biodiesel production by means of biocatalyzed reactions. Lipases production from alternative sources, enzymes immobilization, as well as the use of alternative reaction media, such as supercritical fluids, have appeared as potential alternatives to reach this goal. Besides, the use of alternative feedstock, focusing on agro-industrial residues, seems to be a practical solution to reduce biodiesel production costs while ensuring a minimal environmental impact.

8. References

Adamczak, M., Bornscheuer, U.T., & Bednarski, W. (2009). The application of biotechnological methods for the synthesis of biodiesel. *European Journal of Lipid Science and Technology*, Vol.111, pp. 808-813, ISSN 1438-9312

Adlercreutz, P. (1996). Modes of using enzymes in organic media. In: *Enzymatic reactions in organic media*, Koskinen, A.M.P., & Klibanov, A.M., (Eds.), pp. 9-42, Blackie Academic & Professional, ISBN 075140259, Glasgow, Scotland

Agência Nacional do Petróleo, Gás Natural e Biocombustíveis (2005). *Lei 11.097 de 13 de janeiro de 2005*. Brasília, Brazil, Agência Nacional do Petróleo, Gás Natural e Biocombustíveis

Agência Nacional do Petróleo, Gás Natural e Biocombustíveis (2009). *Oil, Natural Gas and Biofuels Statistical Yearbook*. Rio de Janeiro, Brazil, Ministry of Mining and Energy, National Agency of Petroleum, Natural Gas and Biofuels. ISSN 2177-0271. 78p. Retrieved from <http://www.anp.gov.br>

Agência Nacional do Petróleo, Gás Natural e Biocombustíveis (2011). *Boletim mensal de biodiesel*. Brasília, Brazil, Agência Nacional de Petróleo, Gás Natural e Biocombustíveis, January 2011. 9p. Retrieved from <http://www.anp.gov.br>

Akoh, C.C., & Min, D.B. (2008). *Food lipids: chemistry, nutrition, and biotechnology* (3ed.). CRC Press, ISBN 1420-046632, Boca Ratón, USA. 914p

Akoh, C.C., Chang, S.-W., Lee, G.-C., & Shae, J.-F. (2007). Enzymatic approach to biodiesel production. *Journal of Agricultural and Food Chemistry*, Vol.55, pp. 8995-9005, ISSN 1520-5118

American Society for Testing and Materials (1989). *Standard specification for biodiesel fuel (B100) blend stock for distillate fuels*. ASTM Standards D-6751

Associação Brasileira dos Exportadores de Frango (2011). *Annual report 09-10*. Brazil, Associação Brasileira dos Exportadores de Frango. 31p. Retrieved from <http://www.abef.com.br/Relatorios_Anuais.php>

Batistella, C.B., Moraes, E.B., Maciel-Filho, R., & Wolf Maciel, M.R. (2002). Molecular distillation process for recovering biodiesel and carotenoids from palm oil. *Applied Biochemistry and Biotechnology*, Vol.98-100, 1149-1159, ISSN 0273-2289

Berchmans, H.J., & Hirata, S. (2008). Biodiesel production from crude *Jatropha curcas* L. seed oil with a high content of free fatty acids. *Bioresource Technology*, Vol.99, pp. 1716-1721, ISSN 0960-8524

Boey, P.-L., Maniam, G.P., Hamid, S.A., & Ali, D.M.H. (2011). Crab and cockle shells as catalysts for the preparation of methyl esters from low free fatty acid chicken fat. *Journal of the American Oil Chemists´ Society*, Vol.88, No.2, pp. 283-288, ISSN 1558-9331

Bournay, L., Casanave, D., Delfort, B., Hillion, G., & Chodorge, J.A. (2005). New heterogeneous process for biodiesel production: a way to improve the quality and the value of the crude glycerin produced by biodiesel plants. *Catalysis Today*, Vol.106, No.1-4, pp. 190-192, ISSN 0920-5861

Cambon, E., Bourlieu, C., Salum, T.F.C., Piombo, G., Dubreucq, E., & Villeneuve, P. (2009). Ability of *Vasconcellea x heilbornii* lipase to catalyse the synthesis of alkyl esters from vegetable oils. *Process Biochemistry*, Vol.44, pp. 1265-1269, ISSN 1359-5113

Castilho, L.R., Polato, C.M.S., Baruque, E.A., Sant'Anna Jr., G.L., & Freire, D.M.G. (2000). Economic analysis of lipase production by *Penicillium restrictum* in solid-state and submerged fermentations. *Biochemical Engineering Journal*, Vol.4, pp. 239-247, ISSN 1369-703X

Cerveró, J.M., Coca, J., & Luque, S. (2008). Production of biodiesel from vegetable oils. *Grasas y Aceites*, Vol.59, No.1, pp. 76-83, ISSN 1988-4214

Charoenchaitrakool, M., & Thienmethangkoon, J. (2011). Statistical optimization for biodiesel production from waste frying oil through two-step catalyzed process. *Fuel Processing Technology*, Vol.92, pp. 112-118, ISSN 0378-3820

Cheetham, P.S.J. (1995). The applications of enzymes in industry, In: *Handbook of enzyme biotechnology* (3rd ed.), Wiseman, A., (Ed.), pp. 419-552, ISBN 0-13-382920-0, New York, USA

Chen, Y., Xiao, B., Chang, J., Fu, Y., Lv, P., & Wang, X. (2009). Synthesis of biodiesel from waste cooking oil using immobilized lipase in fixed bed reactor. *Energy Conversion and Management*, Vol.50, pp. 668-673, ISSN 0196-8904

Chiou, B.-S., El Mashad, H. M., AvenaBustillos, R. J., Dunn, R. O., Bechtel, P. J., McHugh, T. H., Imam, S. H., Glenn, G. M., Orts, W. J., & Zhang, R. (2008). Biodiesel from waste salmon oil. *Transactions of the ASABE*, Vol.51, No.3, pp. 797-802, ISSN 0001-2351

Conselho Nacional de Política Energética (2009). *Resolução CNPE nº 6 de 16/9/2009*. Brasília: Conselho Nacional de Política Energética

Costa Neto, P.R., Caro, M.S.B., Mazzuco, L.M., & Nascimento, M.G. (2004). Quantification of soybean oil ethanolysis with ^{1}H NMR. *Journal of the American Oil Chemists´ Society*, Vol.81, No.12, pp. 1111-1114, ISSN 1558-9331

Cunha, M.E., Krause, L.C., Moraes, M.S.A, Faccini, C.S., Jacques, R.A., Almeida, S.R., Rodrigues, M.R.A, & Caramão, E.B. (2009). Beef tallow biodiesel produced in a pilot scale. *Fuel Processing Technology*, Vol.90, No.4, pp. 570-575, ISSN 0378-3820

Cvengros, J., Paligová, J., & Cvengrosová, Z. (2006). Properties of alkyl esters based on castor oil. *European Journal of Lipid Science and Tehnology*, Vol.108, pp. 629-635, ISSN 1438-9312

Dalla Rosa, C., Morandim, M. B., Ninow, J. L., Oliveira, D., Treichel, H., & Oliveira, J.V. (2008). Lipase-catalyzed production of fatty acid ethyl esters from soybean oil in compressed propane. *The Journal of Supercritical Fluids*, V.47, pp. 49-53, ISSN 0896-8446

Dalla Rosa, C., Morandim, M.B., Ninow, J.L., Oliveira, D., Treichel, H., & Oliveira, J.V. (2009). Continuous lipase-catalyzed production of fatty acid ethyl esters from soybean oil in compressed fluids. *Bioresource Technology*, Vol.100, pp. 5818-5826, ISSN 0960-8524

Damstrup, M.L., Jensen, T., Sparso, F.V., Kiil, S.Z., Jensen, A.D., & Xu, X. (2006). Production of heat-sensitive monoacylglycerols by enzymatic glycerolysis in *tert*-pentanol: process optimization by response surface methodology. *Journal of the American Oil Chemists´ Society*, Vol.83, No.1, pp. 27-33, ISSN 1558-9331

Dors, G. (2011). *Etanólise enzimática do óleo de palma visando à produção de biodiesel em sistema contínuo*. Doctorate thesis in Chemical Engineering. Florianópolis, Brazil, Federal University of Santa Catarina, 233p. Retrieved from <www2.enq.ufsc.br/teses/d096.pdf> (In Portuguese)

Du, W., Xu, Y., Liu, D., & Zeng, J. (2004). Comparative study on lipase-catalyzed transformation of soybean oil for biodiesel production with different acyl acceptors. *Journal of Molecular Catalysis B: Enzymatic*, Vol.30, pp. 125-129, ISSN 1381-1177

Dunford, N.T. (2004). Utilization of supercritical fluid technology for oil and oilseed processing. In: *Nutritionally enhanced edible oil and oilseed processing*, Dunford, N.T., & Dunford, H.B., (Eds.), pp. 100-116, AOCS Press, ISBN 1-893997-49-9, Champaign, USA

FAO (2010). *The state of world fisheries and aquaculture 2010*. Rome, Food and Agriculture Organization of the United Nations (FAO). FAO Fisheries and Aquaculture Department. 218p. Retrieved from <http://www.fao.org/fishery/sofia/en>

Feltes, M. M. C., Dors, G., Santos, A.F., Furigo Junior, A., & Ninow, J.L. (2009). Lipase-catalyzed synthesis of biodiesel via ethanolysis of chicken fat from processing by-products. *New Biotechnology*, Vol.25S, p. S277, ISSN 1871-6784

Feltes, M.M.C. (2006). *Estudo da síntese química e enzimática de triglicerídios estruturados a partir de óleo de peixe*. Master thesis in Food Engineering. Florianópolis, Brazil, Federal University of Santa Catarina, 121p. Retrieved from <http://www.tede.ufsc.br/teses/PEAL0066.pdf>. (In Portuguese)

Feltes, M.M.C., Correia, J.F.G., Beirão, L.H., Block, J.M., Ninow, J.L., & Spiller, V.R. (2010). Alternativas para a agregação de valor aos resíduos da industrialização de peixe.

Revista Brasileira de Engenharia Agrícola e Ambiental, Vol.14, No.6, pp. 669-677, ISSN 1415-4366

Ferrari, R.A., Oliveira, V.S., & Scabio, A. (2005). Biodiesel de soja - taxa de conversão em ésteres etílicos, caracterização físico-química e consumo em gerador de energia. *Química Nova*, Vol.28, No.1, pp. 19-23, ISSN 1678-7064

Foglia, T.A., & Villeneuve, P. (1997). *Carica papaya* latex-catalyzed synthesis of structured triacylglycerols. *Journal of the American Oil Chemists´ Society*, Vol.74, No.11, pp. 1447-1450, ISSN 1558-9331

Freitas, L., Da Rós, P.C.M., Santos, J.C., & Castro, H.F. (2009). An integrated approach to produce biodiesel and monoglycerides by enzymaticinterestification of babassu oil (*Orbinya* sp). *Process Biochemistry*, Vol.44, pp. 1068-1074, ISSN 1359-5113

Fukuda, H., Kondo, A., & Noda, H. (2001). Biodiesel fuel production by transesterification of oils. *Journal of Bioscience and Bioengineering*, Vol.92, No.5, pp. 405-416, ISSN 1389-1723

Gamba, M., Lapis, A.A.M, & Dupont, J. (2008). Supported ionic liquid enzymatic catalysis for the production of biodiesel. *Advanced Synthetic & Catalysis*, Vol.350, pp. 160-164, ISSN 1615-4169

Griebeler, N., Polloni, A.E., Remonatto, D., Arbter, F., Vardanega, R., Cechet, J.L., Di Luccio, M., Oliveira, D., Treichel, H., Cansian, R.L., Rigo, E., & Ninow, J.L. (2009). Isolation and screening of lipase-producing fungi with hydrolytic activity. *Food Bioprocess and Technology*, DOI 10.1007/s11947-008-0176-5, ISSN 1935-5149

Gunstone, F.D. (1999). Enzymes as biocatalysts in the modification of natural lipids. Review. *Journal of the Science of Food and Agriculture*, Vol.79, pp. 1535-1549, ISSN 1097-0010

Gunstone, F.D., & Harwood, J.L. (2007). Occurrence and characterization of oils and fats, In: *The Lipid Handbook* (3rd ed.), Gunstone, F.D., Harwood, J.L., & Dijkstra, A.J. (Eds.), pp. 37-142, CRC Press, ISBN 0-849396-88-3, Boca Ratón, USA

Hájek, M., & Skopal, F. (2010). Treatment of glycerol phase formed by biodiesel production. *Bioresource Technology*, Vol.101, No.9, pp. 3242-3245, ISSN 0960-8524

Halling, P. (2002). Enzymic conversions in organic and other low-water media, In: *Enzyme catalysis in organic synthesis: a comprehensive handbook* (2nd ed.), Drauz, K., & Waldmann, H., (Eds.), pp. 259-285, Wiley-VCH, ISBN 3-527-29949-1, Weinheim, Germany

Hildebrand, C., Dalla Rosa, C., Freira, D.M.G., Destain, J., Dariva, C., Oliveira, D., & Oliveira, J.V. (2009). Fatty acid ethyl esters production using a non-commercial lipase in pressurized propane medium. *Ciência e Tecnologia de Alimentos*, Vol.29, No.3, pp. 603-608, ISSN 1678-457X

Hirose, Y. (2002). Production and isolation of enzymes, In: *Enzyme catalysis in organic synthesis: a comprehensive handbook* (2nd ed.), Drauz, K., & Waldmann, H., (Eds.), pp. 41-66, Wiley-VCH, ISBN 3-527-29949-1, Weinheim, Germany

Hoque, E., Singh, A., & Chuan, Y.L. (2010). Biodiesel from low cost feedstocks: The effects of process parameters on the biodiesel yield. *Biomass and Bioenergy*, Vol.35, No.4, pp. 1582-1587, ISSN 0961-9534

Iso, M., Chen, B.X., Eguchi, M., Kudo, T., & Shrestha, S. (2001). Production of biodiesel fuel from triglycerides and alcohol using immobilized lipase. *Journal of Molecular Catalysis B: Enzymatic*, Vol.16, pp. 53–58, ISSN 1381-1177

Knothe, G. (2008). "Designer" biodiesel: optimizing fatty ester composition to improve fuel properties. *Energy Fuels*, V.22, No.2, pp. 1358–1364, ISSN 1520-5029

Knothe, G., & Dunn, R.O. (2005). Biodiesel: an alternative diesel fuel from vegetable oils or animal fats, In: *Industrial uses of vegetable oils*, Erhan, S.Z., (Ed.), pp. 42-89, AOCS Press, ISBN 1-893997-84-7, Champaign, USA

Krause, L.C. (2008). *Desenvolvimento do processo de produção de biodiesel de origem animal.* Doctorate thesis in Chemistry. Porto Alegre, Brazil, Federal University of Rio Grande do Sul. 130p. Retrieved from <http://hdl.handle.net/10183/14362> (In Portuguese)

Kula, M.-R. (2002). Introduction, In: *Enzyme catalysis in organic synthesis: a comprehensive handbook* (2nd ed.), Drauz, K., & Waldmann, H., (Eds.), pp. 1-39, Wiley-VCH, ISBN 3-527-29949-1, Weinheim, Germany

Kulkarni, M.G., & Dalai, A.K. (2006). Waste cooking oils as an economical source for biodiesel: a review. *Industrial Engineering and Chemical Research*, Vol.45, pp. 2901-2913, ISSN

Kumar, G., Kumar, D., Poonam, Johari, R., & Singh, C.P. (2011). Enzymatic transesterification of *Jatropha curcas* oil assisted by ultrasonication. *Ultrasonics Sonochemistry*, In Press, Accepted Manuscript, ISSN 1350-4177

Kurki, A., Hill, A., & Morris, M. (2010). Biodiesel: the sustainability dimensions, In: *ATTRA – National Sustainable Agriculture Information Service*, 10.04.2011, Available from: <http://www.attra.org/attra-pub/PDF/biodiesel_sustainable.pdf>

Laane, C., Boeren, S., Vos, K., & Veegek, C. (1987). Rules for optimization of biocatalysis in organic solvent. *Biotechnology and Bioengineering*, Vol.30, pp. 81–87, ISSN 1097-0290

Lands, W.E.M. (2005). Overall supply of n-3 fatty acids, In: *Fish, Omega-3 and Human Health* (2nd ed.), Lands, W.E.M., pp. 171-185, AOCS Press, ISBN 1-893997-81-2, Champaign, USA

Lanza, M., Priamo, W.L., Oliveira, J.V., Dariva, C., Oliveira, D. (2004). The effect of temperature, pressure, exposure time, and depressurization rate on lipase activity in $SCCO_2$. *Applied Biochemistry and Biotechnology*, Vol.113-116, pp. 181-187, ISSN 1559-0291

Lee, K.-T., Foglia, T.A., & Chang, K.-S. (2002). Production of alkyl ester as biodiesel from fractionated lard and restaurant grease. *Journal of the American Oil Chemists´ Society*, Vol.79, No.2, pp. 191-195, ISSN 1558-9331

Liu, K. (2004). Soybeans as a powerhouse of nutrients and phytochemicals, In: *Soybeans as functional foods and ingredients*, Liu, K., (Ed.), pp. 1-22, AOCS Press, ISBN 1-893997-33-2, Champaign, USA

Meher, L.C., Vidya Sagar, D., & Naik, S.N. (2006). Technical aspects of biodiesel production by transesterification—a review. *Renewable and Sustainable Energy Reviews*, Vol.10, pp. 248–268, ISSN 1364-0321

Meneghetti, S.M.P., Meneghetti, M.R., Wolf, C.R., Silva, E.C., Lima, G.E.S., Coimbra, M.A., Soletti, J.I., & Carvalho, S.H.V. (2006). Ethanolysis of castor and cotton seed oil: a systematic study using classical catalysts. *Journal of the American Oil Chemists´ Society*, Vol.83, No.9, pp. 819-822, ISSN 1558-9331

Miao, X., & Wu, Q. (2006). Biodiesel production from heterotrophic microalgal oil. *Bioresource Technology*, Vol.97, pp. 841-846, ISSN 0960-8524

Ministério da Pesca e Agricultura (2010). *Boletim estatístico da pesca e aqüicultura Brasil 2008-2009*. Brasília, Brazil, Ministério da Pesca e Agricultura, 101p. Retrieved from <http://www.mpa.gov.br>

Moser, B.R. (2009). Biodiesel production, properties, and feedstocks. *In Vitro Cellular Development and Biology - Plant*, V.45, pp. 229-266, ISSN 1475-2689

Nandi, S., Gangopadhyay, S., & Ghosh, S. (2005). Production of medium chain glycerides from coconut and palm kernel fatty acid distillates by lipase-catalyzed reactions. *Enzyme and Microbial Technology*, Vol.36, pp. 725-728, ISSN 0141-0229

Ndiaye, P.M., Franceschi, E., Oliveira, D., Dariva, C., Tavares, F.W., Oliveira, J.V. (2006). Phase behavior of soybean oil, castor oil and their fatty acid ethyl esters in carbon dioxide at high pressures. *Journal of Supercritical Fluids*, Vol.37, pp. 29-37, ISSN 0896-8446

Nelofer, R., Ramanan, R.N., Rahman, R.N.Z.R.A., Basri, M., Ariff, A.B. (2010). Sequential optimization of production of a thermostable and organic solvent tolerant lipase by recombinant *Escherichia coli*. *Annals of Microbiology*, DOI 10.1007/s13213-010-0170-9

Nielsen, P.M., Brask, J., & Fjerbaek, L. (2008). Enzymatic biodiesel production: technical and economical considerations. *European Journal of Lipid Science and Technology*, Vol.110, pp. 692-700, ISSN 1438-9312

Oliveira, D., & Oliveira, J.V. (2001). Enzymatic alcoholysis of palm kernel oil in *n*-hexane and $SCCO_2$. *Journal of Supercritical Fluids*, Vol.19, pp. 141-148, ISSN 0896-8446

Oliveira, D., Di Luccio, M., Faccio, C., Dalla Rosa, C., Bender, J.P., Lipke, N., Menoncin, S., Amroginski, C., & Oliveira, J.V. (2004b). Optimization of enzymatic production of biodiesel from castor oil in organic solvent medium. *Applied Biochemistry and Biotechnology*, Vol.113-116, pp. 771-789, ISSN 0273-2289

Oliveira, D., Feihrmann, A.C., Rubira, A.F., Kunita, M.H., Dariva, C., & Oliveira, J.V. (2006). Assessment of two immobilized lipases activity treated in compressed fluids. *Journal of Supercritical Fluids*, Vol.38, pp. 373-382, ISSN 0896-8446

Oliveira, D., Oliveira, J.V., Faccio, C., Menoncin, S., & Amroginski, C. (2004a). Influência das variáveis de processo na alcoólise enzimática de óleo de mamona. *Ciência e Tecnologia de Alimentos*, Vol.24, No.2, pp. 178-182, ISSN 1678-457X

Pramanik, K. (2003). Properties and use of jatropha curcas oil and diesel fuel blends in compression ignition engine. *Renewable Energy*, Vol.28, pp. 239-248, ISSN 0960-1481

Ramadhas, A.S., Jayaraj, S., & Muraleedharan, C. (2005). Biodiesel production from high FFA rubber seed oil. *Fuel*, Vol.84, pp. 335-340, ISSN 0016-2361

Ramani, K., Kennedy, L.J., Ramakrishnan, M., & Sekaran, G. (2010). Purification, characterization and application of acidic lipase from *Pseudomonas gessardii* using beef tallow as a substrate for fats and oil hydrolysis. *Process Biochemistry*, Vol.45, pp. 1683-1691, ISSN 1359-5113

Rathore, V., Madras, G. (2007). Synthesis of biodiesel from edible and non adible oil in supercritical alcohols and enzymatic synthesis in supercritical carbon dioxide. *Fuel*, V.86, pp. 2650 - 2659, ISSN 0016-2361

Reaney, M.J.T., Hertz, P.B., & McCalley, W.W. (2005). Vegetable oils as biodiesel, In: *Bailey´s Industrial Oil and Fat Products* (6th ed.), Shahidi, F., pp. 223-257, John Wiley & Sons, ISBN 9780471678496, Hoboken, USA

Rigo, E., Polloni, A., Remonatto, D., Arbter, F., Menoncin, S., Oliveira, J.V., Oliveira, D., Treichel, H., Ninow, J.L., Kalil, S.J., & di Luccio, M. (2010). Esterification activity of novel fungal and yeast lipases. *Applied Biochemistry and Biotechnology*, Vol.162, pp. 1881-1888, ISSN 1559-0291

Robles-Medina, A., González-Moreno, P.A., Esteban-Cerdán, L., & Molina-Grima, E. (2009). Biocatalysis: towards ever greener biodiesel production. *Biotechnology Advances*, Vol.27, pp. 398-408, ISSN 0734-9750

Rodrigues, R. C., Volpato, G., Ayub, M. A. Z., & Wada, K. (2008a). Lipase-catalyzed ethanolysis of soybean oil in a solvent-free system using central composite design and response surface methodology. *Journal of Chemical Technology and Biotechnology*, V.83, pp. 849-854, ISSN 1097-4660

Rodrigues, R. C., Volpato, G., Wada, K., & Ayub, M. A. Z. (2008b). Enzymatic synthesis of biodiesel from transesterification reactions of vegetable oils and short chain alcohols. *Journal of the American Oil Chemists´ Society*, Vol.85, pp. 925-930, ISSN 1558-9331

Rokhina, E. V., Lens, P., & Virkutyte, J. (2009). Low-frequency ultrasound in biotechnology: state of the art. *Trends in Biotechnology*, Vol.27, No.5, pp. 298-306, ISSN 0167-7799

Rousseau, D., & Marangoni, A.G. (2002). The effects of interesterification on the physical properties of fats, In: *Physical properties of lipids*, Marangoni, A.G., Narine, S.S., (Eds.), pp. 479-564, Marcel Dekker, ISBN 0-8247-00058, New York, USA

Salum, T.F.C., Villeneuve, P., Baréa, B., Yamamoto, C.I., Côcco, L.C., Mitchell, D.A., & Krieger, N. (2010). Synthesis of biodiesel in column fixed-bed bioreactor using the fermented solid produced by *Burkholderia cepacia* LTEB11. *Process Biochemistry*, Vol.45, pp. 1348-1354, ISSN 1359-5113

Santos, F. F. P., Malveira, J. Q., Cruz, M. G. A, & Fernandes, F. A. N. (2010). Production of biodiesel by ultrasound assisted esterification of *Oreochromis niloticus* oil. *Fuel*, Vol.89, No.2, pp. 275-279, ISSN 0016-2361

Schuchardt, U., Sercheli, R., & Vargas, R.M. (1998). Transesterification of vegetable oils: a review. *Journal of the Brazilian Chemical Society*, Vol.9, No.1, pp. 199-210, ISSN 1678-4790

Sheldon, R.A. (1996). Large-scale enzymatic conversions in non-aqueous media. In: Koskinen, A.M.P., & Klibanov, A.M., (Eds.), *Enzymatic reactions in organic media*. Blackie Academic & Professional, pp. 266-307, ISBN 075140259, Glasgow, Scotland

Shieh, C.-J., Liao, H.-F., & C.-C. Lee. (2003). Optimization of lipase-catalyzed biodiesel by response surface methodology. *Bioresource Technology*, Vol.88, pp. 103-106, ISSN 0960-8524

Silva, C., Weschenfelder, T.A., Rovani, S., Corazza, F.C., Corazza, M.L., Dariva, C., & Oliveira, J.V. (2007). Continuous production of fatty acid ethyl esters from soybean oil in compressed ethanol. *Industrial & Engineering Chemistry Research*, Vol.46, pp. 5304-5309, ISSN 1520-5045

Soccol, C.R., & Vandenberghe, L.P.S. (2003). Overview of applied solid-state fermentation in Brazil. *Biochemical Engineering Journal*, Vol.13, pp. 205-218, ISSN 1369-703X

Soumanou, M.M., & Bornscheuer, U.T. (2003). Lipase-catalyzed alcoholysis of vegetable oils. *European Journal of Lipid Science and Technology*, Vol.105, pp. 656-660, ISSN 1438-9312

Tan, K. T., Lee, K.T., & Mohamed, A. R. (2010). A glycerol-free process to produce biodiesel by supercritical methyl acetate technology: An optimization study via Response Surface Methodology. *Bioresource Technology*, Vol.101, No.3, pp. 965-969, ISSN 0960-8524

Teoh, C.H. (2002). The palm oil industry in Malaysia: from seed to frying pan. Technical report, WWF Switzerland. 145p

Tongboriboon, K., Cheirsilp, B., & H-Kittikun, A. (2010). Mixed lipases for efficient enzymatic synthesis of biodiesel from used palm oil and ethanol in a solvent-free system. *Journal of Molecular Catalysis B: Enzymatic*, Vol.67, pp. 52-59, ISSN 1381-1177

Treichel, H., Oliveira, D., Mazutti, M.A., Di Luccio, M., & Oliveira, J.V. (2010). A review on microbial lipases production. *Food and Bioprocess Technology*, Vol.3, pp. 182-196, ISSN 1935-5130

United States Department of Agriculture (2011). Oilseeds: world markets and trade, In: *United States Department of Agriculture - Foreign Agricultural Service*, 10.04.2011, Available from: <http://www.fas.usda.gov/psdonline/circulars/oilseeds.pdf>

Van Gerpen, J., Shanks, B., Pruszko, R., Clements, D., & Knothe, G. (2004). *Biodiesel production technology*, National Renewable Energy Laboratory, Golden, USA

Vargas, G. D. L. P., Treichel, H., Oliveira, D., Beneti, S. C., Freire, D.M.G., & Di Luccio, M. (2008). Optimization of lipase production by *Penicillium simplicissimum* in soybean meal. *Journal of Chemical Technology and Biotechnology*, Vol.83, pp. 47–54, ISSN 1097-4660

Vicente, G., Coteron, A., Martinez, M., & Aracil, J (1998). Application of the factorial design of experiments and response surface methodology to optimize biodiesel production. *Industrial Crops and Products*, Vol.8, pp. 29-35, ISSN 0926-6690

Villeneuve, P. (2007). Lipases in lipophilization reactions. *Biotechnology Advances*, Vol.25, pp. 515-536, ISSN 0734-9750

Villeneuve, P., Muderhwa, J.M., Graille, J., & Haas, M.J. (2000). Customizing lipases for biocatalysis: a survey of chemical, physical and molecular biological approaches. *Journal of Molecular Catalysis B: Enzymatic*, Vol.9, pp. 113-148, ISSN 1381-1177

Volpato, G., Rodrigues, R.C., Heck, J.X., & Ayub, M.A.Z. (2008). Production of organic solvent tolerant lipase by *Staphylococcus caseolyticus* EX17 using raw glycerol as substrate. *Journal of Chemical Technology and Biotechnology*, Vol.83, pp. 821–828, ISSN 1097-4660

Watanabe, Y., Shimada, Y., Sugihara, A., & Tominaga, Y. (2002). Conversion of degummed soybean oil to biodiesel fuel with immobilized *Candida antarctica* lipase. *Journal of Molecular Catalysis B: Enzymatic*, Vol.17, pp. 151-155, ISSN 1381-1177

Weber, N., & Mukherjee, K.D. (2008). Lipid biotechnology. In: *Food lipids: chemistry, nutrition and biotechnology*, Akoh, C.C., & Min, D.B., (Eds.), pp. 707-766, CRC Press, ISBN 978-1-4200-4663-2, Boca Ratón, USA

Wolski, E., Menusi, E., Remonatto, D., Vardanega, R., Arbter, F., Rigo, E., Ninow, J.L., Mazutti, M.A., di Luccio, M., Oliveira, D., & Treichel, H. (2009). Partial characterization of lipases produced by a newly isolated *Penicillium* sp. in solid state and submerged fermentation: a comparative study. *LWT- Food Science and Technology*, Vol.42, pp. 1557-1560, ISSN 0023-6438

Yazdani, S.S., & Gonzalez, R. (2007). Anaerobic fermentation of glycerol: a path to economic viability for the biofuels industry. *Current opinion in biotechnology*, Vol.18, pp. 213-219, ISSN 0958-1669

Zanotto, S.P., Romano, I.P., Lisboa, L.U.S., Duvoisin Jr., S., Martins, M.K., Lima, F.A., Silva, S.F., Albuquerque, P.M. (2009). Potential application in biocatalysis of mycelium-bound lipases from Amazonian fungi. *Journal of the Brazilian Chemists Society*, Vol.20, No.6, pp. 1046-1059, ISSN 0103-5053

3

Biodiesel Fuel Production by Enzymatic Transesterification of Oils: Recent Trends, Challenges and Future Perspectives

Nevena Luković, Zorica Knežević-Jugović and Dejan Bezbradica
Faculty of Technology and Metallurgy, University of Belgrade
Serbia

1. Introduction

Liquid fuels have been used for many years as the most dominant and basic fuel for motor engines. However, declining fossil fuel resources as well as the tendency for developing new renewable biofuels have shifted the interest of the society towards finding novel alternative fuel sources. Biodiesel (monoalkyl esters of long-chain fatty acids) has a great potential as an alternative diesel fuel. From an environmental point of view it shows clear advantages over conventional fuel: it comes from renewable sources, and hence does not contribute to new carbon dioxide emission, it is biodegradable, its combustion products have reduced levels of particulates, sulphur oxides, carbon oxides, nitrogen oxides, and therefore, significantly reduces pollution (Al-Zuhair, 2007; Salis et al., 2005). One of the advantages of biodiesel in comparison to other biofuels is that biodiesel can be pumped, stored and handled using the same infrastructure employed for conventional diesel fuel (Robles-Medina et al., 2009). Also, major advantage of biodiesel as an alternative fuel is that its energy content is similar to conventional fuels, so it can be used either on its own or mixed with conventional diesel fuel, with no need of altering existing engines (Bozbas, 2005). European countries have recognized need for alternative fuels and issued the Directive on the Promotion of the use of biofuels and other renewable fuels for transport (2003/30/EC). The Directive stipulates that EU countries should replace 5.75% of fossil fuels with alternative, biofuels until 31. December 2010. This Directive has been amended by Directive 2009/28/EC which also promotes the usage of energy from renewable sources (aims at achieving a 20% share of energy from renewable sources in the EU's final consumption of energy by 2020). In 2005, the estimated world production of biodiesel was 2.92 million tones of which 87% was obtained in EU. More importantly, between 2000 and 2005 world production increased threefold, indicating that share of biodiesel in global fuel production will significantly increase in future (Mousdale, 2008).
There are several possible processes for biodiesel synthesis: pyrolysis, the use of microemulsions and transesterification. Though pyrolysis, due to reduced viscosity, gives good quality fuel it still produces more biogasoline than biodiesel fuel. Fuel produced by microemulsion creates engine performance problems (Fukuda et al., 2001; Ma and Hanna, 1999). Conventionally, biodiesel is produced by transesterification of triacylglycerol (TAG) and short chain alcohols, commonly methanol, in the presence of an acid or an alkaline

catalyst (Marchetti et al., 2007). By this process the flashpoint of biodiesel is lowered and the cetane number is improved (Parawira, 2009). Transesterification is a three-step consecutive reaction, in which diglycerides and monoglycerides are formed as an intermediate compounds (Figure 1. and Figure 2.). Three moles of biodiesel and one mole of glycerol are produced for every mole of TAG that undergoes completely conversion. It is a reversible reaction and accordingly, excess alcohol is used to shift the equilibrium to the products side.

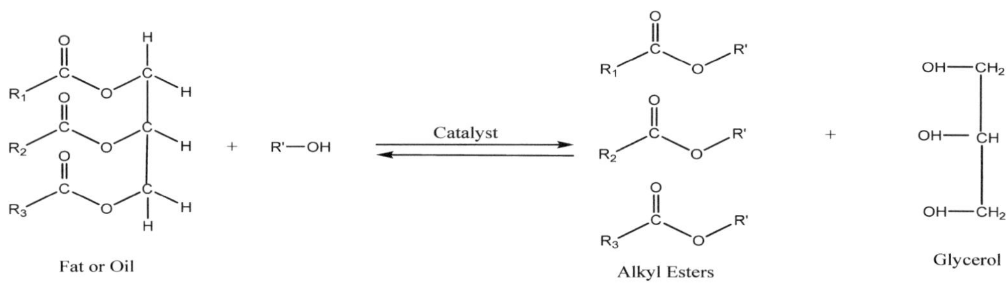

Fig. 1. General equation of transesterification of triacilglycerols with alcohol

This approach of the biodiesel synthesis has been used on industrial scale for decades, but there are several problems associated with this kind of production. Complex removal of catalyst, excessive energy requirements, recovery of glycerol, undesirable side reactions and the cost of the refined feedstock are the major drawbacks for such chemical process. Not only that the preliminary processing of feedstock is required (to reduce soap formation), but waste water must be treated as well (Fjerbaek et al., 2008). There have been several attempts to overcome these problems and to develop the alternative methodologies for biodiesel production. Nowadays, the biological production of biodiesel with lipases has received great consideration and it is undergoing a rapid development.

1.Triglyceride (TG) + R'OH $\xrightleftharpoons{\text{Catalyst}}$ Diglyceride (DG) + $R'COOR_1$

2. Diglyceride (DG) + R'OH $\xrightleftharpoons{\text{Catalyst}}$ Monoglyceride (MG) +$R'COOR_2$

3. Monoglyceride (MG) + R'OH $\xrightleftharpoons{\text{Catalyst}}$ Glycerol (GL) + $R'COOR_3$

Fig. 2. Transesterification of triacilglycerols with alcohol- Three consecutive and reversible reactions

The main advantages of employment of lipases as biocatalyst are mild reaction conditions and easy recovery of glycerol without purification or chemical waste production and production of very high purity product. In addition, free fatty acids content in the oil can be completely converted to methyl esters, with no soap formation, and thus increasing the biodiesel yield and reducing the costs for fuel purification. This characteristic of the enzymes allows the usage of materials with high free fatty acids (FFA) or high water content such as non-edible oils, waste cooking oils and industrial waste oil. Transesterification of triglycerides using lipases is considered to be one of the most efficient processes of biodiesel production from waste oil. It is clear that production process for alkaline method is more complex than enzymatic procedure as it is shown in Figure 3. and Figure 4.

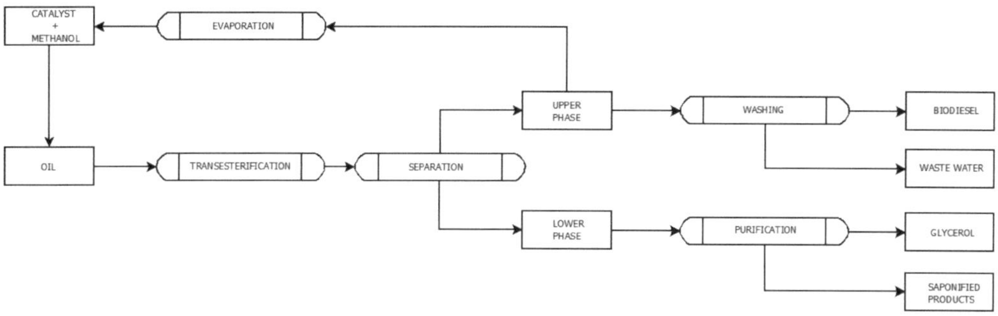

Fig. 3. Biodiesel production by alkali process

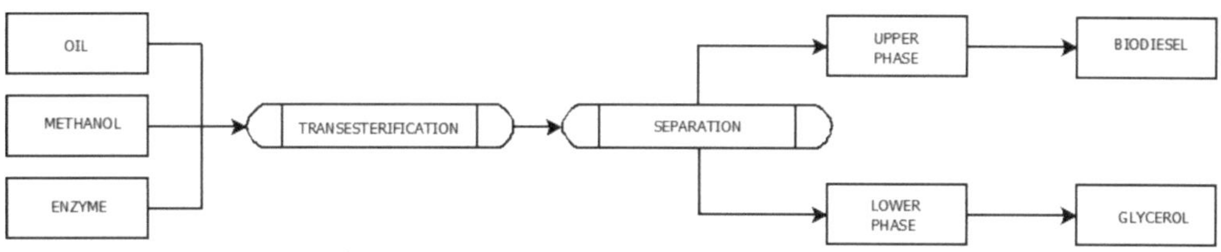

Fig. 4. Enzymatic biodiesel production

Despite numerous advantages, enzymatic processes have drawbacks such as: low reaction rate, enzyme cost for industrial scale use in comparison to alkali catalyst, low enzyme stability in the presence of excess methanol (Bajaj et al., 2010; Fjerbaek et al., 2008). In order to make enzymatic transesterification competitive on industrial scale there are several issues that have to be addressed: solvent engineering, lipases immobilization, selection of acyl acceptor, and selection of the reactor system. In this chapter we will try to review offered solutions to the current problems and to look for new perspectives in the industrial application of enzymatic biodiesel synthesis.

2. Lipases as biocatalysts in biodiesel synthesis

Lipases (triacylglycerol hydrolase, EC 3.1.1.3.) are enzymes that catalyze the hydrolysis of carboxylic ester link in the triacylglycerol molecule to form free fatty acids, di- and monoglycerides and glycerol. Although their natural function is to catalyze hydrolysis of ester links, they can also catalyze the esterification, the creation of this link between alcohol hydroxyl groups and carboxyl groups of carboxylic acids. Therefore, since they can catalyze hydrolysis, alcoholysis, esterification and transesterification they have a wide spectrum of biotechnological applications.

Lipases are also highly specific as chemo-, regio- and enantioselective catalysts. Thanks to direct evolution and protein engineering it is possible to enhance catalytic potential of lipases and to 'tailor' them to specific application and process conditions, enabling further expansion of their industrial use (B van Beilen and Li, 2002; Kirk et al., 2002). Among lipases of plant, animal and microbial origins, most commonly used are microbial lipases. They have numerous advantages over lipases from animals and plants. Using microorganisms it is possible to achieve a higher yield of enzymes, and to genetically manipulate the producing strain in obtaining a low-cost lipase with desired properties for the conversion of natural fats and oils into biodiesel. In addition, the enzymatic yield is independent of potential seasonal variations and it is possible to achieve rapid growth of microorganism in low-cost culture media (Gupta et al., 2004).

Lipases	Oil	Acyl acceptor	Solvent	Time	Yield	Reference
Candida antarctica B	Waste cooking palm oil	Methanol	*tert*-butanol	4h	79.1%	Halim et al. (2009)
Thermomyces lanuginosa	Soybean oil	Ethanol	*n*-hexane/solvent free	10h	70%-100%	Rodrigues et al. (2010)
Candida rugosa *Pseudomonas fluorescens* *Mucor javanicus* *Burkholderia cepacia*	*Jatropha* seed oil	Ethanol	solvent-free	8h	98%	Shah and Gupta (2007)
Candida antarctica	Cotton seed oil	Methanol	*tert*-butanol	24h	97%	Royon et al. (2007)
Rhizomucor miehei *Penicilium cyclopium*	Soybean oil	Methanol	solvent-free	12h	68%-95%	Guan et al. (2010)
Candida antarctica B	Soybean oil	Methyl acetate	solvent-free	14h	92%	Du et al. (2004)
Thermomyces lanuginosa	Sunflower oil Soybean oil Waste cooking oil	Methanol	solvent-free	24h	90%-97%	Dizge et al. (2009)
Candida antarctica	Sunflower oil	Methyl acetate	solvent-free	12h	>95%	Ognjanovic et al. (2009)
Candida antarctica	Cotton seed oil	Methanol, Propanol, Butanol, Amyl alcohol	solvent-free	7h	91.5%	Kose et al. (2002)
Candida antarctica	Rapeseed oil	Methanol	solvent-free	24h	91.1%	Watanabe et al. (2007)
Thermomyces lanuginosa *Candida antarctica*	Rapeseed oil	Methanol	*tert*-butanol	12h	95%	Li et al. (2006)
Candida antarctica	Jatropha seed oil Karanj oil Sunflower oil	Ethyl acetate	solvent-free	12h	>90%	Modi et al. (2007)

Lipases	Oil	Acyl acceptor	Solvent	Time	Yield	Reference
Rhizomucor miehei	Sunflower oil	Methanol	n-hexane	24h	>80%	Soumanou and Bornscheuer (2003)
Thermomyces lanuginosa	Sunflower oil	Methanol	solvent -free	24h	>90%	
Pseudomonas fluorescens						
Candida sp. 99-125	Waste cooking oil	Methanol	*n*-hexane	10h	91.08%	Chen et al. (2009)
Penicillium expansum	Waste oil	Methanol	*t*-amyl alcohol	24h	92,80%	Li et al. (2009)
Bacillus subtilis	Waste cooking oil	Methanol	solvent-free	72h	>90%	Ying and Chen (2007)
Enterobacter aerogenes	Jatropha seed oil	Methanol	tert-butanol	48h	94%	Kumari et al. (2009)
Criptococcus SP-2	Rice bran oil	Methanol	solvent-free	120h	80,20%	Kamini and Iefuji (2001)
Burkholderia cepacia	Soybean oil	Methanol	solvent-free	90h	>80%	Kaieda et al. (2001)

Table 1. Biodiesel production with various lipases

Reaction system for lipase is a two-phase system consisting of aqueous phase with dissolved enzyme and an organic phase with dissolved substrate. Lipases have a specific reaction mechanism due to the fact that active site of the enzyme is covered by amphiphilic peptide loop that acts like a lid. This disables the substrate molecule to bind to the enzyme active site, resulting with negligible activity of lipases in aqueous solutions (in the absence of an interphase or organic solvent). When contact occurs with a lipid/water interface, lid undergoes a conformational rearrangement which renders the active site accessible to the substrate (Schmid and Verger, 1998).

A large number of lipases from different sources have been utilized for biodiesel synthesis as shown in Table 1. *Candida antarctica* B lipase has been the most studied enzyme for biodiesel production in various reaction systems. It catalyzes acyl transfer reactions of various oils and acyl acceptors (alcohols or esters) showing high stability in organic solvents and broad substrate specificity. However, the lipase from *Pseudomonas* has also many advantages considering enzyme stability in a water-containing system in the presence of methanol. Kaieda et al. (2001) reported that the lipase from *P. cepacia* has much higher methanol resistance than those from *Rhizopus oryzae, P. roqueferoti, C. lipolytica* and even *C. antarctica*, making it more attractive for use as a biocatalyst in methanolysis reaction processes. Soumanou and Bornscheuer (2003) confirmed that lipase from *P. fluorescens* showed the greatest methanol resistance among the tested lipases, and a rather high yield of biodiesel (>90%) was obtained even at 4.5 M equivalents of methanol in the solvent free system. Other lipases also have been reported to be an efficient biocatlyst for the methanolysis reaction in the solvent-free system. For example, Kamini and Iefuji (2001) found that a crude lipase from the yeast *Cryptococcus* spp. S-2 efficiently catalyzed the methanolysis of vegetable oil even in a water-containing system without an organic solvent and in a single step, avoiding the stepwise addition of methanol. In general, lipases produced from yeast and fungi are most commonly used due to their low price and accessibility, although several bacterial lipases such as those from Pseudomonas sp., *Enterobacter aerogenes, Bacillus subtilis*, and others have been thoroughly investigated because of their superior methanol resistance (Al-Zuhair, 2007; Jaeger and Eggert, 2002; Noureddini et al., 2005). The main characteristic of lipases for biodiesel production from triacylglycerols is that they should be nonstereospecific so that all tri-, di- and monoglycerides can be converted to fatty acid alkyl esters. It is also an imperative that they should catalyze the esterification of free fatty acids (Fjerbaek et al., 2009). In addition, commercially available lipases (Lipozyme, Novozyme) appear to be more suitable catalysts for large scale application, since the use of free *Pseudomonas* or other lipases requires the additional immobilization step (Kose et al., 2002; Samukawa et al., 2000; Watanabe et al., 2001; Soumanou and Bornscheuer, 2003; Ognjanović et al., 2008).

2.1 Lipases immobilization

One of the main obstacles for industrial application of lipases is the high cost of biocatalysts. Therefore, immobilization of lipases, which allows their reusability, is a necessity to make them more attractive for industrial biodiesel processes. The aim of immobilization is to enhance lipases properties such as thermostability and activity in non-aqueous media, and to improve handling, recovery and recycling of biocatalyst. Recycling of immobilized enzymes greatly reduces the cost of the production, so the most promising immobilization supports and methods could make the enzymatic biodiesel production competitive to chemical processes.

By definition, the immobilization of enzymes is localization or confinement of an enzyme on to a solid support or on a carrier matrix. There are a number of supports that can by used for immobilization and the selection depends on several factors: thermal stability, chemical durability, mechanical strength, lipase type, type of the reaction system, ease of regeneration, loading capacity and cost (Jegannathan and Abang, 2008). Generally, immobilization is accompanied by changes in enzyme activity, optimum pH, temperature and stability. The productivity of an immobilized system is evaluated through both, activity and stability of a biocatalys. Often, the loss of enzyme activity is compensated by enhanced operational stability, which in overall makes the immobilization method a viable procedure.

Methods for enzyme immobilization can be classified as physical adsorption, entrapment, covalent bonding and encapsulation, each with its advantages and disadvantages. Different immobilization techniques of lipases used as catalysts in biodiesel production are summarized in Table 2.

Adsorption is the most widely used method for lipase immobilization primarily because procedure is easy, conditions are mild and the method itself is cheap (Hilal et al., 2006; Nie et al., 2006; Yagiz et al., 2007). This technique involves no toxic chemicals, supports can be regenerated for repeated use, and there is no major activity loss. Various supports have been used such as: textile membrane, alumina, ceramics, sepharose, sepadex, cellulose, hydrotalcite, different type of zeolites, celite, silica gel, polyethylene, polypropylene and others. The nature of support strongly dictates catalytic properties of the lipase such as activity, selectivity, stability (Knežević et al. 2004). Yang et al. explored the influence of pore diameter of nonpolar and weakly polar resin on degree of immobilization and reported that the degree of immobilization increases with increasing pore diameter. Also, the higher lipase activity was obtained employing nonpolar resins (Yang et al. 2006). It has been documented that lipase activity is generally higher with hydrophobic support. At low ionic strength, hydrophobic areas surrounding the active site are adsorbed on hydrophobic support, leaving stabilized the open form of the lipase (Mateo et al. 2007; Petkar et al. 2006).

Shah and Gupta immobilized *Burkholderia cepacia* on celite and concluded that the immobilized lipase performed better than free enzyme, under the same conditions. The immobilized enzyme gave a higher biodiesel yield and, in general, immobilized enzymes are known to give better catalytic performances in non-aqueous media (Shah and Gupta, 2007). However, reusability of immobilized lipases has been reported to decrease after four cycles. Since the enzymes are attached on the surface of support by weak forces (van der Walls, hydrophobic interactions, hydrogen bonds) they are easily stripped off. This leads to inevitable loss of catalytic activity which is not caused by enzyme inactivation. This is the main reason why the immobilization of lipases by adsorption is not the best solution for industrial application.

Immobilization method	Carier used	Source of lipases	Oil	Acyl acceptor	Yield	Reference
Adsorption	Celit	*Burkholderia cepacie*	Jatorpha oil	Ethanol	98%	Shah and Gupta (2007)
Adsorpiton	Cotton membrane	*Candida* sp. 99-125	Waste oil	Methanol	92%	Nie et al. (2006)
Adsorption	Hydrotalcite and zeolites	*Thermomyces lanuginosus*	Waste oil	Methanol	92,80%	Yagiz et al. (2007)
Adsorption	Nonpolar resin	*Candida* sp. 99-125	Soybean oil	Methanol	97,30%	Yang et al. (2006)
Adsorption	Toyonite 200-M	*Burkholderia fluorescens*	Safflower oil	Methanol		Iso et al. (2011)
Adsorption	Polypropylene support, Accurel MP 100	*Burkholderia cepacia*	Mahua oil	Ethanol	96%	Kumari et al. (2007)
Cross-linked enzyme aggregates (CLEAs)	None	*Burkholderia cepacia*	Mahua oil	Ethanol	92%	Kumari et al. (2007)
Protein-coated microcrystals (PCMCs)	None	*Burkholderia cepacia*	Mahua oil	Ethanol	99%	Kumari et al. (2007)
Entrapment	Hydrophobic sol-gel support	*Burkholderia cepacie*	Soybean oil	Methanol, ethanol	65%	Noureddini et al. (2005)
Entrapment	Phyllosilicate sol-gel matrix	*Burkholderia cepacie*	Restaurant grease	Methanol, ethanol	87-95%	Hsu et al. (2002)
Adsorption	Macropouros polypropylene	*Burkholderia fluorescens Burkholderia cepacie*	Soybean oil	Methanol	98%	Salis et al. (2008)
Entrapment	Phyllosilicate sol-gel matrix	*Burkholderia cepacia*	Restaurant grease	Ethanol	96%	Hsu et al. (2004)
Encapsulation	Silica aerogels	*Burkholderia cepacia Candida antarctica*	Sunflower oil	Methyl acetate		Orcaire et al. (2006)

Immobilization method	Carier used	Source of lipases	Oil	Acyl acceptor	Yield	Reference
Covalent bond	Aldehyde-Lewatit	*Thermomyces lanuginosus*	Soybean oil	Ethanol	75%-100%	Rodrigues et al. (2010)
Covalent bond	Silica -PVA composite	*Burkholderia cepacia*	Babassu oil	Ethanol	98%	Freitas et al. (2009)
Covalent bond	Novel microporous polymeric matrix (basically copolimer of styrene cross-linking with divinilbenzene)	*Thermomyces lanuginosus*	Sunflower oil	Methanol	63,80%	
			Soybean oil		55,50%	Dizge et al. (2009)
			Waste cooking oil		50,90%	
Covalent bond	Chitin activated with hexamethylenediamine (HEMDA)	*Candida rugosa*	Butyric acid	Butanol	187	Gomes et al. (2004)
Covelent bond	Amino-functionalized magnetic nanoparticles	*Thermomyces lanuginosa*	Soybean oil	Methanol	94%	Xie and Ma (2009)
Covelent bond	Silica-PVA composite	*Burkholderia fluorescens*	Palm oil	Ethanol	98%	Moreira et al. (2007)

Table 2. Different techniques of the immobilization of lipases used in biodiesel synthesis

Lipases immobilized by entrapment are more stable than physically adsorbed lipase. Entrapment of lipase includes capture of the lipase within a matrix of a polymer. The porose nature of the matrix allows the retention of the enzyme within the support, also enabling substrate and product diffusion. This method uses a relatively simple procedure and a number of natural and organic supports: alginate, agarose, gelatin, phyllosilicate sol-gel matrix. A variety of methods have been documented. For example, Meunier and Legger developed a procedure for entrapment of lipase from *Candida antarctica* on sol-gel supported on diatomaceous earth (Meunier and Legger, 2010). Although the loading of the enzymes and the activity of immobilized enzymes were high, the low conversion of triacylglycerol was noticeable. The ester yield in the reaction of triolein and methanol using lipase from *C. antarctica* immobilized in particles obtained via sol-gel process was around 60%. The low conversions are result of a poor diffusion and enzyme leakage which are a common problem in gel entrapped systems. However, recent advances in materials synthesis have allowed the preparation of an efficient immobilized lipase system suitable for designing a bioreactor for the continuous production of biodiesel. Namely, Hsu et al. (2002) have developed an original procedure to entrap lipase from *P. cepacia* (PS-30) within a phyllosilicate sol-gel matrix (IM PS-30) (Hsu et al. 2002; 2004). Lipase prepared in this way exhibit remarkable stability under normally denaturing conditions. Namely, the immobilized IM PS-30 lipase has shown to be more active biocatalyst and gave higher conversion to ester than both commercial *C. antarctica* lipase supported on a macroporous acrylic resin and *P. cepacia* lipase supported physically on granulated silica. The immobilized *Burkholderia cepacia* was used in transesterification of soybean oil and a conversion of around 95% after 48 h of reaction was achieved (Hsu et al., 2002).

Encapsulation is the confinement of enzyme within a porous membrane forming a bilayer (Jegannathan et al. 2008). In a way, it provides a cage, which prevents enzyme leaching making the resulting immobilized enzyme a highly reusable biocatalyst. Encapsulation can be carried out by using natural polymers, such as alginate and carrageenan, synthetic polymers (photo-cross linkable resins and polyurethane polymers), acrylic polymers, hydrogels, microemulsion based gels and those obtained by sol–gel methods (Yadav and Jadhav, 2005). For the use in biodiesel synthesis lipases from *Burkholderia cepacia* and *C. antarctica* were encapsulated in silica aerogels reinforced with silica quartz fiber felt and dried by the CO_2 supercritical technique (Orcaire et al. 2006). Under optimum reaction conditions final conversion of 56% was achieved and the operational stability of the immobilized lipase was high. However, with the increase of the substrate concentration limitation by diffusion occurs. The solution is to produce an encapsulated enzyme with smaller size to overcome the mass transfer problems, and to use the enzymes of highest purity (Jegannathan et al. 2008).

The main advantage of covalent immobilization is the irreversible bonding of the lipase to the support matrix, thereby preventing leaching of the enzyme in the reaction system. The immobilization by covalent bonding is the most complex one. Since the supports don't have the reactive groups but hydroxyl, amino, amide and carboxy groups, they have to be activated for immobilization of enzymes. Therefore, covalent immobilization usually consists of two stages: activation or modification of the support and attachment of the enzyme on the activated support. Activation of the support implies chemical reaction between support and an activating agent, creating new (usually electrophilic) groups on its surface that show a great affinity towards functional groups (usually nucleophilic) of the enzyme. According to the presence of certain groups, appropriate method of chemical activation is chosen. It is very important that the immobilization of lipase by covalent attachment to a support should involve only functional groups of the lipase that are not essential for its catalytic activity (Knežević et al. 2006; Prlainović et al. 2011). When it comes to supports with carboxy groups, enzymes are immobilized on the polymers through their amino groups, forming a peptide bond, with the presence of carbodiimide as a carboxy activator group. Covalent immobilization via the carboxy group of supports is often used in immobilization of lipase on the polymer derivates of acrylic acid. It must be noted that these active derivatives enable mild covalent coupling of lipase. Polymers containing amino groups can be activated by the introduction of the diazo groups. Enzymes are then easily linked through their α- or ε-amino groups and less by sulfhydryl, hydroxyl, imidazole or carboxy groups. One of the most commonly used methods of activating supports with amino groups is with glutaraldehyde. Recently different authors have focused their attention on developing a large number of supports containing epoxy groups. These groups can react with various nucleophilic groups of amino acids residues (amino, hydroxyl or sulfhydryl group) at very mild conditions: neutral pH and temperatures between 4 and 25°C (Mateo et al. 2000; Bezbradica et al. 2009). In this way, a vast number of very strong bonds are formed, and a considerable increase of stability is achieved due to multiple bonding. In addition, polymers containing epoxy groups can be easily modified by other activating agents such as glutaraldehyde, if the lipase activity decreases due to the formation of undesirable interactions between the enzyme and supports due to short distance (Bayramoglu et al. 2005). Immobilization methods are highly specific and should be optimized for a particular lipase support system. For the use in biodiesel synthesis, lipase from *Thermomyces lanuginosus* has been stabilized by coupling the chemical amination of the enzyme surface to the multipoint covalent attachment of the modified enzyme to glyoxylagarose beads (Rodrigues et al. 2010). The immobilized lipase proved to be very stable and active in transesterification reaction. It presented high reaction activity in the presence of excess alcohol under the optimal conditions. This method is an excellent step in reducing the cost of biodiesel by reducing the cost of enzyme, making it competitive with the chemical process. Lipases from *Thermomyces lanuginosus* and *Pseudomonas fluorescens* were immobilized by multipoint covalent attachment on Toyopearl AF-amino-650 resin. Glutaraldehyde, glycidol and epicchlolohydrin were used as activating agents and better catalytic properties were found when the support was activated with glycidol and epichlorohydrin. *T. lanuginosus* immobilized on glyoxyl-resin obtained derivates with the highest hydrolytic activity and thermal stability, almost 30 times more stable than soluble lipase. This lipase was successfully used for transesterification of palm oil with methanol, with the yield ranging from 93.5 to 100% (Mendes et al., 2011).

2.2 Pretreatment of lipase

Pretreatments of lipase by various methods have a goal of improving enzyme activity, enzyme stability and, especially important for biodiesel synthesis, to improve methanol tolerance. Researches have shown that when immobilized enzyme is successfully pretreated enzyme activity is increased and the ability of the enzyme to resist deactivation by methanol is higher, thereby enabling multiple reuses of enzymes for industrial production. Generally, pre-treatment reagents can be classified as: 1) substrates or their analogues; 2) organic solvents; 3) salts; 4) enzyme lyoprotectants such as crown ethers (Lu et al. 2010). Treatments with different reagents have, more or less, the same goal, keeping the conformation of lipases in its active form, by making the conformational change of active site from closed to open form. The most commonly used method is the usage of substrates or their analogues as pre-treatment reagents. They have the role of enhancing enzyme activity in organic solvents by molecular imprinting. The formed complex enzyme-imprinter keeps the enzyme in its active conformation and therefore retains higher activity in the nonaqueous environment as compared to the nonimprinted enzyme (Rich et al. 2002). Lu et al. pretreated immobilized lipase form *Candida* sp. 99-125 with methanol solutions within volume concentration range 10-20%. This treatment enhanced the enzyme activity and methanol tolerance in three step methanol addition and even in one step methanol addition. However, there were no enhancements for both, the initial rates and the equilibrium biodiesel yields with lipase pretreated with short chain alcohols: *n*-propyl alcohol, *n*-butanol, isopropyl alcohol, *t*-butanol, isobutyl alcohol. The most promising results were achieved when the lipase was treated with 1mM solution of $CaCl_2$ and $MgCl_2$., since lipase activity, methanol tolerance and operational stability were drastically improved. It is assumed that salts could incorporate with the protein to form a more stable molecule, which could resist conformational change induced by high methanol concentration (Lu et al. 2010). Lipases from different origins show to have distinct properties and one activation methods might not be versatile for other lipases. The immobilized lipase from *C. antarctica* was pretreated, immersed, in alcohols: isopropanol, 2-butanol, *t*-butanol. The activity of the commercial immobilized enzyme, Novozyme 435, increased about tenfold in comparison to the enzyme not subjected to any pre-treatment and the methyl ester yield was about 7 to 10 times higher (Chen and Wu, 2003). Smukawa et al. achieved 97% biodiesel yield in methanolysis of plant oil after 3.5 hours with preincubated Novozym 435. Methanolysis progressed much faster when Novozym 435 was preincubated in methyl oleate for 0.5h and subsequently in soybean oil for 12h (Samukawa et al. 2000). It is clear that these procedures can significantly increase productivity of industrial enzyme preparations and make enzymatic processes for industrial biodiesel production economically feasible.

3. Key operational variables

A number of different factors influence the enzymatic biodiesel synthesis: oil source, reaction temperature, choice of acyl acceptors, acyl acceptors to oil molar ratio, amount of water in the system or water activity, and presence of organic solvent in the mixture. Optimal parameters for enzymatic transesterification vary depending on the origin and type of lipase, type of oil source, and reactor type.

Reaction temperature may vary from 23 to 50°C. Optimal temperature for methanolysis of sunflower oil is 50°C when *T. lanuginose* is used as biocatalyst, but when *R. miehei* is used in the same reaction, temperature optimum lower than 40 °C has been reported (Soumanou and Bornscheuer, 2003). In general, increasing the temperature leads to an increase of the reaction rate of biodiesel production. When the optimum is reached, further increase of temperature, leads to decreased catalytic activity of the enzyme due to denaturation and inactivation. The researches have shown that immobilization of enzymes shift temperature optimum to higher values in comparison to free enzymes. It seems that immobilization provides a more rigid external backbone for lipase molecule, leading to the increase of the temperature optima and higher reaction rates (Al-Zuhair, 2007).

One of the key factors of enzymatic ester synthesis is the water content in the system. Lipases need an optimal small amount of water to maintain the activity in the organic media. Nevertheless, increased water concentration has an unfavourable effect on the equilibrium conversion, since it promotes reverse reaction of hydrolysis. The amount of water in the system should be a compromise between minimizing hydrolysis and maximizing lipase activity for the transesterification reaction and it should be determined for a particular reaction system (Chowdary, 2002; Noureddini et al., 2005). Many studies have shown that immobilized enzymes show highest activity in low water system. Tamalampudi et al. showed that, in biodiesel synthesis using lipase from *C. antarctica* (CAL-B), the rate of methanolysis decreased with the increase of the water content, reaching the FAME content of 75% when no water was added in the system. Similar results were achieved using the same lipase in the transesterification of sunflower oil, where yield of over 90% was achieved in an anhydrous reaction medium (Ognjanović et al., 2009). It has been shown that many immobilized lipases contain sufficient amount of water to preserve the catalytic conformation. Another problem occurs when waste oils are used, since the content of free fatty acids (FFA) is increased compared to refined oils. The esterification of FFA releases water, which can shift the reaction equilibrium towards ester hydrolysis. In these cases, molecular sieves are used for the control of water activity and increase ester yields by removing water produced by esterification. However, when lipases are immobilized on hydrophilic support, the molecular sieves are not needed, since, in that case, they have a negligible impact on methyl esters yield (Hsu et al., 2002).

There are many possible raw materials with a potential to obtain biodiesel. Generally, the main feedstock for biodiesel production can be divided in: 1) Vegetable oils such as sunflower oil (Dizge et al., 2009; Modi et al., 2007), soybean oil (Guan et al., 2010; Rodrigues et al., 2010), rapeseed oil (Li at al., 2006; Watanabe et al., 2007), jatropha oil (Shah and Gupta, 2007; Tamalampudi et al. 2008), cotton seed oil (Royon et al.,2007); 2) Animal fats such as tallow, lard (Da Cunha et al.,, 2009; Ngo et al. 2008); 3) Waste cooking oils and industrial waste oils (Chen et al., 2009; Halim et al., 2009).The fatty acid composition of animal fats is not favourable for biodiesel production, since they contain predominantly saturated fatty acids. The limited supplies of animal fats and high melting points of obtained esters, which require addition of very powerful freeze protectants, make animal fats substrate of minor importance. On the other hand, vegetable oils are renewable and virtually inexhaustible energy source. Depending on availability of raw materials in different countries, various oils have been used: soybean oil is predominantly used in the United States, rapeseed oil and sunflower oil are the main feedstock for biodiesel synthesis in many European countries. Coconut oil and palm oil are used in Asian countries (Demirbas, 2009). Figure 5. and Figure 6. shows the usage of oils for biodiesel production as well as world production by feedstock in 2007 (International Grains Council, 2007).

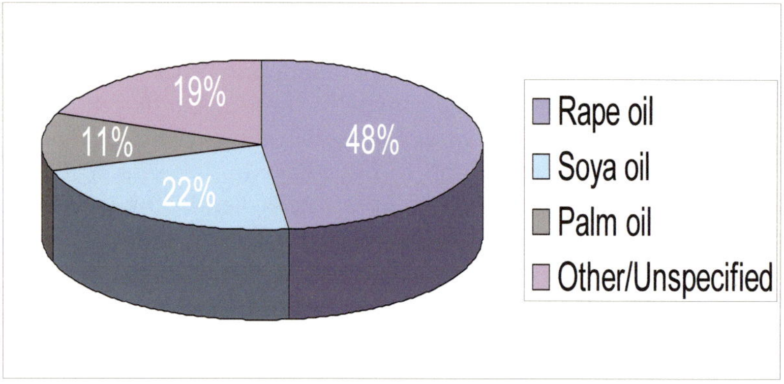

Fig. 5. Vegetable oil in use for biodiesel production

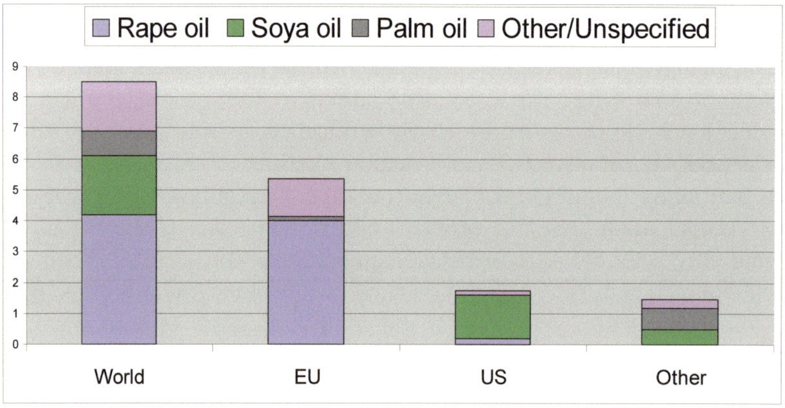

Fig. 6. Biodiesel world production by feedstock

However, edible oils are not in surplus supply and the cost of oil sources accounts for a large part in biodiesel production. In order to make the biodiesel production viable the solution is to develop a production based on waste cooking oils where no competition with food production takes place. But the amount of waste oils alone is not sufficient to meet demands. The optimal solution is to use non-edible oils which can not be used for human consumption because of the presence of some antinutritional factors, or toxic components. The most suitable oils are those from crops with the highest productivity pre hectare, low production cost and that can grow on waste land, such as *Jatropha* oil or rapeseed oil (Shah and Gupta, 2007; Tan et al., 2010). It is important to know that the quality and the properties of biodiesel are greatly influenced by the fatty acid composition of fats and oils used for its synthesis. Fatty acid composition for different vegetable oils is shown in Table 3.

Oil	C 16:0	C 16:1	C 18:0	C 18:1	C 18:2	C 18:3	C 20:0	C 20:1	Others	Ratio SFA/UFA
Almond	6,5	0,5	1,4	70,7	20			3,5	0,9	7,9/91,2
Borage	12,9	0,2	4,3	19,1	39	18,7	0,3		2	17,5/82,5
Corn	11,7		1,9	25,2	60,5	0,5	0,2			13,8/86,2
Cotton seed	28,3		0,9	13,3	57,5					29,2/70,8
Jatropha	16,4	1	6,2	37	39,2		0,2			22,8/77,2
Olive	11,8	1,5	2,7	74,1	8,5	0,7	0,4	0,3		14,9/85,1
Palm	42,6	0,3	4,4	40, 5	10,1	0,2			1,9	47/51,1
Canola	3,5		0,9	64,4	22,3	8,2			0,7	4,4/94,9
Soybean	11,4		4,4	20,8	53,8	9,3	0,3			16,1/83,9
Sunflower	7,1		4,7	25,5	62,4		0,3			12,1/87,9

Table 3. Fatty acid composition for different vegetable oils (Robles-Medina et al., 2007)

Biodiesel produced from oils that have high content of unsaturated fatty acids show higher pour and cloud point properties, making it appropriate for both cold and warm weather. However, the quality of such biodiesel is lower since it has lower cetane number and combustion temperature (Demirbas, 2008; Ghaly, 2010). Then again, biodiesel produced from oils with long chain fatty acids (longer than 18 carbons) has high cetane number and combustion temperature, but low cloud and pour point and greater viscosity. So the choice of feedstock should be a compromise between the unsaturation and the length of fatty acid chain (Robles-Medina et al., 2009).

3.1 Solvent engineering and acyl acceptor selection

The majority of enzymatic syntheses of biodiesel are performed in organic solvents. The yield of biodiesel is greatly influenced by the type of organic solvent present in the reaction system. Immobilized lipases showed high degree of efficiency in the presence of non-polar solvents (Soumanou and Bornscheuer, 2003). The polar, less hydrophobic solvents are not suitable for biocatalytic processes since they can distort water microlayer around the enzyme influencing its native structure, thereby, leading to denaturation. It has been shown that the highest biodiesel yield with lipase from *C. antarctica* was achieved with *n*-hexane as solvent. The lowest yields were obtained with polar solvents such as acetone (Nie et al. 2006). High yields were achieved in the transesterification reaction with immobilized lipase from *Pseudomonas fluorescens* with 1, 4-dioxane as solvent (Iso et al. 2001). *Tert*-butanol is also commonly used solvent for biodiesel enzymatic synthesis. It has been shown that the presence of *t*-butanol significantly reduces the negative effects caused by both methanol and glycerol, due to *t*-butanol ability to dissolve both methanol and glycerol. In the transesterification of rapeseed oil with Lipozyme TI LM and Novozyme 435 as catalyst, and under optimal condition, 90% yield was achieved (Li et la. 2006). *Tert*-butanol was used as a solvent in the enzymatic production of biodiesel from cotton oil with lipase from *C. antarctica*, in the batch reactor and a packed bed reactor. In both cases, the yield was over 90% (Royon et al. 2007).

However, despite the promising results, use of solvents should be avoided because of their toxicity and flammability, damaging effects on the environment and consequential requirement for solvent removal. From the economic point of view, the use of organic solvents is unfavorable due to the necessity of their removal from the final product which further increases production cost. Thus, to enable enzymatic processes to be competitive, enzymatic solvent-free systems have been developed. It has been shown that there are many benefits in using the solvent-free system in comparison with the organic solvent system, including the costs reduction and the improvement in the process control (Kose et al. 2002; Selmi and Thomas, 1998). The main drawback in industrial implementation of lipases into the solvent-free biodiesel synthesis is rather low enzyme stability in the presence of excess methanol, since several studies reported that a high methanol concentration could lead to serious inactivation of lipase (Kose et al. 2002; Royon et al. 2007). Methanol is the most popular alcohol used in transesterification process because of its relatively low price in comparison to other alcohols. In a typical methanolysis reaction, the reaction mixture consists of two phases due to low solubility of methanol in oil, leading to the inactivation of enzyme and the decreased fatty acid methyl ester yield. To minimize the enzyme inactivation, the solvent-free reaction systems with a stepwise addition of methanol have been developed. The most common way is a three step addition of methanol in accordance with the reaction dynamics, enabling conversions as high as 98.4% after 48 h (Shimada et al., 1999). Several other studies confirmed that the three-step addition of methanol is superior to the one-step addition (Shimada et al., 2002; Watanabe et al., 2001, 2002). A group of researches has performed the methanolysis of plant oils by lipase immobilized from *C. antarctica* in batch conditions with two-step (Watanabe et al., 2000) and three-step (Watanabe et al., 1999) addition of methanol (50% or 33% of stoichiometric quantity required) with a yield of 95% and 98.4%. However, operational stability of lipases in repeated cycles of methanolysis is not very high, indicating that an important parameter affecting the economic feasibility of the process is the biocatalysts stability and reusability over an extended period of time. Biocatalyst susceptibility towards methanol shifted interest of scientific community into selecting new acyl acceptors for transesterification reaction.

Several studies have focused their attention on branched and long chain alcohols. Experiment showed that increase of the number of carbon atoms increased the cetane number as well as heat content of the fuel. Also, fatty acid esters of secondary or branched-chain alcohols can be used as fuel additives since they decrease the solidification point, and consequently, the high cloud point and pour point (Salis et al. 2005; Watanabe et al. 2007). Kose et al. dealt with the alcoholysis of cotton seed oil with primary and secondary alcohols by using Novozyme 435 lipase. They analyzed the effect of alcohol types on alcoholysis on cotton seed oil indicating that the highest yield was obtained with isoamyl alcohol (Kose et al., 2002). Ognjanović et al. investigated the influence of methanol, 2-propanol and *n*-butanol on biodiesel synthesis. Operational stability of lipase from *C. antarctica* was investigated in a three-step addition of alcohol in a solvent-free system. With all three acyl acceptors a high initial yield was achieved but lipase exhibited poor activity during the repeated experiments. The replacement of methanol with less polar alcohols resulted in only slight increase in retained activity, but the significant inactivation of lipase still occurs. This might be due to the inactivation effect caused by alcohol and the negative effect caused by byproduct glycerol adsorbed on the surface of the immobilized lipase (Ognjanović et al., 2009). Byproduct glycerol is hydrophobic and insoluble on oil, so it is easily adsorbed onto the surface of the immobilized lipase, also exhibiting a negative effect on lipase activity and operational stability (Du et al., 2004). Most recent studies are focusing their interest on implementing methyl acetate and ethyl acetate as an acyl acceptors in biodiesel synthesis (Modi et al., 2007; Xu et al., 2005). It has been shown that methyl acetate has no negative effect on lipase activity. The usage of methyl acetate also eliminates the risk of deactivation of enzyme by glycerol, since no glycerol is produced in the reaction. In this reaction, instead of glycerol, the byproduct is triacetin, which has a greater value than glycerol. It can be used as a fuel additive, as an antiknock agent which can reduce engine knocking in gasoline, and to improve cold and viscosity properties of biodiesel (Du et al., 2004; Wang et al., 2006). Since methyl acetate has no negative effect on enzyme stability a stepwise addition of methyl acetate is not needed. This greatly reduces the reaction time and simplifies the procedure. Ognjanović et al. reported the transesterification of sunflower oil with methyl acetate by using the immobilized *C. antarctica* (Novozyme 435) in the solvent-free system (Ognjanović et al., 2009). They have explored the substrate ratio on biodiesel production and concluded that equilibrium conversion increases as methyl acetate concentration increases. A large excess of methyl acetate is required in order to shift the interesterfication in the forward direction. The highest methyl ester yield of 98% was obtained at 12:1 molar ratio of methyl acetate to oil, at 45°C and 3% of enzyme based on oil weight. The operational stability of lipase was found to be constant over 7 repeated cycles (200h), whereas the activity of immobilized enzyme decreased rapidly when methanol was used, and after the third cycle it was below 5%. Du et al. also used methyl acetate in the transesterification of soybean oil utilizing Novozym 435 as biocatalyst. This research also showed that in contrary to what occurred when using methanol, large excess of methyl acetate showed no inhibitory effect on lipase (Du et al., 2004). They achieved conversion of 92%, and there was no loss of the enzyme activity even after being continuously used for 100 bathes. Ethyl acetate has been used as acyl acceptor for interesterification of jatropha and sunflower oil with Novozym 435 as biocatalyst. The operational stability of lipase was unchanged for 12 cycles, whereas under the same conditions, when ethanol was used as acyl acceptor, lipase lost all its activity after 6 cycles (Modi et la., 2007).

4. Reactor configuration and industrial application

Important aspect of lipase implementation in biodiesel industrial production is development of an efficient immobilized reactor system. On a laboratory scale most commonly used systems are stirred batch reactors. The advantage of this reactor system is easy handling and control, as well as the simple configuration of the reactor. There are numerous studies on enzymatic transesterification in various reactor configurations with the emphasis on determining the best operational parameters, but there isn't sufficient information on scale-up systems (Du et al., 2007; Talukder et al., 2009). Packed bed reactor (PBR) systems are frequently applied for the large-scale enzymatic reactions. They enable continuous production with high efficiency, and reduced shear denaturation of the biocatalyst. PBR are more suitable for long lasting and industrial production since it allows the reuse of enzyme without the need of a prior separation and the immobilized enzyme is subjected to fewer shear stress, in comparison to stirred batch reactors,. The main advantage of packed bed reactors over fluidized bed reactors or continuous stirred tank reactor is the higher conversion per weight of catalyst. Basic construction of packed bed reactors (on a laboratory scale) is a glass column, packed with immobilized lipases with the reaction mixture placed in thermostat water bath, and pumped upward through the column by a peristaltic pump. Attention has turned to determining the operational and hydrodynamic conditions that maximize the yield in an industrially feasible immobilized system. Chen et al. reported an efficient system for continual production of biodiesel fuel by methanolysis of waste cooking oil (WCO) using immobilized *Candida* lipase in a three-step fixed bed reactor (Chen et al., 2009). They have explored the effect of lipase, solvent, water, temperature, operational stability and flow of the reaction mixture on the synthesis of biodiesel. The reaction was carried with hexane as a solvent, since waste cooking oil was not completely dissolved in methanol. Solvent improved the transfer of the substrate and the products, and reduced the toxic effect of methanol on immobilized lipases by decreasing methanol concentration. A crucial parameter of packed bed reactors are the amount of enzyme and the flow rate. The effect of flow rate on the enzyme efficiency is not straightforward. When the flow rate is low, methanol is in prolonged contact with the enzyme, which can lead to reduced enzyme activity by denaturation. On the other hand, too high flow rates reduce time of contact between the lipase and substrate, significant portion of substrates fail to bind at the enzyme active site, leading to inevitable decrease in biodiesel yield. The optimum reaction conditions were 25:15:10:100 weight ratio of lipase/hexane/water/WCO, temperature of 45°C and reactant flow of 1, 2mlmin^{-1}. The achieved yield was 91.08%, however, it decreased to 76.74% over 100h. There are two possible reasons for such poor stability: the influence of glycerol adsorbed on the surface of the immobilized lipase and the negative effect of methanol on the enzyme. Glycerol must be eliminated in a timely manner during the process in order to achieve better stability. Another solution is to use a different solvent since glycerol is insoluble in n-hexane, so it remains in the reactor adsorbed onto the enzymatic support, inevitably leading to decrease in enzymatic activity.

Packed bed reactor has been tested in laboratory scale by Severac et al. for biodiesel production from high-oleic sunflower oil using *t*-butanol as co-solvent for butanol. Initial substrate concentrations and their molar ratio were optimized by using Novozym 435 as biocatalyst. The crucial point of control is the partition of polar compounds between the reaction medium and the supported enzyme (Severac et al., 2011). Since *tert*-butanol has the ability to dissolve both, polar and nonpolar compound, it is a perfect solvent for this production. The best results were obtained for an initial oil concentration of 500mM and a molar ratio of oil/butanol, 1:5, with estimated productivity of 13,8 tons year^{-1} kg Novozym 435 with 96.5% yield. It is important to notice that system operated without the loss of activity for more than 50 days (over 1200h), making this procedure an effective approach for biodiesel synthesis. One crucial difference between these systems is the choice of catalyst: *Candida* sp. 99-125 immobilized on textile cloth versus commercial immobilized *C. antarctica* on macroporous acrylic resign. The support for *C. antarctica* is hydrophobic and limits the adsorption of polar components presented in the medium such as glycerol in the continuous transesterification conditions (Severac et al., 2011). The selection of catalyst needs to be a compromise between activity, stability, and price. It seems that, although the catalytic life of immobilized *Candida* sp. 99-125 is shorter, the cost of catalyst is significantly lower making it rather competitive for industrial use.

Similar PBR was optimized by Hailm et al. for continuous synthesis of biodiesel from waste cooking oil with methanol using *t*-butanol as solvent. The packed bed reactor consisted of two series of Pyrex column where the substrate mixture was fed upwards through the column using a peristaltic pump (Halim et al. 2009). With Novozyme 435 as biocatalyst, the FAME yield was 80% and the operational stability was longer than 120h. Halim et al. investigated the effect of mass transfer. It seems that at low velocity of substrate the mass transfer could dominate the yield, whereas at high velocities the reaction could dominate the yield. By analyzing two proposed models it was concluded that transesterification reaction of waste cooking oil in a continuous packed bed reactor occurs under mass transfer control conditions. It was suggested that mass transfer rate can be increased using conventional methods, such as enhancement of active area per unit volume of immobilized enzyme.
It was previously reported that, due to internal mass transport limitation, packed bed reactor don't seem to be an option for solvent free enzymatic FAME production (Fjerbaek et al., 2008). There are several problems to address: due to the high viscosity of solvent free systems pressure drop becomes significant. In order to minimize the pressure drop packed bed reactors need to operate at low flow velocities, the size of carrier must be increased or solvent must be added. With increasing carrier particle diameter the pressure drop decreases, but the internal mass transfer rate in biocatalyst decreases, which can affect the overall reaction rate. Despite these apparent limitations several PBR system have been developed for solvent free reactions (Hajar et al., 2009; Hama et al., 2007; Shimada et al., 2002; Ognjanović et al., 2009). The usage of solvents only increases production costs as they have to be removed and purified for recycling. Shimada et al. developed methanolysis of vegetable oil in a PBR packed with *C. antarctica* lipase. Since the influence of methanol in a solvent free system is more detrimental on lipase activity, stepwise addition of methanol is required. In this case glycerol was separated by sedimentation after each step. The conversion reached 90% and a three step reaction was used for 100 days (Shimada et al. 2002). Enzymatic methanolysis of canola oil in the solvent-free system was studied in a PBR using Novozyme 435 with small pieces of loofa (Hajar et al., 2009). Loofa was used for several reasons: it prevents enzyme compression especially for viscous flows, due to structure porosity it facilitates oil passing through the reactor, reduces the accumulation of glycerol and provides better enzyme distribution. As it was previously reported, *C. antarctica* lipase was inactivated by adding more than 1.5 molar equivalents of methanol to the oil in a solvent-free system. Stepwise addition of methanol was required and the appropriate volume of methanol was added at three stages: at the beginning of the reaction and after 24 and 48h. The achieved yield of methyl ester in this PBR configuration was above 95% and biocatalyst retained constant activity for more than 400h. Hajar et al., using response surface methodology, also investigated the influence of flow rate. Fluid dynamics and operating regime of the reactor system need to be investigated with special concern when solvent-free systems are applied. At lower flow rates mass transfer of the oil to the surface of lipase is decreased. Due to insufficient agitation rate hydrophilic layer of methanol and glycerol is formed around lipase. By increasing flow rates exposure of lipase to substrate is improved, thus improving the biodiesel yield. The feasibility of another PBR construction for solvent free system was investigated with methyl acetate as acyl acceptor (Ognjanović et al., 2009). Based on the best results achieved in a batch reactor (3% of the immobilized lipase based on the oil weight, and 12:1 methyl acetate to oil molar ratio) the packed bed reactor was constructed to perform the transesterification of sunflower oil with

commercial *C. antarctica* B lipase immobilized on acrylic resin. The kinetics in the packed bed reactor system proved to have a slightly better profile than in a batch stirred tank reactor. Almost complete conversion of the oil to methyl esters (96.25%) was achieved in 8 hours. The yield of biodiesel remained the same for eight cycles (93.6 ± 3.75%) showing that the enzyme did not loose its activity even in a large excess of methyl acetate. With a proper selection of the reactor configuration and the operating conditions, the immobilized enzyme might offer a great prospect to ensure the high conversion values, improved enzyme stability, and thus prolonged operation times and cost-effective industrial enzymatic processes.
A very interesting and important study was conducted by Sotoft et al. They carried out the process simulation and economical evaluation of an enzymatic biodiesel production plant (Sotoft et al., 2010). Since there is, to date, only one plant in the world that uses enzymatic transesterification of biodiesel these kinds of studies are very important to make a correct evaluation of the industrial potential of a particular process. The study evaluated several important and relevant scenarios for enzyme catalyzed biodiesel production processes. Enzymatic biodiesel production from rapeseed oil and methanol has been investigated for solvent-free and cosolvent production processes. Study included two different production scales (8 and 200 mil.kg biodiesel/year) and different enzyme prices: current price 762.7€/kg enzyme and assumed price of enzyme in the future 7.627€/kg enzyme. The process simulations were carried out in Aspen Plus and Aspen Icarus Process Evaluator. Based on the simulations, the solvent free process is viable for a larger scale production of 200 mil.kg biodiesel/year with the current enzyme price. Also the continuous operation is the only realistic option with stepwise addition of methanol. Byproduct sale is also an important part of process economy. For a solvent free operation the cost of raw materials is distributed as: 50% enzyme, 47% oil and 3% methanol. The influence of enzyme cost is lower for cosolvent operation due to the improved enzyme performance. The estimated product price was 0.73-1.49€/kg biodiesel with the current enzyme price and 0.05-0.75€/kg with the enzyme price for a future for solvent-free processes.

5. Conclusion

The production and consumption of biodiesel will inevitably rise in future due to low environmental impact, ease of handling, and possibility of use without need for major adjustments of existing engines of motor vehicles. Although the majority of biodiesel manufacturers still employ a base-catalyzed process, biocatalytic methods based on activity of various microbial lipases offer several advantages, such as mild reaction conditions and high selectivity of biocatalysts in comparison with chemical catalysts. These advantages have good prospects to cause significant increase of economical feasibility of biodiesel production, since mild conditions decrease energy costs, while high selectivity leads to avoiding of unwanted by-products formation, simpler downstream processing, and easier separation of valuable by-product - glycerol. The main obstacle of biocatalytic process application on industrial level is high costs for enzyme formulations, but simultaneous efforts in optimization of different parameters of biocatalytic process led to significant increase of productivity. Various microbial lipase producers with strong affinity towards catalysis of methanolysis reaction and low susceptibility towards presence of methanol have been identified during previous two decades. The stability of lipases has been further increased by pre-treatment of lipases, selection of adequate immobilization procedures or use of whole-cell biocatalysts, leading to prolonged activity of biocatalyst. Detrimental effects of methanol on lipase activity opened new area of investigation - selection of alternative acyl-acceptors, such as higher or branched alcohols, and esters. Investigations in this area led not only to prolonged stability of biocatalyst, but to additional valuable by-products and even biodiesel of improved fuel properties. Additionally, the reduction of oil costs has been performed by development of processes with waste oil and further increase of productivity was achieved by application of novel reactor designs. It can be concluded that promising novel findings obtained in optimization of lipase-based processes of biodiesel production indicate that biocatalysis has great potential in biodiesel synthesis.

6. Acknowledgment

The authors are grateful to the Ministry of Science, Republic of Serbia for financial support (Project III 46010).

7. References

Al-Zuhair, S. (2007). Production of biodiesel: possibilities and challenges. *Biofuels, Bioproducts and Biorefining*, Vol. 1, (July 2007), pp. 57-66, ISSN 1932-1031

B van Beilen, J. & Li, Z., (2002). Enzyme technology: an overview. *Current Opinion in Biotechnology*, Vol. 13, No. 4, (August 2002), pp. 338-344, ISSN 0958-1669

Bajaj, A., Lohan P., Jha P.N. & Mehrotra R. (2010). Biodiesel production through lipase catalyzed transesterification: An overview. *Journal of Molecular Catalysis B: Enzymatic*, Vol. 62, No. 1, (October 2009), pp. 9-14, ISSN 1381-1177

Bayramoglu, G., Kaya, B. & Anca, M.Y. (2005) Immobilization of *Canida rugosa* lipase onto spacer-arm attached poly(GMA-HEMA-EGDMA) microspheres. *Food Chemistry*, Vol. 92, No. 2, (September 2005), pp. 261-268, ISSN 0308-8146

Bezbradica D., Mijin D., Mihailović M. & Knežević-Jugović Z. (2009) Microvawe-assisted immobilization of lipase from *Candida rugosa* on Eupergit supports, *Journal of Chemical Technology and Biotechnology* Vol. 84, No. 11, (November 2009), pp. 1642-1648, ISSN 0268-2575

Bozbas, K. (2008). Biodiesel as an alternative motor fuel: Production and policies in the European Union. *Renewable and Sustainable Energy Reviews*, Vol. 12, No. 2, (February 2008), pp. 542-552, ISSN 1364-0321

Chen, J-W. & Wu, W-T. (2003) Regeneration of immobilized *Candida antarctica* lipase for transesterification, *Journal of Bioscience and Bioengineering*, Vol. 95, No. 5, (January 2003), pp. 466-469, ISSN1389-1723

Chen, Y., Xiao, B., Chang, J., Fu, Y., Lv, P. & Wang, X. (2009). Synthesis of biodiesel from waste cooking oil using immobilized lipase in fixed bed reactor, *Energy Conversion and Management*, Vol. 50, No. 3, (March 2009), pp. 668-673, ISSN 0196-8904

Chowdary, G.V. & Prapulla, S.G. (2002). The influence of water activity on the lipase catalyzed synthesis of butyl butyrate by transesterification, *Process Biochemistry*, Vol. 38, No. 3, (November 2002), pp. 393-397, ISSN 1359-5113

Da Cunha, M.E., Krause, L.C., Moraes, M.S.A., Faccini, C.S., Jacques, R.A., Almeida, S.R., Rodrigues, M.R.A. & Caramao, E.B. (2009) Beef tallow biodiesel produced in a pilot scale, *Fule Processing Technology*, Vol. 90, No. 4, (April 2009), pp. 570-575, ISSN 0378-3820

Demirbas, A. (2008) Relationships derived from physical properties of vegetable oil and biodiesel fuels, *Fuel,* Vol. 87, No. 8-9, (July 2008), pp. 1743-1748, ISSN 0016-2361

Demirbas, A. (2009). Progress and recent trends in biodiesel fuels, *Energy Conversion and Management,* Vol. 50, No. 1, (January 2009), pp. 14-34, ISSN 0196-8904

Dizge, N., Aydiner, C., Imer, D.Y., Bayramoglu, M., Tanriseven, A. & Keskinler, B. (2009) Biodiesel production from sunflower, soybean and waste cooking oils by transesterification using lipase immolbilized onto a novel microporous polymer, *Bioresource Technology,* Vol. 100, No. 6, (March 2009), pp. 1983-1991, ISSN 0960-8524

Du, W., Xu, Y., Liu D. & Zeng, J. (2004). Comparative study on lipase-catalyzed transformation of soybean oil for biodiesel production with different acyl acceptors, *Journal of Molecular Catalysis B: Enzymatic,* Vol. 30, No. 3-4, (August 2004), pp. 125-129, ISSN 1381-1177

Du, W., Liu, D., Li, L. & Dai, L. (2007) Mechanism exploration during lipase-mediated methanolysis of renewable oils for biodiesel production in a *tert*-butanol system, *Biotechnology progress,* Vol. 23, No. 5, (Septemeber 2007), pp. 1087- 1090, ISSN 8756-7938

Fjerbaek, L., Christensen, K.V. & Norddahl, B. (2009). A review of the current state of biodiesel production using enzymatic transesterification. *Biotechnology and Bioengineering,* Vol. 102, No. 5, pp. 1298-1315, ISSN 0006-3592

Freitas, L., Da Ros, P.C.M., Santos, J.C. & de Castro, H.F. (2009) An integrated approach to produce biodiesel and monoglycerides by enzymatic interestification of babassu oil (*Orbinya* sp). *Process Biochemistry,* Vol. 44, No. 10, (October 2009), pp. 1068-1074, ISSN 1359-5113

Fukuda, H., Kondo A. & Noda H. (2001). Biodiesel fuel production by transesterification of oils. *Journal of Bioscience and Bioengineering,* Vol. 92, No. 5, (September 2001), pp. 405-416, ISSN 1389-1723

Ghaly, A.E., Dave, D., Brooks, M.S. & Budge, S. (2010) Production of biodiesel by enzymatic transesterification: Review, *American Journal of Biochemistry and Biotechnology,* Vol. 6, No. 2, pp. 54-76, ISSN 1553-3468

Gomes, F.M., Pereira, E.B. & Castro, H.F., (2004). Immobilization of lipase on chitin and its use in nonconventional biocatalysis. *Biomacromolecules,* Vol. 5, No. 1, (January 2004), pp. 17-23, ISSN 1525-7797

Guan, F., Peng, P., Wang, G., Yin, T., Peng, Q., Huang, J., Guan, G. & Li, Y. (2010). Combination of two lipases more efficiently catalyzes methanolysis of soybean oil for biodiesel production in aqueous medium, *Process Biochemistry,* Vol. 45, No. 10, (October 2010), pp. 1667-1682, ISSN 1359-5113

Gupta, R., Gupta, N. & Rathi P. (2004). Bacterial lipases: an overview of production, purification and biochemical properties. *Applied Microbiology and Biotechnology,* Vol. 64, No. 6, (February 2004), pp. 763-781, ISSN 0175-7598

Hajar, M., Shokorollahzadeh, S., Vahabzadeh, F. & Monazzami, A. (2009) Solvent-free methanolysis of canola oil in a packed-bed reactor with use of Novozyme 435 plus loofa, *Enzyme and Microbial Technology,* Vol. 45, No. 3, (September 2009), pp. 188-194, ISSN 0141-0229

Halim, S.F.A., Kamaruddin, A.H. & Fernando, W.J.N. (2009). Continuous biosynthesis of biodiesel from waste cooking palm oil in a packed bed reactor: Optimization usig response surface methodology (RSM) and mass transfer studies. *Bioresource Technology,* Vol. 100, No. 2, (January 2009), pp. 710-716, ISSN 0960-8524

Hama, S., Yamaji, H., Fukumizu, T., Numata, T., Tamalampudi, S., Kondo, A., Noda, H. & Fukuda, H. (2007) Biodiesel-fuel production in a packed-bed reactor using lipase-producting *Rhizopus oryzae* cells immobilized within biomass support particles, *Biochemical Engineering Journal,* Vol. 34, No. 3, (June 2007), pp. 273-278, ISSN 1369-703X

Hilal, N., Kochkodan, V., Nigmatullin, R., Goncharuk, V. & Al-Khatib, L. (2006). Lipase-immobilized biocatalytic membranes for enzymatic esterification: Comparison of various approaches to membrane preparation, *Journal of Membrane Science,* Vol. 268, No. 2, (January 2006), pp. 198-207, ISSN 0376-7388

Hsu, A-F., Jones, K., Foglia T.A. & Marmer, W.N. (2002). Immobilized lipase-catalysed production of alkyl esters of restaurant grease as biodiesel. *Biotechnology and Applied Biochemistry,* Vol. 36, No. 3, (December 2002), pp. 181- 186, ISSN 0885-4513

Hsu, A-F., Jones, K., Foglia, T.A. & Marmer, W.N. (2004). Continuous production of ethyl esters of grease using an immobilized lipase. *Journal of American Oil Chemical Society,* Vol. 81, No. 8, (August 2004), pp. 749-752, ISSN 0003-021X

Iso, M., Chen, B., Eguchi, M., Kudo, T., Shrestha S. (2001). Production of biodiesel fuel from triglycerides and alcohol using immobilized lipase. *Journal of Molecular Catalysis B: Enzymatic,* Vol. 16, No. 1, (November 2001), pp. 53-58, ISSN 1381-1177

Jaeger, K.E. & Eggert, T. (2002). Lipases in biotechnology. *Current Opinion in Biotechnology,* Vol. 13, No. 4, (August 2002), pp. 390-397, ISSN 0958-1669

Jegannathan, K.R. & Abang, S., (2008). Production of biodiesel using immobilized lipase-A critical review. *Critical Reviews in Biotechnology,* Vol. 28, No. 4, pp. 253-264, ISSN 0738-8551

Kaieda, M., Samukawa, T., Kondo, A. & Fukuda, H. (2001). Effect of methanol and water contents on production of biodiesel fuel from plant oil catalyzed by various lipases in a solvent-free system. *Journal of Bioscience and Bioengineering,* Vol. 91, No. 1, pp. 12-15, ISSN 1389-1723

Kamini, N.R. & Iefuji, H. (2001). Lipase catalyzed methanolysis of vegetable oils in aqueous medium by *Cryptococcus* spp. S-2. *Process Biochemistry,* Vol. 37, No. 4, (December 2001), pp. 405-410, ISSN 1359-5113

Kirk, O., Borchert, T.V. & Fuglsang, C.C. (2002). Industrial enzyme application. *Current Opinion in Biotechnology,* Vol. 13, No. 4, (August 2002), pp. 345-351, ISSN 0958-1669

Knežević, Z., Šiler-Marinković, S. & Mojović, Lj. (2004). Immobilized lipases as practical catalysts. *Acta Periodica Technologica,* Vol. 35, pp. 151-164, ISSN 1450-7188

Knežević, Z., Milosavić, N., Bezbradica, D., Jakovljević, Ž. & Prodanović, R. (2006) Immobilization of lipase from *Candida rugosa* on Eupergit C supports by covalent attachment, *Biochemical Engineering Journal,* Vol. 30, No. 3, (June 2006), pp. 269-278, ISSN 1369-703X

Kose, O., Tuter, M. & Aksoy, H.A. (2002). Immobilized *Candida antarctica* lipase-catalyzed alcoholysis of cotton seed oil in a solvent-free medium. *Bioresource Technology,* Vol. 83, No. 2, (June 2002), pp. 125-129, ISSN 0960-8524

Kumari, A., Shah, S. & Gupta, M.N. (2007). Preparation of Biodiesel by Lipase-Catalyzed Transesterification of High Free Fatty Acid Containing Oil from Madhuca indica. *Energy and Fuels* Vol. 21, No. 1, (January 2007), pp. 368-371, ISSN 0087-0624

Kumari, A., Mahapatra, P., Garlapati, V.K. & Banerjee, R. (2009) Enzymatic transesterification of Jatropha oil, *Biotechnology for Biofuels,* Vol. 2, No. 1, (January 2009), pp. 1-7, ISSN 1754-6834

Li, L., Du W., Liu, D., Wang, L. & Li, Z. (2006) Lipase-catalyzed transesterification of rapeseed oils for biodiesel production with a novel organic solvent as the reaction medium. *Journal of Molecular Catalysis B: Enzymatic,* Vol. 43, No. 1-4, (December 2006), pp. 58-62, ISSN 1381-1177

Li, N., Zong, M. & Wu, H. (2009) Highly efficient transformation of waste oil to biodiesel by immobilized lipase from *Penicillium expansum. Process Biochemistry,* Vol. 44, No. 6, (June 2009), pp. 685-688, ISSN 1359-5113

Lu, J., Deng, l., Zhao, R., Zang, R., Wang, F. & Tan, T. (2010) Pretreatment of immobilized *Candida* sp. 99-125 lipase to improve its methanol tolerance for biodiesel production. *Journal of Molecular Catalysis B: Enzymatic,* Vol. 62, No. 1, (January 2010), pp. 15-18, ISSN 1381-1177

Ma, F. & Hanna, M.A. (1999). Biodiesel production: a review. *Bioresource Technology,* Vol. 70, No. 1, (Octobar 1999), pp. 1-15, ISSN 0960-8524

Marchetti, J.M., Miguel, V.U. & Erazzu, A.F. (2005). Possible methods for biodiesel production. *Renewable and Sustainable Energy Reviews,* Vol.11, No. 6, (August 2005), pp. 1300-1311, ISSN 1364-0321

Mateo, C., Abian, O., Fernandez-Lafuente, R. & Guisan, J.M. (2000) Increase in conformational stability of enzymes immobilized on epoxy-activated supports by favoring additional multipoint covalent attachment. *Enzyme and Microbial Technology,* Vol. 26, No. 7, (April 2000), pp. 509-515, ISSN 0141-0229

Mateo, C., Palomo, J.M., Fernandez-Lorente, G., Guisan, J. & Fernandez-Lorente, R. (2007) Improvement of enzyme activity, stability and selectivity via immobilization techniques. *Enzyme and Microbial Technology,* Vol. 40, No. 6, (May 2007), pp. 1451-1463, ISSN 0141-0229

Mendes, A., Giordano, R., Giordano, R. & de Castro, H. (2011). Immobilization and stabilization of micribial lipases by multipoint covalet attachment on aldehyde-resin affinity: Application of the biocatalysts in biodiesel synthesis. *Journal of Molecular Catalysis B: Enzymatic,* Vol. 68, No. 1, (January 2011), pp. 109-115, ISSN 1381-1177

Meunier, S. & Legge, R. (2010) Evaluation of diatomaceous earth as a support for sol-gel immobilized lipase for transesterification. *Journal of Molecular Catalysis B: Enzymatic,* Vol. 62, No. 1, (January 2010), pp. 54-58, ISSN 1381-1177

Modi, M. K., Reddy J.R.C., Rao, B.V.S.K. & Prasad, R.B.N. (2007) Lipase-mediated conversion of vegetable oils into biodiesel usig ethyl acetate as acyl acceptor. *Bioresource Technology,* Vol. 98, No. 6, (April 2007), pp. 1260-1264, ISSN 0960-8524

Moreira, A.B.R., Perez, V.H., Zanin, G.M. & Castro, H.F. (2007) Biodiesel Synthesis by Enzymatic Transesterification of Palm Oil with Ethanol Using Lipases from Several Sources Immobilized on Silica–PVA Composite. *Energy & Fuels,* Vol. 21, No. 36, (November 2007), pp. 3689–3694, ISSN 0887-0624

Mousdale, D. M. (2008) *Biofuels – Biotechnology, Chemistry, and Sustainable Development*, CRC Press, ISBN-13: 978-1-4200-5124-7, United States of America

Ngo, H.L., Zafiropoulos, N.A., Foglia, T.A., Samulski, E.T. & Lin, W. (2008) Efficient two-step synthesis of biodiesel from greases, *Energy and Fuels,* Vol. 22, No. 1, (January 2008), pp. 626-634, ISSN 0887-0624

Nie, K., Xie, F., Wang, F. & Tan, T. (2006) Lipase catalyzed methanolysis to produce biodiesel: Optimiyation of the biodiesel production, *Journal of Molecular Catalysis B: Enzymatic,* Vol. 43, No. 1-4, (December 2006), pp. 142-147, ISSN 1381-1177

Noureddini, H., Gao, X. & Philkana, R.S. (2005). Immobilized *Pseudomonas cepacia* lipase for biodiesel fule production from soybean oil, *Bioresource Technology,* Vol. 96, No. 7, (May 2005), pp. 769-777, ISSN 0960-8524

Ognjanović, N., Šaponjić, S., Bezbradica, D. & Knežević-Jugović, Z. (2008). Lipase-catalyzed biodiesel synthesis with different acyl acceptors. *Acta Periodica Technologica,* Vol. 39, pp. 161-169, ISSN 1450-7188

Ognjanović, N., Bezbradica, D. & Knežević, Z. (2008). Optimization of the production of biodiesel by a commercial immobilized lipase in a solvent-free system using a response surface methodology. *Journal of the Serbian Chemical Society,* Vol. 73, No. 2, pp. 147-156, ISSN 0352-5139

Ognjanović, N., Bezbradica, D. & Knežević-Jugović, Z. (2009). Enzymatic conversion of sunflower oil to biodiesel in a solvent-free system: Process optimization and immobilized system stability. *Bioresource Technology,* Vol. 100, No. 21, (November 2009), pp. 5146-5154, ISSN 0960-8524

Orcaire, O., Buisson, P. & Pierre, A. (2006) Application of silica aerogel encapsulated lipases in the synthesis of biodiesel by transesterification reaction. *Journal of Molecular Catalysis B: Enzymatic,* Vol. 42, No. 3-4, (November 2006), pp. 106-113, ISSN 1381-1177

Parawira, W. (2009). Biotechnological production of biodiesel fuel using biocatalysed transesterification: a review. *Critical Reviews in Biotechnology,* Vol. 29, No. 2, (September 2009), pp. 82-93, ISSN 0738-8551

Petkar, M., Lali, A., Caimi, P. & Daminati, M. (2006) Immobilization of lipases for non-aqueous synthesis, *Journal of Molecular Catalysis B: Enzymatic,* Vol.39, No. 1-4, (May 2006), pp. 83-90, ISSN 1381-1177

Prlainović, N., Knežević-Jugović, Z., Mijin, D. & Bezbradica, D. (2011) Immobilization of lipase from *C. rugosa* on Sepabeads: the effect of lipase oxidation by periodates. *Bioprocess Biosystems Engineering* (in press, DOI 10.1007/s00449-011-0530-2)

Rich, J., Mozahev, V., Dordick, J., Clark, D. & Khmelnitsky, Y. (2002) Molecular imprinting of enzymes with water insoluble ligands for nonaqueous biocatalysis. *Journal of the American Chemical Society,* Vol. 124, No. 19, (May 2002), pp. 5254-5255, ISSN 0002-7863

Robles-Medina, A., Gonzales-Moreno, P.A., Esteban-Cerdan, L. & Molina-Grima, E. (2009) Biocatalysis: Towards ever greener biodiesel production. *Biotechnology Advances,* Vol. 27, No. 4, (October 2008), pp. 398-408, ISSN0734-9750

Rodrigues, R.C., Pessela, B.C.C., Volpato, G., Fernandez-Lafuente, R., Guisan J.M. & Ayub, M.A.Z. (2010). Two step ethanolysis: A simple and efficient way to improve the enzymatic biodiesel synthesis catalyzed by an immobilized-stabilized lipase from *Thermomyces lanuginosus*. *Process Biochemistry,* Vol. 45, No. 8, (August 2010), pp. 1268-1273, ISSN 1359-5113

Royon, D., Daz, M., Ellenrieder, G. & Locatelli, S. (2007). Enzymatic production of biodiesel from cotton seed oil using t-butanol as a solvent. *Bioresource Technology,* Vol. 98, No. 3, (February 2007), pp. 648-653, ISSN 0960-8524

Salis, A., Pinna, M., Monduzzi, M. & Solina, V. (2005) Biodiesel production from triolein and short chain alcohols through biocatalysis. *Journal of Biotechnology,* Vol. 119, No. 3, (September 2005), pp. 291-299, ISSN 0168-1656

Salis, A., Pinna, M., Monduzzi, M. & Solina, V. (2008) Comparison among immobilised lipases on macroporous polypropylene toward biodiesel synthesis. *Journal of Molecular Catalysis B: Enzymatic,*Vol. 54, No. 1-2, (July 2008), pp. 19-26, ISSN 1381-1177

Samukawa, T., Kaieda, M., Matsumoto, T., Ban, K., Kondo, A., Shimada, Y., Noda, H. & Fukuda, H. (2000) Pretreatment of immobilized *Candida antarctica* for biodiesel fuel production from plant oil. *Journal of Bioscience and Bioengineering,* Vol. 90, No. 2, (May 2000), pp. 180-183, ISSN 1389-1723

Schmid, R.D. & Verger, R. (1998) Lipases: interfacial enzymes with attractive applications. *Angewandte Chemie International Edition,* Vol. 37, No. 12, (December 1998), pp. 1608-1633, ISSN 1433-7851

Selmi, B. & Thomas, D., (1998) Immobilized lipase-catalyzed ethanolysis of sunflower oil in a solvent-free medium. *Journal of the American Oil Chemists' Society,* Vol. 75, No. 8, (August 1998), pp. 691-695, ISSN 0003-021X

Severac, E., Galy, O., Turon, F., Monsan, P. & Marty, A. (2011) Continuous lipase-catalyzed production of esters from crude high-oleic sunflower oil. *Bioresource Technology,* Vol. 102, No. 8, (April 2011), pp. 4954-4961, ISSN 0960-8524

Shah S. & Gupta M.N., (2007) Lipase catalyzed preparation of biodiesel from *Jatropha* oil in a solvent free system. *Process Biochemistry,* Vol. 42, No. 2, (March 2007), pp. 409-414, ISSN 1359-5113

Shimada, Y., Watanabe, Y., Samukawa, T., Sugihara A. & Noda, H. (1999) Conversion of vegetable oil to biodiesel usin immobilized *Canida antarctica* lipase. *Journal of the American Oil Chemists' Society,* Vol. 76, No. 7,(July 1999), pp. 789-793, ISSN 0003-021X

Shimada, Y., Watanabe, Y., Sugihara, A. & Tominanga, Y., (2002) Enzymatic alcoholysis for biodiesel fuel production and application of the reaction to oil processing. *Journal of Molecular Catalysis B: Enzymatic,* Vol.17, No. 3-5, (June 2002), pp. 133-142, ISSN 1381-1177

Sotoft, L.F., Rong, B.G., Christensen, K.V. & Norddahl, B. (2010) Process simulation and economical evaluation of enzymatic biodiesel production plant. *Bioresource Technology,* Vol. 101, No. 14, (July 2010), pp. 5266-5274, ISSNv0960-8524

Soumanou, M. & Bornscheuer, U.T. (2003). Improvement in lipase-catalyzed synthesis of fatty acid methyl esters from sunflower oil. *Enzyme and Microbial Technology,* Vol. 33, No. 1, (July 2003), pp. 97-103, ISSN 0141-0229

Talukder, M.R., Wu, J.C., Nguyen, T.B.V., Fen, N.M. & Melissa, Y.L.S. (2009) Novozyme 435 for production of biodiesel from unrefined oil: Comparison of the methanolysis methods. *Journal of Molecular Catalysis B: Enzymatic,* Vol.60, No. 3-4, (October 2009), pp.106-112, ISSN 1381-1177

Tamalampudi, S., Talukder, M.R., Hama, S., Numata, T., Kondo, A. & Fukuda, H. (2008). Enzymatic production of biodiesel from Jatropha oil: A comparative study of immobilized-whole cell and commercial lipases as a biocatalyst. *Biochemical Engineering Journal,* Vol. 39, No. 1, (April 2008), pp. 185-189, ISSN 1369-703X

Tan, T., Lu, J., Nie, K., Deng, L. & Wang, F. (2010) Biodiesel production with immobilized lipase: A review. *Biotechnology Advances,* Vol. 28, No. 5, (September-October 2010), pp. 628-634, ISSN 0734-9750

Wang, L., Du, W., Liu, D. & Dai, N. (2006). Lipase-catalyzed biodiesel production from soybean oil deodorizer distillate with absorbent present in *tert*-butanol system. *Journal of Molecular Catalysis B: Enzymatic,* Vol.43, No. 1-4, (December 2006), pp. 29-32, ISSN 1381-1177

Watanabe, Y., Samukawa, T., Sugihara, A., Noda, H., Fukuda, H. & Tominaga, Y. (1999) Conversion of vegetable oil to biodiesel using immobilized *Candida antarctica* lipase. *Journal of the American Oil Chemists' Society,* Vol. 76, No. 7, (July 1999), pp. 789-793, ISSN 0003-021X

Watanabe, Y., Shimada, Y., Sugihara, A., Noda, H., Fukuda, H. & Tominaga, Y. (2000) Continuous production of biodiesel fuel from vegetable oil using immobilized *Candida antarctica* lipase. *Journal of the American Oil Chemists'Society,* Vol. 77, No. 4, (April 2000), pp. 355-360, ISSN 0003-021X

Watanabe, Y., Shimada, Y., Saguhara, A. & Tominaga, Y. (2001) Enzymatic conversion of waste edible oil to biodiesel fuel in a fixed bed reactor, *Journal of the American Oil Chemists' Society,* Vol. 78, No. 7, (July 2001), pp. 701-707, ISSN 0003-021X

Watanabe, Y., Shimada, Y., Sugihara, A. & Tominaga, Y. (2002) Conversion of degummed soybean oil to biodiesel fuel with immobilized *Candida antarctica* lipase, *Journal of Molecular Catalysis B: Enzymatic,* Vol.17, No. 3-5, (June 2002), pp. 151-155, ISSN 1381-1177

Watanabe, Y., Pinsirodom, P., Nagao, T., Yamauchi, A., Kobayashi, T., Nishida, Y., Takagi, Y. & Shimada, Y. (2007) Conversion of acid oil by-produced in vegetable oil refining to biodiesel fuel by immobilized *Candida antarctica* lipase. *Journal of Molecular Catalysis B: Enzymatic,* Vol.44, No. 3-4, (March 2007), pp. 99-105, ISSN 1381-1177

Xie, W. & Ma, N. (2009). Immobilized Lipase on Fe_3O_4 Nanoparticles as Biocatalyst for Biodiesel Production. *Energy & Fuels,* Vol. 23, No. 3, (January 2009), pp. 1347-1353, ISSN 0887-0624

Xu, Y., Du, W. & Liu, D. (2005) Study on the kinetics of enzymatic interesterification of triglycerides for biodiesel production with methyl acetate as the acyl acceptor. *Journal of Molecular Catalysis B: Enzymatic,* Vol. 32, No. 5-6, (March 2005), pp. 241-245, ISSN 1381-1177

Yadav, G. & Jadhav, S. (2005) Synthesis of reusable lipases by immobilization on hexagonal mesoporous silica and encapsulation in calcium alginate: Transesterification in non-aqueous medium. *Microporous and Mesoporous Materials,* Vol. 86, No. 1-3, (November 2005), pp. 215-222, ISSN 1387-1811

Yagiz, F., Kazan, D. & Akin, A.N. (2007) Biodiesel production from waste oils by using lipase immobilized oh hydrotalcite and zeolites. *Chemical Engineering Journal,* Vol. 134, No. 1-3, (November 2007), pp. 262-267, ISSN 1385-8947

Yang, G., Tian-Wei, T., Kai-li, N. & Fang, W. (2006). Immobilization of lipase on macroporous resin and its application in synthesis of biodiesel in low aqueous media. *Chinese Journal of Biotechnology,* Vol. 22, No. 1, (January 2006), pp. 114-118

Ying, M. & Chen, G. (2007). Study on the Production of Biodiesel by Magnetic Cell Biocetalyst Based on Lipase-Producing Bacillus subtilis, *Applied Biochemistry and Biotechnology*, Vol. 137-140, No. 1-12, (April 2007), pp. 793- 804, ISSN 0273-2289

4

Engine Test of Bio-Diesel Manufactured from Waste Cooking Oil and Reward Preferential Benefit Analyses for Its Promotion

Jai-Houng Leu
Yu-Da University
Taiwan, R.O.C.

1. Introduction

The main problem for current biodiesel is high cost. According to the statistics, 70% of manufacturing cost of biodiesel is from feedstock. So using cheep feedstock and raise the transerification rate to reduce cost is the key for the manufacturing biodiesel. But there are following shortcomings for adopting chemical method to synthesize biodiesel: complicated manufacturing process, excess of alcohol; the follow-up process must have corresponding alcohol retrieval device, high energy-consuming, dark colour because the quality of the unsaturated fatty acid is easy to change under high temperature; difficult to retrieve esterified products, high cost; there is waste alkali liquid emission in the manufacturing process.

To solve above-mentioned problems, some researcher have begun to use bio-enzyme to synthesize biodiesel, that is, use animal and plant fatty and low carbon alcohol to conduct transerification reaction through fatty acid and manufacture corresponding fatty acid: methyl and ethyl. The biodiesel synthesized by enzyme method has moderate condition, that is, low transerification rate, 40% to 60% in general. At present time, the fatty acid is effective in the esterificaiton or transerification of long chain fatty alcohol and low esterification rate for short chain fatty alcohol such as methanol and ethanol. Besides, short chain alcohol is poisonous to enzyme that the enzyme life is short. The difficulty to retrieve its by-products glycerol and water constrains the product. The glycerol is poisonous to immobilized enzyme that shortens the life of immobilized enzyme. So it is necessary to develop new immobilization method for fatty enzyme and esterification process to manufacture high quality and low cost biodiesel.

Developing biodiesel by biocatalyst transerificaiton is one of new process that is developing in the world. Using biocatalyst to conduct the esterification production of biodiesel could essentially be divided into extracellular enzyme and intracellular enzyme. There already had many researches on the extracellular reaction. Akoh et al., Who had tested many fatties to dissolve enzyme, found that dissolving enzyme had the best effectiveness [1]. Knothe et al (2000) also offered pre-processing method for fatty ester composition [2]. Wu, W.T. et al. also conducted researches on the enhancement of activity of enzyme in commercialized enzyme Novozyme435 [3]. The biggest shortcoming of extracellular reaction is that the enzyme has to go through the process of isolation and purification that is complicated in

process and the enzyme is usually not stable and expensive. Therefore, many researches use entire cell to catalyze [4]. The research by Ban et al pointed out that using entire cell for immobilization of biocatalyst had big potential for production of biodiesel in biomass support particles, (BSPs) [5].
In Germany, the cars manufactured by Mercedes-Benz, BMW, VW and Audi auto manufacturers are allowed to use bio-diesel without the need of the modification of engines. Because the use of bio-diesel enjoys can enjoy tax-exempt incentive measures, the retail price at present time is about 1.45 marks/litre whereas the price of diesel is 1.60 marks/litre, so bio-diesel has excellent competitiveness comparing to general diesel. The tests of engines of using bio-diesel showed that the waste gas emission indicators not only satisfy present Europe's No. 2 Standard, but also satisfy Europe's tougher No. 3 Standard that will be effective soon. Due to its excellent environment protection property, bio-diesel enjoys the tax preferential policy by government and its price is lower than general diesel. Besides using as fuel diesel, fatty acid methyl esters (bio-diesel) has value in widespread applications to chemical industries [6].
In order to enhance engine efficiency and decrease pollution emission, the relative experiments with mixed fuels were involved in this paper. For example, the bio-diesel derived from waste cooking oil was mixed with petrochemical diesel in addition to the accelerant so as to increase the viscosity of mixed fuel and decrease pollutant emission was added to mixed fuel to promote engine combustion efficiency [7].
Biodiesel, no matter what feedstock and manufacturing process are, is superior to petrochemical diesel in terms of environment protection and human health. So it has been paid more and more attention. In 2000, the world's total production of biodiesel was about 800 thousand tons. But in 2007, the production exceeded three million tons. Most of the incremental quantity was contributed by European countries, Austria, Belgium, France, Germany and Italy in particular. Those countries have been sparing no efforts to the promotions of biodiesel including enacting incentive measures and tax reduction regulations, stipulating national standards of biodiesel and conducting road test and waste gas emissions.
Until today, most of the production of biodiesel come from Europe, top of the whole world with more than 80%, and the production still continue to grow; on the application, in addition to government cars, the metropolitan public traffic vehicles, private cars, trucks, agricultural machines have been shifted to use biodiesel. The main reason of the success is that the direct and indirect incentives and subsidies by European governments and calls for environment protection from the people have contributed to the success of promotion of biodiesel and become leadership in this field [8].

2. Manufacture of biodiesel by transerification reaction with alkali process from waste cooking oil

At present time, the fatty acid is effective in the esterificaiton or transerification of long chain fatty alcohol and low esterification rate for short chain fatty alcohol such as methanol and ethanol. Besides, short chain alcohol is poisonous to enzyme that the enzyme life is short. The difficulty to retrieve its by-products glycerol and water constrains the product. The glycerol is poisonous to immobilized enzyme that shortens the life of immobilized enzyme. So it is necessary to develop new immobilization method for fatty enzyme and esterification process to manufacture high quality and low cost biodiesel. The comparisons of benefit and defect between Alkali process and enzyme process of biodiesel transerification are discussed in Table 1.

Process / Item	Alkali process	Enzyme process	Remark
Reaction temperature	60~70°C	30~40°C	Additive amount of energy depends on temperature
Free fatty acid	Saponified matter	Methyl esters	NaOH with fatty acid may produce saponified matter
Water content of material	Interference reaction	No impact	Water moisture of alkali process reaction can produce caking that has negative impact on following processing.
Methyl esters yield	Average	High	Yield means methyl esters rate per material
Glycerol yield	Difficult	Easy	Glycerol is additional product, and can increase extra income.
Methyl esters purification	Repeated water washing	None	Waste washing water drainage, and additional wastewater treatment is required.
Catalyst cost	Cheap	Very expensive	Increase ferment use time can reduce cost.

Table 1. Comparison between Alkali Process and Enzyme Process

The high acid value oil in waste cooking oil generally composes of impurities, free fatty acid, polymer, resolver, etc. that are not good to transerification and have to be pre-processed. The methods we could consider are physical refining and methanol pre-erification. The oil feedstock we used was Uni-President soybean oil, waste cooking oil from restaurants and waste cooking oil from Yang-Yang Oil Retrieval Company.

The Iodine value is almost the same in Table 2, so the modulation of the Iodine value is not necessary. It showed in Table 2 that we could judge the degree of free fatty acid of the oil from acid value. The higher acid value is that the higher degree of free fatty acid will be. The results of the test of acid value showed that the acid value of Uni-President soybean oil was 0.04488 (mg KOH/g), the acid value of waste cooking oil from restaurants after modulation was 0.26928(mg KOH/g), waste cooking oil from Yang-Yang Oil Retrieval Company was 1.929 (mg KOH/g). It showed that the content of the free fatty acid in the waste cooking oil from Yang-Yang Oil Retrieval Company was higher. The reason is that the waste cooking oil retrieved by Yang-Yang Oil Retrieval Company was the mixed oil from restaurants that usually had higher acid value. This could be acted as referenced basis for the acid value of retrieved waste cooking oil. The iodine value of all the tested feedstock is almost the same under 8% difference. So, the modulation must be focused on the acid value for the feedstock from waste cooking oil.

If Uni-President soybean oil was used as raw material, NaOH catalyst was decreased, and methanol was increased, the saponified substances obviously decreased and can be stirred uniformly after transerification. The tap funnel produced obvious layers after 10 minutes. The methyl ester at upper layer was easy to separate from the glycerol at the lower layer. The crude methyl ester yield increased, and crude methyl ester volume yield was up to 97%. But the residue methanol was not removed from the crude methyl ester volume. After removal, the crude methyl ester volume required correction. After distillation, the methyl ester volume was 395ml, and the methyl ester volume yield was 79.0%. The loss is due to water washing.

When the waste cooking oil from restaurants after modulation was used as raw material, no saponified substance appeared after transerification. The tap funnel had obvious layers after 10 minutes. The methyl ester at upper layer was easy to separate from the glycerol at the lower layer. The crude methyl ester volume was 503 ml, and the crude methyl ester yield was up to 100.6%. After distillation, the methyl ester volume was 400ml, and the methyl ester volume yield was 80.0%. The loss is due to water washing.

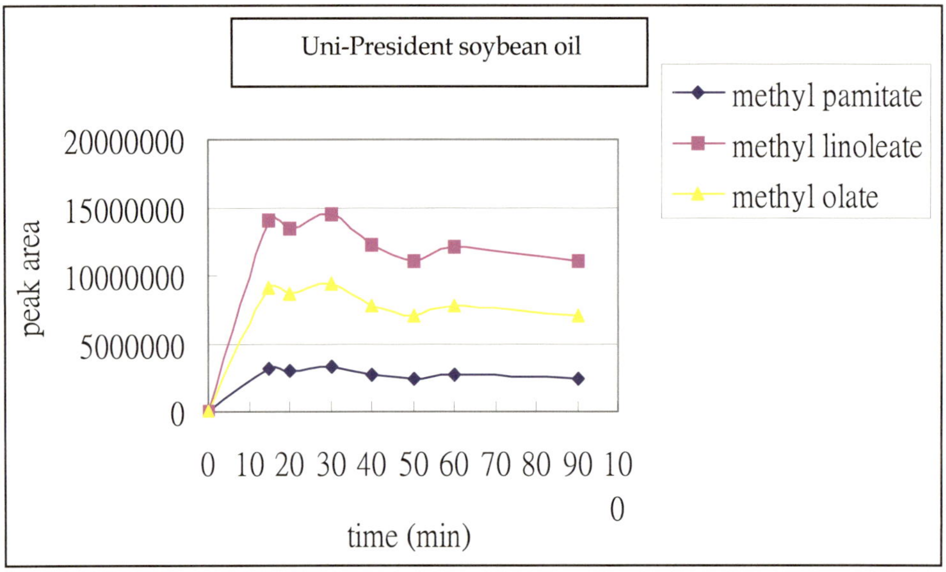

Fig. 1. Uni-President soybean oil transerification reaction curve

If using waste cooking oil retrieved by Yang-Yang Oil Retrieval Company, no saponified substance appeared after transerification. The tap funnel had obvious layers after 10 minutes. The crude methyl ester volume was 528 ml, and the crude methyl ester yield was up to 97.77%. After distillation, the methyl ester volume was 512ml, and methyl ester volume yield was 94.81%. The loss caused by water washing reduced more, and low methyl ester volume was obtained shown in Fig.1 and Fig.2.

From the above experiments, the acid value of waste cooking oil must be modulated down to 0.26928(mg KOH/g) for the self-manufacture of the biodiesel to keep the normal processes of the transerification Reaction with Alkali. The colour of waste cooking oil from Yang-Yang Oil Retrieval Company and restaurants after modulation was deeper than the Uni-President soybean oil before and after transerification. GC-MASS analysis was conducted for products from transerification, chromatography area and time was used to plot reaction curve. The reaction balance can be achieved after 15 minutes for all the tested feedstock in Fig. 1 and Fig. 2.

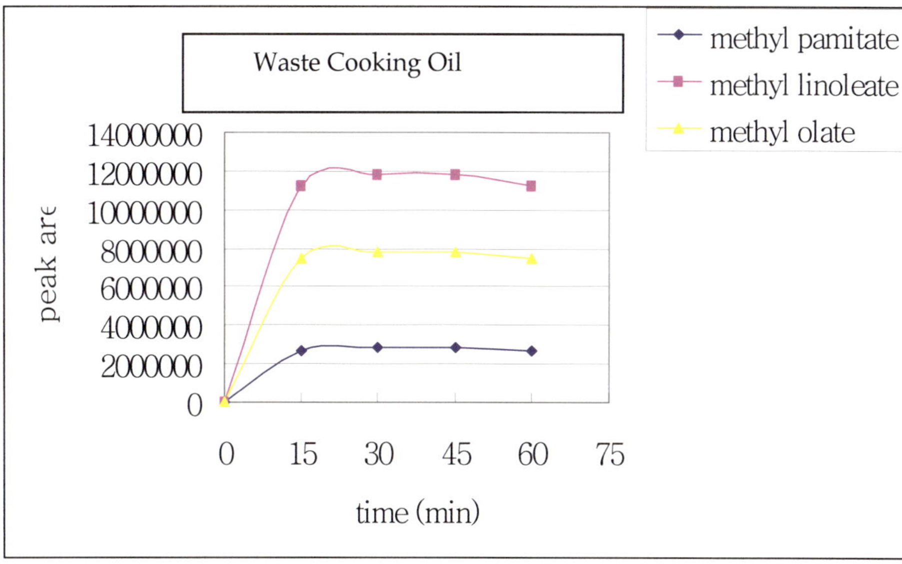

Fig. 2. Transerification reaction curve of waste cooking oil from Yang-Yang Oil Retrieval Company

The production of the methyl linoleate was more over the one of the methyl pamitate during the processes for the manufacture of biodiesel from the waste cooking oil.

Name of Oil	Uni-President soybean oil	Waste cooking oil from Restaurants (after modulation)	Waste cooking oil Retrieved by Yang-Yang Oil Retrieval Company
Iodine Value (g I2 / 100g)	135.91	142.32	145.67
Acid Value (mg KOH / g)	0.04488	0.26928	1.929
Viscosity (cP,22.5°C)	58.6	65.7	67,3
Density (g / ml,25°C)	0.918	0.918	0.908

Table 2. Analyses of the Property of Feedstock Oil for Manufacturing Biodiesel

Item	Unit	Results of Test			Test Method	ASTM Biodiesel Standard
		PD	B20	B100		
Density @15.5°C	G/ml	0.84	0.85	0.88	ASTM D4050	0.85
Flash Point	°C	87	92	150	ASTM D93	100 min.
Contents of Sulphur 1	Wt%	0.0258	0.0200	<0.001	ASTM D2622	0.05 max.
Power Viscosity@40°C	CST	3.460	3.517	4.136	ASTM D445	1.9-6.0
Water and Sediment	Vol.%	0.00	0.00	0.00	ASTM D1796	0.00
Corrosiveness	3hr (50°C)	1a	1a	1a	ASTM D130	No.3b max.
Ash	Wt%	0.003	0.002	0.003	ASTM D482	0.002
Aromatic Hydrocarbons	Wt%	24.5	*	*	ASTM D5186	none
Cetane Index2	none	53.9	53.7	47.3	ASTM D976	53.7
Rams bottom Carbon Residue3	Wt%	0.051	0.033	0.035	ASTM D524	0.033
Carbon Contents	Wt%	86.71	86.49	76.53	ASTM D5291	86.5
Hydrogen Content	Wt%	13.28	13.5	11.72	ASTM D5291	13.5
Oxygen Content	Wt%	0.01	0.01	11.75		none
Total Thermal Value	Cal/g	10990	10625	9490	ASTM D240	10630
Net Thermal Value	Cal/g	10317	9941	8896	ASTM D240	9940

Note: * means the tests that could not be completed by analytic instruments under SOP of Refinery Institute of China Petroleum Company. The number 1, 2, 3 were tested by Material and Chemical Research Laboratories, Industrial Research Institute for the research.

Table 3. Analyses of Fuel Properties of Petroleum Diesel, B20 and B100

The high flash point and low sulphur content exists in the manufactured biodiesel than petroleum biodiesel. The cetane index and heating value of the manufactured biodiesel is a litter low than petroleum biodiesel. And, the properties of the manufactured biodiesel meet the standard of the biodiesel almost.

3. Engine application of bio-diesel manufactured from waste cooking oil

The fuel in the test included petrochemical diesel (PD), self-manufactured pure biodiesel (B100) through transerification reaction of soybean oil and mixed biodiesel that is mixed by above-mentioned two fuels with volume ratios of 4:1. Then we sent these fuels to Refining Institute of China Petroleum Company, Taiwan for analyses and tests.

The engine test was conducted at Alternative Inspection Centre of Ford Company, Taoyuan County, Taiwan. The facilities of this centre are listed in Table 4. We used the equipment of the centre to test the engines of garbage trucks (provided by Cleaning Department of Chungli City. Test was conducted during the maintenance of the trucks). The regulations of the engine are listed in Table 5. According to CNS11644 and CNS11645, we conducted pollution tests of non-loading instant speed up, full load fixed speed. Meanwhile, we used waste gas analysis instruments (listed in Table 6) to test the ingredients of gas emission by engines under full loading, including CO, CO2, SO2, NOX and brake horsepower. Besides,

because when the garbage truck is often in idling state, so we conducted the gas emission test under the idling state and test brake horsepower with car body power meter. Pollution degree is defined as follows:
(CO emission concentration)/ (CO2 emission concentration + CO emission concentration)

Item	Specification
Vehicle body dynamometer	SCHENCK 364/L/D/1200
Smoke meter	KOMYO ST-100
Tachometer	AVL 873

Table 4. Testing Instrument

Item	Specification
Engine brand	INTERNATIONAL
Vehicle type	4900, 4x2, Turbo 7636C.C, A5
Vehicle mileage	270743KM
Engine type	In-line six-cylinder
Starting date	1992
Max power	182.7kW@2600rpm
Vehicle weight	15,000kg

Table 5. Garbage Truck Engine Data

Equipment	Brand and Model	Analyser principle	Measurement range	Accuracy
SO2 analyser	Siemens LTRAMAT6E	NDIR	0~1000ppm	±1.0% of Range
CO2 analyser	Siemens LTRAMAT21P	NDIR	P~20%	±1.0% of Range
CO analyser	Siemens ULTRAMAT6E	NDIR	0~3000ppm	±0.5% of Range
O2 analyser	AMETEK FCA/Rack Mt.	ZrO2	0~20.9%	±1.5% of Range
NOX analyser	Siemens ULTRAMAT23	NDIR	0~1000ppm	±1.0% of Range

Table 6. Exhaust-Gas Analyzer

3.1 Test of brake horsepower

The brake horsepower tests of engine were conducted in using petrochemical diesel, B20 and B100 under the state of full loading. Under full loading, increase loading step by step, the rotational speed from 2600rpm is reduced to 2500, 2400, 2300 and finally 2200rpm. The output brake horsepower by the engine in using premium diesel, B20 and B100 were shown as Fig. 3.Calculated by average values, comparing to premium diesel, the brake horsepower output by B20 is reduced by 8%, and, B100 is reduced by 10%.

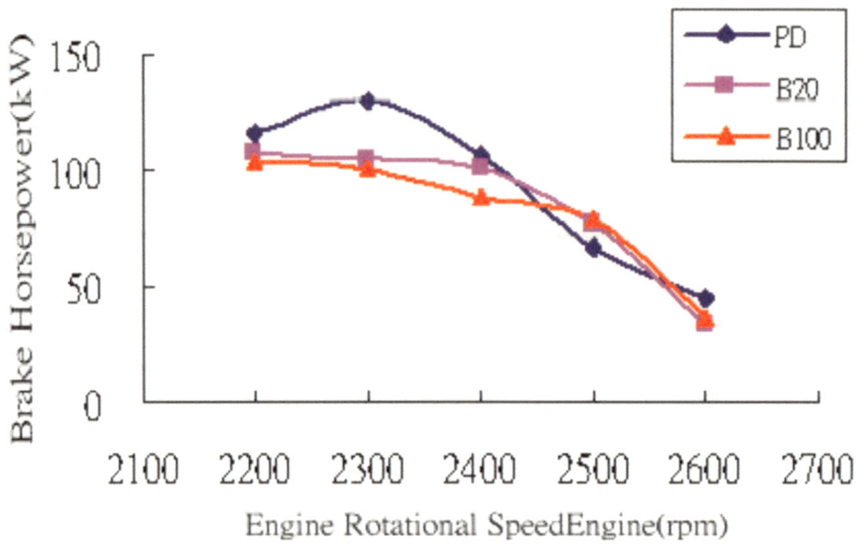

Fig. 3. Engine Horsepower Output by Three Fuels under Full Loading

3.2 Test of pollutant emissions by engine

Reorganized concentration of waste flue such as carbon monoxide, sulfur dioxide and nitrogen oxide compound emitted by engine under the state of full load and idle as Table 7.

Under the test of full load, when used B20 and B100, the concentrations of carbon monoxide emitted by engine were reduced 14% and 51% respectively comparing petrochemical diesel. Under the test of idle, when used petrochemical diesel, B20 and B100, the concentrations of carbon monoxide emitted by engine were almost the same.

Under the test of full load, when used B20 and B100, the concentrations of sulphur dioxide were reduced 68% and 73% respectively comparing petrochemical diesel,. Under the test of idle, the concentration of sulphur dioxide emitted by B100 was the lowest; the concentrations of sulphur dioxide emitted by petrochemical diesel and B20 were almost the same.

Under the test of full load, when used B20 and B100, the concentration of nitrogen oxide compound increased 1% and 13% respectively comparing petrochemical diesel,. Under the test of idle, the concentration of nitrogen oxide emitted by B100 was the highest; the concentrations of nitrogen oxide emitted by petrochemical diesel and B20 were almost the same.

Under the test of full load, the load increased gradually. When the engine speed decreased from 2680rpm to 2360rpm and to 2100rpm, B100 had the lowest pollution degree, followed by B20, and the premium grade diesel had the highest pollution degree, see Fig. 4. In terms of average value, as compared to premium grade diesel, the pollution degree of B20 can be reduced by 8%; the pollution degree of B100 can be reduced by 69%.

Under the test of idle, pollution degree of premium grade diesel, B20 and B100 was 7%, 3% and 2% respectively, see Fig. 5. Under the test of free acceleration, pollution degree of the premium grade diesel was close to that of B20, and the pollution degree was 31.3% and 38.6% respectively. B100 had the lowest pollution degree, 19.3%, see Fig. 6.

RPM	CO_2(%)	CO(ppm)	O_2(%)	SOx(ppm)	NOx(ppm)
PD					
2600	8.1	200	10	18	256
2340	9.1	282	8	46	520
2080	9.5	316	7	57	517
Idle	1.6	98	18	17	275
B20					
2600	6.7	174	12	18	333
2340	9.3	257	8	25	453
2080	9.5	255	8	29	525
Idle	1.7	103	19	13	263
B100					
2600	6.8	124	12	10	331
2340	9.1	133	9	10	547
2080	9.6	131	8	12	582
Idle	1.8	108	18	12	334

Note> The idle is at 630 rpm

Table 7. Emission Test under the State of Full Load

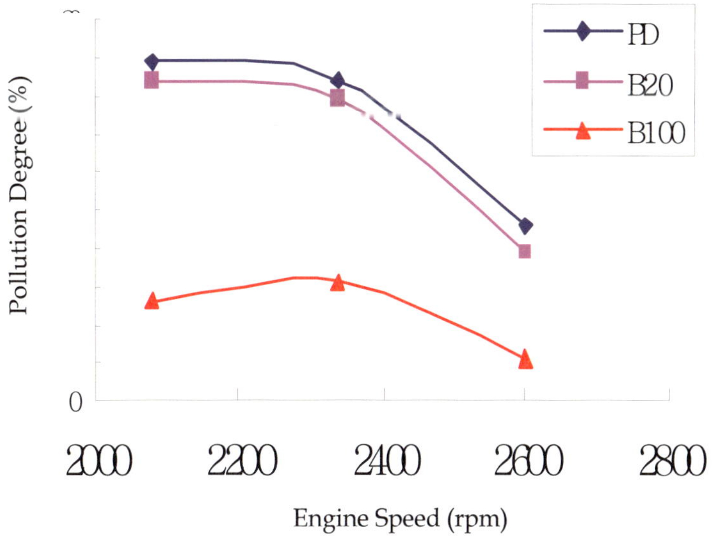

Fig. 4. Pollution Degree Test under Full Load

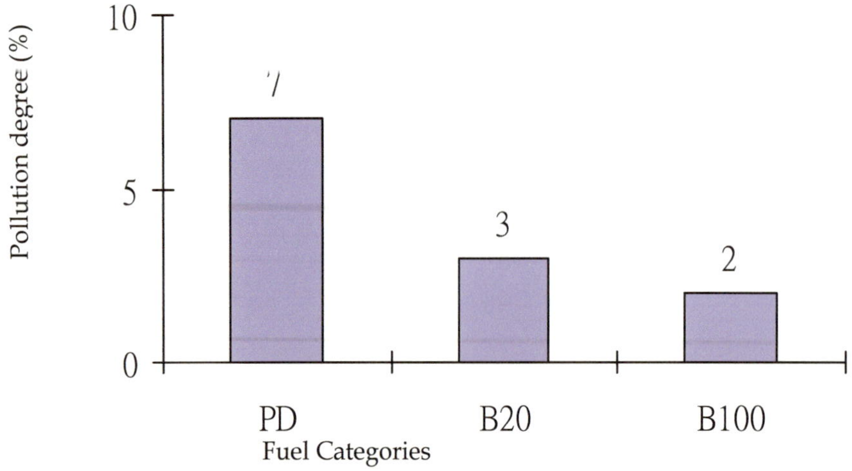

Fig. 5. Pollution Degree Test under Idle State

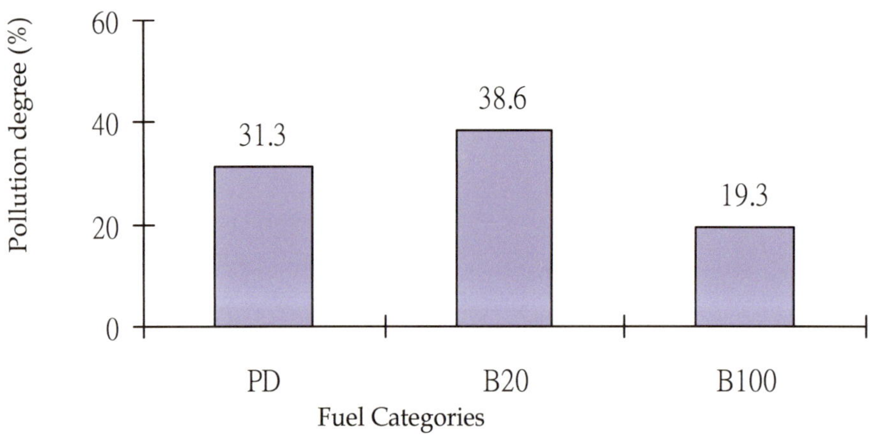

Fig. 6. Pollution Degree Test under Free Acceleration

4. The analyses of reward preferential benefit for biodiesel promotion in Taiwan by financial IRR and PB estimation

Through economic research, the study aims to understand the potential and feasibility of promotions and application of biodiesel in Taiwan. It is hoped that the results of research could provide the operators with references in investment and the government with references in working out regulations and incentives. With reference to the establishment and operation of factories abroad and conditions in Taiwan, we evaluated the feasibility of establishing biodiesel factory through economic analyses to provide operators with the base in planning and establishing factories in Taiwan.

The article collected foreign related literature and analyzed the cost and benefits in three parts:
(1) Net Present Value Method: After initial investment, the yearly flowing-in amount of net cash was conversed into present value according to capital cost (minimum investment reward). After deducting the present value of initial investment, the difference is net present value (NPV). When NPV>0, we could accept the investment plan; if NPV<0, we should abandon the plan. Under consideration of inflation, the formula is as follows:

$$NPV=\sum_{t=1}^{n}\frac{[(R_t-C_t)-T_r(R_t-C_t)+D_t]}{[(1+r)^t \times (1+np)]}-CE_0 \quad (1)$$

Rt: The total income at the end of the t period, the value is the monetary value at the end of the t period.
Ct: The total cost at the end of the t period, the value is the monetary value at the end of the t period.
Tr: Tax rate of income for enterprise
Dt: Depreciation
I: Capital Cost
P: Inflation Rate
Ceo: Initial investment cost
(2) Internal Reward Method: After investment, the discount rate of total of every year's flowing-in amount of net cash that is conversed into present value equivalent to total initial investment is internal reward rate (IRR). When IRR >1 (capital cost), we could accept the investment plan, but if IRR<1, we should abandon the plan. Under consideration of inflation, the formula is as follows:

$$NPV-CE_0=0$$

$$NPV=\sum_{t=1}^{n}\frac{[(R_t-C_t)-T_r(R_t-C_t)+D_t]}{[(1+r)^t \times (1+np)]} \quad (2)$$

Note. The definition of symbols is the same as net present value method and in which r is the IRR.
Discount and Repayment of Principal Method: It refers the time that the initial total investment could be repaid through operation. When the net flowing amounts of cash in every period (i.e. the results of total income deducted by total expenses of the period) are equal, the time of discount and repayment of principal (PB) is results of initial investment amount being divided by the total of each period's discount net cash flowing amount. On the other hand, when and if each period's discount net cash flowing amounts are equal, then we need to use interpolation to obtain PB.
In the research, we took the example of an Ireland's 3000 tons/year biodiesel factory to conduct benefit analyses to explore the unit input cost of different feedstock. The related investment and operation cost came from Ireland's literature [2]. We inducted as follows:
Mould Factory Capacity Planning: 1) Oil-Squeezing Factories (waste eating oil does not need the equipment): with 30% oilseed transform ratio, the average oilseed investment cost is NT$1,415/ton (or NT$1.25/litre). If we wish to provide mixed biodiesel factories with 25% tea oil, then we will need a tea oil oil-squeezing factory with capacity of 750 tons/year, the investment cost is NT$5.12 Millions; if we want to supply a 100% biodiesel factory, we

have to invest NT$ 21.2 millions for a tea oil oil-squeezing factory with capacity of 3,000 tons/year; 2) biodiesel refinery: with oil transform ratio 95% for vegetable oil and 87% for waste eating oil, the total investment for a factory with the capacity of 3,000 tons is about NT$38,4 millions, average oil-squeezing cost is NT$2,625/ton (or NT$2.31/litre. Operation hours: 24 hours per day and 330 days per year; Operation pattern: lot production pattern. Three different feedstock: 100% tea oil camellia, mixed (25% tea oil and 75% waste eating oil), 100% waste oil.

(3) Discount Rate (necessary reward rate): 20%, pound: NT $ dollars = 1:45

Furthermore, the subject of the research was small mould factory with output of 1,275 kiloliter one year. We also conducted evaluation of economic benefits of hardware equipment and price incentives according to the instalment cost of hardware equipment and operation cost provided by Pacific Biodiesel, Inc. The analyses were conducted with self-developed software [9].

4.1 The Analyses of unit output cost of different feedstock

This research took an Ireland's biodiesel mould factory with output 3,000 tons/year as an example to compare unit output cost of three different feed stocks (tea oil-camellia, mixed, waste eating oil):

From Table 8, we could learn that the feedstock was 60% to 70% of production cost. If we raise the additive ratios of the waste eating oil or all replace with waste eating oil, the unit cost could be reduced 12.5% to 17%. Therefore, the biodiesel with waste eating oil as feedstock could not only reduce production cost but also had benefit of environment protection, and would never be affected by the fluctuation of prices of agricultural products.

Unit: NT dollars/litre

Category of Feedstock / Items of Cost	100% Tea Oil	Mixed (25% Tea Oil and 75% Waste Eating Oil)	100% Waste Eating Oil
* Feedstock Cost	12.02	9.74	8.98
Operation Cost	3.29	3.29	3.29
Investment Cost**	2.31	2.31	2.31
Unit Cost **	17.6	15.4	14.6

Note: *Feedstock Cost: Tea Oil includes agricultural products purchase cost and oil-squeezing cost; waste eating oil means charges fro cleaning, but not included purchase cost for waste eating oil.
** Investment cost: included only investment cost of biodiesel refinery. The investment cost of oil-squeezing factory had been included in feedstock cost.
***Annual production: 3000 tons (3000÷0.8827=3,400 kilolitres)

Table 8. Comparison and Analyses of Unit Cost of Different Feedstock

4.2 The analyses of benefits of incentives for biodiesel mould factory in Taiwan

For the present development of biodiesel in Europe, the size of small mould factory is annual output of 5,500 to 13,500 kilolitres. This research will target at mould factories with annual output of 1,275. According to the instalment cost of hardware equipment and operation cost calculated by Pacific Biodiesel, Inc., we conducted the evaluation of economic benefits of the incentives for hardware equipment and price. This article will take the popularized gasoline and diesel prices as the benchmark for comparing with biodiesel (13.3 to 15.4 NT Dollars/litre of published tag prices, but here we used it as the lowest price to conduct analyses).

After analyzing unit output cost of different of different feedstock in Table 1, we could learn that the feedstock cost was 60% to 70% of production cost. There was significant difference in the effects on the production cost between oil plants and waste eating oil as feedstock. Using waste eating oil as feedstock could reduce 17% of production cost. Therefore, for the planting conditions of land in Taiwan, it seems more feasible to use low-cost waste eating oil as feedstock. The research will take waste eating oil and the oil of cooking residue as subjects to conduct related analyses.

CASE1 to 3 are interrelationship between the retrieval time limit of equipment subsidy ratio, internal reward rates and oil purchase prices of biodiesel. Reorganizing the data of CASE1 to 3, we could obtain results as Table 9 and 10. When the feedstock is waste eating oil (the purchase cost is zero) and oil of cooking residue, because of the reduction of feedstock and perhaps the income of disposal charges, the operators could have economic benefit without any subsidy. But, when the purchase cost of waste eating oil is NT$5/kg, the operator will suffer loss and could not be able to have economic benefit with ordinary diesel sales price 13.3/litre as oil price. From Table 10, we could learn that since waste eating oil (Case 1 - purchase cost was zero), and the oil of cooking residue (Case 3) could have economic benefit without any subsidy, Table 9 only explores the price and equipment subsidy under a situation in which the purchase cost > 0 (Case 2). With 12%, 15% and 20% as reasonable reward rates for the operation of investors. Use interpolation method to calculate demand of oil purchase price and retrieval time limit of the goal of each reward rate; if we adopts price subsidy policy, then the oil purchase prices demand should deduct the sales price NT$13.3/litre of the ordinary fossil diesel, the difference is the demand of price subsidy.

From Table 9, we could learn that when the purchase cost of waste eating oil is NT8/kg (including purchase price NT$5/kg and cleansing and transportation charges NT$3/kg). the oil price of break-even point (IRR=12%) is NT$14.6/kg; Retrieval time limit is 21.8 years. If we take current ordinary fossil diesel price NT$13.3/litre as oil purchase price, then the subsidy demand of single price is NT$1.3/litre, if not consider price subsidy, then the equipment subsidy ratio will be as high as 90% to make up difference with ordinary fossil diesel sales price. However, if we consider equipment subsidy ratio in couple with price subsidy (mixed subsidy), then we could reduce demand of price subsidy and equipment subsidy ratio. For example, when IRR=12%, equipment subsidy 50%, demand of price subsidy is only NT$0.5/litre, retrieval time limit could be reduced to eight years.

If we want to encourage intention of investment and raise inventory reward rate (IRR), for example 15~20%, then the demand of relative non-subsidy oil price will also be raised to NT$16.9/kg, but the retrieval time limit will reduce to 6.4 years and the subsidy for the price difference will be increased to NT$3.6/litre. If we only use equipment subsidy, it is impossible to make up the difference even if we use 100% subsidy. So we should have price subsidy in the same time to reach the goal of raising investment reward rate. For example, when IRR=20%, equipment subsidy 50%, then the price subsidy may only need NT$2/litre and retrieval time limit could reduce to five years. Summarizing the results of analyses of Table 9 as follows:

1. If 12%～20% are reasonable internal reward rate for operation, the oil purchase price should be between NT$14.6 to 16.9/litre.

2. If the tag price of ordinary fossil diesel is NT$13.3/liter, the difference of purchase price is NT$1.3 to 3.6 after deducting NT$13.3/litre, the average difference is NT$2.5/litre.
3. If we couple with using equipment subsidy ratio, for example, 59%, then the difference of oil purchase will be between NT$0.5/litre to NT$2.1/litre, the average is NT$1.3/litre, reducing about half, comparing to the difference by using single price subsidy.

Using Case 2 cost benefit evaluation of model factory of output 1.275 kilolitres to further analyze the effects of financial subsidies such as hardware and equipment subsidy and low-interest loan on unit production cost.

Item		IRR=12%*	IRR=15%	IRR=20%
No Subsidy	Demand of Oil Price under No Subsidy (NT$/litre)	14.61	15.34	16.92
	Retrieval Time Limit (year)	21.8	15.7	6.4
Single Subsidy	Demand of Price Subsidy** (NT$/litre)	1.3	2.0	3.6
	Equipment Subsidy Ratio (%)	90	>100	>100
Mixed Subsidy	Equipment Subsidy Ratio (%)	30	30	30
	Demand of Price Subsidy (NT$/litre)	1.1	1.4	2.9
	Retrieval Time Limit (year)	10.8	9.8	5.3
	Equipment Subsidy Ratio (%)	40	40	40
	Demand of Price Subsidy	0.8	1.3	2.5
	Retrieval Time Limit (year)	9.6	8.3	5.1
	Equipment Subsidy Ratio (%)	50	50	50
	Demand of Price Subsidy	0.5	1.1	2.1
	Retrieval Time Limit (year)	8.2	7.1	5.0

Note: *IRR=12% is break-even point
** Demand of Price Subsidy=Demand of Oil Price under No Subsidy−13.3
** Ordinary fossil diesel sales price NT$13.3/litre

Table 9. Analyses of Various Reward Rates and Price Subsidy and Equipment Subsidy of Oil Purchase Price

Oil purchase price	Retrieval time limit (year)--Attached Figure 1				Internal Reward Rate（%）--Attached Figure 2			
NT$ /Liter	No Subsidy	30%Subsidy	40%Subsidy	50%Subsidy	No Subsidy	30%Subsidy	40%Subsidy	50%Subsidy
10	50.0	50.0	50.0	50.0	0.0%	0.0%	0.0%	0.0%
11	50.0	50.0	50.0	50.0	0.0%	0.0%	0.5%	2.1%
12	50.0	34.1	16.1	10.3	7.9%	11.5%	13.0%	15.0%
13	14.9	7.9	6.5	5.3	13.7%	17.0%	18.3%	19.8%
14	8.0	5.2	4.4	3.6	17.2%	20.2%	21.6%	23.2%
15	5.8	3.9	3.3	2.7	19.5%	22.8%	24.2%	25.9%
16	4.6	3.1	2.7	2.2	21.5%	24.9%	26.3%	28.1%
17	3.8	2.6	2.2	1.9	23.2%	26.7%	28.2%	30.0%
18	3.3	2.3	1.9	1.6	24.7%	28.2%	29.8%	31.6%
19	2.9	2.0	1.7	1.4	26.0%	29.6%	31.2%	33.0%
20	2.6	1.8	1.5	1.2	27.2%	30.9%	32.5%	34.3%

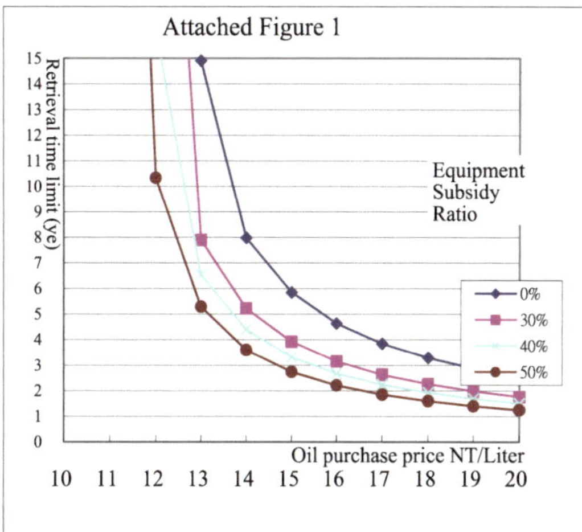

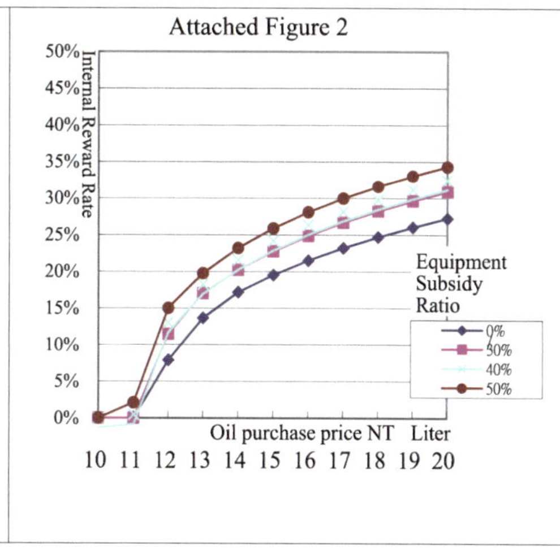

Fixed Conditions : Specification of Equipment : 1, 275kl/Annual Fuel Oil Output, 8 hours operation per day, 300 work days per year

Initial Investment Cost in Equipment: NT$20, 770, 000

Discount Time Limit: 10 years

Discount Rate: 12%

The oil price is different from the measured benchmark of general diesel

The feedstock of biodiesel is waste eating oil, the purchase price per kilogram is NT$0

Case 1. The feedstock is waste eating oil, the purchase price per kilogram is NT$0. The interrelationship between subsidy ratio of biodiesel system equipment, retrieval time limit, internal reward rate and oil purchase proces.

Oil purchase price NT$ /Liter	Retrieval time limit (year)--Attached Figure 3				Internal Reward Rate（%）--Attached Figure 4			
	No Subsidy	30%Subsidy	40%Subsidy	50%Subsidy	No Subsidy	30%Subsidy	40%Subsidy	50%Subsidy
11	50.0	50.0	50.0	50.0	0.0%	0.0%	0.0%	0.0%
12	50.0	50.0	50.0	50.0	0.0%	0.0%	0.0%	0.0%
13	50.0	50.0	50.0	50.0	0.0%	0.0%	0.0%	0.0%
14	50.0	50.0	33.2	14.8	7.1%	9.8%	11.3%	13.0%
15	19.3	8.8	7.3	5.8	13.6%	16.1%	17.3%	18.7%
16	8.6	5.6	4.7	3.8	17.7%	19.5%	20.8%	22.4%
17	6.2	4.1	3.5	2.9	20.2%	22.1%	23.5%	25.1%
18	4.8	3.3	2.8	2.3	22.3%	24.3%	25.7%	27.4%
19	4.0	2.7	2.3	1.9	24.1%	26.1%	27.6%	29.4%
20	3.4	2.3	2.0	1.6	25.7%	27.7%	29.3%	31.0%
21	2.9	2.0	1.7	1.4	27.2%	29.2%	30.7%	32.5%
22	2.6	1.8	1.5	1.3	28.5%	30.5%	32.0%	33.8%
23	2.3	1.6	1.4	1.1	29.7%	31.6%	33.2%	35.0%
24	2.1	1.5	1.3	1.0	30.8%	32.7%	34.2%	36.1%
25	1.9	1.3	1.2	1.0	31.8%	33.6%	35.2%	37.1%

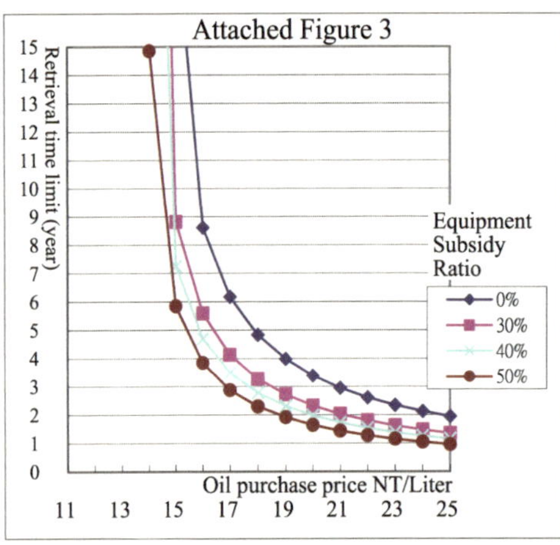

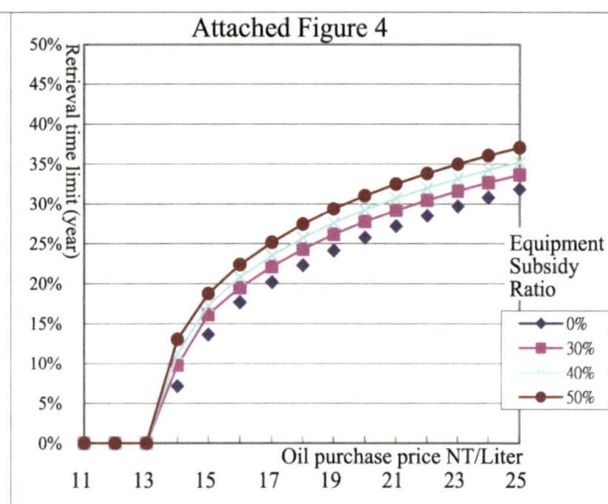

Fixed Conditions : Specification of Equipment : 1, 275kl/Annual Fuel Oil Output, 8 hours operation per day, 300 work days per year

Initial Investment Cost in Equipment: NT$20, 770, 000

Discount Time Limit: 10 years

Discount Rate: 12%

The oil price is different from the measured benchmark of general diesel

The feedstock of biodiesel is waste eating oil, the purchase price per kilogram is NT$5

Case 2. The feedstock is waste eating oil, the purchase price per kilogram is NT$5. The interrelationship between subsidy ratio of biodiesel system equipment, retrieval time limit, internal reward rate and oil purchase proces.

Oil purchase price NT$ /Liter	Retrieval time limit (year)--Attached Figure 5				Internal Reward Rate（%）--Attached Figure 6			
	No Subsidy	30% Subsidy	40% Subsidy	50% Subsidy	No Subsidy	30% Subsidy	40% Subsidy	50% Subsidy
6	50.0	50.0	50.0	50.0	0.0%	0.0%	0.0%	0.0%
7	50.0	50.0	50.0	50.0	0.0%	0.0%	0.6%	2.5%
8	50.0	50.0	18.9	11.1	8.0%	11.9%	13.6%	16.0%
9	17.1	8.4	6.9	5.6	13.9%	17.7%	19.2%	20.9%
10	8.5	5.5	4.6	3.8	17.6%	21.0%	22.6%	24.5%
11	6.2	4.1	3.5	2.9	20.0%	23.7%	25.3%	27.4%
12	4.9	3.3	2.8	2.3	22.1%	25.9%	27.6%	29.7%
13	4.1	2.8	2.4	1.9	23.8%	27.8%	29.6%	31.8%
14	3.5	2.4	2.0	1.7	25.4%	29.5%	31.3%	33.6%
15	3.0	2.1	1.8	1.5	26.8%	31.0%	32.9%	35.2%
16	2.7	1.8	1.6	1.3	28.1%	32.3%	34.3%	36.6%
17	2.4	1.7	1.4	1.2	29.3%	33.6%	35.5%	37.9%
18	2.2	1.5	1.3	1.1	30.4%	34.7%	36.7%	39.1%
19	2.0	1.4	1.2	1.0	31.4%	35.8%	37.8%	40.2%
20	1.9	1.3	1.1	0.9	32.3%	36.8%	38.8%	41.3%

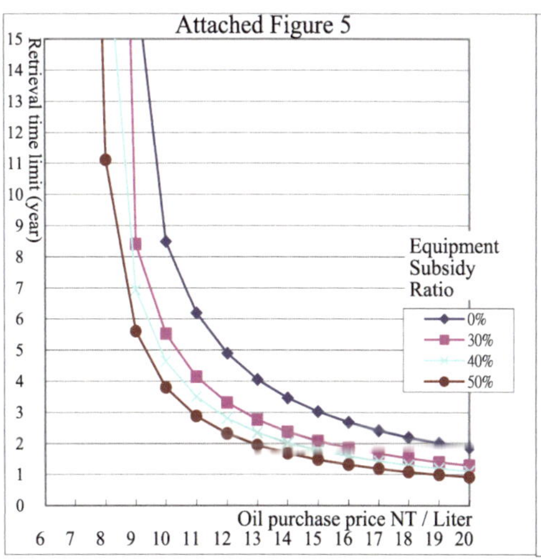

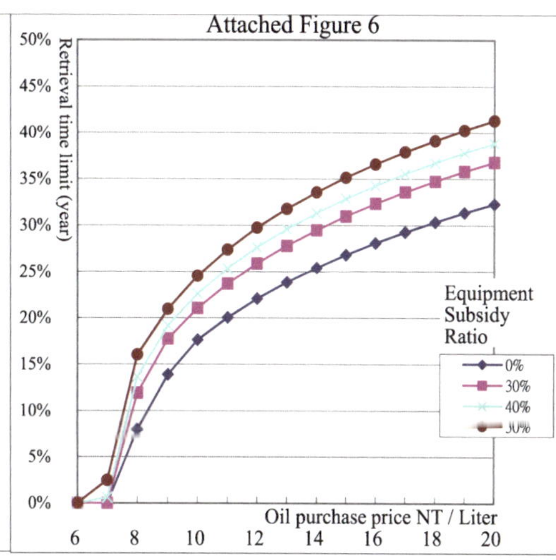

Fixed Conditions: Specification of Equipment: 1.275kl/Annual Fuel Oil Output, 8 hours operation per day,
300 work days per year
Initial Investment Cost in Equipment: NT$21, 120, 000
Discount Time Limit: 10 years
Discount Rate: 12%
The oil price is the same as the measured benchmark of general diesel
The feedstock of biodiesel is waste eating oil, the purchase price per ton is NT$900

Case 3. The feedstock is cooking residue, the purchase price per kilogram is NT$900. The interrelationship between subsidy ratio of biodiesel system equipment, retrieval time limit, internal reward rate and oil purchase proces.

Table 10. The Evaluation of 1275 k litre / year Biodiesel Model Factory Subsidy Benefit

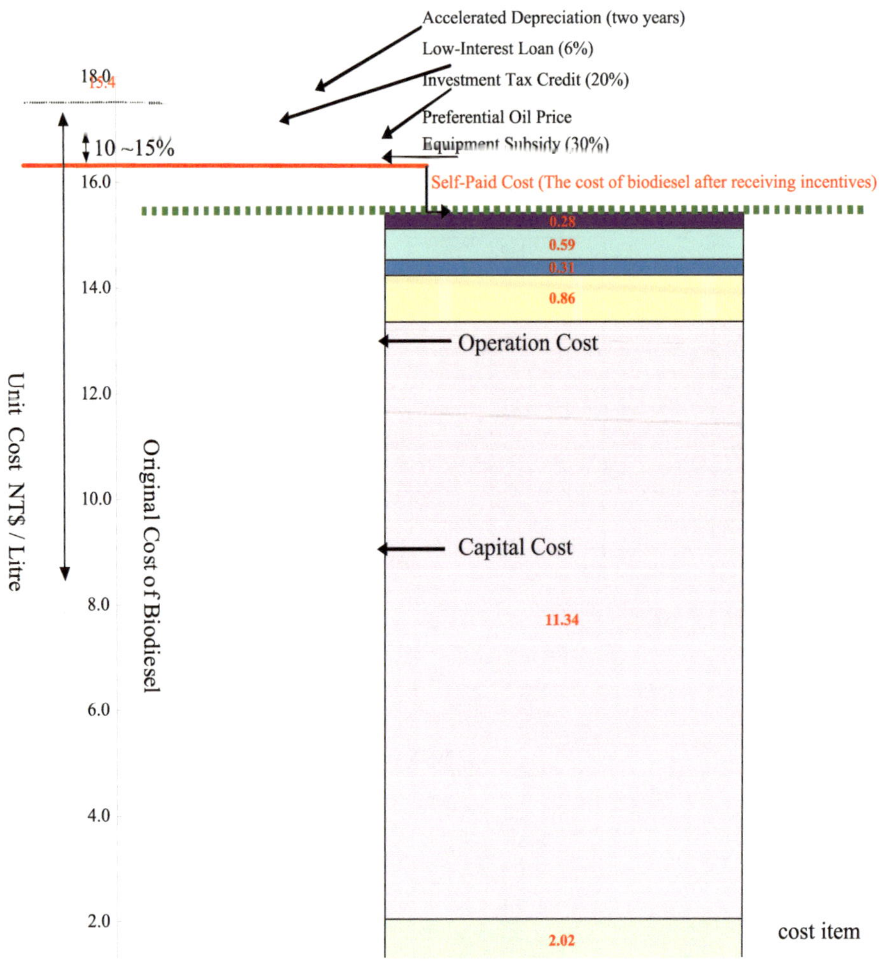

Fig. 7. Analyses of reward preferential benefit for biodiesel

From Fig. 7, we could learn that the unit production cost of biodiesel in Case 2 (purchase cost >0) is NT$15.4/litre. Under no subsidy, equipment investment (expenditure) cost is NT%2.88/litre, operation cost is NT$12.52/litre; if we consider the equipment subsidy 30%, then equipment investment (expenditure) cost could reduce NT$0.86/litre; if we consider financial subsidies, such as investment tax credit (20%), low-interest loan (6%) and speed up discount (two years), then the total financial subsidy could help to reduce the unit production cost by NT$1.18/litre. Plus equipment subsidy, the total subsidies could reduce unit production cost by NT$2.04/litre.

Comparing the model factory of output 1.275 kilolitres (Case 2) with the above-mentioned Irish model factory (uses 100% waste eating oil as feedstock) of output 3,000 tons (about 3,400 kilolitres), we obtain the unit production cost of biodiesel under current production conditions in Taiwan: NT$15.4/litre, higher than the unit production cost of the factory in Ireland NT$14.6/litre. The primary difference comes from the fact that in Taiwan, in addition to the cleansing and transportation cost, the cost of using waste eating oil has to include purchase price cost whereas in Ireland, they only include cleansing and transportation cost. The other reason is unit equipment investment cost: NT$2.88/litre in Taiwan is higher than the Irish NT$2.31/litre. Through comparing unit production cost of model factories in two countries, we may estimate the unit production cost of biodiesel is between NT$14-16/litre while unit equipment investment cost is between NT$2.3-2.9/litre [10].

From promotion point of view, if the goal market is focused on the transportation utilization of the whole urban regions and public transport roads in Taiwan, the annual consumption quantity is 330 thousand KLOE of B20 bio-diesel; the annual market benefit is 4.3 billion NT dollars for the manufacture of bio-diesel, in addition, the annual CO2 reduction quantity is about 880 thousand ton in Taiwan. Furthermore, the competition abilities were analyzed; the strategy trends were proposed, and, the policy issues were suggested from feedstock supply, manufacture & marketing sales point of views for the future bio-diesel application in Taiwan.

3. Conclusion

The production of the methyl linoleate was more over the one of the methyl pamitate during the processes for the manufacture of biodiesel from the waste cooking oil. The reaction balance can be achieved after 15 minute same as the feedstock soybean oil. The crude methyl ester volume was 528 ml, and the crude methyl ester yield was up to 97.77%. After distillation, the methyl ester volume was 512ml, and methyl ester volume yield was 94.81%. The loss caused by water washing reduced more, and low methyl ester volume was obtained.

The process of manufacturing biodiesel by waste cooking oil needs pre-processing to adjust. In the transerification of fat, we have to strictly control the impurities, water and acid value in oil. The acid value of waste cooking oil must be modulated down to 0.26928(mg KOH/g) for the self-manufacture of the biodiesel to keep the normal processes of the transerification reaction with alkali. And, the properties of the manufactured biodiesel meet the standard of the biodiesel almost.

Using bio-diesel to conduct engine test, the results showed that under the test of idle, when used petrochemical diesel, B20 and B100, the concentrations of carbon monoxide emitted by engine were almost the same. Under the test of full load, there was significant effect of reduction of pollution, particularly when use B100 to replace petrochemical diesel. But the reduction of brake horsepower of engines and the increase of oil-consumption signified that the effects are limited to the improvement of environment. Comparing petrochemical diesel, under the test of full load, when used B20 and B100, the concentration of carbon monoxide emitted by engine reduced 14% and 51% respectively and the concentration of sulphur dioxide reduced 68% and 73% respectively, the concentration of nitrogen oxide compound increased 1% and 13% respectively, the smoke reduced 8% and 69% respectively; brake horsepower reduced 8% and 10% respectively; oil-consumption increased 9.4%and 11.3% respectively.

If we take current ordinary fossil diesel price NT$13.3/litre as oil purchase price, then the subsidy demand of single price is NT$1.3/litre; if not consider price subsidy, then the equipment subsidy ratio will be as high as 90%to make up difference with ordinary fossil diesel sales price. However, if we consider equipment subsidy ratio in couple with price subsidy (mixed subsidy), then we could reduce demand of price subsidy and equipment

subsidy ratio. For example, when IRR=12%, equipment subsidy 50%, demand of price subsidy is only NT$0.5/litre, retrieval time limit could be reduced to around eight years.
If we want to encourage intention of investment and raise investment reward rate (IRR), for example 15~20%, then the demand of relative non-subsidy oil price will also be raised to NT$16.9/kg, but the retrieval time limit will reduce to 6.4 years and the subsidy for the price difference will be increased to NT$3.6/litre. If we only use equipment subsidy, it is impossible to make up the difference even if we use 100% subsidy. So we should have price subsidy in the same time to reach the goal of raising investment reward rate. For example, when IRR=20%, equipment subsidy 50%, then the price subsidy may only need NT$2/litre and retrieval time limit could reduce to five years.

4. References

[1] Akoh C.C., Chang S.W., Lee G.C. and Shaw J.F.(2007).Enzymatic Approach to Biodiesel Production. *Journal of Agricultural and Food Chemistry*, Vol.55, No.22 (2007), pp.8995-9005, ISSN: 0021-8561

[2] Knothe, G.(2008).Designer Biodiesel: Optimizing Fatty Ester Composition to Improve Fuel Properties.*Energy & Fuels,* Vol.22(2008),pp.1358–1364, ISSN: 0887-0624

[3] Wu, W. T. and Chen, J. W.(2001).Method of preparing lower alkyl fatty acid ester and in particular biodiesel.US Patent No. 6398707

[4] Marchetti, J. M. And Errazu, A. F.(2008).Comparison of different heterogeneous catalysts and different alcohols for the esterification reaction of oleic acid.*Fuel*, Vol.87(2008), pp.3477–3480, ISSN: 0016-2361

[5] Ban, K., Hama, S., Nishizuka, K., Kaieda, M., Mastumoto, T., Kondo, A., Noda, H., and Fukuda, H.(2002).Repeated Use of Whole-Cell Biocatalysts Immobilized within Biomass Support Particles for Biodiesel Fuel Production. *J. Molec. Catal. B: Enzymatic*, Vol.17 (2002), pp.157-165, ISSN: 03045102

[6]Wu,Y.P.andChang,C.T.(2010).Combustion Characteristics of waste-oil produced biodiesel/diesel fuel blends.*Fuel*,Vol.86(2010), pp.1772–1780, ISSN: 0016-2361

[7] Leu J.H. and Huang M. Y.(2009). Alternative Fuel Test Study of an Diesel Engine Fed by Synthetic Fuel Derived from Waste Bilge Oil.*Journal of Marine Science and Technology*,Vol.17, No.1(2009), pp.7-12,ISSN 1023-2796

[8] Leu,J.H.(2011).The Analyses of Reward Preferential Benefit for Biodiesel Promotion in Taiwan by Financial IRR and PB Estimation.*Journal of Interdisciplinary Mathematics*, Vol.7, No.3(2011), pp.1-17,ISSN 0972-0502

[9] Wu,Li-fong and Su,Ay(2002).The Electronic Plasma FusionTechnology, Benefits and Policy Evaluation for Metropolitan Garbage Incinerators Ash. Mater Thesis , Yun Zi Univerversity,Taiwan.

[10] USA(2000),Available from http://www.biodiesel.org/fuelfactsheet.htm

5

Waste Animal Fats with High FFA as a Renewable Energy Source for Biodiesel Production - Concept, Experimental Production and Impact Evaluation on Air Quality

Francisc Popescu and Ioana Ionel
"Politehnica" University of Timisoara
Romania

1. Introduction

This chapter presents one method to obtain the renewable "biodiesel" energy source thru animal fats transesterification with high free fatty acid content. The authors tested different methods to produce biodiesel from waste animal fats. The purpose of research was to use the byproduct from the meat industry, the wastes in form of fats with high free fatty acid contend (FFA), fats that cannot be used in other sectors, like supplement in animal food, due to high contamination and FFA content.

The research underlines a correlation between social-economic development and air quality in an urban area. We also intend to improve the knowledge in the field of alternative fuels, field not well covered in Romania, especially if the waste management and potential are in sight. The relevance of this research theme in the knowledge development is related to the fact that studies regarding environment protection are not an objective but a dynamic process that requires adequate instruments and concrete actions in the frame of a coherent legislative frame. (Popescu, 2009)

The chapter will follow three directions: first will describe the research conducted in order to obtain an "receipt" to produce the biodiesel from waste animal fats; second will describe the pilot installation constructed to improve the biodiesel production receipt and third will describe some research conducted in order to prove (or validate) the positive impact on the urban environment if the biodiesel in used in stead of regular diesel.

In a broad sense, *energy conversion* is the capacity to promote changes and/or actions (heating, motion, etc.), and *biomass* includes all kinds of materials that were directly or indirectly derived not too long ago from contemporary photosynthesis reactions, such as vegetal matter and its derivatives: wood fuel, wood-derived fuels, fuel crops, agricultural and agro-industrial by-products, and animal by-products. *Bioenergy* is the word used for energy associated to biomass, and *biofuel* is the bioenergy carrier, transporting solar energy stored as chemical energy. (Ionel, 2008) *Biofuels* can be considered a renewable source of energy as long as they are based on sustainable biomass production. The style will adjust your fonts and line spacing. Improving the efficiency of the existing power-plants using fossil fuels, the use of renewable fuels and renewable energy sources and the increased use of nuclear power are all considered to be important means of reducing greenhouse-gas emissions. One possibility to reduce greenhouse-gas emissions is to substitute biomass for coal in energy units or to add in a restrictive amount biofuel to fossil fuel, in internal engine combustion systems. The use of biomass in energy plants or of liquid biofuels in engines also offers advantages associated with emissions, other than reducing greenhouse-gas emissions. These are due to the low sulphur and low nitrogen contents of biomass. Also it is worth to notice that liquid biofuels are of interest as well, despite the present debate of not affecting price and production of food products. Not at least supplementary problems that must be solved are attested (emission of CO, unburned hydrocarbons, etc.). There is evidence that power generation from biomass is an attractive technology that utilizes agricultural residual waste. It is evident from research that efficient co-firing of biomass with coal can be achieved with minimal modifications to existing coal-fired boilers; as well existing engines might run, with minimum investment costs, using a blended fossil flue with biofuel. (Ionel 2008)

Among the available alternative energy sources, including hydro, solar, wind etc. to mitigate greenhouse emissions, using biofuels is the only carbon-based sustainable option. It is increasingly understood that 1st–generation biofuels (produced primarily from food crops such as grains, sugar beet and oil seeds) are limited in their ability to achieve targets for oil-product substitution, climate change mitigation, and economic growth. Their sustainable production is under review, as is the possibility of creating undue competition for land and water used for food and fibre production. A possible exception that appears to meet many of the acceptable criteria is ethanol produced from sugar cane. The cumulative impacts of these concerns have increased the interest in developing biofuels produced from non-food biomass.

With exception of hydroelectricity and nuclear energy, the major part of all energy quantities consumed worldwide comes from petroleum, charcoal and natural gas. However, these sources are limited, and will be exhausted by the end of this century. Thus, looking for alternative sources of energy is of vital importance.

Vegetable oils and animal fats are a renewable and potentially inexhaustible source of energy with an energetic content close to diesel fuel. The physical characteristics of fatty acid ester (bio-diesel) resulted in the end of the transesterification are very close to those of diesel fuel and the process is relatively simple. Furthermore, the methyl or ethyl esters of fatty acids can be burned directly in unmodified diesel engines, with very low deposit formation.

Transesterification is the general term used to describe the important class of organic reaction where an ester is transformed into another through interchange of the alkoxy moiety. When the original ester is reacted with an alcohol, the transesterification process is called alcoholysis.

The general transesterification reaction equation is:

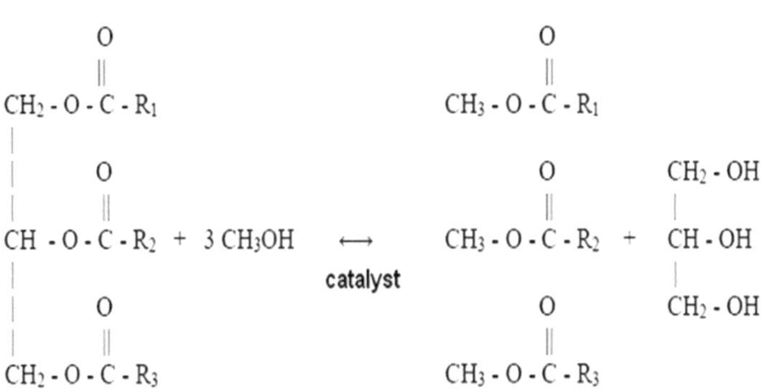

(1)

To meet rising energy demands and compensate for diminishing petroleum reserves, fuels such as bio-diesel and bio-ethanol are in the forefront of alternative technologies. It is well known that transport is almost totally dependent on fossil-, particularly petroleum-, based fuels such as gasoline, diesel fuel, liquefied petroleum gas, and natural gas. An alternative fuel to petro-diesel must be technically feasible, economically competitive, environmentally acceptable, and easily available. Accordingly, the viable alternative fuel for compression-ignition engines is bio-diesel.

Thus, it is obvious that the transesterification is an equilibrium reaction and the transformation occurs essentially by mixing the reactants. However, the presence of the catalyst (typically a strong acid or base) accelerates considerably the adjustment of the equilibrium. In order to achieve a high yield of the ester, the alcohol has to be used in excess. Several aspects, including the type of catalyst (alkaline or acid), alcohol/fat molar ratio, temperature, purity of the reactants (mainly water content) and free fatty acid content have an influence on the curse of the transesterification.

One popular process for producing bio diesel from fats/oils is transesterification of triglyceride by methanol (methanolysis) to make methyl esters of straight-chain fatty acids. The purpose of the transesterification process is to lower the viscosity of oil. The transesterification reaction proceeds well in the presence of some homogeneous catalysts such as potassium hydroxide (KOH), sodium hydroxide (NaOH) and sulphuric acid or heterogeneous catalysts such as metal oxides or carbonates. (Popescu, 2009)

The transesterification process of the fresh fats and vegetable oils is an easy one with no complications. It is more difficult to control the transesterification process when the raw materials are waste fats with animal origin and very high content of free fatty acids. The scope of this paper is to provide an receipt, based on authors experiences, to succeed in the transformation of waste fats to bio-diesel and to use the energy contents of waste fats, that otherwise must be incinerated/destroyed in special conditions with high costs, as an renewable energy source.

2. Experiments to obtain a "receipt" for biodisel production from waste fats

In this paragraph the properties of 11 old fats sample provided by the same manufacturer will be presented. It may be interesting to mention that the fats were originally produced to be used as a food in animal farms. The qualities of the fats are below the in force in Germany at the time of experiments, GROFOR - that stipulates the acceptable limits for the quality of animal food - and that is a very good reason and argument to try to use the fats for bio-diesel production. Measurements have been made for: iodine number (DGF-Methoden C-V 11b), sulfuric ash (DIN 51 575), total contamination (DIN 51419-A), peroxide number, kinematic viscosity (DIN 51 562 part 1) and free fatty acid (DIN 51 558 part 1). In Table 1 a synthesis of the old fat proprieties used in experiments are given.

In figures 1 and 2 the total contamination and free fatty acid number are presented in correlation with the admissible limits according to GROFOR norms for animal foods.

The fat must be heated until it becomes liquid and then might be filtered. Two stages (acid-catalyzed stage and base-catalyst stage) should be accomplished. (Trent, 2002)

For a successful reaction the oil must be free of water. There are two common methods of removing the water content:

- Boiling the water off: heating to 100 °C. As the heat raises the water, it separates out and falls to the bottom. The water is then drained out to avoid steam explosion. The temperature (> 100 °C) should be maintained until no bubbles are rising any more.

- Settling the water out: This method saves energy. The fats must be heated to 60 °C and to keep this temperature for 15 minutes. After this one pours the fats into a settling tank and lets it settle for at least 24 hours. The water will separate and fall to the bottom.

Sample	Iodine number	Acid number	Total contamination [mg/kg]	Viscosity [mm²/s]	Sulfuric Ash [%]	FFA [%]	Peroxide number
S1	64,76	21,63	12227	44,136	0,6839	10,87	58,4
S2	66,12	23,56	10709	43,084	0,4328	11,85	2,9
S3	69,7	16,59	12028	43,264	0,4344	8,34	9,1
S4	66	17,93	9385	42,666	0,3332	9,02	1,5
S5	58,7	19,51	10598	49,592	0,3674	9,8	177,8
S6	61	27,89	9026	41,373	0,3465	14,02	146,1
S7	57	41,59	11050	49,59	0,6218	20,91	116,9
S8	61	37,58	7177	44,978	0,475	18,89	1,4
S9	53,4	32,84	9351	46,278	0,4934	16,5	13,3
S10	66,7	29,99	6243	41,819	0,2988	15,08	28,9
S11	58,9	22,3	7426	43,569	0,3715	11,21	14,5

Table 1. Main properties / characterization of fats used in experiments

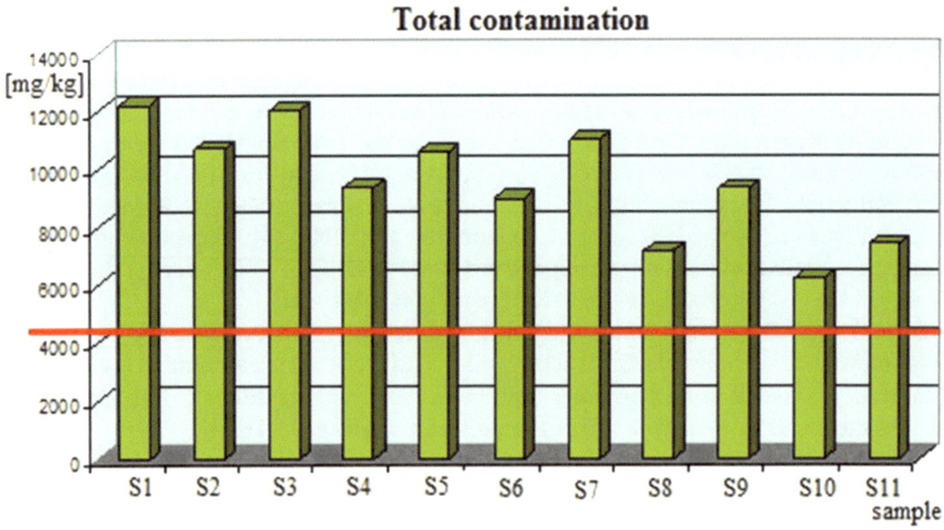

Fig. 1. Total contamination of the probes used in experiments, expressed in mg of contaminants in one kg of waste fats. (e.g. bone residues)

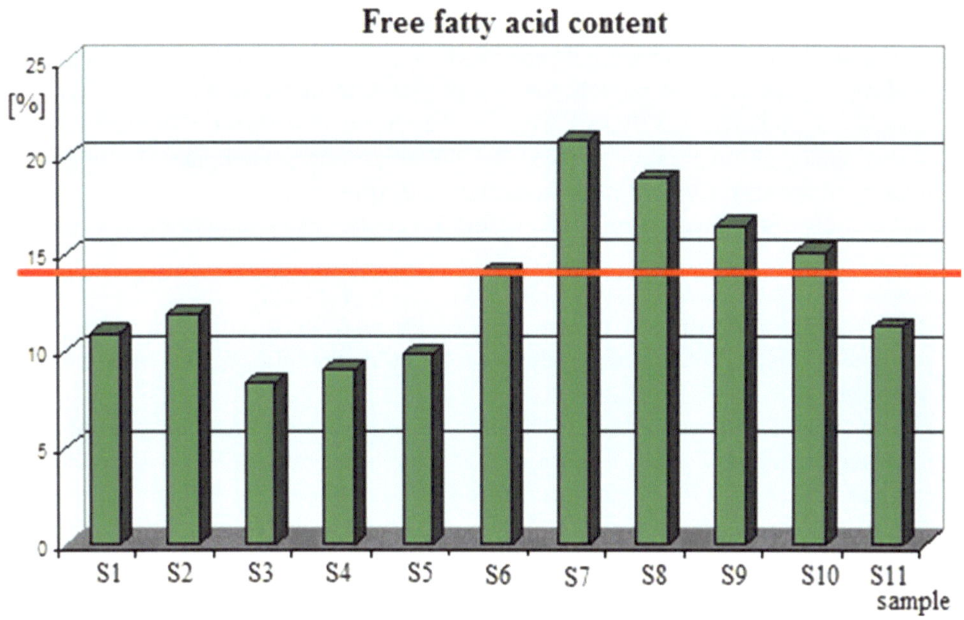

Fig. 2. Free fatty acid (FFA) content of the probes used in experiments.

The following strategy/receipt must then be fulfilled, step-by-step, for accomplishing successfully the fats preparation. First described is the acid-catalyzed stage, giving the basic acid stage reaction.

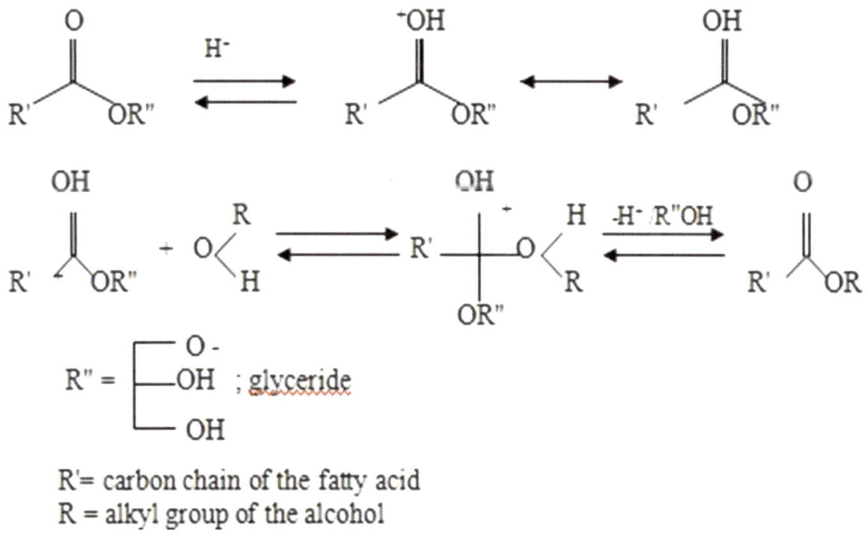

(2)

Bellow is given the First stage (acid-catalyzed stage), in detail:

- Measuring the volume of oil/fats to be processed (preferably in liters).
- Heating the oil/fats to 55 °C - all solid fats must be melted.
- Measure out the methanol and add 0.1 liter of methanol for each liter of oil/fats (10 % by volume). Add the methanol to the heated oil.

- Mixing for 5 minutes; the mixture will become murky because of solvent change (methanol is a polar compound, oil is strongly non-polar and a suspension will occur).
- Adding for each liter of oil/fats the quantity of 1 milliliter of 95-97 % sulfuric acid (H_2SO_4). A graduated eyedropper, a graduated syringe or a pipette should be used.
- Mixing gently at low rotation speed (rpm), while keeping the temperature at 55 °C. The rotation of the stirrer should not exceed 500 to 600 rpm.
- Maintaining the temperature at 55 °C for 50 minutes then stop heating. Continuing stirring.
- Preparation of the sodium methoxide that consists of adding 0.1 liter of methanol for each liter of oil/fat (10 % by volume) and 3.1 grams of 99 % pure sodium lye (NaOH) per liter of oil/fat. Finally mixing the lye into the methanol until the lye is completely dissolved.
- Pouring half of the prepared methoxide into the mixture after 1.5 hours (oils only or oils plus fats) or 2 hours (for fats that are solid at room temperature). This will stop the acid-catalyzed reaction and prevent ester back splitting. Mixing for 5 more minutes, then stop.
- Allowance to the mixture to settle for 6 to 12 hours, then draining off the glycerin. (the brown or dark brown compound at the bottom).

Bellow is given the Second stage (base-catalyzed stage), in detail:

- Heating the mixture to 55 °C. Make sure that any remaining room-temperature solid fats are melted.
- Adding the second half of the prepared sodium methoxide to the heated mixture and start mixing at the same low speed of not more than 500 to 600 rpm. Mix for 1 hour.
- Allow settling for 6 to 12 hours.
- Drain off the glycerin. The biodiesel is now obtained.

Final stage is represented by the washing out the biodiesel and the pH of the FAME (fatty acid methyl ester) should be known. The receipt comprises the following steps: Put the FAME in one vessel with ½ water or the same quantity of water as the FAME to wash. The FAME and water must have the same temperature (room temperature). The water pH must have as many units under 7 as the FAME pH is above 7. Use strong vinegar to obtain lower pH for water. Use compressed air to create bubbles in the vessel ho contains the FAME/water mixture. Let it bubble for up to 6 hours. The bubbles will carry the water up. When this water falls down again, it washes the soaps and surplus methanol out of the FAME and the vinegar neutralizes the remaining lye. After settling for 12 hours the water will fall to the bottom, turning completely white and the bio-diesel will look much lighter in color now (Figure 3).

In Table 2 are presented the chemicals (CH_3OH - methanol, H_2SO_4 - sulphuric acid, NaOH - sodium hydroxide, KOH - potassium hydroxide) and their concentration values used in the tests for producing the bio-diesel.

The most successful test was No 6, realized with an extra-step with KOH. The viscosity of the bio-diesel obtained was 5.3 mm^2/s. Also test No 7 was a partial success. It should be mentioned that for the last step (KOH) for No 6 it was used just 188.28 g collected at the end of the second step. Figure 3 is a view over samples from biodiesel No 6. Note the lighter colors and no depositions.

Implementing the process described above results in the process flow shown in the figure 4.

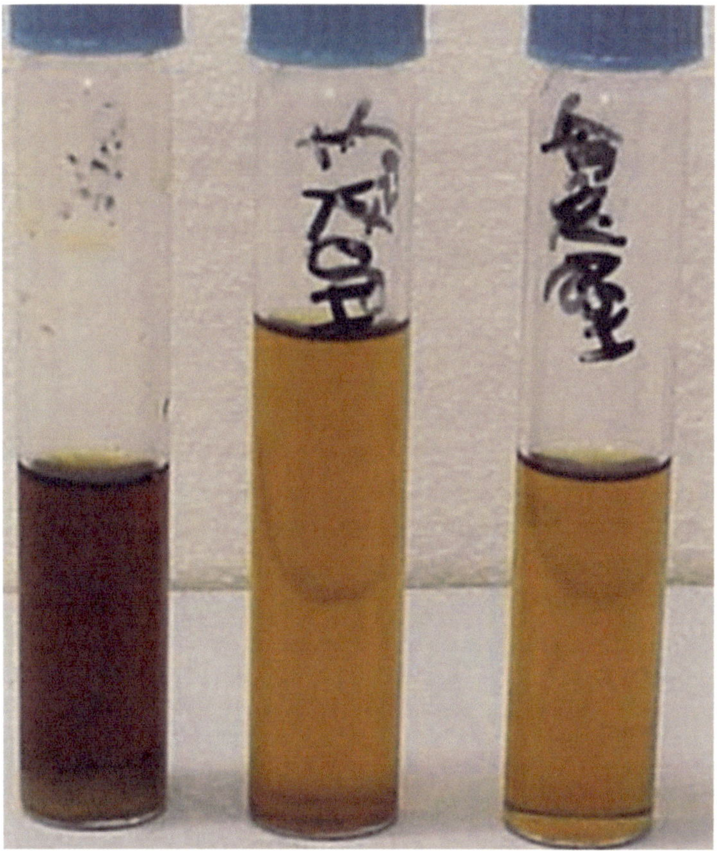

Fig. 3. Biodiesel sample after second stage (left), before (middle) and after washing (right)

Test No	Fats (ml)	CH_3OH (ml)	H_2SO_4 (ml)	NaOH (g)	CH_3OH (ml)	Stage I (h)	Stage II (h)	KOH (g)	CH_3OH (g)	Reaction time (min)
1	500	50	0,5	3,5	50	2	1			
2	500	50	1	3,5	50	2	1			
3	500	50	1	5	50	2	1			
4	500	60	1	4	50	5,3	1			
5	260	25	1	1	25	2	1			
6	250	40	2	1,5	40	2	1	2,53	26,35	25
7	250	40	2,5	2,65	40	2	1			

Table 2. Chemicals needed for bio-diesel preparation

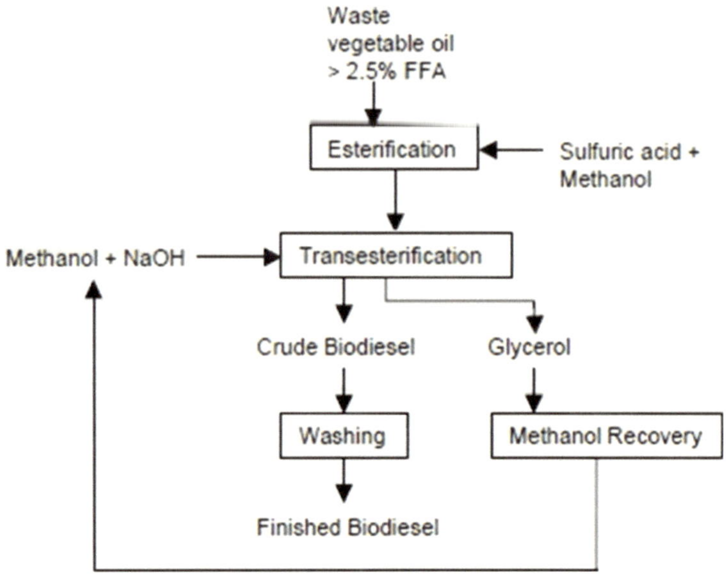

Fig. 4. Biodiesel production flow-chart

Based on the experiments described above and on the derived flow chard of the process, the pilot station presented in figure 5 was built at "Politehnica" University of Timisoara. The control of the installation and transesterification process is fully automated and design to use waste animal fats from meat processing industry. The ration between raw fats introduced and biodiesel produced is up to ~ 50% for waste fats with the FFA content higher than 12% and is getting higher as the FFA content in the raw material decreases. The installation capacity is 20 liters of fats and the processing time is from 3 to 10 hours, depending on the FFA content of the fats

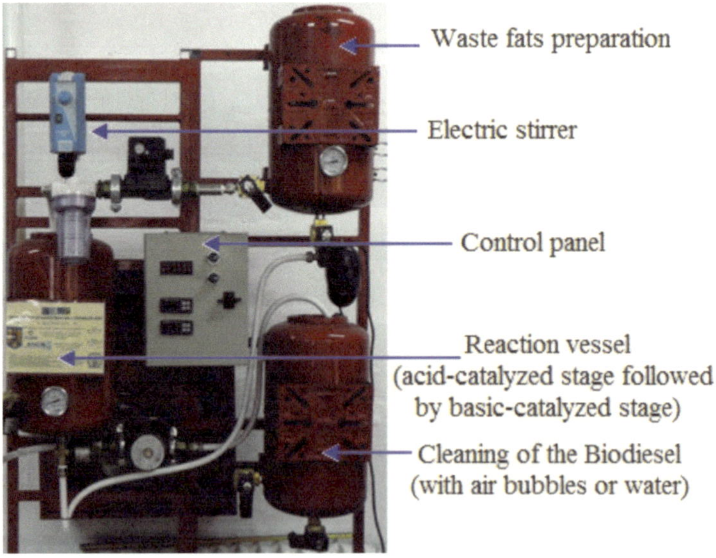

Fig. 5. Pilot installation for Biodiesel production from waste fats with high FFA content.

3. Advantages in the use of biodiesel in an typical urban vehicles fleet

Air pollution is a major concern for all nations, with a higher or lower development level. The rapid increase of the industry sector and urban development had generated substantial quantities of substances and poisonous materials, witch are, mostly evacuated in the atmosphere. The human society was not willing to recognize that the environment has only a limited capacity to process all this waste, without major changes. As a consequence we are able to observe disfunctionalities in the health degree, deterioration of flora and fauna, materials, buildings, in parallel with the lost of natural resources. Each of us is a polluter but also a victim of pollution. (Ghani, 2009)

The scope of this paragraph focuses on the utilization of the bio-diesel in a comparative study regarding air quality impact of its utilization as main fuel in a real fleet in urban area, in comparison to the present situation, when fossil diesel is used.

Further examples of how air quality might be determined by numerical modeling in order to establish if new industrial sources (that are also polluting sources) might affect the quality of air. One knows that the total air concentration is a sum of the natural pollution and the induced anthropic pollution. In order to modernize a city it is necessary to build new industrial area that are simultaneously also polluting. By numerical simulation the strategy might be optimized. (Silva, 2010)

In the next paragraph and figures is presented the simulation in the case that all diesel engines are using the bio-diesel fuel, for one major Timisoara crossroad, M. The values for emission factors both for diesel and bio-diesel fuels are extracted from Corinair database and traffic structure and meteorological date were measured *in situ*. The CORINAIR - CORe INventory of AIR emissions - was a project performed since 1995 by the then European Topic Centre on Air Emissions under contract to the European Environment Agency. The aim is to collect, maintain, manage and publish information on emissions into the air, by means of a European air emission inventory and database system. This concerns air emissions from all sources relevant to the environmental problems of climate change, acidification, eutrophication, tropospheric ozone, air quality and dispersion of hazardous substances.

The software used was CALROADS with modeling code CALINE 4 (US EPA approved) and the areas studied were Michelangelo crossroad and Marasti crossroad, both major polluting area point in the city of Timisoara. The traffic surveillance and cars characterizations have been conducted for the time frame considered in both crossroads. The medium number of cars in both intersections was around 45 000 vehicles/day (about 30% diesel powered vehicles). The diesel emission factors have been taken from Corinair database.

In figure 6 an example of road traffic structure measured in the Michelangelo crossroad is given. The figure shows the traffic structure over one hour period from a typical day in a time frame with medium traffic intensity.

The road traffic measurements and fleet characterization has been done over a week period, the results for total road traffic in the Michelangelo crossroad is given in table 3.

In figure 6 the road traffic structure is given for one hour period, for the Michelangelor crossroad, as the concentrations resulted after the simulations are one hour mean values.

The formulas used to calculate the emission factors for CO, NOx and particles is in accordance to Corinair methodology and the estimated results for diesel and biodiesel are calculated. (EMEP, 2007)

Day		Road traffic [veh. tot.]
15.05.2007	Thursday	49430
16.05.2007	Wednesday	50601
17.05.2007	Thursday	50077
18.05.2007	Friday	50990
19.05.2007	Sunday	41256
20.05.2007	Saturday	32105
21.05.2007	Monday	50589
22.05.2007	Thursday	48883
23.05.2007	Wednesday	40101

Table 3. Measured data for total road traffic in Michelangelo crossroad.

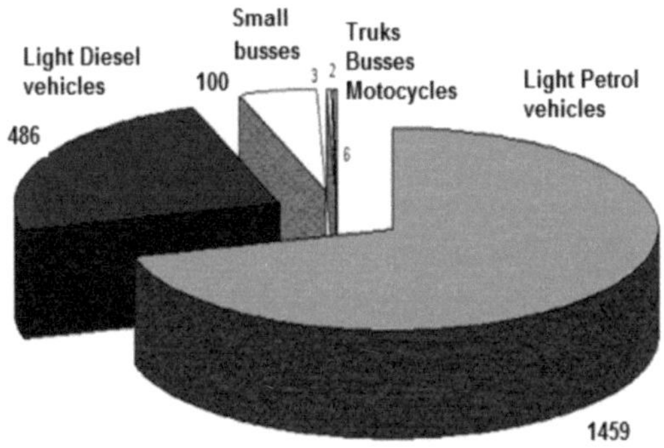

Fig. 6. Road traffic structure for one hour period in Michelangelo crossroad

$$E_{CO} = 5.41301 \cdot V^{-0.574} \quad [g/km] \tag{3}$$

$$E_{NOx} = 1.331 - 0.018 \cdot V + 0.000133 \cdot V^2 \quad [g/km] \tag{4}$$

$$E_{PM} = 0.45 - 0.0086 \cdot V + 0.000058 \cdot V^2 \quad [g/km] \tag{5}$$

In the formulas above V is the mean speed of the vehicles. The calculated results for a typical mean vehicle speed in Michelangelo crossroad is about 30 [km/h] are presented. Emissions factors for diesel fuel, for each vehicle, are:

$$E_{CO}^{\ diesel} = 0.7683 \quad [g/km] \tag{6}$$

$$E_{NOx}^{\ diesel} = 0.91 \quad [g/km] \tag{7}$$

$$E_{PM}^{\ diesel} = 0.2442 \quad [g/km] \tag{8}$$

Emission factors for biodiesel fuel (B100 blend), for each vehicle, are:

$$E_{CO}^{\ biodiesel} = 0.30732 \quad [g/km] \tag{6}$$

$$E_{NOx}^{\ biodiesel} = 0.93 \quad [g/km] \tag{7}$$

$$E_{PM}^{biodiesel} = 0.09768 \quad [g/km] \tag{8}$$

The most relevant results are presented in figures 8, 9 and 10 and in table 4 the maximum values obtained for considered pollutants, after dispersion, are given.

Pollutant	Fuel	Maximum obtained value
CO	Diesel	2.60 [ppm]
	Biodiesel	0.70 [ppm]
NOx	Diesel	0.02 [ppm]
	Biodiesel	0.02 [ppm]
Particles	Diesel	28 [μg/m^3]
	Biodiesel	11.70 [μg/m^3]

Table 4. Maximum values obtained for considered pollutants, after simulation.

4. Conclusion and discussions

With exception of hydroelectricity and nuclear energy, the major part of all energy is produced from petroleum, charcoal and natural gas. However, these sources are limited, and will be exhausted by the end of this century. Thus, looking for alternative sources of energy is of vital importance.

Vegetable oils and animal fats are renewable and potentially inexhaustible sources of energy with an energetic content close to diesel fuel. The physical characteristics of fatty acid ester (bio-diesel) resulted in the end of the transesterification are very close to those of diesel fuel and the process is relatively simple. Furthermore, the methyl or ethyl esters of fatty acids can be burned directly in unmodified diesel engines, with very low deposit formation.

It is quite visible from Figures 7, 8 and 9 the advantages of using bio-diesel in diesel engines and that if all diesel engines will use bio-diesel fuel the reduction of emitted CO and particle concentration is up to 50%.

As it is known the crude oil world reserves will not last forever and it is necessary to obtain alternatives fuels for piston engines. Another aspect is the more decent impact of these fuels on the environment. The search for alternative fuels is on wide spread in all developed countries.

The advantage of bio-diesel as an alternative fuel relies first of all on its physical properties, similar with those of diesel fuel so that bio-diesel can be used directly in diesel engines with no essential modification. Second advantage is the very low impact on environment, and the third advantage is the accessibility of the breeding materials like old animal fats and a large variety of vegetable oils.

The tests described above are a real success and a step forward in developing new alternative fuels for piston engines.

This paper brings in attention an example of possible strategy in order to meet the general EU strategy concerning the RES utilization and percentage support for the total energy consumption by 2020. Also, gives an example on how numerical simulations with software approved by international institutions can be used by local urban and environment authorities in order to control pollutants emissions in urban or industrial areas, with an emphasis on sustainable and eco-friendly urban development. The advantage of numerical simulations is given by fast results, low cost and possibility to evaluate different pollution reduction scenarios.

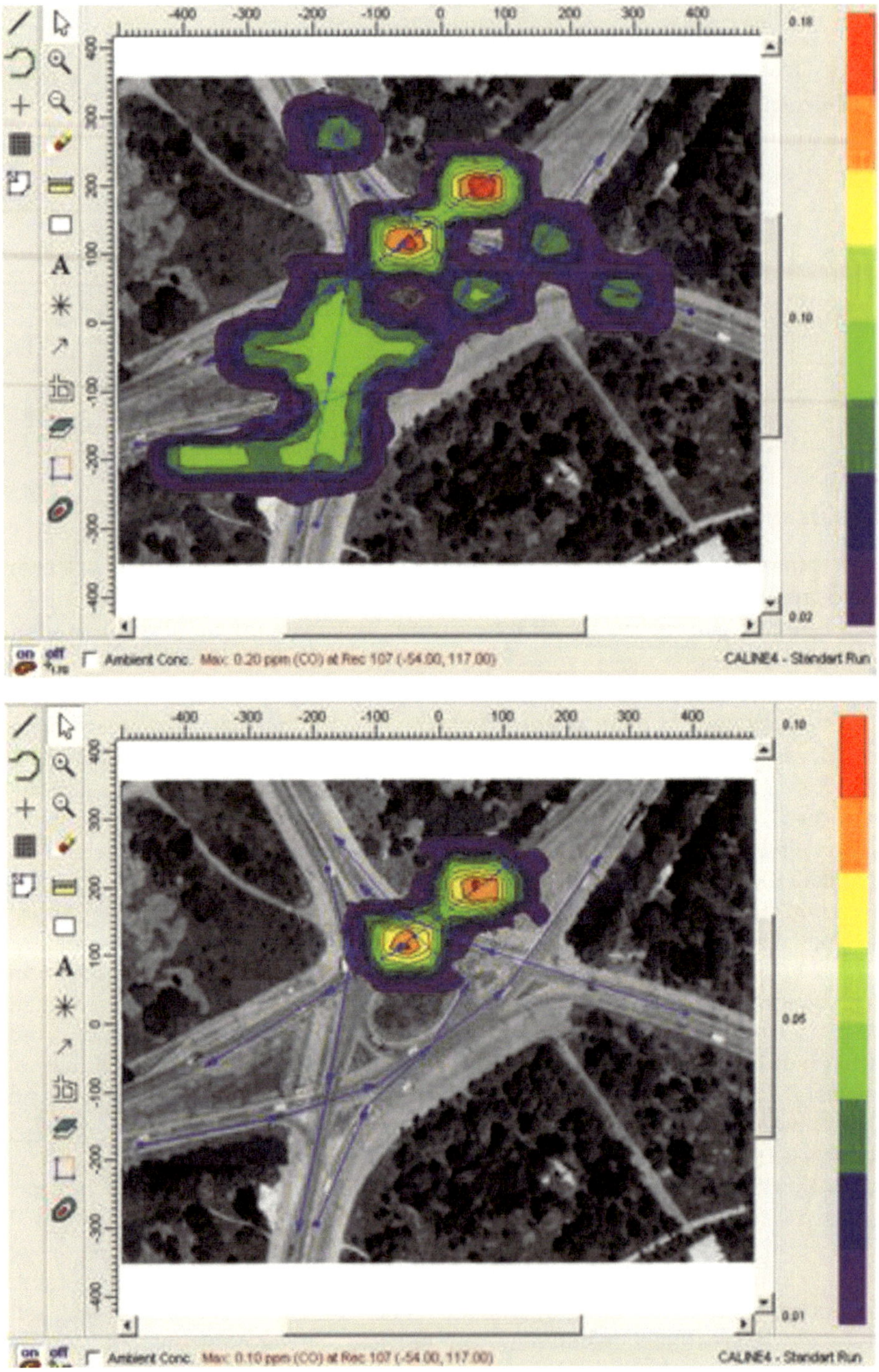

Fig. 7. CO concentrations for the Michelangelo crossroad: diesel (*up*) versus bio-diesel (*down*), where red color represents highest values and violet lowest values. The maximum resulted value for CO concentration in ambient air is in red, in the figure bottom.

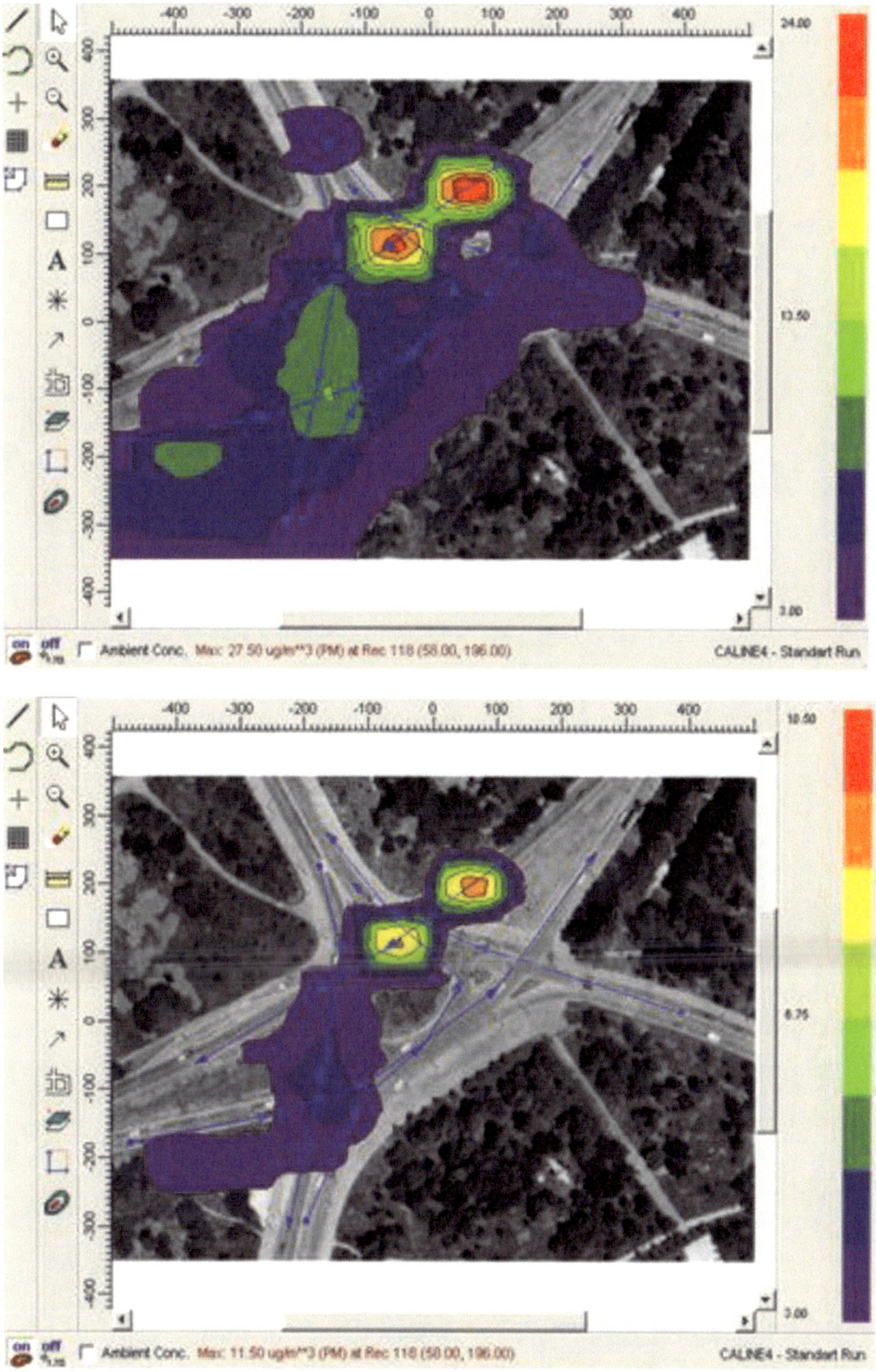

Fig. 8. Particle (PM) concentrations for the Michelangelo crossroad: diesel (*up*) versus bio-diesel (*down*), where red color represents highest values and violet lowest values. The maximum resulted value for PM concentration in ambient air is in red, in the figure bottom.

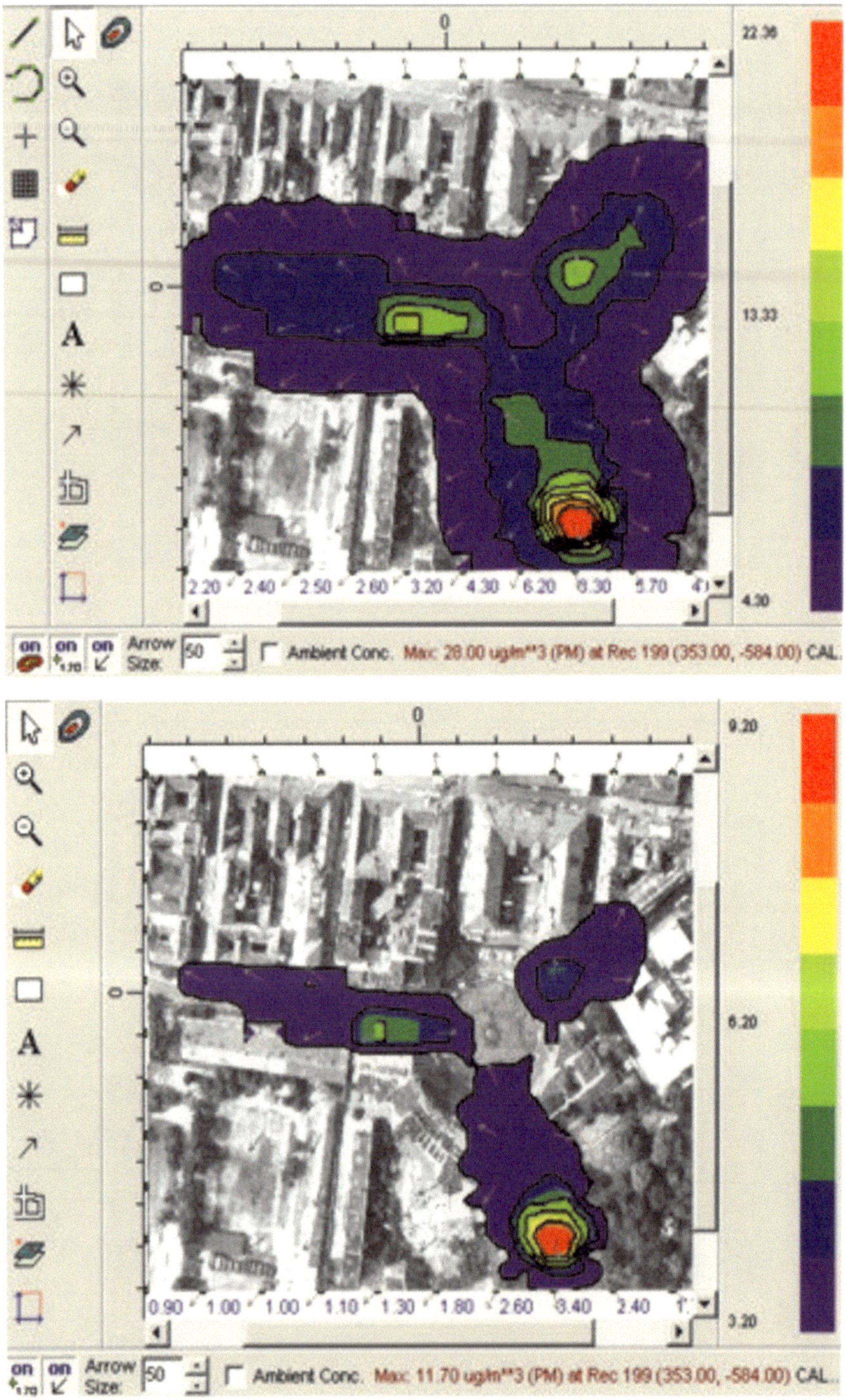

Fig. 9. Particles concentrations for the Marasti crossroad: diesel (*up*) versus bio-diesel (*down*), where red color represents highest values and violet lowest values. The maximum resulted value for PM concentration in ambient air is in red, in the figure bottom.

Worldwide, there is a growing interest in the use of solid, liquid and gaseous biofuels for energy purposes. There are various reasons for this, such as:

1. Political benefits (for instance, the reduction of the dependency on imported oil);
2. Employment creation - biomass fuels create up to 20 times more employment than coal and oil; and
3. Environmental benefits such as mitigation of greenhouse gas emissions, reduction of acid rain and soil improvements.

The advantage of biofuel as an alternative fuel relies first of all on its physical properties, similar with those of diesel fuel so that biofuels (biodiesel or buthanol) can be used directly in diesel engines with no essential modification. Second advantage is the very low impact on environment, and the third advantage is the accessibility of the breeding materials like old animal fats and a large variety of vegetable oils.

In this respect, knowledge exchange as well as the creation of conductive market mechanisms and legislation is essential for a more widespread introduction of biomass and biofuels energy systems.

5. Acknowledgment

All tests presented where successfully made by the authors in the frame of several projects of the hosts, "Politehnica" University of Timisoara and T.U. München with the financial support of Alexander von Humboldt Foundation, Romanian National Council for Scientific Research and Balkan Environmental Association for which deep thanks are expressed.

6. References

Popescu, F. (2009). Advantages in the use of bio diesel in an urban fleet. Case study: major crossroads in Timisoara city. *Journal of Environmental Protection and Ecology*, Vol.10, 2009, pp. 182-191, ISSN 1311-5065

Ionel I.; Popescu F.; Lontis N.; Trif-Tordai G. & Russ W. (2009). Co-combustion of fossil fuel with bio fuel in small cogeneration systems, between necessity and achievements. *Proceedings of the 11th WSEAS International Conference on Sustainability in Science Engineering*, pp. 352-357, ISBN 978-960-474-080-2, Timisoara, Romania, Vol.2, 2009

Silva L.T.; Mendes F.G. & Ramos A.R. (2010). Urban air dispersion model of a mid-sized city. Validation methodology. *WSEAS Transaction on Environment and Development*, Issue 1, Vol. 6., 2010, pp. 1-10, ISSN 1790-5079

Trent W.R. (2002). Process of treating fatty glycerides, U.S. patent ofice, No. 462.370

EMEP (2006) EMEP/CORINAIR Emission Inventory Guidebook, Technical report No 11, Published by European Environment Agency, 2006, http://www.eea.europa.eu/publications/EMEPCORINAIR4/

Popescu F.; Ionel I. & Ungureanu C.; (2009). Air Quality Monitoring in the City of Timisoara. *Journal of Environmental Protection and Ecology*, Vol. 10, 2009, 1-13, ISSN 1311-5065

Ionel I.; a.o. (2008). Method for determination of an emission factor for a surface source, *Optoelectronics and Advances Materials – Rapid Comunications*, Vol. 2, No. 12, 2008, 851-854, ISSN 1842-6573

Ghani W.; Alias A.; Savory R. & Cliff K. (2009). Co-combustion of agricultural residues with coal in a fluidised bed combustor, *Waste Management, Vol.* 29, 2009, pp. 767–773, ISSN 0956-053X

*** DIN 51419-A, DIN 51575, DIN 51562-Teil 1.

Part 2

Bioethanol

6

Delignification Process of Agro-Industrial Wastes an Alternative to Obtain Fermentable Carbohydrates for Producing Fuel

Oscar Sánchez[1], Roció Sierra[1] and Carlos J. Alméciga-Díaz[2]
[1]*Universidad de Los Andes, Chemical Engineering Department*
[2]*Pontificia Universidad Javeriana*
Institute for the Study of Inborn Errors of Metabolism, Bogotá
Colombia

1. Introduction

Fossil fuels, mainly petroleum, coal, and natural gas, were the main energy sources for most industries during the 20th century, and are still the most important feedstocks to produce energy in the world. "Currently, the world energy market worth around 1.5 trillion dollars is dominated by fossil fuels" (Goldemberg, 2006). However, these sources are not longer regarded sustainable, and their availability is much lower. Shafiee and Topal (2009) predicted that oil, coal and gas reserves will last around 35, 107 and 37 years for petroleum, coal and natural gas, respectively. In addition their combustion involve environmental issues such as global warming due to greenhouse gas emissions (Naik et al., 2010). Therefore, the interest for questing sustainable and environmental energy sources has risen in the last two decades; giving origin to the production of fuels from renewable feedstocks, such as biomass. These feedstocks are commonly divided in three categories: wood, residues from agricultural, industrial or domestic origin, and energy crops from dedicated farming (Bringezu et al., 2007).

The use of these renewable resources to produce fuel has created two different generations in biofuel production. The first biofuel generation is based on grain or food sources, and is constituted mainly by ethanol, fatty acid methyl ester (FAME), and pure plant oil (PPO) (Bringezu et al., 2007). However, environmental impacts, energy efficiency, and eutrophycation have limited the production of these first generation biofuels. In addition, land competition of energy crops with food crops has arisen a fuel-versus-food debate exacerbated by the increase of food prices, particularly those of maize, wheat, sugar beet, cassava, sweet sorghum, sugarcane, oilseed rape, soybean and oil palm (Börjesson & Tufvesson, 2011; Vries et al., 2010). For example, the United States Department of Energy reported that in 2005 the bioethanol production reached about 15 billion litters for which nearly 36 MMt of maize, 13% of total US maize crops, were used. These values were estimated to double in 2010 by the Renewable Fuel Association (Cassman & Liska, 2007).

The second generation of biofuels is produced from non-grain and non-food sources such as lignocellulosic sources and algae biomass (Naik et al., 2010; Simmons et al., 2008).

Lignocellulosic feedstocks include agro-industrial by-products, perennial grasses, vegetable and wood residues. They can be burned to produce heat and electricity and also be used to obtain liquid fuels (Naik et al., 2010). Wastes and by-products from agro-industrial processes such as coconut shells, rice husks, sugarcane bagasse, corncob and corn stover among many others, are abundantly produced in the world daily and have modest if any applications. These wastes and by-products are rich sources of cellulose and hemicellulose, which constitute important substrates in fermentative processes directed to biofuel production. However, as opposed to sugarcane juice or maize starch, these substrates are not readily available. The structural carbohydrates in the plant cell wall are wrapped up in lignin, which is an inert polymer that protects the plant and consequently constitutes an important barrier to fermentation. Therefore, a very effective way —but not the only option— to significantly increase biomass digestibility is lignin degradation or separation (delignification). The operation is aimed to increase the digestibility of constituent sugars through increment in gross material pore size (Sierra et al., 2008). It is challenging due to the recalcitrance of lignin and may require expensive chemicals and relatively high temperatures and pressures for acceptable reaction rates. Otherwise, at mild conditions (i.e. use of microorganisms or purified enzymes) it takes long times. Other ways to increase lignocelluloses digestibility include partial to total solubilization of hemicelluloses, and separation acetyl groups that link hemicellulose and lignin (Zhu et al., 2008). Crystallinity reduction of cellulose fibrils is sought because low crystallinity results in more reactivity; however, after delignification and sugar degradation with chemicals, an increased crystallinity is usually observed (Chang & Holtzapple, 2000). This increase is attributed to a preferential degradation of amorphous cellulose and less ordered crystalline forms during chemical pretreatment. A common method to obtain a significant reduction in crystallinity is sudden release of reactor vapor pressure. This operation is known as steam explosion.

The operations aimed to turn lignocellulose digestible through either of the mechanisms described above are widely known as "pretreatments". They typically start with size reduction by chipping and grinding. In addition to being a rate-limiting step, a chemical pretreatment increases the cost of bioethanol production due to the high-energy requirements of heating and mechanical size reduction. Energy consumption during size reduction of wood may surpass 0.1 – 0.4 MJ/kg, which is the required energy consumption to achieve sensible net energy output from wood to ethanol production (Kumar et al., 2009).

Chemical separation of lignin and carbohydrates can be achieved through the use of acids, alkalies, and solvents, which promote selective solubilisation of either component. If acidic, carbohydrates solubilise; if alkaline, lignin degrades and solubilises (Mosier et al., 2005); if with solvents -widely known as organosolv pretreatment- carbohydrates solubilise (Zhao et al., 2009). Chemical processes may not be as selective as biological processes but may represent advantages related to required time, scalability, and process control.

Biological delignification can be conducted using either microorganisms, which produce a set of enzymes that work synergically, or purified enzymes. The most widely used microorganisms are fungi from the *Basidiomysetes* family. Nevertheless, bacteria from *Pseudomonas, Flavobacteria, Xanthomonas, Bacillus, Aeromonas* and *Cellulomonas* strains can also decompose lignin and its derivatives. Biological lignin degradation can be conducted by culturing the microorganism in submerged, semisolid or solid cultures where enzymes such as lignin peroxidase, xylanase, laccase, and manganese peroxidase (among others) perform selective lignin degradation.

In this chapter, the reactions that lead to delignification of feedstocks in several different choices of chemical and biological pretreatments are discussed. In addition, other compositional changes during pretreatment are briefly reviewed. Although some basic chemical aspects related to lignocelluloses compositional analysis is presented here, it is out of the scope of this chapter to make a deep chemical discussion on this subject. All throughout the text, the reader is referred to several recent reviews and key articles for more detailed discussions.

2. Generalities of lignocellulosic compounds

Lignocellulose is the major renewable organic matter with an estimated production of about 200 x 10^9 tons per year (Reddy & Yang, 2005). This is mainly constituted by lignin and cell wall polysaccharides such as cellulose, pectins, and hemicelluloses, which may be valuable substrates for biofuels production. Lignin is found in the cell wall and sometimes within woody tissue in quantities that vary widely depending on the type of plant, the part of the plant, and its age. Klason lignin is reported to range between 8 and 22% for herbaceous crops and between 19 to 30% for woody crops (Hatakeyama & Hatakeyama, 2005). Lignins are complex natural polymers resulting from oxidative coupling of, primarily, 4-hydroxyphenylpropanoids. The monomers conforming lignin are p-coumaril, conyferil, and sinapyl alcohols. They differ from each other in the degree of methoxylation (Fig. 1). The proportion in which they are present in lignin varies widely depending on the type of plant. These monomers produce p-hydroxyphenyl, guaiacyl, and syringyl phenylpropanoid units (Fig. 1), which are capable of generating electron delocalized radicals that couple at various sites (Boerjan et al., 2003). These monomers can form under different chemical routes dimmers or participate in lignifications. Some basic linkages between lignin monomers are illustrated in Fig. 1. For a deeper discussion refer to Boerjan et al. (2003).
Cellulose is constituted entirely by glucan chains linked by β(1→4) bonds which interact with each other via hydrogen bonds (Keegstra, 2010). Hemicelluloses on the other hand, are constituted by backbones of glucose β(1→4) linked, mannose, and xylose with an equatorial configuration. Consequently, hemicelluloses include xylans (mainly), glucans, mannans, glucomannans, and xyloglucans (Scheller & Ulvskov, 2010).

2.1 Physic-chemical properties of lignin

Complete studies and reviews of the chemistry and physic-chemical properties of lignin were reported in the ACS symposium 397 (Glasser & Sarkanen, 1989), and by few authors (Agarwal & Atalla, 2010; Hatakeyama & Hatakeyama, 2010a; 2010b; Hatakeyama & Hatakeyama, 2005; Ralph & Landucci, 2010; Schmidt, 2010). A brief summary of these revisions is presented here. Readers seeking a greater detail should consult the aforementioned references.
Lignin molecular mass ranges from 10^3 and 10^5 depending on plant species and variations. The reported values may also be influenced by extracting or processing methods and analytic techniques. Soluble lignin presents a discrete maximum absorbance at 205 or 280 nm, with an extinction coefficient (ε) varying according to the lignin source and the used solvent. It also presents a characteristic fluorescence with a broad maximum emission ranging from 400 to 500 nm, depending on the excitation wavelength. Lignin is an amorphous polymer and presents a broad distribution of molecular arrangements, with

intra- and intermolecular distances that can vary from 0.40 to 0.98 nm. The glass transition of lignin in solid state does not present first order thermodynamic transitions; and at temperatures lower than the decomposition temperature, lignin can adopt a glassy or rubbery state. It has been shown that the heat capacity and intermolecular distance increase at the glass transition temperature (T_g). Also, T_g increases linearly with increments in lignin molecular weight, and strongly depends on the thermal history of the lignin sample. For example, T_g in kraft lignin is close to 100 °C, while it varies between 70 to 85 °C for hydrolysis lignin. Since the solidification rate affects the enthalpy of glassy lignin, enthalpy of the glassy state decreases when a lignin sample is slowly cooled from the melt. On the other hand, if the sample is quenched, the enthalpy increases since molecular chains are more randomly frozen than those of slowly glassified samples. Then, in samples below T_g the measured enthalpy of glassy polymers decreases as a function of time. This is called enthalpy relaxation, which is monitored through heat capacity change at glass transition.

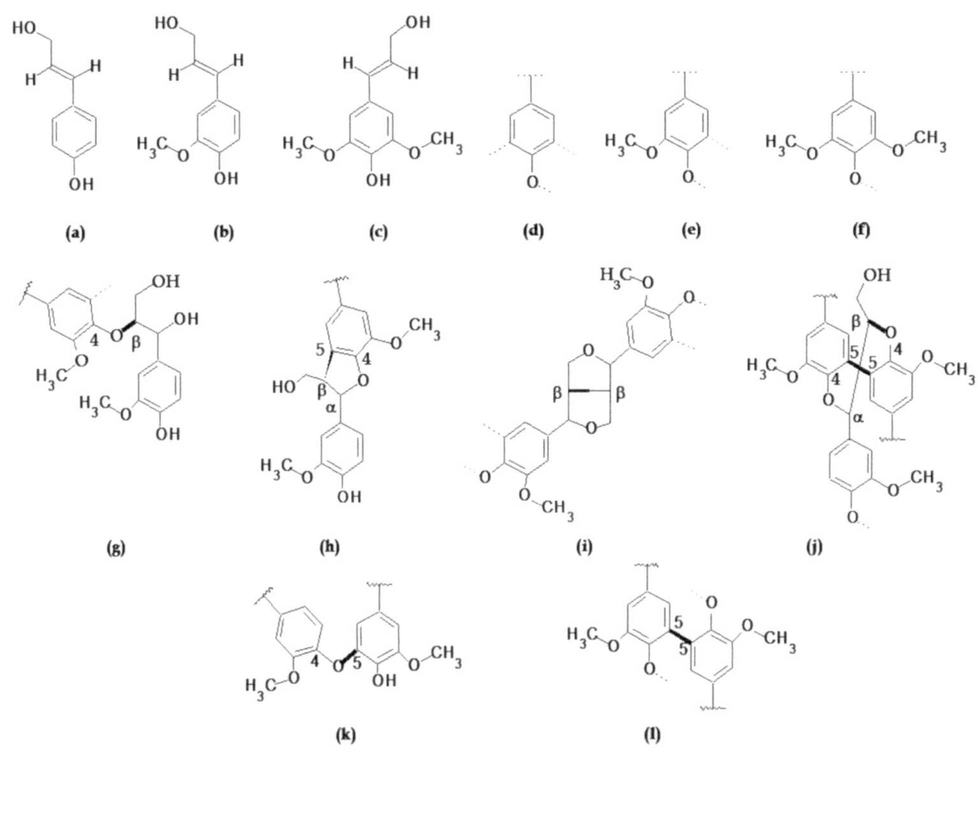

Fig. 1. Lignin monomers p-coumaril (a), conyferil (b) and sinapyl (c) alcohols, and their units inside lignin p-hydroxyphenyl (d), guaiacyl (e), and syringyl (f). Major structural units of lignin and presence in softwood (sw) and hardwood (hw), (g) β-O-4 unit (β- aryl ether), sw: 50 % and hw: 60 %; (h) β-5 (α-O-4) unit (phenylcoumaran), sw: 11% and hw 6 %; (i) β-β unit (resinol), sw and hw ~ 2 %; (j) [5-5/β-O-4(α-O-4)] unit (dibenzodioxocin); (k) 4-O-5 unit (biphenyl ether), sw and hw ~ 5 %; (l) 5-5 unit (biphenyl), sw: 18 % and hw: 10 % (Boerjan et al., 2003; Dimmel, 2010).

2.2 Chemical composition of agro-industrial wastes and by-products

Knowledge of chemical components of agro-industrial wastes produced worldwide should precede any attempt for fuel production. Although it is difficult to establish the major agro-industrial crops and their by-products generated worldwide, this work presents in Table 1 a selection of agro-industrial by-products based on their use for bioethanol production, and major crops reported by the FAO (2008) for the most populated countries in the word (China, India, US, Indonesia, Brazil, Bangladesh, Nigeria, Russia and Japan). This is according to the Population Reference Bureau (PRB, 2010).

3. Biological delignification

The production of alcohols from lignocellulosic materials involve two steps: 1) the hydrolysis of macromolecules (cellulose, hemicellulose, lignin and xylose) to release the carbohydrates that are further depolymerized to produce free sugars, and 2) the fermentation of these free sugars (Martinez et al., 2009; Perez et al., 2002). Among the lignocellulose constituent macromolecules, lignin is the most recalcitrant due to its amorphous hydrophobic heteropolymeric nature (Martinez et al., 2009). Although chemical/thermal processes allow good delignification levels, biodelignification has mild reaction conditions, higher product yields, few side reactions, less energy demand, and less reactor requirements to resist pressure and corrosion, which makes the bio-delignification a promissory pre-treatment for the production of biofuels. However, bio-delignification takes much longer than chemical or thermal processes, usually 8-12 weeks (Yu et al., 2010a), showing the importance of exploring new microorganisms, and the improving of culture conditions. Tables 2 and 3 summarize some fungi and bacteria used for bio-delignification of lingocellulosic materials.

3.1 Delignifying enzymes

Bio-delignification can be produced by the action of lignin peroxidase (LiP), manganese peroxidase (MnP), laccase, and versatile peroxidase (VP). Although up to now it has not been described a microorganism producing all the ligninolytic enzymes, the presence of two or more enzymes have been associated with higher delignification levels than those observed with the isolated enzymes showing the synergistic action of this set of enzymes (Costa et al., 2002; Gonçalves et al., 1998; Kannan et al., 1990).

3.1.1 Lignin peroxidase

LiP [EC 1.11.1.14] has been mainly isolated from *Phanerochaete chrysosporium*. Nevertheless, LiPs have been also found in other white- and brown-rot fungi strains, *Aspergilllus* strains, and bacteria such as *Acinetobacter calcoaceticus*, *Streptomyces viridosporus* and *Streptomyces lividans* (Crawford et al., 1993; Ghodake et al., 2009; Zerbini et al., 1999). LiPs are glycosilated enzymes of about 340 amino acids with a molecular weight between 38 and 50 kDa, a single heme, and two calcium ions (Hammel & Cullen, 2008; Sinclair et al., 1992). LiPs isozymes of *P. chrysosporium* are encoded by a family of 10 closely related and well-characterized genes (Hammel & Cullen, 2008), while in *Trametes versicolor* the LiP gene is arranged into a gene cluster encoding for two LiP and one MnP (Johansson & Nyman, 1996), and bacterial LiP seems to be encoded only for one gene (Wang et al., 1990).

By-product	Country	L[c]	G	X	A	C	H	Reference
Corn storver	US	13.3	31.9	18.9	2.8			(Templeton et al., 2009)
Cotton stalk	China	19.75				35.60	21.88	(Zhang et al., 2010)
Sugarcane baggase	Brazil	22.8	43.4	24.4	2.0			(Carrasco et al., 2010; Guo et al., 2009)
Oil palm (empty fruit bunches)	Colombia	4.15				46.77	17.92	(Piarpuzán et al., 2011)
Silvergrass	Europe [a,b]	7.6-11.5				43.0-52.2	24.8-33.9	(Hodgson et al., 2010)
Switchgrass (*Panicum sp.*)	US	21.4	35.0	21.8	3.5			(Garlock et al., 2011)
Rice straw	China	7.2-12.8				30.3-38.2	19.8-31.6	(Jin & Chen, 2007)
Wheat straw	Canada	18.15	38.27	18.75	1.5			(Tamaki & Mazza, 2010)
Oat straw	India	15.5				36.0	48.5	(Pandey & Pandey, 2002)
Barley straw	Russia	22.2				41.3	19.5	(Kocheva et al., 2008)
Rye straw	Russia	20.5				49.0	17.4	(Kocheva et al., 2008)
Soya stalks	Russia	25.4				37.6	17.3	(Torgashov et al., 2010)
Coconut husk	Malaysia	32.8				44.2	12.1	(Khalil et al., 2006)
Eucalyptus wood	China	26.2	44.9	11.4				(Yu et al., 2010c)
Sunflower stalks	India	17.5				38.5	33.5	(Sharma et al., 2002)
Sorghum stover	China	14.3	27.3	13.1	1.4			(Li et al., 2010)
Cassava baggase	Thailand	2.2	19.1	4.2	1.4			(Kosugi et al., 2009)
Olive wood	Spain	20.4				34.4	20.3	(Ruiz et al., 2006)
Eucalyptus wood	Brazil China	27.1 26.2	51.49 44.9	13.11 11.4	0.47			(Brito et al., 2008) (Yu et al., 2010c)
Construction wood waste	Korea	22.7-25.3		6.3-9		45.4-51.2		(Cho et al., 2011)
Pinus wood	US	26.8		6.6	1.6	43.6		(Frederick Jr et al., 2008)

[a] Samples were taken in three countries: Germany, Denmark and Sweden. [b] Variations are due to the *Miscanthus* species and harvest season. [c] Lignin composition refers to total acid- soluble and insoluble lignin. N.D: Not determined.

Table 1. Composition of major agro-industrial by-products use in delignification processes to obtain fermentable carbohydrates (L, lignin; G, glucan; X, xylan; A, arabian or arabinosyl substituents; C, cellulose, and H, hemicellulose).

Strain	Culture Conditions			L* (%)	Substrate	Remarks	Reference
	Process	Temp. (°C)	Time (d)				
Aspergillus niger	SS	55	0.08	20	Cellulose pulp	Crude extract containing xylanases	(Betini et al., 2009)
Aspergillus niveus	SS	55	0.08	20	Cellulose pulp	Crude extract containing xylanases	(Betini et al., 2009)
Aspergillus ochraceus	SS	55	0.08	20	Cellulose pulp	Crude extract containing xylanases	(Betini et al., 2009)
Ceriporiopsis subvermispora	SS	27	30	20	Sugarcane bagasse	Xylanase and MnP production	(Costa et al., 2005)
	SS	25	60	N.S.	Paddy straw	Cellulose lost.	(Sharma & Arora, 2010)
	SS	28	42	32	Corn stover	Improvement in sugar hydrolysis	(Wan & Li, 2010)
Echinodontium taxodii	Sm	25	30	42	Corn straw	Enhanced efficiency of alkaline/oxidative treatment	(Yu et al., 2010b)
Euc-1	SS	28	35	55	Wheat straw	Laccase, MnP and LiP production	(Dias et al., 2010)
Gleophyllum trabeum	SS	27	12	0	Wood chips	68% of carbohydrate degradation	(Fissore et al., 2010)
Gonoderma lacidum	Sm	25	30	55	Corn straw	Enhanced efficiency of alkaline/oxidative treatment	(Yu et al., 2010b)
Irpex lacteus	SS	28	35	55	Wheat straw	MnP and LiP production	(Dias et al., 2010)
	SS	28	15	12	Corn stalks	Combination with mild alkaline pretreatment allowed 80% lignin reduction	(Yu et al., 2010a)
Oxysporus sp.	Sm	25	60	69	Pomace from olive oil	Laccase production	(Haddadin et al., 2002)
Panos tigrinus strains	SS	30	10	4 to 7	Sugarcane bagasse	Expression of MnP, LiP and lacasse	(Gonçalves et al., 1998)
Penicillium oxalicum + Pleurotus osteatus	SS	N.S.	0.125	21	Wood pulp	Synergistic action of crude extract containing xylanase and laccase	(Dwivedi et al., 2010)
Phanerochaete chrysosporium	Sm	39	14	34	Cotton stalks	Improvement in carbohydrate availability	(Shi et al., 2009)
	SS	39	10	21	Cotton stalks	Improvement in carbohydrate availability	(Shi et al., 2009)
Phlebia brevispora	SS	25	60	N.S.	Paddy straw	Cellulose lost.	(Sharma & Arora, 2010)

Strain	Culture Conditions			L*	Substrate	Remarks	Reference
	Process	Temp. (°C)	Time (d)	(%)			
Phlebia floridensis	SS	25	60	N.S.	Paddy straw	Cellulose lost.	(Sharma & Arora, 2010)
Phlebia radiate	SS	25	60	N.S.	Paddy straw	Cellulose lost.	(Sharma & Arora, 2010)
Pycnoporus cinnabarinus	SS	24	30	12	Sugarcane bagasse	No cellulose degradation	(Meza et al., 2006)
Pycnoporus sanguineus	Sm	30	21	25	Wheat straw		(Lu et al., 2010)
	Sm	28	1	71	Wheat straw	Native and recombinant laccase	(Lu et al., 2010)
	Sm	30	21	27	Corn strover		(Lu et al., 2010)
	Sm	28	1	57	Corn strover	Native and recombinant laccase	(Lu et al., 2010)
Pynnoporus cinnabarinus	SS	30	15	12	*Prosopis juliflora*	20% increment in sugar release	(Gupta et al., 2010)
	SS	30	15	8	*Lantana camara*	20% increment in sugar release	(Gupta et al., 2010)
Trametes versicolor	Sm	25	30	52	Corn straw	Enhanced efficiency of alkaline/oxidative treatment	(Yu et al., 2010b)

Table 2. Fungus strains used for the delignification of lignocelullose materials. L* = Lignin Lost, Sm = Submerged Fermentation, SS = Solid-state fermentation, MnP = Manganese peroxidase, LnP = Lignin peroxidase , N.S. = Non specified.

Strain	Culture Conditions			L	Substrate	Remarks	Reference
	Process	Temp. (°C)	Time (d)	(%)			
Bacillus macerans	SS	28	45	50	Cotton saw		(Singh et al., 2008)
Bacillus pamilus	Sm	55	2	8	Kraft-pulp	Cellulase-free	(Kaur et al., 2010)
Bacillus sp.	Sm	35	2	56	Pulp and paper mill effluent	Xylanase and MnP production	(Mishra & Thakur, 2010)
Cellulomonas cartae	SS	28	45	50	Cotton saw		(Singh et al., 2008)
Cellulomonas uda	SS	28	45	50	Cotton saw		(Singh et al., 2008)
Zymomonas mobilis	SS	28	45	33	Cotton saw		(Singh et al., 2008)

Table 3. Bacteria strains used for the delignification of lignocelullose materials. L* = Lignin Lost, Sm = Submerged Fermentation, SS = Solid-state fermentation, MnP = Manganese peroxidase, LnP = Lignin peroxidase , N.S. = Non specified.

Due to the complex nature of lignin, the delignification process must be carried out extracellulary, and the LiP is found in all the peripheral regions of the fungal cell cytoplasm in association with the cell membrane, fungal cell wall, and extracellular slime materials (Daniel et al., 1989). In addition, in solid wood LiP is detected in low concentrations associated with both superficial and degradation zones within secondary cell walls undergoing fungal attack, while in liquid cultures (i.e. submerged-state fermentation) a much greater level of extracellular peroxidase activity is associated with wood fragments degraded by fungus. Although LiP is able to oxidize aromatic compounds with a high redox potentials by single electron abstraction and is the most efficient lignin-degrading enzyme (Piontek et al., 2001), *in vitro* the majority of lignin-derived preparations actually experience overall polymerization after LiP exposure (Sarkanen et al., 1991). LiP activity is controlled by the ionization degree suggesting that specifics amino acids residues play a role in lignin binding, and that this enzyme is capable of oxidizing lignin directly at the protein surface by a long-range electron transfer process (Johjima et al., 1999).

LiPs degrade lignin through H_2O_2 using a mechanism that resembles that used for other peroxidades (Hammel & Cullen, 2008; Martinez et al., 2005; Wong, 2009). The first step involves the formation of a high redox potential oxo-ferryl intermediate (compound I, LiP-I) as result of the reaction of the heme cofactor with H_2O_2. In a second step two consecutive $1e^-$ reductions are carried out (Fig. 2): (i) a $1e^-$ reduction of LiP by a reducing substrate yields to compound II (LiP-II) and a substrate radical cation, and (ii) a $1e^-$ reduction that returns the enzyme to the ferric oxidation state, completing the catalytic cycle (Wong, 2009). The radical cation of the substrate produced in this cycle undergoes rearrangements and non-enzymatic degradations (Fig. 2), which in turn leads to a set of reactions that results in lignin depolymerization (Wong, 2009). Mutagenesis studies have shown that LiP has two distinct substrate interaction sites: (i) the classical heme edge, which degrades lignin-derived compounds and dyes, and (ii) the glutamine 146 site, which is implicated in catalysis of lignin-derived compounds.

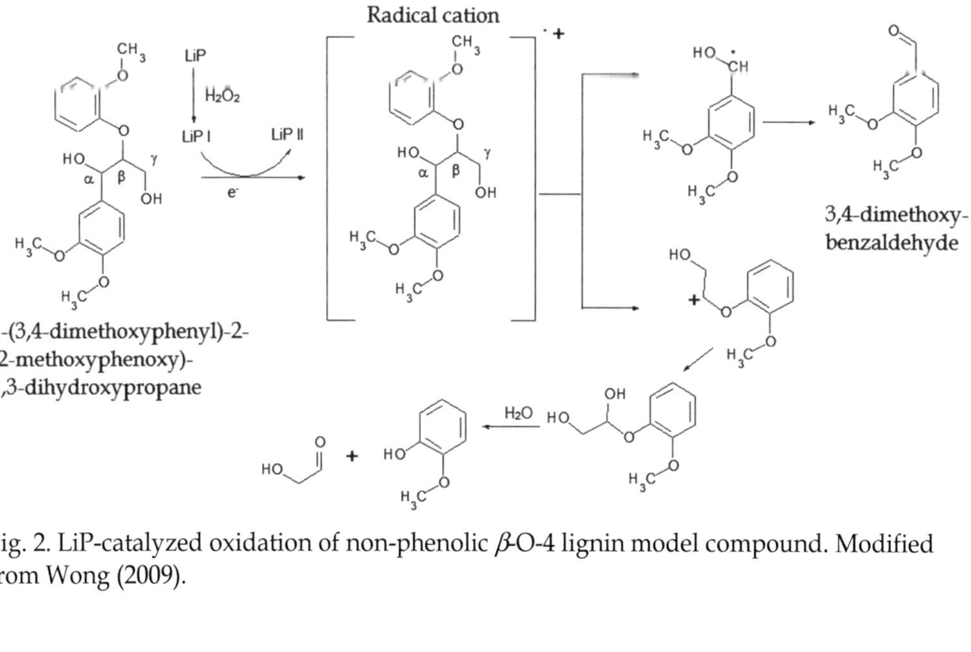

Fig. 2. LiP-catalyzed oxidation of non-phenolic β-O-4 lignin model compound. Modified from Wong (2009).

LiP genes have been cloned and expressed in *Phanerochaete sordida, Trichoderma reesei, Pichia pastoris, Pichia methanolica* and *Escherichia coli,* obtaining recombinant enzymes with similar properties to those observed in native enzymes (Nie et al., 1998; Saloheimo et al., 1989; Sugiura et al., 2009; Wang & Wen, 2009). A recent work showed that by coupling a directed evolution technique with a rapid colorimetric screening method, it was possible to obtain a recombinant LiP with improved H_2O_2 stability, polychlorinated phenol degradability, and kinetic properties (Ryu et al., 2008).

3.1.2 Manganese peroxidase

The MnPs [EC 1.11.1.13] share many of the characteristics previously discussed for LiPs. They are glycosilated, extracellular heme-containing enzymes of about 350 amino acids, a molecular weight around 40 kDa, and the same peroxidase catalytic cycle of LiP (Martinez et al., 2009; Martinez et al., 2005). MnP has been isolated mainly from basidiomycetes including *P. chrysosporium, Schizophyllum sp., Ceriporiopsis subvermispora, Panus tigrinus, Lentinula edodes, Nematoloma frowardii, Bjerkandera adusta, T. versicolor,* and *Dichomitus squalens,* among others. The presence of MnP has been also reported in an *Aspergillus terreus* strain (Kanayama et al., 2002). The enzyme production is specie- and strain-dependent, with an important role of the carbon source, the lignocellulosic substrate and the presence of aromatic compounds (Elisashvili & Kachlishvili, 2009). Catalytic cycle of MnP is similar to that of LiP (Fig. 3): (i) compound I (MnP-I), a Fe(IV)-oxo-porphyrin radical cation, is produced by reaction of enzyme and H_2O_2, (ii) Mn^{2+} reduces compound I to compound II, producing Mn^{3+}, and (iii) the resting enzyme is regenerated (Wong, 2009). This Mn^{3+} oxidizes phenolic substrates in a second-order reaction, producing phenoxy-radicals, which in turn led to a set of reactions that result in lignin depolymerization (Perez et al., 2002) (Fig. 3).

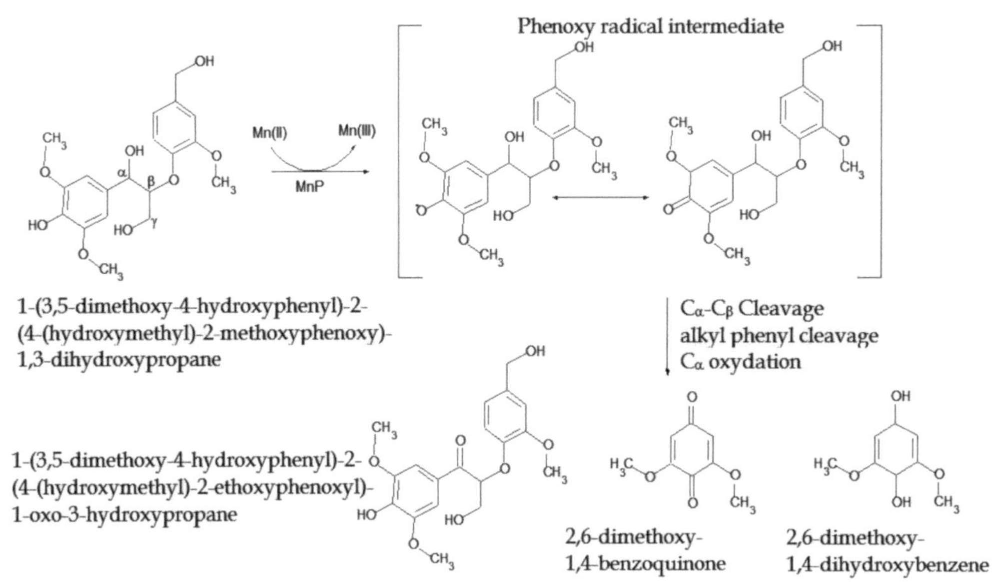

Fig. 3. MnP-catalyzed oxidation of phenolic aryglycerol β-aryl ether lignin model compound. Modified from Wong (2009).

The MnP has a single binding site for Mn^{2+} near the heme, from which two Mn^{3+} equivalents are obtained at the expense of one H_2O_2 equivalent in a reaction requiring oxalate or another appropriate manganese chelator (Timofeevski & Aust, 1997; Wariishi et al., 1992). However, MnP can also catalyze a manganese-dependent disproportionation of H_2O_2 in the absence of manganese chelator, which may protect the enzyme from inactivation by H_2O_2 under limited free oxalate conditions (Timofeevski & Aust, 1997). Reaction of Mn^{3+} with H_2O_2 is catalyzed by Cu^{2+}, which explains the inhibition of MnP by Cu^{2+} (Aitken & Irvine, 1990). Active site and crystal structure analysis of *P. chrysosporium* MnP, the most studied MnP enzyme, have shown that arginine 42 and acidic amino acids (i.e glutamic or aspartic acid) at positions 35, 45, 39 and 179 are involved in Mn^{2+} binding (Sundaramoorthy et al., 2010; Whitwam et al., 1997), while serine 172 seems to been involved in heme binding (Ambert-Balay et al., 2000). MnP encoding genes have been cloned from several white-rot fungi strains, where up to three different genes have been reported each one encoding for a specific isoform (Alvarez et al., 2009; Martinez et al., 2009; Martinez et al., 2005). The expression of MnP genes is mainly regulated by Mn via a growth-stage-specific and concentration-dependent mechanism (Brown et al., 1991). Cu^{2+}, Zn^{2+}, Ag^{+}, Cd^{2+}, Se, H_2O_2, ethanol, sodium arsenite, and 2,4-dichlorophenol, as well as heat shock regulate the MnP gene expression, while the co-administration with Mn significantly enhances the MnP production (Alvarez et al., 2009; Catal et al., 2008).
MnP genes have been homo- or heterologous expressed in *P. chrysosporium*, *P. ostreatus*, *P. pastoris*, *Aspergillus oryzae*, and *Zea mays*, producing recombinant enzymes with similar kinetic and stabilities properties to those observed for native enzymes (Clough et al., 2006; Gu et al., 2003; Jiang et al., 2008).

3.1.3 Laccase

Laccases [EC 1.10.3.2] are glycosilated multicopper enzymes, with wide molecular weights ranging from 50 to 140 kDa (Gaitan et al., 2011; Wong, 2009). Unlike LiP and MnP, laccases are widely distributed in nature and can be found in plants, fungi, bacteria and insects (Dwivedi et al., 2011; Riva, 2006). Although transgenic plants over-expressing laccase genes have been used for energy production, phytoremediation, and alteration in phenolic metabolism (Riva, 2006), most of the delignification process have been carried out by using microorganisms. In fungi, where they play a role in the host-pathogen interaction during the first step of infection (Punelli et al., 2009), laccases have been found in *Trametes versicolor, Trametes pubescens, Pycnoporus cinnabarinus, Myceliophthora thermophila, Pleurotus eryngii, Pleurotus ferulae, Pleurotus astreatus, Cerrena unicolor, Cyathus bulleri, Lentinula edodedes*, and *Agaricus blazei*, among others. In bacteria the enzyme has been reported in *Azospirrullum lipoferum, Marinomonas medierranea, Streptomyces griseus, Haloferax volcanii*, and *Bacillus subtilis*. Laccases are monomeric or homopolymeric enzymes (Riva, 2006; Wong, 2009), and the cellular localization depends on the organism, with plant and fungal laccases as extracellular enzymes, while an intracellular localization is observed in most of the bacterial laccases (Diamantidis et al., 2000; Riva, 2006).
Laccase catalyze the monoelectric oxidation of phenolic and nonphenolic substrates to their corresponding reactive radicals in a reaction mediated by the four Cu atoms located at the catalytic core. During the substrate oxidation there is a reduction of one oxygen molecule to two water molecules and four radicals (Arora & Sharma, 2010). The four copper atoms are arranged in three different groups: Type-1 or blue Cu, Type-2 or normal Cu and Type-3 or

coupled binuclear Cu centers. These Cu atoms produce the lignin degradation in a three-step reaction: (i) the Type-1 Cu is reduced by oxidizing lignin, (ii) the electron is transferred from Type-1 Cu to Type-2 and Type-3 Cu cluster, and (iii) oxygen is reduced to water at Type-2 and Type-3 Cu centres (Dwivedi et al., 2011; Riva, 2006). Laccases can oxidize a wide range of substrates including polyphenols, methoxysubstituted phenols, and aromatic diamines, among others, through Cα-Cβ cleavage, alkyl-aryl cleavage or C_α oxidation (Fig. 4) (Arora & Sharma, 2010; Wong, 2009). Since laccases use oxygen instead of the H_2O_2 uses for peroxidases, the toxic effect of this compound in cell viability can be avoided giving to the laccases an important role in early stages of delignification (Sterjiades et al., 1993). Due to the large size of laccases and the highly complex structure of lignin, in most cases it cannot be degraded directly, requiring the presence of intermediate substrates (i.e. chemical mediators), which oxidized radicals are able to induce the depolymerization of complex substrates (Riva, 2006).

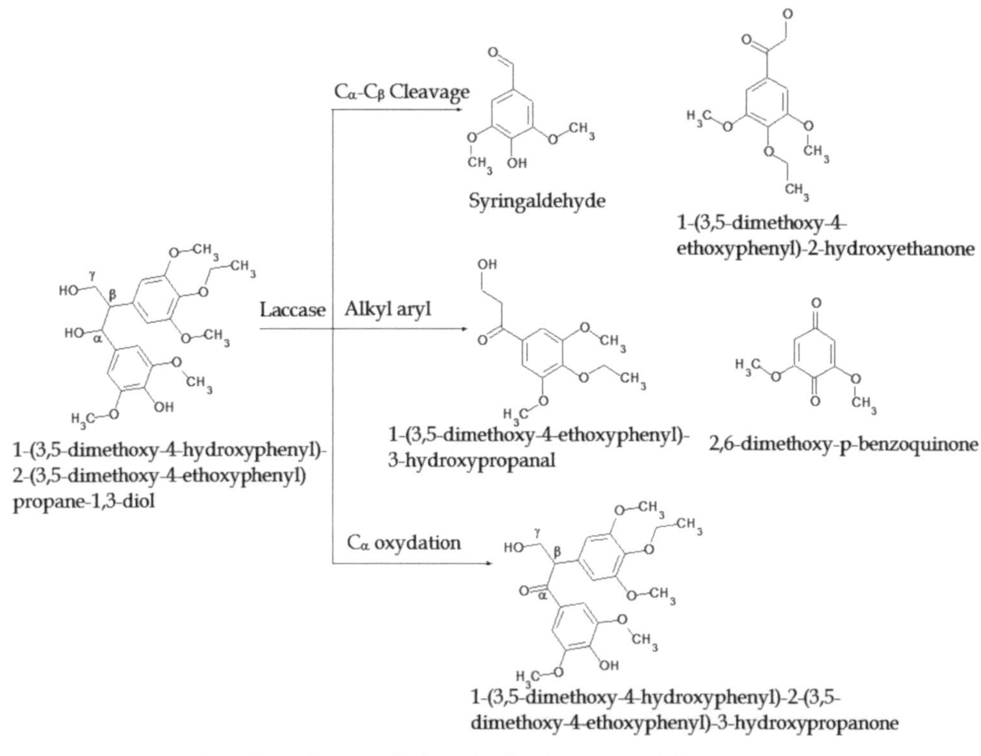

Fig. 4. Laccase-catalyzed oxidation of phenolic β-1 lignin model compound. Modified from Wong (2009).

Laccases are presented as up to six isoenzymes, with different molecular weights, expression profiles, stability, substrate affinity, and encoded by different genes (Dwivedi et al., 2011; Gaitan et al., 2011; Xiao et al., 2006). Laccase gene expression is mainly up regulated by Cu^{2+} addition, although other potent inducers are Mn^{2+}, Fe^{3+}, heavy metals, 2,6-dimethoxy-1,4-benzoquinone, H_2O_2, caffeine, amphotericin B, syringic acid, tannic acid, Tween 80, soybean oil, aromatic compounds, and microclimatic changes (i.e. lower temperature and osmotic pressure) (Dekker et al., 2007; Galhaup et al., 2002; Xiao et al.,

2006). On the other hand, laccase production can be down-regulated or inhibited by proline, urea, glucose Hg^{2+}, Mg^{2+}, Ca^{2+}, Sn^{2+}, Ba^{2+}, Co^{2+}, Cd^{2+}, Mn^{2+}, and Zn^{2+}, fatty acids, sulfydryeagents, hydroxyglycine, kojic acid, EDTA, l-cysteine, dithiothreitol, glutathione, thiourea, and cationic quaternary ammonium detergents; by chelating the Cu atoms, by modifying amino acid residues or by causing conformational changes in the glycoprotein (Dekker et al., 2007; Dwivedi et al., 2011; Galhaup et al., 2002).

Laccase genes have been cloned and expressed in *Pichia pastoris, P. methanolic, Kluyveromyces lactis, Aspergillus niger, Coriolus versicolor,* and *E. coli,* with significantly higher levels and similar or even better kinetic and stability profiles than those observed for native enzymes (Guo et al., 2006; Hong et al., 2007; Ranieri et al., 2009; Rodriguez et al., 2008; Salony et al., 2008).

3.2 Fungi delignification

During the last two decades white-rot fungi strains have shown to produce the more efficient delignification (Martinez et al., 2009; Vicuña, 2000; Wong, 2009) *Phanerochaete chrysosporium* is the most studied white-rot fungi used in biodelignification, with lignin reduction levels ranging from 11 to 73% and incubation times between 10 to 60 days. However, the differences among the substrates (i.e. rice straw, cotton stalks, mustard straw and sugarcane bagasse) and the culture conditions (e.g. solid-state fermentation-SSF, submerged fermentation-SmF, temperature and time), limit the comparison between the studies.

Shi et al. (2009), showed that higher lignin lost was observed in SmF than in SSF delignification of cotton stalks, while a 73% lignin lost of kraft pulp was observed with SmF while no delignification was observed in SSF (Pellinen et al., 1989). Wheat straw and sugarcane bagasse, two important substrates for biofuels production (Kaparaju et al., 2009; Vasquez et al., 2007), showed low lignin degradation by *P. chrysosporium* strains (Dorado et al., 1999; Li et al., 2002). Similar results were observed for sugarcane bagasse delignification with *Panos tigrinus* (Gonçalves et al., 1998), *Phanerochaete sordida* (Li et al., 2002) and *Pycnoporus cinnabarinus* (Meza et al., 2006), showing the recalcitrant nature of this substrate for delignification. However, delignification levels above 50% were observed with *Irpex lacteus* for wheat straw (Dias et al., 2010) and *Phlebia sp.* for sugarcane bagasse (Li et al., 2002; Osono & Takeda, 2006). Significant delignification results have also been reported for *Coriolus versicolor* (Tripathi et al., 2008), *Echinodontium taxodii* (Yu et al., 2010b), *Euc-1* (Dias et al., 2010), *Gonoderma* (Bourbonnais et al., 1997; Haddadin et al., 2002; Tripathi et al., 2008; Yu et al., 2010b), *Oxysporus sp.* (Haddadin et al., 2002), *Phlebia sp.* (Li et al., 2002), *Trametes versicolor* (Yu et al., 2010b) and *Trichoderma reesei* (Singh et al., 2008), with reduction of lignin levels ranging from 40% to 60%. Recent studies have shown that actinomycetes strains (e.g. *Aspergillus*) can induce lignin degradation levels of up to 92%, with similar culture conditions to those used by basidiomycetes (Singh et al., 2008).

Since carbohydrates are needed for the second stage of biofuel production (i.e. sugar fermentation), the degradation during the delignification stage is a side reaction that affects the yield of biofuel production. Hence, a selection of a microorganism with a low cellulose activity is an important task. *Phanerochaete chrysosporium* (Shi et al., 2009), *Pycnoporus cinnabarinus* (Meza et al., 2006), *Echinodontium taxodii* (Yu et al., 2010b), *Euc-1* (Dias et al., 2010), *Irpex lacteus* (Dias et al., 2010; Yu et al., 2010a), and *Pycnoporus sanguineus* (Lu et al., 2010), have shown high lignin degradation specificity, while significant levels of carbohydrate degradation have been reported for *Ceriporiopsis subvermispora* (Wan & Li, 2010), *Gleophyllum trabeum* (Fissore et al., 2010), *Phlebia brevispora* (Sharma & Arora, 2010), *Phlebia floridensis* (Sharma & Arora, 2010), *Phlebia radiata* (Sharma & Arora, 2010), and *Pleurotus sajor-caju* (Kannan et al., 1990). This carbohydrates degradation during biodelignification could also depend on the substrate, since fungi with high lignin selectivity can show carbohydrate lost with certain lignocellulosic materials, as observed for paddy straw that presented cellulose lost with all the white-rot fungi used in the delignification (Sharma & Arora, 2010).

Biodelignification has a direct effect on the availability of sugars for ethanol production, which significantly improves the yield of biofuel production. Although this beneficial effect of delignification is not evaluated in all the studies, the results show that there is no need to produce a 100% delignification since increases above 20% in carbohydrates availability can be obtained with just 8% delignification (Gupta et al., 2010; Shi et al., 2009; Wan & Li, 2010). In addition, the biodelignification has been used before chemical pretreatments (i.e. alkaline/oxidative or mild alkaline) allowing up to 80% lignin reduction (Yu et al., 2010a; Yu et al., 2010b). These results strongly suggest that combination of biological and chemical/physical delignification methods might be a feasible alternative to improves delignification levels and reduces the volume and conditions of the chemical/physical pretreatments, promoting the generation of eco-friendly processes for paper and biofuels companies.

Finally, the use of crude extracts from ligninolytic fungi has shown to be a promissory alternative for delignification. Crude extracts from *Aspergillus niger*, *A. niveus*, and *A. ochraceus*, containging xylanases, allowed a 20% delignification of cellulase pulp in 2 h (Betini et al., 2009), while a crude extract of *Penicillium oxalicum* and *Pleurotus osteatus*, containg xylanase and laccase enzymes, produced a 21% delignification of wood pulp in 3 h (Dwivedi et al., 2010).

3.3 Bacterial delignification

Bacteria are less used microorganism in delignification of lignocellulose materials. Most of the studies use *Bacillus* strains (Table 3), although *Cellulomonas* and *Zymomonas* have been recently reported as promissory ligninolytic bacteria (Singh et al., 2008).

The loss of lignin ranges from 8% to 56%, with similar temperatures and times to those used in fungi-mediated delignification. Bacteria delignification is mainly mediated by extracellular xylanases, although a synergistic effect has been observed by the addition of MnP, pectinase or α-L-arabinofuranosidase (Bezalel et al., 1993; Kaur et al., 2010). However, these high lignin removal levels might be accompanied with high levels of cellulose degradation, as observed for *B. macerans*, *C. cartae*, *C. uda*, and *Z. mobilis* with cellulose reductions ranging from 31% to 51%. One strategy to overcome this issue is the use of cellulase-free extracts (Kaur et al., 2010) or purified enzymes (Bezalel et al., 1993), that allows up to 20% delignification levels within shorter incubations times than those observed with the whole microorganisms.

Even though bacteria have shown significant delignification levels, the main application is in the paper industry on pre-bleaching steps or to reduce toxic compounds from paper mill effluents. However, biodelignification has a direct impact in pulp properties, free-sugars, and reduction in effluent toxicity and in the use of chloride compounds (Kaur et al., 2010; Mishra & Thakur, 2010; Singh et al., 2008).

4. Chemical separation of Lignin and carbohydrates

This section establishes the basis of three types of chemical pretreatment: acidic, alkaline and with solvents. For each, lignin and carbohydrate solubilisation (or degradation) are discussed. It is clear that different types of chemical pretreatments affect biomass differently not only because of differences in the solubilized component (Harmsen et al., 2010), but also because the accessibility attained is different and changes differently as subsequent enzymatic hydrolysis occurs (Kumar & Wyman, 2009).

A few reports giving meaningful comparisons between different types of chemical pretreatment have been obtained (Wyman et al., 2005a; 2005b; Wyman et al., 2009). A more complete coverage is impractical because different authors present important differences in selected substrates, analytical procedures, and/or method to report results. Abatzoglou et al. (1992) discuss the application of a severity factor or severity parameter to account for the effect of acid on lignocelluloses composition and the extent of reaction if acid pretreatment is used. They state that due to the complexity of reactions, kinetic parameters lack mechanistic meaning and are functions of the ranges of experimental conditions used. Severity factors combine into a single parameter the effect of the different operational variables and have been adapted to other types of chemical pretreatments. One widely adopted defining equation (slightly or importantly modified in some later studies) is as follows:

$$R_0 = e^{\left[(T-100)/14.5\right]t} \tag{1}$$

where R_0 is the reaction ordinate, T is the pretreatment temperature in °C, and t is the pretreatment time in min. R_0 is meant to provide a zonal indication of the predominant reactions taking place in lignocelluloses. This end result was empirically motivated.

4.1 Acidic

Two main configurations of acid pretreatment are diluted and concentrated. During dilute acid pretreatment structural carbohydrates are efficiently solubilized and possibly degraded to an extent that depends on the pretreatment conditions. Here, partial hemicelluloses solubilisation is obtained, therefore increasing pore size. Hydrolysis of the remaining hemicelluloses and cellulose is frequently achieved through a subsequent enzymatic hydrolysis stage that uses enzyme cocktails containing cellulases and xylanases (Mosier et al., 2005; Wyman & Lloyd, 2005). The acidic reagent may be liquid hot water (pH of water decreases with temperature), dilute or concentrated hydrochloric, sulfuric, phosphoric, peracetic, oxalic, and maleic acid among others. Virtually any acid (either mineral or organic) may be used, but sulfuric acid is widely preferred, even though recent studies have shown better hemicelluloses yields with other acids (Lee & Jeffries, 2011). Pretreatment time (accounted from the moment the biomass reaches the desired pretreatment temperature) ranges between 1 min and 180 h. Reactor configurations include but are not limited to flow-through, continuous, and batch. Temperatures range between 25 and 200°C at pressures between 1 atm and 15 atm. Initial acid concentration may range between 0.1 and 6%, more typically between 0.7 and 4% (Kumar et al., 2009; Sierra et al., 2008). In batch reactors, total pressure corresponds to the saturation pressure of the mixture at the pretreatment temperature. Typically, the highest temperatures require less time and result in a more extensive cellulose degradation. The dilute sulfuric acid pretreatment can achieve high reaction rates and significantly improves cellulose hydrolysis (Esteghlalian et al., 1997). Its cost is similar or higher than for other pretreatment options (Eggeman & Elander, 2005). Neutralization of pH is necessary for the downstream enzymatic hydrolysis or fermentation processes. Dilute-acid pretreatment is also known to have a negative influence on the enzymatic hydrolysis of biomass. For example, recent studies report the formation of spherical droplets of lignin and/or lignin and carbohydrate complexes on the cellulose surface. These droplets, preferentially formed at temperatures greater than 130°C, have an important negative impact on biomass subsequent enzymatic digestibility (Selig et al., 2007). Some other researches argue that materials that have been subjected to acid hydrolysis can be harder to ferment because of the presence of toxic substances including but not limited to furfural and hydroxymethylfurfural (Galbe & Zacchi, 2002). Furthermore, acid pretreatment results in costly materials of construction, high pressures, neutralization and conditioning of hydrolysate prior to biological steps, slow cellulose digestion by enzymes, and nonproductive binding of enzymes to lignin (Wyman et al., 2005b).

The concentrated acid pretreatment is more widely known as concentrated acid hydrolyisis because it results in complete solubilisation of carbohydrates. Clearly, subsequent enzymatic hydrolysis is not required. This type of pretreatment has a long story (1883 for hydrolysis of cellulose in cotton) with the first industrial plant built in Germany in 1937 using hydrochloric acid. The Arkenol's process, claimed to reduce sugar loss to less than 3%, separating 98% of the acid (Farone & Cuzens, 1996), starts mixing lignocelluloses waste (less than 10% moisture content) with 70%-77% sulfuric acid added at a ratio of 1.25:1 (acid : cellulose+hemicellulose), the temperature is controlled at less than 50°C. Afterwards, water is added to dilute the acid to 20%-30 and hydrolysis at 100°C takes place for an hour. This results in a gel, which is pressed to remove an acid/sugar product stream. Residual solids are subjected to a second hydrolysis step. A chromatographic column is used to separate acid and sugars (Nanguneri & Hester, 1990). The fermentation converts both the xylose and the glucose to ethanol at theoretical yields of 85% and 92%, respectively. A triple effect evaporator is required to reconcentrate the acid (Yancey & Kadam, 1997).

The industrially used concentrated pretreatment must not be confused with the complete solubilisation of carbohydrates widely used for quantification of structural lignin and carbohydrates. The corresponding protocol uses sulfuric acid reagent in two stages, one concentrated at mild temperature (72% w/w, 30°C) and the other diluted at high temperature (4% w/w, 121°C) (Sluiter et al., 2004).

4.1.1 Reactions during acid pretreatment

During acid pretreatment a rather severe solubilisation (and some degradation of monomers) of hemicelluloses occur. Concurrently, a change in the lignin structure takes place. Wood lignin submitted to diluted acid pretreatment (Sannigrahi et al., 2008), as well as wood submitted to steam explosion (Li et al., 2007), and switchgrass submitted to diluted acid pretreatment (Pingali et al., 2010) presented a decrease in the β-O-4 linkages in lignin, which are fragmented during high temperature acid-catalyzed reactions. The extent at which these reactions occur is illustrated in Fig. 5 (Li et al., 2007). Other effects of acid pretreatment on the structure of wood lignin include a decrease in the protonated and oxygenated aromatic carbons per aromatic ring (Sannigrahi et al., 2008).

As a result of the substantial cleavage of β-O-4 linkages, the molecular weight distribution of lignin would be expected to decrease; however, this is not the case. A comprehensive repolymerization (condensation reactions) resulting in an increase in molecular size and a more heterogeneous lignin structure is obtained (Li et al., 2007) (Fig. 6). The simultaneous depolymerization and repolymerization of lignin during an acid pretreatment are undesirable reactions since they will lead to an increase in the heterogeneity of the resulting lignin. In addition, solubility and reactivity properties will be negatively affected. Reactions, shifting the pH-conditions of the steam treatment towards the alkaline side could be one way to reduce the formation of carbonium ions and inhibit severe structural changes in the lignin polymer. Alternatively, the addition of inhibitors for the repolimerization reactions (i.e., reactive phenol, 2-naphthol (Li et al., 2007)) lead to an important delignification with the production of lignin with uniform structure.

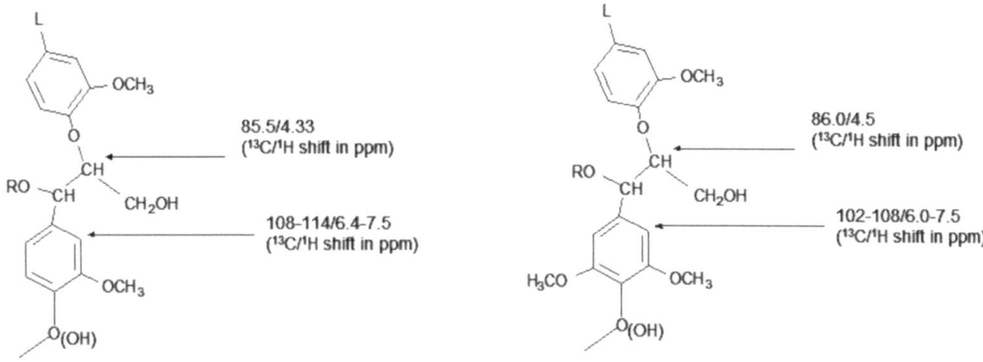

Fig. 5. The β-O-4 structure in lignin together with $^{13}C/^{1}H$ shift values for CH-groups.

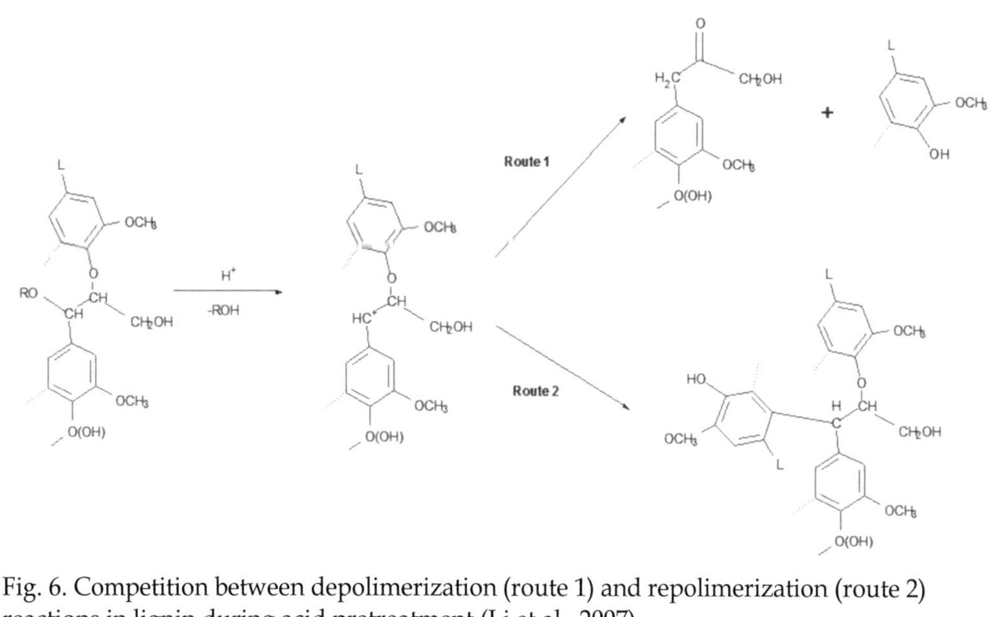

Fig. 6. Competition between depolimerization (route 1) and repolimerization (route 2) reactions in lignin during acid pretreatment (Li et al., 2007).

On the other hand, depolymerization of polysaccharides resulting from steam treatment through cleavage of glucosidic linkages is desirable for conversion of carbohydrates through fermentation processes. However, if monomeric carbohydrates are further degraded into

HMF (from hexoses) and furfural (from pentoses), a true yield loss occurs. Maximum loss of glucose and xylose in HFM and furfural formation have been reported as ~35% and ~5% of original content respectively depending on pretreatment conditions (Farone & Cuzens, 1996) While the carbohydrate content decreases following pretreatment, the proportion of hexose sugars within the carbohydrate fraction increases. Li et al (Li et al., 2005) reported that only about half this fraction was true lignin while the rest comprised "pseudo-lignin" material. This lignin like material is formed by dehydration and repolymerization of polysaccharide degradation products such as furfural, which are formed during high temperature pretreatments.

The discussion presented here is not comprehensive but only fundamental. The presence of products from lignin, glucose and xylose degradation reactions including but not limited to organic acids, phenols and aromatic aldehydes (Chen et al., 2006) is an indication of this fact. In depth studies to elucidate the mechanisms through which these reactions occur is important because all of the degradation products are inhibitors of fermentation and result in an undesired carbohydrates yield loss.

4.1.2 Kinetic modeling acid pretreatment

As discussed in the previous section, all of the main cell wall constituents of lignocelluloses react during acid pretreatment. However, kinetic studies are not available for lignin reactions. The main focus of kinetic modeling work has been hemicelluloses hydrolysis with its consequent degradation. Because xylose is often the main constituent in hemicelluloses, comprising up to 90% of the total hemicellulose dry weight depending on the feedstock (McMillan, 1992), hemicelluloses degradation modeling only accounts for the production of xylose monomers. Three main mechanisms have been proposed and are depicted in Fig. 7. The most adapted mechanism (Fig. 7b) uses a four-step pseudo-first-order irreversible reactions with Arrhenius type constants.

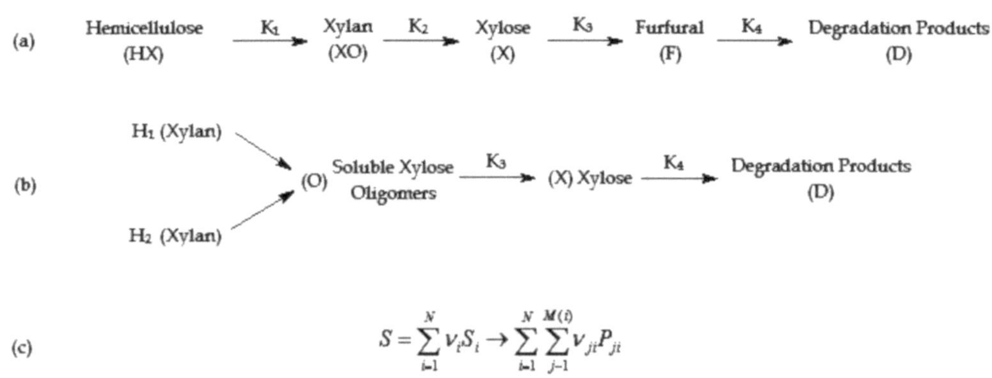

Fig. 7. Mechanisms for degradation of hemicelluloses during acid pretreatment. (a) Accounts for the formation of furfural and from there other degradation products (Morinelly et al., 2009) (b) Accounts for different hemicelluloses hydrolysis reactivities (fast and slow)(Chen et al., 1996; Esteghlalian et al., 1997; Grohmann et al., 1985; Lee et al., 2000; Schell et al., 2003) (c) lumped system approach. A reaction system (S) is composed of N subsystems producing M final products P. The reaction stoichiometric coefficients are ν (Abatzoglou et al., 1992).

Some work also accounts for cellulose degradation reactions as shown in the following equation (Lee et al., 2000):

$$\text{Cellulose} \xrightarrow{k_1} \text{Glucan} \xrightarrow{k_2} \text{Glucose} \xrightarrow{k_3} \text{Degradation products} \quad (2)$$

Rate equations are then obtained for the different reactors and reactions considered. For example, for reactions in Fig. 7a the equations are:

$$\frac{dXH(t)}{dt} = -k_1 XH(t); \frac{dXO(t)}{dt} = -k_1 XH(t) - k_2 XO(t); \frac{dX(t)}{dt} = -k_2 XO(t) - k_3 X(t); \frac{dF(t)}{dt} = -k_3 X(t) - k_4 F(t) \quad (3)$$

For models in Fig. 7b, reported activation energies in k_1 and k_2 varied between 50 and 250 kJ kmol^{-1} depending on the reactor employed and the feedstock used. The highest values were reported for continuous reactors at pilot plant scale using corn stover (Schell et al., 2003) and the lowest for wheat straw in a batch reactor (Grohmann et al., 1985). High activation energies may serve as an indication that the system follows a chemical reaction-limited kinetics rather than a mass transport limited kinetics. Some interesting conclusions from the diverse dilute sulfuric acid kinetic studies are: (i) xylose yields are favored at high temperatures and short times (Chen et al., 1996; Esteghlalian et al., 1997; Grohmann et al., 1985; Lee et al., 2000; Morinelly et al., 2009; Schell et al., 2003); (ii) selectivity defined in terms of either Arrhenius constants ratios or activation energies ratios shows that high temperatures have an enhancing effect on the hemicelluloses breakdown to oligomers (i.e., in Fig. 7a, $E_1/E_2 > 1$), however, the hydrolysis of oligomers is less favored than the formation of degradation products (i.e., in Fig 7a, $E_2/E_3 < 1$) (Morinelly et al., 2009); (iii) high solids concentration is desirable because in addition to help the process economics, less degradation of xylose to furfural is obtained, and these solids act as a barrier that protects monomers from degradation, though the main drawback of is slow reaction rates (Morinelly et al., 2009); (iv) total carbohydrates yields are lower at pilot scale than they are in laboratory scale, demonstrating the importance of pilot plant scale measurements in the scaling up of the process (Schell et al., 2003); (v) countercurrent shrinking bed reactors result in solubilisation of both cellulose and hemicellulase, but selectivity of desired monomers is better than with other reactors because of lower residence times (Lee et al., 2000); and (vi) no direct comparisons among different studies can be made due to important differences in reaction conditions.

4.2 Alkaline

Alkali delignification of lignocellulosic biomass is widely applied at industrial scale. The process is known as soda and kraft pulping and its chemistry has been extensively discussed (Alen, 2000; Klinke et al., 2002; Sjöström, 1981). Frequently, an oxidative agent is applied at high temperatures obtaining an important enhancement of the pretreatment effects. Sodium, potassium, calcium, and ammonium hydroxides are suitable for this type of pretreatment. Among these, calcium hydroxide is the least expensive and can be easily regenerated using the lime Kiln technology; however, at industrial scale, sodium hydroxide is widely preferred (Sanchez, 2007). Lime pretreatment has proven to be a useful method for selectively reducing the lignin content of lignocellulosic biomass without significant loss in carbohydrates, thus realizing an important increase in biodigestibility (Chang et al., 1997; Kim & Holtzapple, 2005; Sierra, 2005; Sierra et al., 2009a; 2009b). With ammonium hydroxide two main processes have been developed: Ammonia Freeze Explosion (AFEX) and Ammonia Recycle Percolation (ARP). In AFEX, biomass is mixed with liquid ammonia and then treated at high-pressure. When pretreatment time is elapsed, the pressure is suddenly released. The combined chemical and physical effects of this pretreatment increase lignocelluloses pore size making it digestible (Shao et al., 2010). There are many adjustable parameters in the AFEX process: ammonia loading, water loading, temperature, time, blow down pressure, and number of treatments (Holtzapple et al., 1991). In ARP the biomass is pretreated with aqueous ammonia in a flow-through column reactor. The liquid flows at high temperature through the reactor column, which has been previously packed with biomass. To prevent flash evaporation the reactor system must be slightly pressurized (Kim et al., 2003). After reaction the solid fraction, rich in cellulose and hemicellulose, is separated from the liquid.

Another pretreatment ranked within this category is referred to as *wet alkaline oxidation* or *catalytic wet oxidation*. The alkali agents reported are NaOH and Na_2CO_3. This is not to be confused with *wet oxidation* pretreatment which does not use an alkali. The oxidative agent may be oxygen, air, or hydrogen peroxide (Klinke et al., 2002). Alkaline wet oxidation uses temperatures above 180°C, pressures in the range 1.4 - 1.6 MPa, and reaction times up to 30 min.

4.2.1 Reactions during alkaline pretreatment

The major effect of alkaline pretreatment is delignification. Alkaline pretreatments successfully increase lignocelluloses digestibility without the production of furfural and methylfurfural (Harmsen et al., 2010; Sierra et al., 2008). Concurrently, acetyl removal is obtained, which is advantageous because acetyl groups inhibit fermentation (Wyman et al., 2009). Alkaline hydrolysis mechanism is based on saponification of intermolecular ester bonds crosslinking xylan hemicelluloses and lignin (Sun & Cheng, 2002).

Although lignin degradation has been the subject of much study over many years, it is far from being completely understood. Complete biomass delignification is difficult because of its location within the deep cell wall, hydrophobicity, physical stiffness, strong polyring bonds of C-O-C, C-C, and the tendency of recondensation (Kim et al., 2003). Fortunately, complete delignification is not required to make biomass fully digestible (Sierra et al., 2009b; Sierra et al., 2010).

Alkaline depolymerization of lignin mostly depends on the cleavage of two types of aryl ether bonds: $C_{aliphatic}$—O—$C_{aromatic}$ and $C_{aromatic}$—O— $C_{aromatic}$ (ordered from least to most stable), which frequently correspond to α- and β-aryl ether bonds (50-70% in wood). Examples of these typical delignification reactions (only OH^- anions involved) are presented in Fig. 8. Oxidative agents importantly enhance the effects of alkaline pretreatments. Oxygen is relatively unreactive; however, in alkaline media it is reduced through the reaction with phenolic hydroxyl groups to superoxide radical ($-O_2\bullet$). The production of these groups requires very basic conditions (pH >12). Reactions involved in alkaline oxidative pretreatments are primarily single-electron (radical) reactions. Delignification reactions involve the formation of several different acids that introduce hydrophilic groups into the lignin structure. Nucleophilic attack also occurs in some extend causing ring opening, which promotes further degradation and solubilization. Condensation products may leave remaining lignin unreactive in the oxidative alkaline media.

Because lignin fractions contain reactive groups, undesirable condensation reactions may occur between lignin entities retarding delignification. This is known to occur mostly in terminal phases of delignification processes and at the unoccupied C-5 position of phenolic units (Sjöström, 1981).

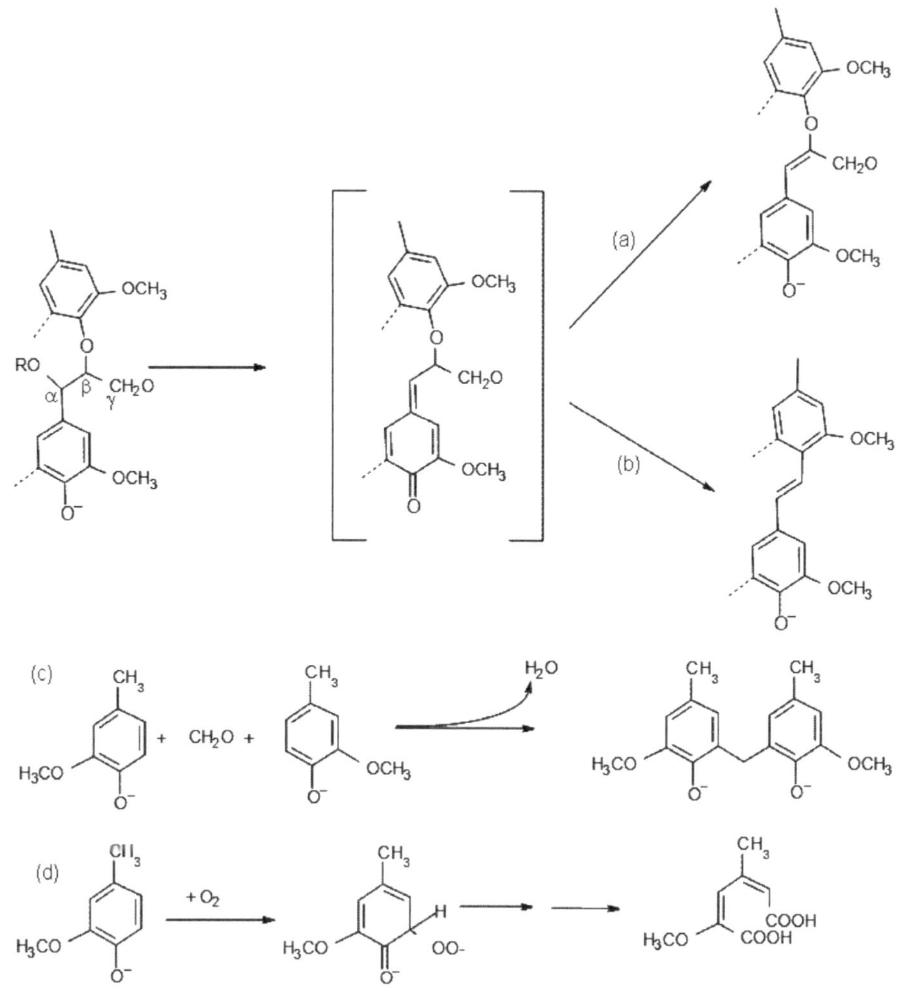

Fig. 8. Lignin degradation reactions in alkaline conditions involving α- and β-aryl ether linkages (a) cleavage of α-aryl ether linkage (b) cleavage of CH_2O group (c) example of a possible condensation reaction. (d) Example of alkaline oxygen degradation of lignin (Gierer & Stockholm, 1985; Guay, 2000).

Unfortunately, alkaline delignification is not completely selective. A fraction of carbohydrates in the presence of alkali and oxygen undergo both oxidation and alkaline degradation generating a complex mixture of products. A wide variety of organic acids are produced as a result of carbohydrates degradation during alkaline pretreatment (Klinke et al., 2002). Due to this fact, an important decrease in pH may take place during pretreatment, particularly if initial alkali concentration is not high. If pretreatment starts with a high initial alkali concentration, pH is approximately constant and first order reaction rates of cellobiose degradation are obtained (Bonn et al., 1985).

4.3.2 kinotic modeling alkaline pretreatment

Kinetic models applied to oxygen bleaching of paper pulp focus on the degradation of polymers, either lignin or carbohydrates. Even though there is complex modeling (Kopelman, 1988), most models separate different moieties that degrade at three different rates: rapid, medium, and slow (de Groot et al., 1994; Susilo & Bennington, 2007). These models were successfully applied to lignin and carbohydrate degradation of lime pretreatment of sugar cane bagasse (Granda, 2005), corn stover (Kim & Holtzapple, 2006), and poplar wood (Sierra et al., 2011a; Sierra et al., 2011b).

Two parallel, first-order reactions accurately represent delignification and carbohydrate degradation. They consider differing reactivities for each biomass component (lignin, cellulose, and hemicellulose). This model successfully represents literature data (Sierra et al., 2011a).

$$-\frac{dY_i}{dt} = k_{if} M_{O_2}{}^{\beta if} Y_{if} + k_{is} M_{O_2}{}^{\beta is} Y_{is} \tag{4}$$

Where

$$k_{ij} = a_{ij} \exp\left(-\frac{E_{ij}}{RT}\right) \tag{5}$$

and

k_{ij} = rate constant for component i (L, G o X) and moiety j (f, s) (min^{-1})

a_{ij} = frequency factor for component i (L, G o X) and moiety j (f, s) (min^{-1})

E_{ij} = activation energy for component i (L, G o X) and moiety j (f, s) (kJ/mol)

R = ideal gas constant (8.314×10^{-3} kJ/(mol·K))

T = absolute temperature (K)

M_{O2} = initial oxygen charge (kg initial oxygen/kg initial dry biomass)

β_{ij} = exponent for component i (L, G o X) and moiety j (f, s) (dimensionless)

The models allow the calculation of selectivity, defined for glucan with equation 6. Glucan and xylan selectivity measure the ability of pretreatment to remove lignin while retaining carbohydrates.

$$S_{dG} \equiv \frac{dY_L/dt}{dY_G/dt} = \frac{a_{Lf}\exp\left(-\frac{E_{Lf}}{RT}\right) M_{O2}^{\beta Lf} Y_{Lf} + a_{Ls}\exp\left(-\frac{E_{Ls}}{RT}\right) M_{O2}^{\beta Ls} Y_{Ls}}{a_{Gf}\exp\left(-\frac{E_{Gf}}{RT}\right) M_{O2}{}^{\beta Gf} Y_{Gf} + a_{Gs}\exp\left(-\frac{E_{Gs}}{RT}\right) M_{O2}{}^{\beta Gs} Y_{Gs}} \tag{6}$$

Using this definition it is possible to calculate the pretreatment conditions that result in a desired biomass composition.

4.3 Organosolv

Organosolv pretreatment is similar to organosolv pulping. The main purpose is lignin separation from the carbohydrates matrix. This pretreatment is effective removing biomass recalcitrance because it achieves almost complete separation of biomass components (Zhao et al., 2009). Three separate fractions are obtained: solid lignin, an aqueous stream, and high purity cellulose. Lignin is recovered unaltered; thus, it has high potential for other applications. In the aqueous phase hemicelluloses are either hydrolysed to xylose monomers or degraded (Johanson et al., 1987).

Some of the most commonly used solvents are low boiling point methanol, ethanol, and acetone. Both alcohols are low cost and miscible in water. At 180 °C, the addition of a catalyst is required (at ~0.2% concentration). Both mineral acids (i.e., hydrochloric, sulfuric and phosphoric) and organic acids (i.e., formic, oxalic, acetylsalicylic, and salicylic) may be used as catalysts to accelerate delignification and xylan degradation (Zhao et al., 2009); however, acids are not required if the pretreatment is conducted at temperatures in the range 190 to 210 °C because liquid hot water is acidic (Duf & Murray, 1996). In alcohol organosolv, the solvent concentration is 50-80% with the highest concentration for hard to pretreat biomass. Usually, pretreatment temperatures for methanol are lower than for ethanol, but ethanol is safer due to its lower toxicity. The Lignol technology uses ethanol as solvent and converts a range of cellulosic feedstocks into several valuable organic chemicals and fuels (Arato et al., 2005). Using low boiling point solvents, such as ethanol, methanol and acetone, organosolv pretreatment requires high pressures because pretreatment runs at high temperatures. To perform at atmospheric pressure, a variety of high boiling point alcohols may be used including but not limited to ethylene glycol, glycerol, and tetrahydrofurfuryl. Organosolv pretreatment with low boiling point alcohols has successfully run at 240 °C for 4 h. Drawbacks include high-energy consumption for solvents recovery and high costs of solvents (Aziz and Sarkanen 1989).

Other kind of organic compounds such as formic, acetic, and peracetic acids, dimethylsulfoxide, ethers, ketone, and phenols are widely used (Thring et al., 1990). When the solvents are organic acids, subsequent enzymatic hydrolysis may not be as successful as with other solvents, probably due to inhibition caused by acetyl groups (Pan et al., 2006). If peracetic acid is used, pretreatment can be conducted at ambient or mild temperatures (e.g., 80°C) for long times and the addition of NaOH may cause biomass swelling. Compared to acid and alkaline pretreatment at the same conditions, peracetic acid is more effective (Zhao et al., 2008).

Due to important advantages obtained through efficient separation of components, organosolv pre-treatments are the most promising options for a biorefinery. Nevertheless, these type of pre-treatments are still very undeveloped because they require extensive and cumbersome washing steps in order to obtain complete solvents removal, and they are energy intensive if high solvents recovery is to be achieved (Zhao et al., 2009). Removal of solvents is not only required for the process economics but also because their presence may be inhibitory for enzymatic hydrolysis and fermentation (Holtzapple & Humphrey, 1984). If high pressures must be applied, the equipment costs increase substantially.

4.3.1 Reactions during organosolv pretreatment

In organosolv pretreatment, the reactions that take place highly depend on the solvent choice. For example, if alcohols are used, OH^{-} attacks the lignin-hemicellulose acid ester bonds at the α carbons of the lignin monomers as well as the ether and 4-O-methylglucoronic bonds. Of these, cleavages of ether linkages are primarily responsible for lignin breakdown. Also, hydrolysis of the glycosidic bonds in hemicelluloses occurs. If an acid catalyst is employed, degradation of lignin and monosaccharides may happen (Chum et al., 1990).

On the other hand, if peracetic acid is used, the hydroxonium ion is present (OH^+) generated from peracetic acid in acidic media. This ion attacks lignin by ring hydroxylation. Peracetic acid is a powerful oxidizing agent; thus, oxidative demethylation and ring opening may occur. Other reactions are displacement of side chains and cleavage of β-arylether bonds.

5. Biotransformation of lignocellulosic material into alternative fuels

The main interest in the use of agro-industrial by-products and agro-waste streams is the release of the fermentable carbohydrates contained into the lignocellulosic matrix. The cost of lignocellulosic feedstocks is currently appealing because it is lower than any other energy sources. For example, crude oil price ranges from $40 to $80 per barrel, generating an energy price that ranges from $7.1 to $14.2 GJ^{-1}, while the energy price from lignocellulosic ranges from $0 to $3 GJ^{-1} (Zhang, 2008). Nevertheless, the actual cost of fuel production from lignocellulosic feedstock can be higher than conventional processes (e.g. ethanol from sugarcane or corn) generating an expensive fuel (Ferris & Joshi, 2010). This can be caused by increased lignocelluloses prices due to higher demand, biomass conditioning and transportation costs, and the cost and yields of the different process stages like pretreatment, saccharification, detoxification and fermentation (Aden et al., 2002). The National Renewable Energy Laboratory (NREL), in its economical evaluation of biotransforming corn stover to ethanol in a plant that uses simultaneous saccharification and fermentation technology, calculated an ethanol selling price close to $1.07 per gallon (+$0.12/-$0.05) in a plant with an installed capacity of 2000 metric tons of lignocellulose per day. They considered the total project investment, the variable operating cost and fixed operating cost for a plant life of 20 years and an ethanol production of 69.5 millions of gallons per year. They showed that the most expensive item in the process is the feedstock cost followed by the pretreatment cost. Further details can be consulted in the NREL report (Aden et al., 2002). Technological alternatives to Simultaneous Saccharification and Fermentation may even result in more favorable prices.

Transformation of highly oxygenated compounds like carbohydrates allows the production of alcohols, carboxylic acids, and esters. Compounds like fragrances solvents, and lubricants may be produced from lignocellulosic feedstocks, and bio-polymers such as polylactic acid and poly(trimethylene terephthalate) that use lactic acid and 1,3-propanediol as precursors, respectively, can be obtained through carbohydrates fermentations (Ragauskas et al., 2006). Among the different chemical compounds that can be used as fuels, alcohols are of a particular interest due to characteristics like a high octane number, broad flammability limits, high flame speeds, and high vaporization heats. These properties have allowed blending ethanol, one of the most important alcohols, with gasoline in mixtures that contain up to 85% ethanol (E85)(Turner et al., 2011). Currently, ethanol and methanol are the most important alcohols used in internal engine combustion. Although these alcohols are obtained after fermentation of the free sugars that result from the depolymerization of carbohydrate polymers, alcohols also can be obtained by direct fermentation of the released carbohydrate polymers. Fig. 9 summarizes the process stages in the widely studied Simultaneous Saccharification and Fermentation Technology.

During pretreatment toxic compounds are released in a concentration that may vary widely. These compounds have been grouped in four categories: (i) sugar degradation products, (ii) lignin degradation products, (iii) compounds derived from cellulose structure, and (iv) heavy metal ions released by corrosion of the hydrolysis equipments. These compounds include but are not limited to furfural and hydroxymethyl furfural formed by the degradation of pentoses and hexoses; acetic, formic, galacturonic and glucuronic acids formed during the hydrolysis of hemicellulose; and aromatic and phenolic compounds, like cinnamaldehyde, p-hydroxybenzaldehyde, and syringaldehyde (Cardona et al., 2010; Mussatto & Roberto, 2004; Sánchez & Cardona, 2008).

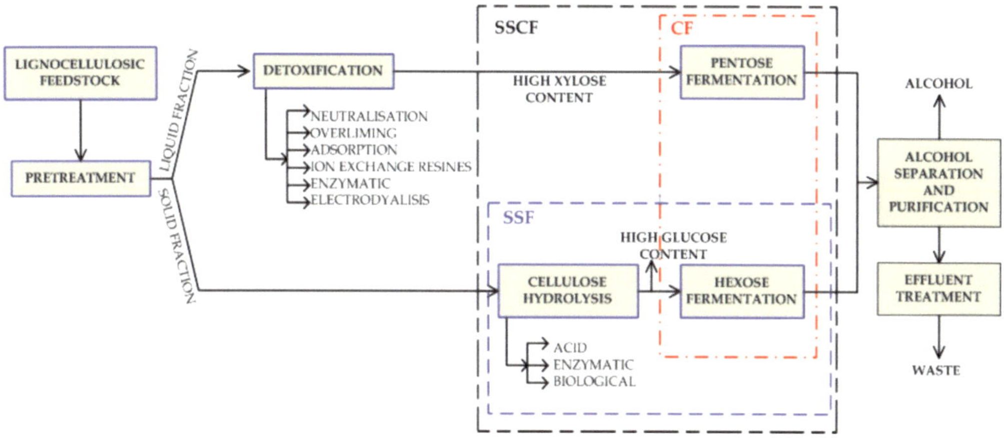

Fig. 9. Process scheme of the lignocellulosic biomass transformation to liquid fuels, mainly alcohols. SSF: simultaneous saccharification and fermentation, CF: co-fermentation, and SSCF: simultaneous saccharification and co-fermentation (Adapted from Cardona et al. (2010))

The detoxification stage, in which most of the inhibitory compounds formed in the pre-treatment stage are removed, is important for the further fermentation stage. This process can be conducted by physical and chemical or biological means. The biological detoxification (BD) may be advantageous because of minimal generation of waste streams, chemical inputs are not required, possible recharging of adsorption resins is not needed, and it is suitable for solid-liquid mixtures (Nichols et al., 2010). The participation of peroxidases and laccases in BD processes has been evaluated. Lacasses have completely removed phenol compounds of low molecular weight by an oxidative polymerization mechanism without varying the acetic acid and furans composition in the hydrolyzate (Chandel et al., 2007). The use of microorganisms like the fungi *Thrichoderma reesei* and *Aspergillus nidulas* to remove acetic and formic acids, furfural and hydroxymethyl furfural, and acid benzoic derivatives has been also reported (Palmqvist & Hahn-Hägerdal, 2000; Yu et al., 2011). Finally, another important but not widely reported method is the adaptation of microorganisms to the hydrolyzate. This method is based in the re-use of the microorganism in successive hydrolizate treatments where the microorganism of one assay is used as inoculum for the next one (Silva & Roberto, 2001). Classical physical-chemical processes are based on toxic compound precipitation, adsorption onto charcoal, diatomaceous earth, ion-exchange resins, and pH adjustment of the extractive (Mussatto & Roberto, 2004). Moroever, new alternatives that use membranes have emerged. One example is the use of reactive membrane extraction in which is avoid the dispersion of one phase in the other reducing the presence of organic compounds, or solvents, in the aqueous phase or hydrolyzate (Simonne et al., 1997).

Carbohydrates depolymerization widely known as hydrolysis or saccharification is carried out by using cellulolytic enzymes complexes constituted by endoglucanases, exoglucanases or cellobiohydrolases, and cellobiases (Bansal et al., 2009; Lee, 1997). Endoglucanases (EC 3.2.1.4, endo-1,4-D-glucanohydrolase) create free chain-ends attacking low crystallinity regions in the cellulose fiber by cleaving α-1,4-glycosidic bonds. Exoglucanases (EC 3.2.1.91, 1,4-b-D-glucan cellobiohydrolase) remove cellobiose units from the free chain-ends providing a further hydrolysis of the molecule. Finally, the cellobiase or β-glucosidase (EC 3.2.1.21) hydrolyses the cellobiose to glucose. Bansal et al. (2009) reviewed the mechanism and mathematical modelling of cellulose hydrolysis by cellulases where the mechanism main steps are: (i) cellulose adsorption onto the substrate via the binding domain, (ii) location of a suitable bond to hydrolysis on the substrate surface, usually the end of the chain or cleavable bond if cellobiohydrolase or endoglucanase is presented, respectively, (iii) formation of enzyme–substrate complex to initiate hydrolysis, (iv) hydrolysis of the α-glycosidic bond and simultaneous forward sliding of the enzyme along the cellulose chain, (v) desorption of cellulases from the substrate or repetition of step iv or steps ii and iii if the catalytic domain detaches from chain, vi) hydrolysis of cellobiose to glucose if α-glucosidase. The main product of the cellulose hydrolysis is glucose, which can be further fermented or chemically transformed. In this stage, the most important parameters to be controlled are reaction time, temperature, pH, enzyme dosage and substrate load (Sánchez & Cardona, 2008).

The last biological stage is fermentation. Streams from the detoxification and/or saccharification stages have a high xylose and glucose concentration. These sugars might be bio-converted to fuels by several microorganisms, including *Candida albicans, Candida shehatae, Kluveromyces fragilis, Kluveromyces marxianus, Pichia stipitis, Saccharomyces cerevisiae, Bacillus subtilis, Bacillus stearothermophilus, Clostridium cellulolyticum, Clostridium beijerinckii, Escherichia coli,* and *Zymomonas mobilis*, which can produce compounds like ethanol, glycerol, butanol, and acetate, among others depending on their metabolic networks. Nevertheless, important differences in the metabolic transformation of xylose a C-5 carbohydrate and glucose a C-6 carbohydrate between bacteria and fungi cause that not all native microorganisms are capable of metabolizing both carbohydrates to ethanol.

It is known that bacteria can transform xylose to xylulose by using xylose isomerase, while most yeasts, fungi, plants, and animals use aldose (xylose) reductase and xylitol dehydrogenase with xylitol as an intermediate (Chiang & Knight, 1960). Nevertheless, if the first reaction step uses NADPH as a cofactor, the reaction is tied to NADPH production, with a subsequent NAD^+ reduction, which can cause a cofactor imbalance under oxygen limitation (Bruinenberg et al., 1983). For example, Toivari (2004) reported an endogenous xylose metabolic pathway in *Saccharomyces cerevisiae* which in aerobic conditions generated a xylitol accumulation into cells caused by lack of NAD^+, but when xylitol is released to the medium the ethanol production yield decreases (Eliasson et al., 2000; Eliasson et al., 2001). Due to these metabolic drawbacks, genetically modified microorganisms capable to ferment both C-5 and C-6 carbohydrates are the focus of attention to improve ethanol production yields. Extensive literature use *S. cerevisiae* as biological model due to its tolerance to media up to 20 $\%^v/_v$ in ethanol among other toxic compounds released during its growth. Also, *Z. mobilis* is amply considered since it has been genetically modified to obtain ethanol by co-fermentation of several substrates like glucose, xylose and arabinose. This topics have been reviewed elsewhere (Kambam & Henson, 2010; Vinuselvi et al., 2011).

For the production of liquid biofuels, mainly ethanol, sequential or simultaneous operational configurations between the hydrolysis or saccharification and fermentation stages have been adopted. Simultaneous saccharification-fermentation (SSF) processes present better ethanol production yields and less energy consumption than sequential hydrolysis-fermentation (SHF) (Sánchez & Cardona, 2008), fermentation time is shorter than SHF, contamination risk with external microorganism is reduced, and presents an important cost reduction since cellulose hydrolysis occurs during glucose fermentation that decreases the inhibition caused by sugars on cellulases. The most important disadvantages of SSF are the difference of optimal temperature for the hydrolysis (45-50°C) and fermentation (28-35°C) stages, the inhibition of the fermenting microorganism and even the cellulose activity caused by the ethanol and the toxic compounds than can be eventually produced during the lignocellulosic biomass pretreatment, and low final concentration of products due to the use of dilute media to obtain suitable rheological properties (Sánchez & Cardona, 2008; Szczodrak & Fiedurek, 1996). The most important advantage of SHF is that hydrolysis and fermentation stages can be carried out at their optimal condition.

The search of new strains or genetically engineered strains that can produce ethanol from xylose allows that through independent fermentations (co-fermentation, CF) of pentoses and hexoses, the ethanol production yield increases (Fig. 9). Nevertheless, the fact that most of the genetically engineered strains to metabolize xylose produce ethanol from glucose has risen in the inclusion of the co-fermentation process to the SSF, which is known as simultaneous saccharification and co-fermentation (SSCF) (Fig. 9). This configuration integrates in only one unit the hydrolysis of cellulose and the fermentation of pentoses. Although economical benefits are evident by using CF and SSCF, challenges emerge because most of xylose-fermenting yeasts (XFY) are not ethanol tolerant; thus, they are prompt to inhibition. New ethanol-tolerant strains should be selected for the CF process, otherwise, joining ethanol removal processes to the SSCF process can help incrementing the ethanol production yield (Chen, 2010).

The direct microbial conversion (DMC) of lignocellulosic biomass to ethanol has also been studied. In DMC, the biodelignification, hydrolysis, and fermentation are done in the same process unit by either one microorganism or a consortium of microorganisms. DMC seems to be an economical process since all the necessary stages for lignocellulosic bioconversion are interconnected. Nevertheless, DMC are slow, and undesired by-products mainly acetic and lactic acids may result when bacteria are used. Another obstacle in DMC is that hydrolysis products such as glucose and cellobiose, can cause inhibition of the cellulase complex limiting the rate of saccharification and final concentration of sugars which result in low ethanol production yields (Mielenz, 2001; Szczodrak & Fiedurek, 1996).

Although most of alternative fuel investigations have focused on biotechnology processes, mainly ethanol production, other chemical compounds of interest useful as fuel, like esters hydrogen, and hydrocarbons can be obtained. In addition, chemical technologies have been rising up as a feasible alternative in which the main goal is to obtain oxygenated hydrocarbons, by removing oxygen from carbohydrates, which can be subsequently transformed to an alternative fuel (Ragauskas et al., 2006).

Despite the fact that biological agents can be involved in all the lignocellulosic transformation stages, few processes employ biological agents in all the stages. More commonly, a combination of biological, chemical, and thermo-chemical processes is employed along the different stages. This results in the production of methanol, butanol, hydrogen, and methane, as important fuel alternatives. Methanol production can be accomplished by (i) biochemical conversion of sugars, (ii) thermo-chemical conversion of biomass by using heat energy and chemical catalyst, (iii) production of syngas from biomass gasification at high temperature and an oxygen-starved environment, and (iv) biomass pyrolysis at high temperature under an oxygen-free environment (Dolan, 2010). Hydrogen production from biomass can be achieved by (i) syngas production and its later transformation by water shift reaction, (ii) autothermal reforming of bio-oils produced by pyrolysis, (iii) aqueous phase reforming of sugars and sugar alcohols, and (iv) biological means, enzymes and microorganisms, in dark fermentations (Tanksale et al., 2010). Methane is produced mainly by anaerobic digestion of the biomass (Li et al., 2011). The catalytic thermochemical processes do not require a delignification stage of biomass, but this is a necessary step to improve the amount of fermentable carbohydrates if the fuel is to be produced by reforming the gasses produced by anaerobic fermentation of these sugars or by any biological means. Thermochemical and chemical catalyst production of these fuels has been reviewed elsewhere (Kalinci et al., 2009; Tanksale et al., 2010).

6. Conclusions

An efficient biorefinery must be capable of using all the components in lignocellulosic wastes resourcefully for the production of fuels in addition to other chemicals and products. Valuable substances contained in these resources must be extracted first. Then, polysaccharides and lignin must be used to produce fuels and other valuable chemicals. Some carbohydrates derived products are: ethanol, carboxylic acids, levulinic acid, glycerol and sorbitol. Lignin, on the other hand, may be used as an energy source through combustion or can be transformed into syngas or bio-oil.

If polysaccharides are to be biologically processed, the lignocelluloses digestibility must be importantly improved. This is achieved by pretreatment, which separate lignin from carbohydrates through chemical or biological means. In some instances, this is achieved by partial delignification or lignin degradation.

Full biological processes are advantageous due to high selectivity and mild reaction conditions, but they are not available yet. Recent studies promote the production of genetically engineered multitask microorganisms able not only to lignocelluloses delignification but also saccharification and fermentation. Additionally, these microorganisms should be marginally inhibited by substrate and product concentration. While this technological dream becomes a scalable economic reality, pretreatments may be carried out by chemical means. In this sense, there is a wide range of possibilities. Advantages and disadvantages of each have been briefly discussed above.

Notable efforts have been conducted to improve ethanol production yields by using lignocellulosic wastes and crops, some of which have been submitted to genetic modifications in the lignin composition. Concurrently, highly desirable integrated processes for non-liquid alternative fuels such as hydrogen are in a very early stage of development. Thus, liquid fuels and products from lignocelluloses can be considered as a bridge between the current highly contaminating petroleum age and the future more sophisticated and clean hydrogen combustion age.

7. References

Abatzoglou, N., Chornet, E. & Belkacemil, K. (1992). Phenomenological kinetics of complex systems: The development of a generalized severity parameter and its application to lignocellulosics fractionation. *Chemical Engineering Sicence*. Vol. 47, No. 5, (April 1992), pp. 1109-1122, ISSN 0009-2509.

Aden, A., Ruth, M., Ibsen, K., Jechura, J., Neeves, K., Sheehan, J. & Wallace, B. (2002). *Lignocellulosic Biomass to Ethanol Process Design and Economics Utilizing Co-Current Dilute Acid Prehydrolysis and Enzymatic Hydrolysis for Corn Stover* (1st ed.), National Renewable Energy Lab, Golden, CO.

Agarwal, U. & Atalla, R. (2010). Vibrational Spectroscopy, In: *Lignin and Lignans*, Heitner, C., Dimmel, D. & Schmidt, J., 103-136, CRC Press, ISBN 978-1-57444-486-5, Boca Raton, FL.

Aitken, M.D. & Irvine, R.L. (1990). Characterization of reactions catalyzed by manganese peroxidase from Phanerochaete chrysosporium. *Archives of Biochemistry and Biophysics*. Vol. 276, No. 2, (February 1990), pp. 405-14, ISSN 0003-9861.

Alen, R. (2000). Basic Chemistry of Wood Delignification., In: *Forest Products Chemistry*, Stenius, P., 59-104, Finnish Paper Engineers' Association, ISBN 9-525-216-039, Helsinki, Finland.

Alvarez, J.M., Canessa, P., Mancilla, R.A., Polanco, R., Santibanez, P.A. & Vicuna, R. (2009). Expression of genes encoding laccase and manganese-dependent peroxidase in the fungus Ceriporiopsis subvermispora is mediated by an ACE1-like copper-fist transcription factor. *Fungal Genetics and Biology*. Vol. 46, No. 1, (January 2009), pp. 104-11, ISSN 1096-0937.

Ambert-Balay, K., Dougherty, M. & Tien, M. (2000). Reactivity of manganese peroxidase: site-directed mutagenesis of residues in proximity to the porphyrin ring. *Archives of Biochemistry and Biophysics*. Vol. 382, No. 1, (October 2000), pp. 89-94, ISSN 0003-9861.

Arato, C., Pye, E. & Gjennestad, G. (2005). The lignol approach to biorefining of woody biomass to produce ethanol and chemicals. *Applied Biochemistry and Biotechnology*. Vol. 123, No. 1-3, (March 2005), pp. 871-882, ISSN 0273-2289.

Arora, D.S. & Sharma, R.K. (2010). Ligninolytic Fungal Laccases and Their Biotechnological Applications. *Applied Biochemistry and Biotechnology*. Vol. 160, No. 6, (March 2010), pp. 1760-1788, ISSN 0273-2289.

Bansal, P., Hall, M., Realff, M.J., Lee, J.H. & Bommarius, A.S. (2009). Modeling cellulase kinetics on lignocellulosic substrates. *Biotechnology Advances*. Vol. 27, No. 6, (November 2009), pp. 833-848, ISSN 0734-9750.

Betini, J.H., Michelin, M., Peixoto-Nogueira, S.C., Jorge, J.A., Terenzi, H.F. & Polizeli, M.L. (2009). Xylanases from Aspergillus niger, Aspergillus niveus and Aspergillus ochraceus produced under solid-state fermentation and their application in cellulose pulp bleaching. *Bioprocess and Biosystems Engineering*. Vol. 32, No. 6, (October 2009), pp. 819-24, ISSN 1615-7605.

Bezalel, L., Shoham, Y. & Rosenberg, E. (1993). Characterization and delignification activity of a thermostable α-L-arabinofuranosidase from Bacillus stearothermophilus.

Applied Microbiology and Biotechnology. Vol. 40, No. 1, (January 1993), pp. 57-62, ISSN 0175-7598.

Boerjan, W., Ralph, J. & Baucher, M. (2003). Lignin Biosynthesis. *Annual Review of Plant Biology*. Vol. 54, No. 1, (June 2003), pp. 519–46, ISSN 1543-5008.

Bonn, G., Binder, H., Leonhard, H. & Bobleter, O. (1985). The alkaline degradation of cellobiose to glucose and fructose. *Monatshefte für Chemie / Chemical Monthly*. Vol. 116, No. 8, pp. 961-971, 0026-9247.

Börjesson, P. & Tufvesson, L.M. (2011). Agricultural crop-based biofuels - resource efficiency and environmental performance including direct land use changes. *Journal of Cleaner Production*. Vol. 19, No. 108–120, (January 2011), ISSN 0959-6526.

Bourbonnais, R., Paice, M.G., Freiermuth, B., Bodie, E. & Borneman, S. (1997). Reactivities of various mediators and laccases with kraft pulp and lignin model compounds. *Applied and Environmental Microbiology*. Vol. 63, No. 12, (December 1997), pp. 4627-32, ISSN 0099-2240.

Bringezu, S., Ramesohl, S., Arnold, K., Fischedick, M., von Geibler, J., Liedtkeand, C. & Schütz, H. (2007). Towards a sustainable biomass strategy. What we know and what we should know. *Wuppertal Institute for Climate, Environment and Energy*. Vol. 163, (June 2007), ISSN 0949-5266.

Brito, J.O., Silva, F.G., Leão, M.M. & Almeida, G. (2008). Chemical composition changes in eucalyptus and pinus woods submitted to heat treatment. *Bioresource Technology*. Vol. 99, No. 18, (December 2008), pp. 8545-8548, ISSN 0960-8524.

Brown, J.A., Alic, M. & Gold, M.H. (1991). Manganese peroxidase gene transcription in Phanerochaete chrysosporium: activation by manganese. *Journal of Bacteriology*. Vol. 173, No. 13, (July 1991), pp. 4101-6, ISSN 0021-9193.

Bruinenberg, P.M., de Bot, P.H.M., van Dijken, J.P. & Scheffers, W.A. (1983). The role of redox balances in the anaerobic fermentation of xylose by yeasts. *European Journal of Applied Microbiology and Biotechnology*. Vol. 18, No. 5, (September 1983), pp. 287–292, ISSN 0175-7598.

Cardona, C.A., Quintero, J.A. & Paz, I.C. (2010). Production of bioethanol from sugarcane bagasse: Status and perspectives. *Bioresource Technology*. Vol. 101, No. 13, (July 2010), pp. 4754-4766, ISSN 0960-8524.

Carrasco, C., Baudel, H.M., Sendelius, J., Modig, T., Roslander, C., Galbe, M., Hahn-Hägerdal, B., Zacchi, G. & Lidén, G. (2010). SO2-catalyzed steam pretreatment and fermentation of enzymatically hydrolyzed sugarcane bagasse. *Enzyme and Microbial Technology*. Vol. 46, No. 2, (February 2010), pp. 64-73, ISSN 0141-0229.

Cassman, K.G. & Liska, A.J. (2007). Food and fuel for all: Realistic or foolish? *Biofuels, Bioproducts, and Biorefining*. Vol. 1, No. 1, (September 2007), pp. 18–23, ISSN 1932-1031.

Catal, T., Liu, H. & Bermek, H. (2008). Selenium induces manganese-dependent peroxidase production by the white-rot fungus Bjerkandera adusta (Willdenow) P. Karsten. *Biological Trace Element Research*. Vol. 123, No. 1-3, (June 2008), pp. 211-7, ISSN 0163-4984.

Chandel, A.K., Kapoor, R.K., Singh, A. & Kuhad, R.C. (2007). Detoxification of sugarcane bagasse hydrolysate improves ethanol production by Candida shehatae NCIM 3501. *Bioresource Technology*. Vol. 98, No. 10, (July 2007), pp. 1947-1950, ISSN 0960-8524.

Chang, V., Burr, B. & Holtzapple, M. (1997). Lime pretreatment of switchgrass. *Applied Biochemistry and Biotechnology*. Vol. 63-65, No. 1, (March 1997), pp. 3-19, ISSN 0273-2289.

Chang, V. & Holtzapple, M. (2000). Fundamental factors affecting biomass enzymatic reactivity. *Applied Biochemistry and Biotechnology*. Vol. 84-86, No. 1, (March 2000), pp. 5-37, ISSN 0273-2289.

Chen, R., Lee, Y. & Torget, B. (1996). Kinetic and modeling investigation on two-stage reverse-flow reactor as applied to dilute-acid pretreatment of agricultural residues. *Applied Biochemistry and Biotechnology*, No. 57/58, (March 1996), pp. 133-146, ISSN 0273-2289.

Chen, S., Mowery, R., Castleberry, V., Walsum, P. & Chamvliss, K. (2006). High-performance liquid chromatography method for simultaneous determination of aliphatic acid, aromatic acid and neutral degradation products in biomass pretreatment hydrolysates. *Journal of Chromatography*, No. 1004, (February 2006), pp. 54-61, ISSN 0021-9673.

Chen, Y. (2010). Development and application of co-culture for ethanol production by co-fermentation of glucose and xylose: a systematic review. *Journal of Industrial Microbiology and Biotechnology*. Vol. 38, No. 5, (May 2010), pp. 581-597, ISSN 1367-5435.

Chiang, C. & Knight, S.G. (1960). The metabolism of D-xylose by moulds. *Nature*. Vol. 188, No. 79–81, (October 1960), ISSN 0028-0836.

Cho, D.H., Shin, S.-J., Bae, Y., Park, C. & Kim, Y.H. (2011). Ethanol production from acid hydrolysates based on the construction and demolition wood waste using Pichia stipitis. *Bioresource Technology*. Vol. 102, No. 6, (March 2011), pp. 4439-4443, ISSN 0960-8524.

Chum, H., Johnson, D. & Black, S. (1990). Organosolv pretreatment for enzymic hydrolysis of poplars. 2. Catalyst effects and the combined severity parameter. *Industrial and Engineering Chemistry Research*. Vol. 29, No. 29, (February 1990), pp. 156-162, ISSN 0888-5885.

Clough, R.C., Pappu, K., Thompson, K., Beifuss, K., Lane, J., Delaney, D.E., Harkey, R., Drees, C., Howard, J.A. & Hood, E.E. (2006). Manganese peroxidase from the white-rot fungus Phanerochaete chrysosporium is enzymatically active and accumulates to high levels in transgenic maize seed. *Plant Biotechnology Journal*. Vol. 4, No. 1, (January 2006), pp. 53-62, ISSN 1467-7652.

Costa, S.M., Gonçalves, A.R. & Esposito, E. (2002). Action of white-rot fungus Panus tigrinus on sugarcane bagasse: Evaluation of selectivity. *Applied Biochemistry and Biotechnology - Part A Enzyme Engineering and Biotechnology*. Vol. 98-100, (March 2002), pp. 357-364, ISSN 0273-2289.

Costa, S.M., Gonçalves, A.R. & Esposito, E. (2005). Ceriporiopsis subvermispora used in delignification of sugarcane bagasse prior to soda/anthraquinone pulping. *Applied Biochemistry and Biotechnology - Part A Enzyme Engineering and Biotechnology*. Vol. 122, No. 1-3, (March 2005), pp. 695-706, ISSN 0273-2289.

Crawford, D.L., Doyle, J.D., Wang, Z., Hendricks, C.W., Bentjen, S.A., Bolton, H., Jr., Fredrickson, J.K. & Bleakley, B.H. (1993). Effects of a lignin peroxidase-expressing recombinant, Streptomyces lividans TK23.1, on biogeochemical cycling and the

numbers and activities of microorganisms in soil. *Applied and Environmental Microbiology*. Vol. 59, No. 2, (February 1993), pp. 508-18, ISSN 0099-2240.
Daniel, G., Nilsson, T. & Pettersson, B. (1989). Intra- and Extracellular Localization of Lignin Peroxidase during the Degradation of Solid Wood and Wood Fragments by Phanerochaete chrysosporium by Using Transmission Electron Microscopy and Immuno-Gold Labeling. *Applied and Environmental Microbiology*. Vol. 55, No. 4, (April 1989), pp. 871-81, ISSN 0099-2240.
de Groot, B., Van Dam, J.E.G., van der Zwan, R.P. & Riet, K.v. (1994). Simplified Kinetic Modelling of Alkaline Delignification of Hemp Woody Core. *Holzforschung*. Vol. 48, No. 3, (March 1994), pp. 207-214, ISSN 0018-3830.
Dekker, R.F., Barbosa, A.M., Giese, E.C., Godoy, S.D. & Covizzi, L.G. (2007). Influence of nutrients on enhancing laccase production by Botryosphaeria rhodina MAMB-05. *International Microbiology*. Vol. 10, No. 3, (September 2007), pp. 177-85, ISSN 1139-6709.
Diamantidis, G., Effosse, A., Potier, P. & Bally, R. (2000). Purification and characterization of the first bacterial laccase in the rhizospheric bacterium Azospirillum lipoferum. *Soil Biology and Biochemistry*. Vol. 32, No. 7, (July 2000), pp. 919-927 ISSN 0038-0717.
Dias, A.A., Freitas, G.S., Marques, G.S., Sampaio, A., Fraga, I.S., Rodrigues, M.A., Evtuguin, D.V. & Bezerra, R.M. (2010). Enzymatic saccharification of biologically pre-treated wheat straw with white-rot fungi. *Bioresourse Thecnology*. Vol. 101, No. 15, (August 2010), pp. 6045-50, ISSN 1873-2976.
Dimmel, D. (2010). Overview, In: *Lignin and Lignans*, Heitner, C., Dimmel, D. & Schmidt, J., 1-10, CRC Press, ISBN 978-1-57444-486-5, Boca Raton, FL.
Dolan, G.A. (2010). *Methanol Production and Utilization* (1st ed.), Blackwell Publishing Ltd., ISBN 9780-470-750-025, Hoboken, NJ.
Dorado, J., Almendros, G., Camarero, S., Martínez, A.T., Vares, T. & Hatakka, A. (1999). Transformation of wheat straw in the course of solid-state fermentation by four ligninolytic basidiomycetes. *Enzyme and Microbial Technology*. Vol. 25, No. 7, (October 1999), pp. 605-612, ISSN 0141-0229.
Duf, S. & Murray, W. (1996). Bioconversion of forest products industry waste cellulosics to fuel ethanol: A review. *Bioresource Technology*. Vol. 55, No. 1, (n.d.), pp. 1-33, ISSN 0960-8524
Dwivedi, P., Vivekanand, V., Pareek, N., Sharma, A. & Singh, R.P. (2010). Bleach enhancement of mixed wood pulp by xylanase-laccase concoction derived through co-culture strategy. *Applied Biochemistry and Biotechnology*. Vol. 160, No. 1, (January 2010), pp. 255-268, ISSN 0273-2289.
Dwivedi, U.N., Singha, P., Pandeya, V.P. & Kumara, A. (2011). Structure–function relationship among bacterial, fungal and plant laccases. *Journal of Molecular Catalysis B: Enzymatic*. Vol. 68, No. 2, (Febrary 2011), pp. 117-128, ISSN 1381-1177.
Eggeman, T. & Elander, R.T. (2005). Process and economic analysis of pretreatment technologies. *Bioresource Technology*. Vol. 96, No. 18, (December 2005), pp. 2019-2025, ISSN 0960-8524.
Eliasson, A., Christensson, C., Wahlbom, C.F. & Hahn-Hagerdal, B. (2000). Anaerobic Xylose Fermentation by Recombinant Saccharomyces cerevisiae Carrying XYL1, XYL2, and XKS1 in Mineral Medium Chemostat Cultures. *Applied and Environmental Microbiology*. Vol. 66, No. 8, (August 1, 2000), pp. 3381-3386.

Eliasson, A., Hofmeyr, J.-H.S., Pedler, S. & Hahn-Hägerdal, B. (2001). The xylose reductase/xylitol dehydrogenase/xylulokinase ratio affects product formation in recombinant xylose-utilising Saccharomyces cerevisiae. *Enzyme and Microbial Technology*. Vol. 29, No. 4-5, (September 2001), pp. 288-297, ISSN 0141-0229.

Elisashvili, V. & Kachlishvili, E. (2009). Physiological regulation of laccase and manganese peroxidase production by white-rot Basidiomycetes. *Journal of Biotechnology*. Vol. 144, No. 1, (October 2009), pp. 37-42, ISSN 1873-4863.

Esteghlalian, A., Hashimoto, A., Fenske, J. & Penner, M. (1997). Modeling and optimization of the dilute sulfuric-acid pretreatment of corn stover, poplar and switchgrass. *Bioresource Technology*. Vol. 59, No. 2-3, (February 1997), pp. 129-136, ISSN 0960-8524.

FAO. (2008). Food and Agricultural commodities production. *http://faostat.fao.org/site/339/default.aspx*.

Ferris, J. & Joshi, S. (2010). Prospects for Ethanol and Biodiesel, 2008 to 2017 and Impacts on Agriculture and Food, In: *Handbook of Bioenergy Economics and Policy*, Khanna, M., Scheffran, J. & Zilberman, D., 91-111, Springer, ISBN 978-1-4419-0369-3, New York.

Fissore, A., Carrasco, L., Reyes, P., Rodríguez, J., Freer, J. & Mendonça, R.T. (2010). Evaluation of a combined brown rot decay-chemical delignification process as a pretreatment for bioethanol production from Pinus radiata wood chips. *Journal of Industrial Microbiology and Biotechnology*. Vol. 37, No. 9, (May 2010), pp. 893-900, ISSN 1367-5435.

Frederick Jr, W.J., Lien, S.J., Courchene, C.E., DeMartini, N.A., Ragauskas, A.J. & Iisa, K. (2008). Co-production of ethanol and cellulose fiber from Southern Pine: A technical and economic assessment. *Biomass and Bioenergy*. Vol. 32, No. 12, (December 2008), pp. 1293-1302, ISSN 0961-9534.

Gaitan, I.J., Medina, S.C., Gonzalez, J.C., Rodriguez, A., Espejo, A.J., Osma, J.F., Sarria, V., Almeciga-Diaz, C.J. & Sanchez, O.F. (2011). Evaluation of toxicity and degradation of a chlorophenol mixture by the laccase produced by Trametes pubescens. *Bioresource Technology*. Vol. 102, No. 3, (February 2011), pp. 3632-5, ISSN 1873-2976.

Galbe, M. & Zacchi, G. (2002). A review of the production of ethanol from softwood. *Applied Microbiology and Biotechnology*. Vol. 59, No. 6, (September 2002), pp. 618-628, ISSN 0175-7598.

Galhaup, C., Goller, S., Peterbauer, C.K., Strauss, J. & Haltrich, D. (2002). Characterization of the major laccase isoenzyme from Trametes pubescens and regulation of its synthesis by metal ions. *Microbiology*. Vol. 148, No. Pt 7, (July 2002), pp. 2159-69, ISSN 1350-0872.

Garlock, R.J., Balan, V., Dale, B.E., Ramesh Pallapolu, V., Lee, Y.Y., Kim, Y., Mosier, N.S., Ladisch, M.R., Holtzapple, M.T., Falls, M., Sierra, R., Shi, J., Ebrik, M.A., Redmond, T., Yang, B., Wyman, C.E., Donohoe, B.S., Vinzant, T.B., Elander, R.T., Hames, B., Thomas, S. & Warner, R.E. (2011). Comparative material balances around leading pretreatment technologies for the conversion of switchgrass to soluble sugars. *Bioresource Technology*. Vol. In Press, (n.d.), ISSN 0960-8524.

Ghodake, G.S., Kalme, S.D., Jadhav, J.P. & Govindwar, S.P. (2009). Purification and partial characterization of lignin peroxidase from Acinetobacter calcoaceticus NCIM 2890 and its application in decolorization of textile dyes. *Applied Biochemistry and Biotechnology*. Vol. 152, No. 1, (January 2009), pp. 6-14, ISSN 1559-0291.

Gierer, J. & Stockholm, S. (1985). Chemistry of delignification. Part 1: General concept and reactions during pulping. *Wood Science and Technology*. Vol. 19, (n.d.), pp. 289-312, ISSN 0043-7719.

Glasser, W.G. & Sarkanen, S. (1989). *Lignin* (American Chemical Society, ISBN 0-8412-1631-2, Washington, DC.

Goldemberg, J. (2006). The promise of clean energy. *Energy Policy*. Vol. 34, (Octuber 2006), pp. 2185–2190, ISSN 0301-4215.

Gonçalves, A.R., Esposito, E. & Benar, P. (1998). Evaluation of Panus tigrinus in the delignification of sugarcane bagasse by FTIR-PCA and pulp properties. *Journal of Biotechnology*. Vol. 66, No. 2-3, (December 1998), pp. 177-185, ISSN 0168-1656.

Grohmann, K., Torget, R. & Himmel, M. (1985). Optimization of dilute-acid pretreatment of biomass. *Biotechnology and Bioengineering Symposium*. Vol. 15, (n.d.), pp. 59-80, ISSN 0572-6565.

Gu, L., Lajoie, C. & Kelly, C. (2003). Expression of a Phanerochaete chrysosporium manganese peroxidase gene in the yeast Pichia pastoris. *Biotechnology Progress*. Vol. 19, No. 5, (September 2003), pp. 1403-9, ISSN 8756-7938.

Guay, D.F. (2000). Mechanisms of Oxidative Degradation of Carbohydrates During Oxygen delignification. I. Reaction of Photochemically generated Hydroxyl Radical with Methyl β-D-Glucoside. *Journal of Wood Chemistry and Technology*. Vol. 20, No. 4, (n.d.), pp. 375-394, ISSN 0277-3813.

Guo, G.-L., Hsu, D.-C., Chen, W.-H., Chen, W.-H. & Hwang, W.-S. (2009). Characterization of enzymatic saccharification for acid-pretreated lignocellulosic materials with different lignin composition. *Enzyme and Microbial Technology*. Vol. 45, No. 2, (August 2009), pp. 80-87, ISSN 0141-0229.

Guo, M., Lu, F., Du, L., Pu, J. & Bai, D. (2006). Optimization of the expression of a laccase gene from Trametes versicolor in Pichia methanolica. *Applied Microbiology and Biotechnology*. Vol. 71, No. 6, (August 2006), pp. 848-52, ISSN 0175-7598.

Gupta, R., Mehta, G., Khasa, Y.P. & Kuhad, R.C. (2010). Fungal delignification of lignocellulosic biomass improves the saccharification of cellulosics. *Biodegradation*. Vol. In press, (August 2010), ISSN 0923-9820.

Haddadin, M.S., Al-Natour, R., Al-Qsous, S. & Robinson, R.K. (2002). Bio-degradation of lignin in olive pomace by freshly-isolated species of Basidiomycete. *Bioresource Technology*. Vol. 82, No. 2, (April 2002), pp. 131-137, ISSN 0960-8524.

Hammel, K.E. & Cullen, D. (2008). Role of fungal peroxidases in biological ligninolysis. *Current Opinion in Plant Biology*. Vol. 11, No. 3, (June 2008), pp. 349-55, ISSN 1369-5266.

Harmsen, P., Huijgen, W., Bermúdez-López, L. & Bakker, R. (2010). *Literature Review of Physical and Chemical Pretreatment Processes for Lignocellulosic Biomass* (1st ed.), ISBN 9789-085-857-570, Netherlands.

Hatakeyama, H. & Hatakeyama, T. (2010a). Lignin Structure, Properties, and Applications, In: *Biopolymers*, Abe, A., Dusek, K. & Kobayashi, S., 1-63, Springer, ISBN 9783-642-13629-0, Berlin.

Hatakeyama, H. & Hatakeyama, T. (2010b). Thermal Properties of Isolated and in situ Lignin, In: *Lignin and Lignans*, Heitner, C., Dimmel, D. & Schmidt, J., 301-319, CRC Press, ISBN 978-1-57444-486-5, Boca Raton, FL.

Hatakeyama, T. & Hatakeyama, H. (2005). Lignin, In: *Thermal Properties of Green Polymers and Biocomposites*, Hatakeyama, T. & Hatakeyama, H., 171-215, Springer, ISBN 1-4020-2354-5, Netherlands.

Hodgson, E.M., Lister, S.J., Bridgwater, A.V., Clifton-Brown, J. & Donnison, I.S. (2010). Genotypic and environmentally derived variation in the cell wall composition of Miscanthus in relation to its use as a biomass feedstock. *Biomass and Bioenergy*. Vol. 34, No. 5, (May 2010), pp. 652-660, ISSN 0961-9534.

Holtzapple, M. & Humphrey, A. (1984). The effect of organosolv pretreatment on the enzymic hydrolysis of poplar. *Biotechnology and Bioengineering*. Vol. 26, No. 7, (July 1984), pp. 670-676, ISSN 0006-3592.

Holtzapple, M., Jun, J., Ganesh, A., Srinivas, P. & Dale, B. (1991). The Ammonia Freeze Explosion (AFEX) Process: A Practical Lignocellulose Pretreatment. *Applied Biochemistry and Biotechnology*. Vol. 28-29, No. 1, (March 1991), pp. 59-74, ISSN 0273-2289.

Hong, Y.Z., Zhou, H.M., Tu, X.M., Li, J.F. & Xiao, Y.Z. (2007). Cloning of a laccase gene from a novel basidiomycete Trametes sp. 420 and its heterologous expression in Pichia pastoris. *Current Microbiology*. Vol. 54, No. 4, (April 2007), pp. 260-5, ISSN 0343-8651.

Jiang, F., Kongsaeree, P., Schilke, K., Lajoie, C. & Kelly, C. (2008). Effects of pH and temperature on recombinant manganese peroxidase production and stability. *Applied Microbiology and Biotechnology*. Vol. 146, No. 1-3, (March 2008), pp. 15-27, ISSN 1559-0291.

Jin, S. & Chen, H. (2007). Near-infrared analysis of the chemical composition of rice straw. *Industrial Crops and Products*. Vol. 26, No. 2, (August 2007), pp. 207-211, ISSN 0926-6690.

Johanson, A., Aaltonen, O. & Ylinen, P. (1987). Organosolv pulping - method and pulp properties. *Biomass*. Vol. 13, No. 1, (n.d.), pp. 45-65, ISSN 0144-4565.

Johansson, T. & Nyman, P.O. (1996). A cluster of genes encoding major isozymes of lignin peroxidase and manganese peroxidase from the white-rot fungus Trametes versicolor. *Gene*. Vol. 170, No. 1, (April 1996), pp. 31-8, ISSN 0378-1119.

Johjima, T., Itoh, N., Kabuto, M., Tokimura, F., Nakagawa, T., Wariishi, H. & Tanaka, H. (1999). Direct interaction of lignin and lignin peroxidase from Phanerochaete chrysosporium. *Proceedings of the National Academy of Sciences of the United States of America*. Vol. 96, No. 5, (March 1999), pp. 1989-94, ISSN 0027-8424.

Kalinci, Y., Hepbasli, A. & Dincer, I. (2009). Biomass-based hydrogen production: A review and analysis. *International Journal of Hydrogen Energy*. Vol. 34, No. 21, (November 2009), pp. 8799-8817, ISSN 0360-3199.

Kambam, P.K.R. & Henson, M.A. (2010). Engineering bacterial processes for cellulosic ethanol production. *Biofuels*. Vol. 1, No. 5, (September 2010), pp. 729-743, ISSN 1759-7269.

Kanayama, N., Tohru, S. & Keiichi, K. (2002). Purification and characterization of an alkaline manganese peroxidase from Aspergillus terreus LD-1. *Journal of Bioscience and Bioengineering*. Vol. 93, No. 4, (April 2002), pp. 405-10, ISSN 1389-1723.

Kannan, K., Oblisami, G. & Loganathan, B.G. (1990). Enzymology of ligno-cellulose degradation by Pleurotus sajor-caju during growth on paper-mill sludge. *Biological Wastes*. Vol. 33, No. 1, (January 1990), pp. 1-8, ISSN 0269-7483.

Kaparaju, P., Serrano, M., Thomsen, A.B., Kongjan, P. & Angelidaki, I. (2009). Bioethanol, biohydrogen and biogas production from wheat straw in a biorefinery concept. *Bioresource Technology*. Vol. 100, No. 9, (May 2009), pp. 2562-8, ISSN 1873-2976.

Kaur, A., Mahajan, R., Singh, A., Garg, G. & Sharma, J. (2010). Application of cellulase-free xylano-pectinolytic enzymes from the same bacterial isolate in biobleaching of kraft pulp. *Bioresource Technology*. Vol. 101, No. 23, (July 2010), pp. 9150-9155, ISSN 1873-2976.

Keegstra, K. (2010). Plant cell walls. *Plant Physiology*. Vol. 154, No. 2, (October 2010), pp. 483-6, ISSN 1532-2548.

Khalil, H.P.S.A., Alwani, M.S. & Omar, A.K.M. (2006). Chemical composition, anatomy, lignin distribution, and cell wall structure of Malaysian plant waste fibers. *BioResources*. Vol. 1 No. 2, (November 2006), pp. 220–232., ISSN 1930-2126.

Kim, S. & Holtzapple, M.T. (2005). Lime pretreatment and enzymatic hydrolysis of corn stover. *Bioresource Technology*. Vol. 96, No. 18, (December 2005), pp. 1994-2006, ISSN 0960-8524.

Kim, S. & Holtzapple, M.T. (2006). Delignification kinetics of corn stover in lime pretreatment. *Bioresource Technology*. Vol. 97, No. 5, (March 2006), pp. 778-785, ISSN 0960-8524.

Kim, T.H., Kim, J.S., Sunwoo, C. & Lee, Y.Y. (2003). Pretreatment of corn stover by aqueous ammonia. *Bioresource Technology*. Vol. 90, No. 1, (October 2003), pp. 39-47, ISSN 0960-8524.

Klinke, H.B., Ahring, B.K., Schmidt, A.S. & Thomsen, A.B. (2002). Characterization of degradation products form alkaline wet oxidation of wheat straw. *Bioresource Technology*. Vol. 82, (March 2002), pp. 15-26, ISSN 0960-8524.

Kocheva, L., Karmanov, A., Mironov, M., Belyi, V., Belyaev, V. & Monakov, Y. (2008). Straw lignins: Hydrodynamic and conformational properties of the macromolecules. *Russian Journal of Applied Chemistry*. Vol. 81, No. 11, (November 2008), pp. 2033-2039, ISSN 1070-4272.

Kopelman, R. (1988). Fractal Reaction Kinetics. *Science*. Vol. 241, No. 4873, (September 1988), pp. 1620-1626, ISSN 0036-8075.

Kosugi, A., Kondo, A., Ueda, M., Murata, Y., Vaithanomsat, P., Thanapase, W., Arai, T. & Mori, Y. (2009). Production of ethanol from cassava pulp via fermentation with a surface-engineered yeast strain displaying glucoamylase. *Renewable Energy*. Vol. 34, No. 5, (May 2009), pp. 1354-1358, ISSN 0960-1481.

Kumar, P., Barrett, D., Delwiche, M. & Stroeve, P. (2009). Methods for pretreatment of lignocellulosic biomass for efficient hydrolysis and biofuel production. *Industrial and Engineering Chemistry Research*. Vol. 48, No. 8, (March 2009), pp. 3713-3729, ISSN 0888-5885.

Kumar, R. & Wyman, C. (2009). Does change in accesibility with conversion depend on both the substrate and the pretreatment technology? *Bioresource Technology*, No. 100, (September 2009), pp. 4193-4202, ISSN 0960-8524.

Lee, J. (1997). Biological conversion of lignocellulosic biomass to ethanol. *Journal of Biotechnology*. Vol. 56, No. 1, (July 1997), pp. 1-24, ISSN 0168-1656.

Lee, J. & Jeffries, T. (2011). Efficiencies of acid catalyst in the hydrolysis of lignocellulosic biomass over a range of combined severity factors. *Bioresource Technology*. Vol. In press, (n.d.), ISSN 0960-8524.

Lee, Y., Zhangwen, W. & Torget, R. (2000). Modeling of countercurrent shringking-bed reactor in dilute-acid total-hydrolysis of lignocellulosic biomass. *Bioresource Technology*. Vol. 71, No. 1, (January 2000), pp. 29-39, ISSN 0960-8524.

Li, B.-Z., Balan, V., Yuan, Y.-J. & Dale, B.E. (2010). Process optimization to convert forage and sweet sorghum bagasse to ethanol based on ammonia fiber expansion (AFEX) pretreatment. *Bioresource Technology*. Vol. 101, No. 4, (February 2010), pp. 1285-1292, ISSN 0960-8524.

Li, J., Gunnar, H. & Góran, G. (2007). Lignin depolymerization/repolymerization and its critical role for deligniWcation of aspen wood by steam explosion. *Bioresource Technology*. Vol. 98, (n.d.), pp. 3061-3068, ISSN 0960-8524.

Li, J., Henriksoon, G. & Gellerstedt, G. (2005). Carbohydrate reactions during high-temperature steam treatment of aspen wood. *Applied Biochemistry and Biotechnology*. Vol. 125, No. 3, (June 2005), pp. 175-188, ISSN 0273-2289.

Li, X., Kondo, R. & Sakai, K. (2002). Biodegradation of sugarcane bagasse with marine fungus Phlebia sp. MG-60. *Journal of Wood Science*. Vol. 48, No. 2, (April 2002), pp. 159-162, ISSN 1435-0211.

Li, Y., Park, S.Y. & Zhu, J. (2011). Solid-state anaerobic digestion for methane production from organic waste. *Renewable and Sustainable Energy Reviews*. Vol. 15, No. 1, (January 2011), pp. 821-826, ISSN 1364-0321.

Lu, C., Wang, H., Luo, Y. & Guo, L. (2010). An efficient system for pre-delignification of gramineous biofuel feedstock in vitro: Application of a laccase from Pycnoporus sanguineus H275. *Process Biochemistry*. Vol. 45, No. 7, (July 2010), pp. 1141-1147, ISSN 1359-5113.

Martinez, A.T., Ruiz-Duenas, F.J., Martinez, M.J., Del Rio, J.C. & Gutierrez, A. (2009). Enzymatic delignification of plant cell wall: from nature to mill. *Current Opinion in Biotechnology*. Vol. 20, No. 3, (June 2009), pp. 348-57, ISSN 1879-0429.

Martinez, A.T., Speranza, M., Ruiz-Duenas, F.J., Ferreira, P., Camarero, S., Guillen, F., Martinez, M.J., Gutierrez, A. & del Rio, J.C. (2005). Biodegradation of lignocellulosics: microbial, chemical, and enzymatic aspects of the fungal attack of lignin. *International Microbiology*. Vol. 8, No. 3, (September 2005), pp. 195-204, ISSN 1139-6709.

McMillan, J. Process for pretreating lignocellulosic biomass: a review, In: *National Renewable Energy Laboratory*, Available from: http://www.osti.gov/energycitations/product.biblio.jsp?osti_id=7171656

Meza, J.C., Sigoillot, J.C., Lomascolo, A., Navarro, D. & Auria, R. (2006). New process for fungal delignification of sugar-cane bagasse and simultaneous production of laccase in a vapor phase bioreactor. *Journal of Agricultural and Food Chemis ry*. Vol. 54, No. 11, (May 2006), pp. 3852-8, ISSN 0021-8561.

Mielenz, J.R. (2001). Ethanol production from biomass: technology and commercialization status. *Current Opinion in Microbiology*. Vol. 4, No. 3, (June 2001), pp. 324-329, ISSN 1369-5274.

Mishra, M. & Thakur, I.S. (2010). Isolation and characterization of alkalotolerant bacteria and optimization of process parameters for decolorization and detoxification of pulp and paper mill effluent by Taguchi approach. *Biodegradation*. Vol. 21, No. 6, (April 2010), pp. 967-978, ISSN 0923-9820.

Morinelly, J., Jensen, J., Browne, M., Co, T. & Shonnard, D. (2009). Kinetic characterization of xylose monomer and oligomer concentrations during dilute acid pretreatment of lignocellulosic biomass from forests and switchgrass. *Industrial and Engineering Chemistry Research*. Vol. 48, No. 22, (September 2009), pp. 9877-9884, ISSN 0888-5885.

Mosier, N., Wyman, C., Dale, B., Elander, R., Lee, Y.Y., Holtzapple, M. & Ladisch, M. (2005). Features of promising technologies for pretreatment of lignocellulosic biomass. *Bioresource Technology*. Vol. 96, No. 6, (April 2005), pp. 673-686, ISSN 0960-8524.

Mussatto, S.I. & Roberto, I.C. (2004). Alternatives for detoxification of diluted-acid lignocellulosic hydrolyzates for use in fermentative processes: a review. *Bioresource Technology*. Vol. 93, No. 1, (May 2004), pp. 1-10, ISSN 0960-8524.

Naik, S.N., Goud, V.V., Rout, P.K. & Dalai, A.K. (2010). Production of first and second generation biofuels: A comprehensive review. *Renewable and Sustainable Energy Reviews*. Vol. 14, (February 2010), pp. 578–597, ISSN 1364-032.

Nanguneri, D. & Hester, R. (1990). Acid/sugar separation using ion exclusion resins: A process analysis and a design. *Separation Science and Technology*. Vol. 25, No. 13-15, (n.d.), pp. 1829-1842, ISSN 1520-5754.

Nichols, N.N., Dien, B.S. & Cotta, M.A. (2010). Fermentation of bioenergy crops into ethanol using biological abatement for removal of inhibitors. *Bioresource Technology*. Vol. 101, No. 19, (October 2010), pp. 7545-7550, ISSN 0960-8524.

Nie, G., Reading, N.S. & Aust, S.D. (1998). Expression of the lignin peroxidase H2 gene from Phanerochaete chrysosporium in Escherichia coli. *Biochemical and Biophysical Research Communications*. Vol. 249, No. 1, (August 1998), pp. 146-50, ISSN 0006-291X.

Osono, T. & Takeda, H. (2006). Fungal decomposition of Abies needle and Betula leaf litter. *Mycologia*. Vol. 98, No. 2, (March 2006), pp. 172-179, ISSN 0027-5514.

Palmqvist, E. & Hahn-Hägerdal, B. (2000). Fermentation of lignocellulosic hydrolysates. I: inhibition and detoxification. *Bioresource Technology*. Vol. 74, No. 1, (August 2000), pp. 17-24, ISSN 0960-8524.

Pan, X., Gilkes, N. & Saddler, J. (2006). Effect of acetyl groups on enzymatic hydrolysis of cellulosic substrates. *Holzforschung*. Vol. 60, No. 4, (n.d.), pp. 398-401, ISSN 0018-3830.

Pandey, P. & Pandey, A.K. (2002). Production of cellulase-free thermostable xylanases by an isolated strain of Aspergillus niger PPI, utilizing various lignocellulosic wastes. *World Journal of Microbiology and Biotechnology*. Vol. 18, No. 3, (April 2002), pp. 281-283, ISSN 0959-3993.

Pellinen, J., Abuhasan, J., Joyce, T.W. & Chang, H.M. (1989). Biological delignification of pulp by Phanerochaete chrysosporium. *Journal of Biotechnology*. Vol. 10, No. 2, pp. 161-170, ISSN 01681656.

Perez, J., Munoz-Dorado, J., de la Rubia, T. & Martinez, J. (2002). Biodegradation and biological treatments of cellulose, hemicellulose and lignin: an overview. *International Microbiology*. Vol. 5, No. 2, (June 2002), pp. 53-63, ISSN 1139-6709.

Piarpuzán, D., Quintero, J.A. & Cardona, C.A. (2011). Empty fruit bunches from oil palm as a potential raw material for fuel ethanol production. *Biomass and Bioenergy*. Vol. 35, No. 3, (March 2011), pp. 1130-1137, ISSN 0961-9534.

Pingali, S., Urban, V., Heller, W., McGaughey, J., O'Neill, H., Foston, M., Myles, D., Ragauskas, A. & Evans, B. (2010). Breakdown of Cell Wall Nanostructure in Dilute Acid Pretreated Biomass. *Biomacromolecules*. Vol. 11, No. 9, (August 2010), pp. 2329-2335, ISSN 1525-7797.

Piontek, K., Smith, A.T. & Blodig, W. (2001). Lignin peroxidase structure and function. *Biochemical Society Transactions*. Vol. 29, No. Pt 2, (May 2001), pp. 111-6, ISSN 0300-5127.

PRB. 2010 World Population Data Sheet, In: *Population Reference Burea*, Available from: http://www.prb.org/pdf10/10wpds_eng.pdf

Punelli, F., Reverberi, M., Porretta, D., Nogarotto, S., Fabbri, A.A., Fanelli, C. & Urbanelli, S. (2009). Molecular characterization and enzymatic activity of laccases in two Pleurotus spp. with different pathogenic behaviour. *Mycological Research*. Vol. 113, No. Pt 3, (March 2009), pp. 381-7, ISSN 0953-7562.

Ragauskas, A.J., Williams, C.K., Davison, B.H., Britovsek, G., Cairney, J., Eckert, C.A., Frederick, W.J., Jr., Hallett, J.P., Leak, D.J., Liotta, C.L., Mielenz, J.R., Murphy, R., Templer, R. & Tschaplinski, T. (2006). The path forward for biofuels and biomaterials. *Science*. Vol. 311, No. 5760, (January 2006), pp. 484-9, ISSN 1095-9203.

Ralph, J. & Landucci, L. (2010). NMR of Lignins, In: *Lignin and Lignans*, Heitner, C., Dimmel, D. & Schmidt, J., 137-243, CRC Press, ISBN 1-57444-486-5, Boca Raton, FL.

Ranieri, D., Colao, M.C., Ruzzi, M., Romagnoli, G. & Bianchi, M.M. (2009). Optimization of recombinant fungal laccase production with strains of the yeast Kluyveromyces lactis from the pyruvate decarboxylase promoter. *FEMS Yeast Research*. Vol. 9, No. 6, (September 2009), pp. 892-902, ISSN 1567-1364.

Reddy, N. & Yang, Y. (2005). Biofibers from agricultural byproducts for industrial applications. *Trends in Biotechnology*. Vol. 23, No. 1, (January 2005), pp. 22-7, ISSN 0167-7799.

Riva, S. (2006). Laccases: blue enzymes for green chemistry. *Trends in Biotechnology*. Vol. 24, No. 5, (May 2006), pp. 219-26, ISSN 0167-7799.

Rodriguez, E., Ruiz-Duenas, F.J., Kooistra, R., Ram, A., Martinez, A.T. & Martinez, M.J. (2008). Isolation of two laccase genes from the white-rot fungus Pleurotus eryngii and heterologous expression of the pcl3 encoded protein. *Journal of Biotechnology*. Vol. 134, No. 1-2, (March 2008), pp. 9-19, ISSN 0168-1656.

Ruiz, E., Cara, C., Ballesteros, M., Manzanares, P., Ballesteros, I. & Castro, E. (2006). Ethanol Production From Pretreated Olive Tree Wood and Sunflower Stalks by an SSF Process, In: *Twenty-Seventh Symposium on Biotechnology for Fuels and Chemicals*, McMillan, J. D., Adney, W. S., Mielenz, J. R. & Klasson, K. T., 631-643, Humana Press, ISBN 978-1-59745-268-7, New York, NY.

Ryu, K., Kang, J.H., Wang, L. & Lee, E.K. (2008). Expression in yeast of secreted lignin peroxidase with improved 2,4-dichlorophenol degradability by DNA shuffling. *Journal of Biotechnology*. Vol. 135, No. 3, (June 2008), pp. 241-6, ISSN 0168-1656.

Saloheimo, M., Barajas, V., Niku-Paavola, M.L. & Knowles, J.K. (1989). A lignin peroxidase-encoding cDNA from the white-rot fungus Phlebia radiata: characterization and expression in Trichoderma reesei. *Gene*. Vol. 85, No. 2, (December 1989), pp. 343-51, ISSN 0378-1119.

Salony, Garg, N., Baranwal, R., Chhabra, M., Mishra, S., Chaudhuri, T.K. & Bisaria, V.S. (2008). Laccase of Cyathus bulleri: structural, catalytic characterization and expression in Escherichia coli. *Biochimica et Biophysica Acta*. Vol. 1784, No. 2, (February 2008), pp. 259-68, ISSN 0006-3002.

Sanchez, D. Recausticizing: Principles and practice, In: *Vector process equipment Inc.*, Available from: http://www.tappi.org/content/events/08kros/manuscripts/2-1.pdf

Sánchez, Ó.J. & Cardona, C.A. (2008). Trends in biotechnological production of fuel ethanol from different feedstocks. *Bioresource Technology*. Vol. 99, No. 13, (September 2008), pp. 5270-5295, ISSN 0960-8524.

Sannigrahi, P., Ragauskas, A. & Miller, S. (2008). Effects of Two-Stage Dilute Acid Pretreatment on the Structure and Composition of Lignin and Cellulose in Loblolly Pine *Bioenegy Research*. Vol. 1, No. 3-4, (December 2008), pp. 205-214, ISSN 1939-1234.

Sarkanen, S., Razal, R.A., Piccariello, T., Yamamoto, E. & Lewis, N.G. (1991). Lignin peroxidase: toward a clarification of its role in vivo. *Journal of Biological Chemistry*. Vol. 266, No. 6, (February 1991), pp. 3636-43, ISSN 0021-9258.

Schell, D., Farmer, J., Newman, M. & McMillan, J. (2003). Dilute-sulfuric acid pretreatment of corn stover in pilot-scale reactor. *Applied Biochemistry and Biotechnology*. Vol. 105, No. 1, (n.d.), pp. 69-85, ISSN 0273-2289.

Scheller, H.V. & Ulvskov, P. (2010). Hemicelluloses. *Annual Review of Plant Biology*. Vol. 61, No. 1, (Junuary 2010), pp. 263-89, ISSN 1545-2123.

Schmidt, J. (2010). Electronic Spectroscopy of Lignins, In: *Lignin and Lignans*, Heitner, C., Dimmel, D. & Schmidt, J., 49-102, CRC Press, ISBN 1-57444-486-5, Boca Raton.

Selig, M.J., Viamajala, S., Decker, S.R., Tuker, M.P. & Himmel, M.E. (2007). Deposition of lignin droplets produced during dilute acid pretreatment of maize stems retards enzymatic hydrolysis of cellulose. *Biotechnology Progress*. Vol. 23, No. 6, (November 2007), pp. 1333-1339, ISSN 8756-7938.

Shafiee, S. & Topal, E. (2009). When will fossil fuel reserves be diminished? *Energy Policy*. Vol. 37, No. 1, (January 2009), pp. 181–189, ISSN 0301-4215.

Shao, Q., Chundawat, S., Krishnan, C., Bals, B., Leonardo, da Costa Sousa, L., Thelen, K., Dale, B. & Balan, V. (2010). Enzymatic digestibility and ethanol fermentability of AFEX-treated starch-rich lignocellulosics such as corn silage and whole corn plant. *Biotechnology for biofuels*. Vol. 3, No. 12, (January 2010), ISSN 1754-6834.

Sharma, R.K. & Arora, D.S. (2010). Changes in biochemical constituents of paddy straw during degradation by white rot fungi and its impact on in vitro digestibility. *Journal of Applied Microbiology*. Vol. 109, No. 2, (February 2010), pp. 679-686, ISSN 1364-5072.

Sharma, S.K., Kalra, K.L. & Grewal, H.S. (2002). Enzymatic saccharification of pretreated sunflower stalks. *Biomass and Bioenergy*. Vol. 23, No. 3, (September 2002), pp. 237-243, ISSN 0961-9534.

Shi, J., Sharma-Shivappa, R.R. & Chinn, M.S. (2009). Microbial pretreatment of cotton stalks by submerged cultivation of Phanerochaete chrysosporium. *Bioresource Technology*. Vol. 100, No. 19, (October 2009), pp. 4388-95, ISSN 1873-2976.

Sierra, R., García, L. & Holtzapple, M. (2011a). Selectivity and delignification kinetics for oxidative short-term lime pretreatment of poplar wood. Part I: constant-pressure. *Biotechnology Progress*. Vol. In press, ISSN 1520-6033.

Sierra, R., Garcia, L., Holtzapple, M. & Laverde, D. (2011b). Selectivity and delignification kinetics for oxidative short-term lime pretreatment of poplar wood. Part II: varying pressure. *Biotechnology Progress*. Vol. In press, ISSN 1520-6033.

Sierra, R., Granda, C.B. & Holtzapple, M. (2009a). Lime Pretreatment, In: *Biofuels: Methods and Protocols*, Mielenz, J. R., 115-124, Humana Press, ISBN 978-1-60761-213-1, New York.

Sierra, R., Granda, C.B. & Holtzapple, M. (2009b). Short-Term Lime Pretreatment of Poplar Wood. *Biotechnology Progress*. Vol. 25, No. 2, (March 2009), pp. 323-332, ISSN 1520-6033.

Sierra, R., Holtzapple, M. & Granda, C.B. (2010). Long-term Lime Pretreatment of Poplar Wood. *AIChE Journal*. Vol. 57, No. 5, (July 2010), pp. 1320-1328, ISSN 0001-1541.

Sierra, R., Smith, A., Granda, C.B. & Holtzapple, M. (2008). Producing Fuels and Chemicals from Lignocellulosic Biomass. *Chemical Engineering Progress*. Vol. 104, No. 8, (August 2008), pp. S10-S18.

Silva, C.J.S.M. & Roberto, I.C. (2001). Improvement of xylitol production by Candida guilliermondii FTI 20037 previously adapted to rice straw hemicellulosic hydrolysate. *Letters in Applied Microbiology*. Vol. 32, No. 4, (April 2001), pp. 248-252, ISSN 1472-765X.

Simmons, B.A., Loque, D. & Blanch, H.W. (2008). Next-generation biomass feedstocks for biofuel production. *Genome Biology*. Vol. 9, No. 242, (December 2008), ISSN 1465-6914.

Simonne, A., Simonne, E., Eitenmiller, R., Mills, H. & Cresman, C. (1997). Could the Dumas Method Replace the Kjeldahl Digestion for Nitrogen and Crude Protein Determinations in Foods? *Journal of the Science of Food and Agriculture*. Vol. 73, No. 1, (n.d.), pp. 39-45, ISSN 1097-0010.

Sinclair, R., Yamazaki, I., Bumpus, J., Brock, B., Chang, C.S., Albo, A. & Powers, L. (1992). Structure of the active site of lignin peroxidase isozyme H2: native enzyme, compound III, and reduced form. *Biochemistry*. Vol. 31, No. 20, (May 1992), pp. 4892-900, ISSN 0006-2960.

Singh, P., Suman, A., Tiwari, P., Arya, N., Gaur, A. & Shrivastava, A.K. (2008). Biological pretreatment of sugarcane trash for its conversion to fermentable sugars. *World Journal of Microbiology and Biotechnology*. Vol. 24, No. 5, (May 2008), pp. 667-673, ISSN 0959-3993.

Sjöström, E. (1981). The chemistry of oxygen delignification. *Paperi ja puu*. Vol. 63, No. 6-7, (n.d.), pp. 438-442, ISSN 0031-1243.

Sluiter, A., Hames, B., Ruiz, R., Scarlata, C., Sluiter, J. & Templeton, D. (2004). *Determination of Structural Carbohydrates and Lignin in Biomass*. (1st edition), National Renewable Energy Laboratory, Golden, CO.

Sterjiades, R., Dean, J., Gamble, G., Himmelsbach, D. & Eriksson, K. (1993). Extracellular laccases and peroxidases from sycamore maple (Acer pseudoplatanus) cell-suspension cultures - Reactions with monolignols and lignin model compounds. *Planta*. Vol. 190, No. 1, (May 1993), pp. 75-87, ISSN 0032-0935.

Sugiura, T., Yamagishi, K., Kimura, T., Nishida, T., Kawagishi, H. & Hirai, H. (2009). Cloning and homologous expression of novel lignin peroxidase genes in the white-rot fungus Phanerochaete sordida YK-624. *Bioscience, Biotechnology, and Biochemistry*. Vol. 73, No. 8, (August 2009), pp. 1793-8, ISSN 1347-6947.

Sun, Y. & Cheng, J. (2002). Hydrolysis of lignocellulosic materials for ethanol production: a review. *Bioresource Technology*. Vol. 83, No. 1, (May 2002), pp. 1-11, ISSN 0960-8524.

Sundaramoorthy, M., Gold, M.H. & Poulos, T.L. (2010). Ultrahigh (0.93A) resolution structure of manganese peroxidase from Phanerochaete chrysosporium: implications for the catalytic mechanism. *Journal of Inorganic Biochemistry*. Vol. 104, No. 6, (June 2010), pp. 683-90, ISSN 1873-3344.

Susilo, J. & Bennington, C.P.J. (2007). Modelling Kappa Number and Pulp Viscosity in Industrial Oxygen Delignification Systems. *Chemical Engineering Research and Design*. Vol. 85, No. 6, (n.d.), pp. 872-881, ISSN 0263-8762.

Szczodrak, J. & Fiedurek, J. (1996). Technology for conversion of lignocellulosic biomass to ethanol. *Biomass and Bioenergy*. Vol. 10, No. 5-6, (July 1996), pp. 367-375, ISSN 0961-9534.

Tamaki, Y. & Mazza, G. (2010). Measurement of structural carbohydrates, lignins, and micro-components of straw and shives: Effects of extractives, particle size and crop species. *Industrial Crops and Products*. Vol. 31, No. 3, (May 2010), pp. 534-541, ISSN 0926-6690.

Tanksale, A., Beltramini, J.N. & Lu, G.M. (2010). A review of catalytic hydrogen production processes from biomass. *Renewable and Sustainable Energy Reviews*. Vol. 14, No. 1, (January 2010), pp. 166-182, ISSN 1364-0321.

Templeton, D., Sluiter, A., Hayward, T., Hames, B. & Thomas, S. (2009). Assessing corn stover composition and sources of variability via NIRS. *Cellulose*. Vol. 16, No. 4, (August 2009), pp. 621-639, ISSN 0969-0239.

Thring, R., Chornet, E. & Qverend, R. (1990). Recovery of a solvolytic lignin: effects of spent liquor/acid volume ratio, acid concentration and temperature. *Biomass*. Vol. 23, (n.d), pp. 289-305, ISSN 0144-4565.

Timofeevski, S.L. & Aust, S.D. (1997). Effects of Mn2+ and oxalate on the catalatic activity of manganese peroxidase. *Biochemical and Biophysical Research Communications*. Vol. 239, No. 3, (October 1997), pp. 645-9, ISSN 0006-291X.

Toivari, M.H., Salusjarvi, L., Ruohonen, L. & Penttila, M. (2004). Endogenous Xylose Pathway in Saccharomyces cerevisiae. *Applied and Environment Microbiology*. Vol. 70, No. 6, (June 2004), pp. 3681-3686, ISSN 0099-2240.

Torgashov, V., Gert, E., Zubets, O. & Kaputskii, F. (2010). A comparative study of isolation conditions, morphology, and properties of cellulose obtained from the stalks of cereals and oil-yielding plants. *Russian Journal of Bioorganic Chemistry*. Vol. 36, No. 7, (December 2010), pp. 838-846, ISSN 1068-1620.

Tripathi, M.K., Mishra, A.S., Misra, A.K., Vaithiyanathan, S., Prasad, R. & Jakhmola, R.C. (2008). Selection of white-rot basidiomycetes for bioconversion of mustard (Brassica compestris) straw under solid-state fermentation into energy substrate for rumen micro-organism. *Letters in Applied Microbiology*. Vol. 46, No. 3, (March 2008), pp. 364-70, ISSN 1472-765X.

Turner, D., Xu, H., Cracknell, R.F., Natarajan, V. & Chen, X. (2011). Combustion performance of bio-ethanol at various blend ratios in a gasoline direct injection engine. *Fuel*. Vol. 90, No. 5, (May 2011), pp. 1999-2006, ISSN 0016-2361.

Vasquez, M.P., da Silva, J.N., de Souza, M.B., Jr. & Pereira, N., Jr. (2007). Enzymatic hydrolysis optimization to ethanol production by simultaneous saccharification and fermentation. *Applied Biochemistry and Biotechnology*. Vol. 137-140, No. 1-12, (April 2007), pp. 141-53, ISSN 1559-0291.

Vicuña, R. (2000). Ligninolysis. A very peculiar microbial process. *Molecular Biotechnology*. Vol. 14, No. 2, (February 2000), pp. 173-6, ISSN 1073-6085.

Vinuselvi, P., Park, J.M., Lee, J.M., Oh, K., Ghim, C.-M. & Lee, S.K. (2011). Engineering microorganisms for biofuel production. *Biofuels*. Vol. 2, No. 2, (February 2011), pp. 153-166, ISSN 1759-7269.

Vries, S.C.d., Ven, G.W.J.v.d., Ittersum, M.K.v. & Giller, K.E. (2010). Resource use efficiency and environmental performance of nine major biofuel crops, processed by first-generation conversion techniques. *Biomass and Bioenergy*. Vol. 34, (May 2010), pp. 588 - 601, ISSN 0961-9534.

Wan, C. & Li, Y. (2010). Microbial delignification of corn stover by Ceriporiopsis subvermispora for improving cellulose digestibility. *Enzyme and Microbial Technology*. Vol. 47, No. 1-2, (July 2010), pp. 31-36, ISSN 0141-0229.

Wang, W. & Wen, X. (2009). Expression of lignin peroxidase H2 from Phanerochaete chrysosporium by multi-copy recombinant Pichia strain. *Journal of Environmental Sciences (China)*. Vol. 21, No. 2, (February 2009), pp. 218-22, ISSN 1001-0742.

Wang, Z.M., Bleakley, B.H., Crawford, D.L., Hertel, G. & Rafii, F. (1990). Cloning and expression of a lignin peroxidase gene from Streptomyces viridosporus in Streptomyces lividans. *Journal of Biotechnology*. Vol. 13, No. 2-3, (February 1990), pp. 131-44, ISSN 0168-1656.

Wariishi, H., Valli, K. & Gold, M.H. (1992). Manganese(II) oxidation by manganese peroxidase from the basidiomycete Phanerochaete chrysosporium. Kinetic mechanism and role of chelators. *Journal of Biological Chemistry*. Vol. 267, No. 33, (November 1992), pp. 23688-95, ISSN 0021-9258.

Whitwam, R.E., Brown, K.R., Musick, M., Natan, M.J. & Tien, M. (1997). Mutagenesis of the Mn2+-binding site of manganese peroxidase affects oxidation of Mn2+ by both compound I and compound II. *Biochemistry*. Vol. 36, No. 32, (August 1997), pp. 9766-73, ISSN 0006-2960.

Wong, D.W. (2009). Structure and action mechanism of ligninolytic enzymes. *Applied Biochemistry and Biotechnology*. Vol. 157, No. 2, (May 2009), pp. 174-209, ISSN 1559-0291.

Wyman, C. & Lloyd, T.A. (2005). Combined sugar yields for dilute sulfuric acid pretreatment of corn stover followed by enzymatic hydrolysis of the remaining solids *Bioresource Technology*. Vol. 96, No. 18, (December 2005), pp. 1967-1977, ISSN 0960-8524.

Wyman, C.E., Dale, B.E., Elander, R.T., Holtzapple, M., Ladisch, M.R. & Lee, Y.Y. (2005a). Comparative sugar recovery data from laboratory scale application of leading pretreatment technologies to corn stover. *Bioresource Technology*. Vol. 96, No. 18, (December 2005), pp. 2026-2032, ISSN 0960-8524.

Wyman, C.E., Dale, B.E., Elander, R.T., Holtzapple, M., Ladisch, M.R. & Lee, Y.Y. (2005b). Coordinated development of leading biomass pretreatment technologies. *Bioresource Technology*. Vol. 96, No. 18, (December 2005), pp. 1959-1966, ISSN 0960-8524.

Wyman, C.E., Dale, B.E., Elander, R.T., Holtzapple, M., Ladisch, M.R., Lee, Y.Y. & Saddler, J.N. (2009). Comparative sugar recovery and fermentation data following pretreatment of poplar wood by leading technologies. *Biotechnology Progress*. Vol. 25, No. 2, (March 2009), pp. 333-339, ISSN 1520-6033.

Xiao, Y.Z., Hong, Y.Z., Li, J.F., Hang, J., Tong, P.G., Fang, W. & Zhou, C.Z. (2006). Cloning of novel laccase isozyme genes from Trametes sp. AH28-2 and analyses of their differential expression. *Applied Microbiology and Biotechnology*. Vol. 71, No. 4, (July 2006), pp. 493-501, ISSN 0175-7598.

Yancey, M. & Kadam, K. (1997). *Biomass to ethanol facility design, cost estimate, and financial evaluation: Volume I.* (1st ed.), National Renewable Energy Laboratory, Gorlden, CO.

Yu, H., Du, W., Zhang, J., Ma, F., Zhang, X. & Zhong, W. (2010a). Fungal treatment of cornstalks enhances the delignification and xylan loss during mild alkaline pretreatment and enzymatic digestibility of glucan. *Bioresource Technology*. Vol. 101, No. 17, (September 2010), pp. 6728-34, ISSN 1873-2976.

Yu, H., Zhang, X., Song, L., Ke, J., Xu, C., Du, W. & Zhang, J. (2010b). Evaluation of white-rot fungi-assisted alkaline/oxidative pretreatment of corn straw undergoing enzymatic hydrolysis by cellulase. *Journal of Bioscience and Bioengineering*. Vol. 110, No. 6, (December 2010), pp. 660-664, ISSN 1389-1723.

Yu, Q., Zhuang, X., Yuan, Z., Wang, Q., Qi, W., Wang, W., Zhang, Y., Xu, J. & Xu, H. (2010c). Two-step liquid hot water pretreatment of Eucalyptus grandis to enhance sugar recovery and enzymatic digestibility of cellulose. *Bioresource Technology*. Vol. 101, No. 13, (July 2010), pp. 4895-4899, ISSN 0960-8524.

Yu, Y., Feng, Y., Xu, C., Liu, J. & Li, D. (2011). Onsite bio-detoxification of steam-exploded corn stover for cellulosic ethanol production. *Bioresource Technology*. Vol. 102, No. 8, (January 2011), pp. 5123-5128, ISSN 0960-8524.

Zerbini, J.E., Oliveira, E.M. & Bon, E.P. (1999). Lignin peroxidase production by Streptomyces viridosporus T7A: nitrogen nutrition optimization using glucose as carbon source. *Applied Biochemistry and Biotechnology*. Vol. 77-79, (March 1999), pp. 681-8, ISSN 0273-2289.

Zhang, G., Zhang, L., Deng, H. & Sun, P. (2010). Preparation and characterization of sodium carboxymethyl cellulose from cotton stalk using microwave heating. *Journal of Chemical Technology and Biotechnology*. Vol. In press, (n.d.), ISSN 1097-4660.

Zhang, Y.H. (2008). Reviving the carbohydrate economy via multi-product lignocellulose biorefineries. *Journal of Industrial Microbiology and Biotechnology*. Vol. 35, No. 5, (May 2008), pp. 367-375, ISSN 1367-5435.

Zhao, X., Chen, K. & Liu, D. (2009). Organosolv pretreatment of lignocellulosic biomass for enzymatic hydrolysis. *Applied Microbiology and Biotechnology*. Vol. 82, No. 5, (April 2009), pp. 815-827, ISSN 0175-7598.

Zhao, X., Zhang, L. & Liu, D. (2008). Comparative study on chemical pretreatment methods for improving enzymatic digestibility of crofton weed stem. *Bioresource Technology*. Vol. 99, No. 9, (June 2008), pp. 3729-3736, ISSN 0960-8524.

Zhu, L., O'Dwyer, J.P., Chang, V.S., Granda, C.B. & Holtzapple, M.T. (2008). Structural features affecting biomass enzymatic digestibility. *Bioresource Technology*. Vol. 99, No. 9, (June 2008), pp. 3817-3828, ISSN 0960-8524.

Performance and Emission Characteristics of Spark Ignition Engine Fuelled with Ethanol and Methanol Gasoline Blended Fuels

Ioannis Gravalos[1], Dimitrios Moshou[2], Theodoros Gialamas[1], Panagiotis Xyradakis[2], Dimitrios Kateris[2] and Zisis Tsiropoulos[1]

[1]Technological Educational Institute of Larissa, Faculty of Agricultural Technology Department of Biosystems Engineering, Larissa

[2]Aristotle University, School of Agriculture, Department of Hydraulics Soil Science and Agricultural Engineering, Thessaloniki

Greece

1. Introduction

Depletion of fossil fuels and environmental pollution has led researchers to anticipate the need to develop bio-fuels. Alcohols are an important category of bio-fuels. Methanol can be produced from coal, biomass or even natural gas with acceptable energy cost. Also, gasification of biomass can lead to methanol, mixed alcohols, and Fischer–Tropsch liquids (Chum and Overend, 2001). Ethanol is produced from sugars (particularly sugar cane) and starch by fermentation. The biomass industry can produce additional ethanol by fermenting some agricultural by-products (Prasad et al., 2007). Lignocellulosic biomass is a potential source for ethanol that is not directly linked to food production (Freudenberger, 2009). Shapouri et al. (1995) showed that the net energy value of corn ethanol has become positive in recent years due to technological advances in ethanol conversion and increased efficiency in farm production. Corn ethanol is energy efficient, as indicated by an energy ratio of 1.24, that is, for every Btu dedicated to producing ethanol, there is a 24-percent energy gain. Goldemberg et al. (2004) demonstrated, through the Brazilian experience with ethanol, that economy of scale and technological advances lead to increased competitiveness of this renewable alternative, reducing the gap with conventional fossil fuels. Consequently alcohols are particularly attractive as alternative fuels because they are a renewable bio-based resource and oxygenated, thereby providing the potential to reduce particulate emissions in spark ignition engines. Kim and Dale (2004) estimated that the potential for ethanol production is equivalent to about 32% of the total gasoline consumption worldwide, when used in E85 (85% ethanol in gasoline) for a midsize passenger vehicle. Such a substitution immediately addresses the issue of reducing our use of non-renewable resources (fossil fuels) and the attendant impacts on climate change, especially carbon dioxide and the resulting greenhouse effect (von Blottnitz and Curran, 2007). The conversion of biomass to bio-fuel has some ecological drawbacks. It is well known that conversion of biomass requires additional energy inputs, most often provided in some form of fossil fuel. Also, agricultural production of biomass is relatively land intensive, and there

is a risk of pollutants entering water sources from fertilisers and pesticides that are applied to the land to enhance plant growth (Pimentel, 2003; Niven, 2005).
The use of alcohol blended with gasoline was a subject of research in the 1980s and it was shown that ethanol and methanol gasoline blends were technically acceptable for existing spark ignition engines. There is a considerable amount of literature relative to various blends of ethanol, methanol and gasoline. Winnington and Siddiqui (1983) studied the effect of using ethanol gasoline blends as a fuel on the performance of spark ignition engines. The Ricardo engine, over the test range of 8:1 to 10:1 compression ratio, showed an average drop in power compared to premium gasoline of 2.5% on blend A and 7.5% on blend B. The specific fuel consumption of the ethanol gasoline blend showed an increase compared to premium gasoline of around 0.5% and 4% on blends A and B, respectively. The Peugeot engine tests showed that the power was down, overall, by around 1% and 2.5% on blends A and B, respectively, and the specific fuel consumption was increased by about 0.5% for blend A and 1% for blend B. El-Kassaby (1993) investigated the effect of ethanol gasoline blends on spark ignition engine performance. The performance tests showed that the engine showed power improvement with ethanol addition, the maximum improvement occurring at the 10% ethanol and 90% gasoline fuel blend. Abdel-Rahman and Osman (1997) carried out performance tests using different percentages of ethanol in gasoline fuel, up to 40%, under variable compression ratio conditions. The results show that the engine showed power improvement with the percentage addition of the ethanol in the fuel blend. The maximum improvement occured at 10% ethanol/90% gasoline fuel blend. Yacoub et al. (1998) quantified the performance and exhaust gas emissions for an engine optimized to operate on C1–C5 alcohol/gasoline blends with matched oxygen content. The performance and exhaust gas emissions characteristics of the blends were quantified by using a single-cylinder spark ignition engine. Lower alcohols (C1, C2 and C3)/gasoline blends showed a wider range of operation relative to neat gasoline. Ethanol/gasoline blends showed the highest knock resistance improvement among all tested blends. On the other hand, higher alcohol (C4 and C5)/gasoline blends showed degraded knock resistance when compared with neat gasoline. Al-Hasan (2002) showed that blending unleaded gasoline with ethanol increased the brake power, torque, volumetric and brake thermal efficiencies and fuel consumption, while it decreased the brake specific fuel consumption and equivalence air-fuel ratio. Wu et al. (2004) tested ethanol/gasoline blended fuel in a conventional engine under various air–fuel equivalence ratios (λ) for its performance and emissions. When air-fuel ratio is slightly smaller than one, maximum torque output and minimum brake specific heat consumption (bshc) are available. Using ethanol/gasoline blended fuels improves torque output. However, bshc does not change noticeably. Yücesu et al. (2006) examined the effect of compression ratio on engine performance and exhaust emissions at stoichiometric air/fuel ratio, full load and minimum advanced timing for the best torque in a single cylinder, four stroke, with variable compression ratio spark ignition engine. The engine torque increased with increasing compression ratio up to 11:1, the increasing ratio was about 8% when compared with 8:1 compression ratio. The highest increasing ratio of engine torque was obtained at 13:1 compression ratio with E40 and E60 fuels, the increment was about 14% when compared with 8:1 compression ratio. Minimum brake specific fuel consumption (BSFC) was obtained at 11:1 compression ratio with E0 fuel. A comparison with 8:1 compression ratio, showed that the BSFC decreased 10% and after 11:1 compression ratio the BSFC increased again. The maximum decrease in BSFC was found to be 15% when E40 was used. Liu et al. (2007) showed that the engine power and torque will decrease when increasing the fraction of methanol in the fuel blends under wide open throttle (WOT)

conditions. However, if spark ignition timing is advanced, the engine power and torque can be improved under WOT operating conditions. Engine thermal efficiency is thus improved in almost all operating conditions. Engine combustion analyses show that the fast burning phase becomes shorter, however, the flame development phase is slightly delayed. Koç et al. (2009) investigated the effects of unleaded gasoline (E0) and unleaded gasoline/ethanol blends (E50 and E85) on engine performance in a single cylinder four-stroke spark-ignition engine at two compression ratios (10:1 and 11:1). The engine speed was changed from 1500 to 5000 rpm at WOT. The results of the engine test showed that ethanol addition to unleaded gasoline increased the engine torque, power and fuel consumption. It was also found that ethanol/gasoline blends allow increasing compression ratio (CR) without knock occurrence. Yücesu et al. (2007) proposed a new approach based on artificial neural network (ANN) to determine the engine torque and brake specific fuel consumption. Ethanol/unleaded gasoline blends (E10, E20, E40 and E60) were tested in a single cylinder, four stroke spark ignition and fuel injection engine. The tests were performed by varying the ignition timing, relative air-fuel ratio (RAFR) and compression ratio at a constant speed of 2000 rpm and at WOT. Maximum brake torque (MBT) timing of the engine showed no significant variation with unleaded gasoline and unleaded gasoline/ethanol blends. When the ignition timing retarded, ethanol blends yielded higher brake torque of the engine than unleaded gasoline. The maximum torque was obtained at 0.9 RAFR for all test fuels for both compression ratios 8:1 and 10:1. The engine torque of ethanol blended fuels was higher than that of E0 obtained at richer working region than stoichiometric air-fuel ratio especially at 10:1 compression ratio. The BSFC varied depending on both engine torque and especially the heating value of the used fuel. The BSFC increased in proportion to the ethanol percentage. From the results of mathematical modelling, the calculated engine torque and specific fuel consumption were obviously within acceptable uncertainty margins. Najafi et al. (2009) proposed the use of ANN to determine the engine power, torque, brake specific fuel consumption, brake thermal efficiency, volumetric efficiency and emission components based on different gasoline/ethanol blends and speeds. Experimental data demonstrated that the use of ethanol/gasoline blended fuels will marginally increase the brake power and decrease the brake specific fuel consumption. It was also found that the brake thermal efficiency and volumetric efficiency tend to increase when ethanol/gasoline blends are used. Analysis of the experimental data by the ANN revealed that there is a good correlation between the ANN-predicted results and the experimental data.

The effects of using ethanol and methanol unleaded gasoline blends on emissions characteristics in spark ignited engine have been investigated by other researchers. Research studies of exhaust emission levels from spark ignited engine are important from different perspectives. The combustion of fuel in an engine generates by-products that we all know as emissions. The four main engine emissions are carbon monoxide (CO), carbon dioxide (CO_2), hydrocarbons (HC), and oxides of nitrogen (NOx) (though others, such as particulates and formaldehyde, are also produced). Gasoline, as a compound hydrocarbon, is not a particularly clean-burning fuel. Alcohols, in comparison, burn nearly pollution-free. Alcohols already contain oxygen integral with the fuel, which can lead to a more homogenous combustion. Alcohols burn with a faster flame speed than gasoline, and they do not contain additional elements such as sulphur and phosphorus. Rajan and Saniee (1983) investigated the characteristics of hydrated ethanol with gasoline as a means of reducing the cost of ethanol/gasoline blends for use as a spark ignition engine fuel. Engine experiments indicate that, at normal ambient temperatures, a water/ethanol/gasoline blend containing up to 6 vol% of water in the ethanol constitutes a desirable motor fuel with power characteristics similar to those of the base gasoline. As a means of reducing the smog causing components of the exhaust gases, such as the oxides of nitrogen and the unburnt hydrocarbons, the water/ethanol/gasoline blend is superior to the base gasoline. Palmer (1986) showed that 10% of ethanol addition to gasoline could reduce the concentration of CO emission up to 30%. Bata et al. (1989) had tested different blend rates of ethanol gasoline fuels in engines, and found that the ethanol could reduce the CO and UHC emissions. Taylor et al. (1996) used four alcohol fuels to blend with gasoline and concluded that adding ethanol can reduce CO, HC and NO emissions. Chao et al. (2000) indicated that using ethanol gasoline blended fuels increases the emission of formaldehyde, acetaldehyde and acetone several times than those from gasoline. Gautam et al. (2000) investigated the emissions characteristics between higher alcohol/gasoline blends and neat gasoline. It was found that the cycle emissions of CO, CO_2 and organic matter hydrocarbon equivalent from the higher alcohol/gasoline blends were very similar to those from neat gasoline. Cycle emissions of NOx from the blends were higher than those from neat gasoline. However, for all the emissions species considered, the brake specific emissions (g/kW h) were significantly lower for the higher alcohol/gasoline blends than for neat gasoline. This was because the blends had greater resistance to knock and allowed higher compression ratios, which increased engine power output. The contribution of alcohols and aldehydes to the overall organic matter hydrocarbon equivalent emissions was found to be minimal. Al-Hasan (2002) investigated the effect of using unleaded gasoline/ethanol blends on a four stroke, four cylinder spark ignition (SI) engine performances and exhaust emission. The results showed that the CO and HC emissions concentrations in the engine exhaust decrease, while the CO_2 concentration increases. Hsieh et al. (2002) investigated the engine performance and pollutant emission of a commercial SI engine using ethanol–gasoline blended fuels with various blended rates (0%, 5%, 10%, 20%, 30%). It was found that with increasing the ethanol content, the RVP of the blended fuels initially increases to a maximum at 10% ethanol addition, and then decreases. Results of the engine test indicated that using ethanol/gasoline blended fuels, CO and HC emissions decrease dramatically as a result of the leaning effect caused by the ethanol addition, and CO_2 emission increases because of the improved combustion. Finally, it was noted that NOx emission depends on the engine operating condition rather than the ethanol content. He et al. (2003) investigated the effect of ethanol blended gasoline fuels on emissions and catalyst conversion efficiencies in a spark ignition engine with an electronic fuel injection system. Ethanol can decrease engine-out regulated emissions. The fuel containing 30% ethanol by volume can drastically reduce engine-out total hydrocarbon emissions (THC) at operating conditions and engine-out THC, CO and NOx emissions at idle speed, but unburned ethanol and acetaldehyde emissions increase. According to Yüksel and Yüksel (2004) one of the major problems for the successful application of gasoline–alcohol mixtures as a motor fuel is the realization of a stable homogeneous liquid phase. To overcome this problem, authors designed a new carburetor. Sixty percent ethanol and forty percent gasoline blend was exploited to test the performance, the fuel consumption, and the exhaust emissions. Experimental results indicated that using ethanol–gasoline blended fuel, the CO and HC emissions decreased dramatically as a result of the leaning effect caused by the ethanol addition, and the CO_2 emission increased because of the improved combustion. Bayraktar (2005) investigated experimentally and theoretically the effects of ethanol addition to gasoline on an SI engine performance and exhaust emissions. Experimental applications have been carried out with the blends containing 1.5, 3, 4.5, 6, 7.5, 9, 10.5 and 12 vol% ethanol. Numerical applications have been performed up to 21 vol% ethanol. Engine was operated with each blend at 1500 rpm for compression ratios of 7.75 and 8.25 and at full throttle setting. Experimental results have shown that among the various blends, the blend of 7.5% ethanol was the most suitable one from the engine performance and CO emissions points of view. However, theoretical comparisons have shown that the blend containing 16.5% ethanol was the most suited blend for SI engines. Jia et al. (2005) investigated emission characteristics from a four-stroke motorcycle engine using 10 vol% ethanol/gasoline blended fuel (E10) at different driving modes on the chassis dynamometers. The results indicate that CO and HC emissions in the engine exhaust were lower with the operation of E10 as compared to the use of unleaded gasoline, whereas the effect of ethanol on NOx emission is not remarkable. Hydrocarbon species except ethanol, acetaldehyde and ethylene emissions were decreased somewhat from ethanol/gasoline blends-fuelled motorcycle engine relative to gasoline-fuelled engine. Additionally, this analysis showed that aromatic compounds and fatty group ones are major compounds in motorcycle engine exhaust. Ceviz and Yüksel (2005) investigated the effects of using ethanol/unleaded gasoline blends on cyclic variability and emissions in a spark-ignited engine. Results of this study showed that using ethanol/unleaded gasoline blends as a fuel decreased the coefficient of variation in indicated mean effective pressure, and CO and HC emission concentrations, while increased CO_2 concentration up to 10 vol.% ethanol in fuel blend.

From the literature review, it is obvious that alcohol gasoline blended fuels can effectively increase the brake power and decrease the emissions without major modifications to the engine design. This chapter was prepared for the purpose of presenting the results of experience to date with a selected list of possible alternative fuels to be used in SI engines.

2. Ethanol and methanol gasoline blended fuels

Methanol and ethanol based liquid fuels can be used as substitutes for gasoline fuels in conventional engines, such as spark ignition engines, without modification to the engines. Several test fuels were used in this study. The first was unleaded gasoline as a base fuel. The others were ethanol and methanol blended gasoline fuels.

2.1 Blend properties

Each fuel has its own set of combustion-related properties. These properties change the engine performance and emission characteristics. Laboratory tests were then carried out using ASTM tests standards to determine the combustion-related properties. A list of fuel properties that compares ethanol and methanol gasoline blended fuels is given in Table 1. It shows heat of combustion, Reid vapour pressure (RVP), research octane number (RON), density at 15.5 °C and distillation temperature including initial boiling temperature (IBT), 10%, 50%, 90% distillation temperatures and final distillation temperature.

Ethanol (ethyl alcohol) C_2H_5OH is a clear, colorless liquid with a characteristic, agreeable odor. Ethanol is an alcohol, a group of chemical compounds whose molecules contain a hydroxyl group, -OH, bonded to a carbon atom. Ethanol melts at -114.1 °C, boils at 78.5 °C, and has a density of 0.789 g/mL at 20°C (De Caro et al., 2001). The heating value of ethanol is lower than that of gasoline. Table 1 further indicates that the heating value of the blended fuel will decrease with the increase of the ethanol content. RON increases with the increase of ethanol concentration. Compared to unleaded gasoline, RON of the blended fuels is increased by 3.5, 8.6 and 14.1, respectively. Therefore, ethanol is an excellent additive for preventing engine knock and improving engine performance where high octane requirements exist. Despite the improved octane performance of ethanol/gasoline blends, engine driveability is generally degraded. Cold starting is more difficult because of the added heat of vaporization in blends. Hot starting is complicated because of increased volatility, which leads to potential vapor locking conditions (Sinor and Bailey, 1993). Adding 10vol.%, 20vol.% and 30vol.% ethanol to gasoline increase the RVP of the base gasoline of about 24.53 kPa, 19.61 kPa and 18.31 kPa, respectively. The increase in vapor pressure for low level blends of ethanol is caused by dilution of hydrogen bonding between ethanol molecules in the final blend. When ethanol is diluted in gasoline, the hydrogen bonding effect is reduced and the ethanol molecules behave more like their low molecular weight would indicate, resulting in increased volatility (Bailey, 1997). Adding ethanol also modifies the distillation curve of the gasoline. It can also be observed that the addition of ethanol to gasoline increases IBT, but the rates of 10%, 50%, 90% and final distillation temperatures decrease. Ethanol forms a minimum boiling azeotrope with gasoline, causing the distillation curve to be depressed between 10vol.% and 90vol.% distilled points.

Methanol CH_3OH, which is also called methyl alcohol, is colorless, odorless, water-soluble liquid. It freezes at -97.8 °C, and boils at 64.6 °C. It is miscible with water in all proportions, and spillages are rapidly dispersed. Blends with between 6.7 % and 36 % of air are flammable. The auto-ignition temperature of methanol is 467 °C, which is high compared with 222 °C for gasoline. This may account for the high octane number, 106, of methanol; a typical gasoline has an octane number of 90 to 100. Although methanol is not the cheapest fuel, its properties make it competitive with the other fuels. Methanol used as an additive or substitute for gasoline could immediately help to solve both energy and air pollution problems (Reed and Lerner, 1973). The heating value of methanol is less than that of gasoline and ethanol so that its blends contain less MJ/kg. Methanol gasoline blends cause slight but significant decrease in efficiency of the engine. Methanol has also a high RON, which increases with the increase of methanol concentration. Compared to unleaded gasoline, RON of the methanol blended fuels is increased by 3.4, 9.6 and 13.6, respectively. This number reflects the fact that the blending of methanol with gasoline is a very effective method of increasing the octane number of the fuel. Moreover, the result of this effect demonstrates the elimination of knocking. It's possibility of replacing anti-knock additives in gasoline with a low percentage of methanols in a blend, helping to minimize air pollution (Yamamoto, 1972). Adding 10vol.%, 20vol.% and 30vol.% methanol to gasoline increases the RVP of the base gasoline at about 22.43 kPa, 31.58 kPa and 33.74 kPa, respectively. Because of the disruption of hydrogen bonding in methanol when it is blended with a hydrocarbon, the vapor pressure of a blend of methanol and gasoline deviates greatly from ideal behavior, exhibiting a much higher vapor pressure than would be expected. This excess vapor pressure can lead to vapor problems (driveability problems), difficulties with hot starts, stalling, hesitation, and poor acceleration (Ceci1, 1974). Several solutions to these problems have been proposed (Fitch and Kilgroe, 1970; Adelman, 1972). It is possible to add high vapor pressure liquids or gases such as butane either generally or preferably during cold start situations. Either gasoline or LPG could be injected at cold starts to accomplish the same effect. Aside from the cold start problem, the performance of the methanol fuelled engine has been shown to be equivalent to a gasoline fuelled engine. Adding methanol also modifies the distillation curve of the gasoline. It can also be observed that the addition of methanol to gasoline increases IBT, 90% and final distillation temperature rates, but 10% and 50%, decrease.

Property item	**Test fuel**							**Test method**
	Gasoline	**E10**	**E20**	**E30**	**M10**	**M20**	**M30**	
Heat of combustion (MJ/kg)	44.133	42.447	40.672	38.673	41.615	38.233	36.247	ASTM D340
Reid vapour pressure (kPa)	35.00	59.53	54.61	53.31	57.43	66.58	68.74	ASTM D323
Research octane number	84.8	88.3	93.4	98.9	88.2	94.4	98.4	ASTM D2699
Density at 15.5°C (kg/l)	0.7678	0.7760	0.7782	0.7794	0.7692	0.7707	0.7734	ASTM D1298
Distillation temperature (°C)								
IBT	38.5	39.5	40.3	40.7	43.2	43.7	44.5	ASTM D86
10 vol.%	57.2	52.3	55.4	55.7	48.2	50.4	51.3	
50 vol.%	93.5	71.8	71.6	72.5	81.0	79.7	81.6	
90 vol.%	156.0	143.7	143.1	142.7	165.1	164.8	164.7	
End point	181.7	176.1	176.6	176.5	206.2	206.3	206.7	

Table 1. Properties of different ethanol and methanol gasoline blended fuels.

2.2 Testing procedure

The performance and emission characteristics of the spark ignition engine running on ethanol and methanol blended with gasoline were evaluated and compared with neat gasoline fuel. Apparatuses used in the present study were an engine, a dynamometer and an exhaust analyzer. The schematic diagram of the experimental set-up is shown in Figure 1. A single cylinder, carburetted, four-stroke, spark ignition non-road engine (type Bernard moteures 19A), was chosen. Non-road gasoline engines differ from automotive engines in several technical specifications. Because of these design differences, the effects of alcohol/gasoline blended fuel changes on performance and emission characteristics from non-road gasoline engines are quite different from the effects of alcohol/gasoline blended fuel changes on performance and emissions from automotive gasoline engines. This engine had a 56 mm bore and a 58 mm stroke (total displacement 143 cm^3). Its rated power was 2.2 kW. The ignition system was composed of the conventional coil and spark plug arrangement with the primary coil circuit operating on a pulse generator unit. The engine was coupled to a hydraulic dynamometer. Exhaust gases were sampled from the outlet and then were measured on line by an exhaust analyzer Bosch.

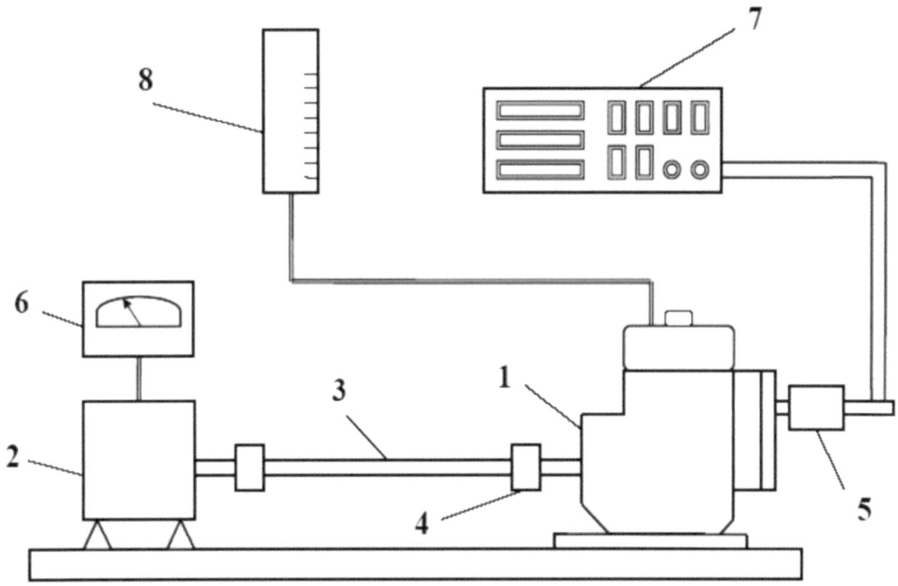

Fig. 1. The schematic diagram of the experimental set-up. (1) Engine, (2) Dynamometer, (3) Shaft, (4) Flywheel, (5) Exhaust pipe, (6) Dynamometer control unit, (7) Gas analyzer, and (8) Fuel measurement system.

A series of experiments were carried out using gasoline, and various ethanol/methanol blends. The test blends were prepared just before starting the experiment to ensure that the fuel mixture was homogeneous and prevent the reaction of ethanol with water vapor. The engine was started and allowed to warm up for a period of 20–30 min. Before running the engine with a new fuel blend, it was allowed to run for sufficient time to consume the remaining fuel from the previous experiment. All the blends were tested under varying engine speed conditions. Engine tests were performed at maximum to idling rpm engine speed. The required engine load was obtained through the dynamometer control. The engine speed, fuel consumption, and load were measured, while the brake power, brake torque and brake specific fuel consumption (BSFC) were computed. For each experiment, three runs were performed to obtain an average value of the experimental data.

After the engine reached a stable working condition (steady state), emission parameters such as CO, CO_2, HC, and NO_x from an online exhaust gas analyser were recorded. CO, CO_2, HC, and NO_x emissions reached average values of the acquired data within 20s for each stable operating condition. The concentration of each gas was measured continuously by digital data acquisition. The exhaust gas temperature was monitored during the experiments to ensure that the engine was in a steady state condition.

2.3 Engine performance characteristics

The results of the brake power, torque, and specific fuel consumption for ethanol and methanol gasoline blended fuels at different engine speeds are presented here.

Fig. 2a shows the influence of ethanol gasoline blended fuels on engine brake power. When the ethanol content in the blended fuel was increased, the engine brake power slightly increased for all engine speeds. However, the brake power of gasoline was slightly lower than that of E10–E30, especially for low engine speeds (e.g., 1000 rpm). With an increase in ethanol percentage, the density of the blend and the engine volumetric efficiency increased and this caused an increase in power. A similar behaviour has been reported by almost all investigators on various types of engines and conditions (Al-Hasan, 2002; Bayraktar, 2005). Fig. 2b shows the effect of different ethanol gasoline blended fuels on engine torque. The increase of ethanol content increased slightly the torque of the engine. The brake torque of gasoline was lower than those of E10–E30, especially for low engine speeds. Due to the addition of ethanol the octane number raised. Therefore, antiknock behaviour improved and allowed a more advanced timing that result in higher combustion pressure and thus higher torque (Agarwal, 2007). From the experimental results, the brake specific fuel consumption (BSFC) was calculated in order to understand the variations of fuel consumption in the test engine using different ethanol gasoline blended fuels. The BSFC (g/kWh) is defined as the ratio of the rate of fuel consumption (g/h) and the brake power (kW). Fig. 2c indicates the variations of the BSFC for different ethanol gasoline blended fuels under various engine speeds. As shown in this figure, the BSFC decreased as the ethanol percentage increased. Also, a slight difference exists between the BSFC using pure gasoline and using ethanol gasoline blended fuels. As engine speed increases reaching 1600 rpm, the BSFC decreases reaching its minimum value. This is due to the increase in brake thermal efficiency (Najafi et al., 2009).

Fig. 3a shows the effect of methanol gasoline blended fuels on engine brake power. With an increasing fraction of methanol engine power slightly decreased for all engine speeds. The brake power of gasoline was higher than those of M10–M30, especially for high engine speeds (e.g., 2500 rpm). Fig. 3b shows the influence of methanol gasoline blended fuels on engine torque. The increase of methanol content decreased slightly the torque of the engine. The brake torque of gasoline was higher than those of M10–M30. Fig. 3c indicates the variations of the BSFC for methanol gasoline blended fuels under various engine speeds. As shown in this figure, the BSFC increased as the methanol percentage increased. Also, a slight difference exists between the BSFC while using gasoline and while using methanol gasoline blended fuels. As engine speed increased reaching 1600 rpm, the BSFC decreased reaching its minimum value. However, if the spark ignition time is advanced by 2° without any further optimizations, under WOT full load operation conditions, the engine power shows almost no reduction and BSFC can be decreased as well (El-Emam and Desoky, 1984; Liu et al., 2007).

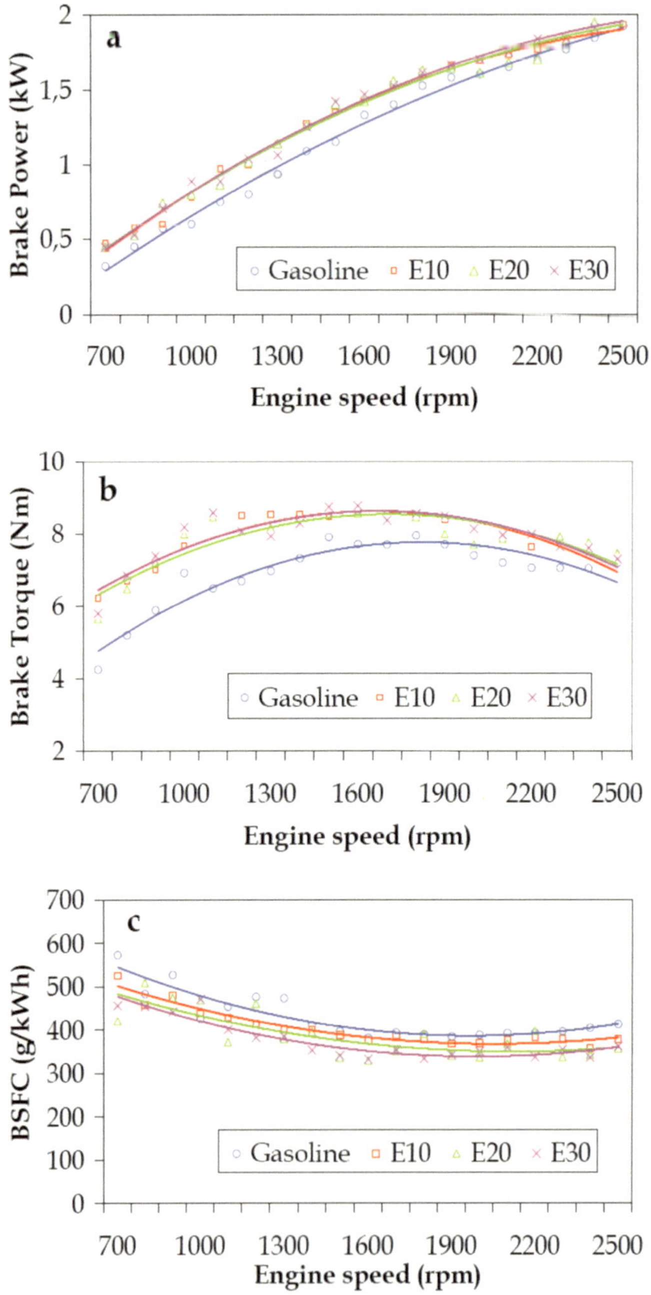

Fig. 2. Experimental results of engine performance characteristics using different ethanol gasoline blended fuels under various engine speeds. (a) Brake power, (b) Brake torque, and (c) Brake specific fuel consumption.

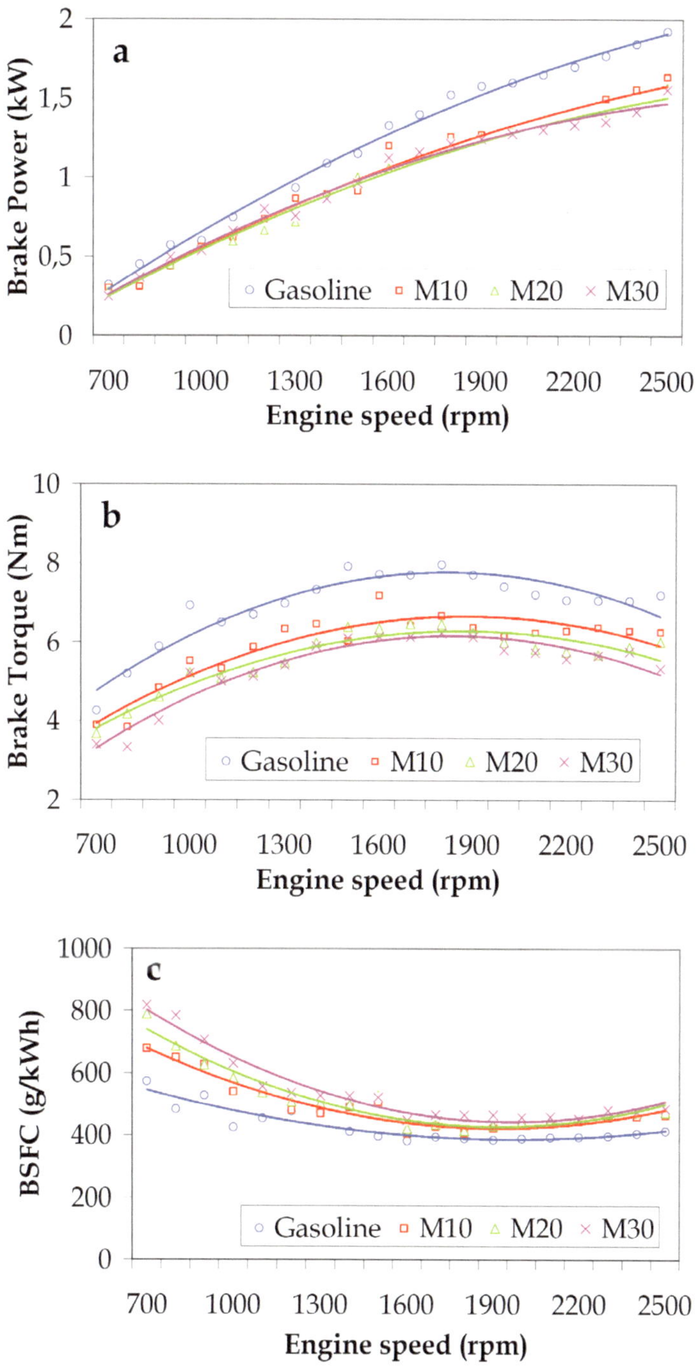

Fig. 3. Experimental results of engine performance characteristics using different methanol gasoline blended fuels under various engine speeds. (a) Brake power, (b) Brake torque, and (c) Brake specific fuel consumption.

The influences of ethanol and methanol addition to unleaded gasoline on SI engine performance characteristics at variable engine speeds are illustrated in Figs. 4 & 5. As shown in Figs. 4a and 4b the brake power and torque slightly decreased as the percentage of ethanol increased for all engine speeds. In fig. 4c, the BSFC decrease continued until the percentage of ethanol reached 40%. Above this point, BSFC started to increase.

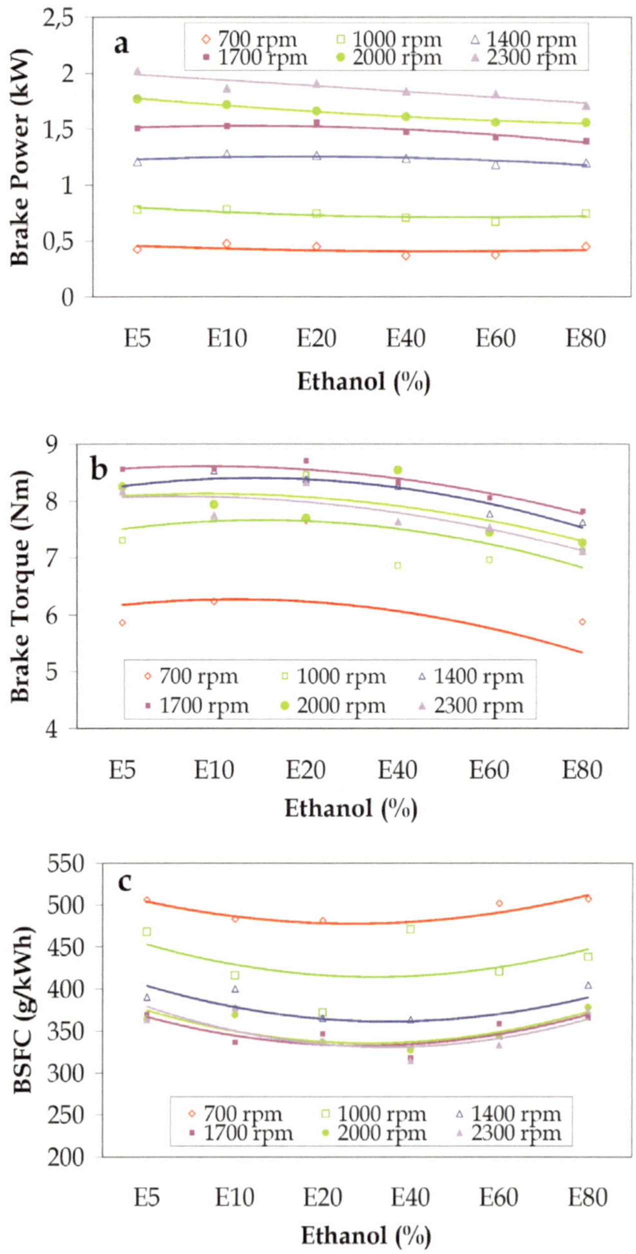

Fig. 4. The influence of ethanol addition on the engine performance characteristics. (a) Brake power, (b) Brake torque, and (c) Brake specific fuel consumption.

As shown in Figs. 5a, 5b and 5c brake power, brake torque and BSFC characteristics have opposite line tendency between lower and higher engine speeds. These characteristics increased as the percentage of methanol increased for lower engine speeds (700-1400 rpm), while the characteristics slightly decreased for higher engine speeds (1700-2300 rpm).

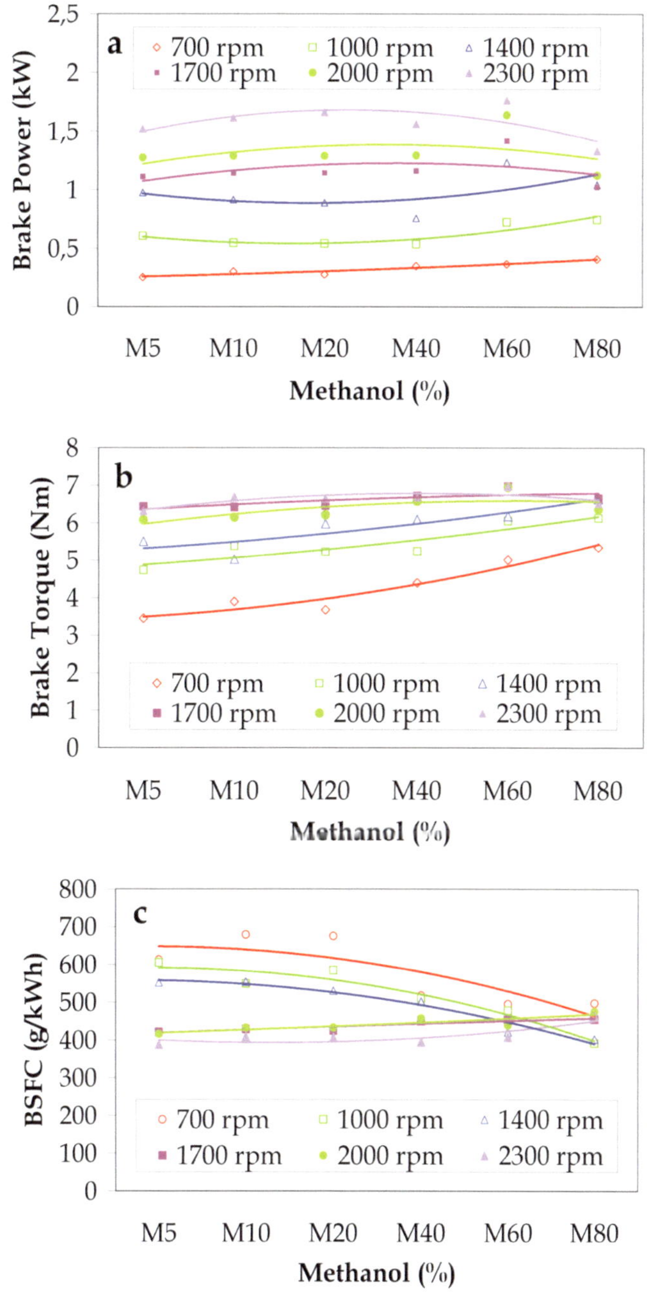

Fig. 5. The influence of methanol addition on the engine performance characteristics. (a) Brake power, (b) Brake torque, and (c) Brake specific fuel consumption.

Fig. 6 shows the comparison between brake power characteristics under different ethanol and methanol blended fuels and engine speeds. One can see that ethanol/gasoline blends have significant higher brake power values than methanol/gasoline blends until the percentage of these blends reaches 40% for lower engine speeds (700-1400 rpm).Beyond this point, both brake power characteristics start to converge. For higher engine speeds (1700-2300 rpm), brake power characteristics are converging until the percentage of the blends reaches 60 %, while beyond this percentage start to diverge. This is due to the influence of the combustion-related properties of the blended fuels. Fig. 7 shows the comparison between BSFC characteristics under different ethanol and methanol blended fuels and engine speeds. One can see that ethanol/gasoline blends have significant lower BSFC values than methanol/ gasoline blends. For lower engine speeds, BSFC characteristics values are converging, while BSFC characteristics values are diverging for higher engine speeds.

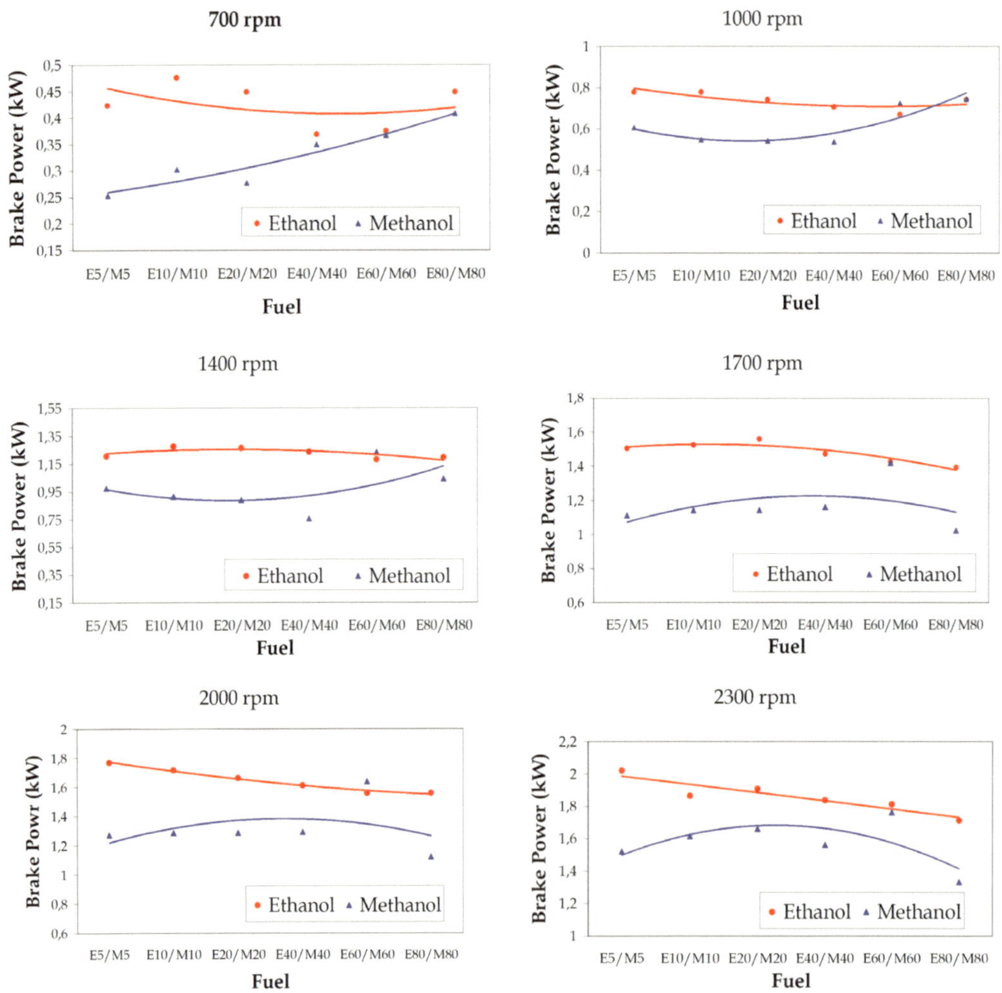

Fig. 6. Comparison of brake power characteristics using different ethanol and methanol gasoline blended fuels.

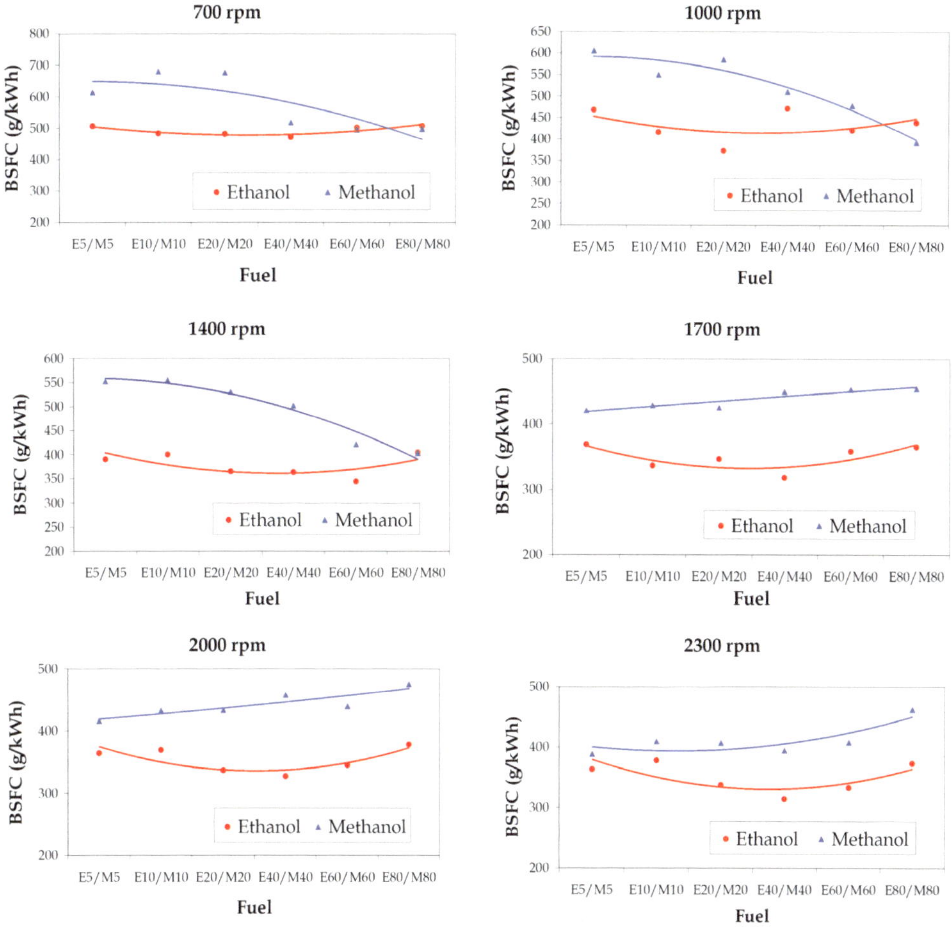

Fig. 7. Comparison of brake specific fuel consumption using different ethanol and methanol gasoline blended fuels.

Adding ethanol to gasoline will lead to improved performance characteristics in a spark ignition non-road engine with low efficiency. It was experimentally demonstrated that adding 30% ethanol to the blend led to an increase in the engine brake power, torque and decreased the BSFC. These findings broadly concur with those of previous studies. The brake power and brake torque of methanol gasoline blends are lower than those of gasoline for all engine speeds. Also, the BSFC of methanol blend fuels is higher than that of gasoline. Therefore, it was shown that the addition of moderate amounts of methanol to gasoline should not appreciably affect the performance characteristics of an unmodified spark ignition engine. This effect is attributed to the following factors: a) the lower heating value per unit mass of methanol and b) the lower stoichiometric air-fuel ratio of methanol gasoline blends.

2.3 Engine emission characteristics

To investigate the effect of different ethanol/methanol gasoline blended fuels on exhaust emissions, results of the engine test at 2000 rpm with full throttle valve opening were selected for comparison, as shown in Fig. 8.

CO is a toxic gas that is the result of incomplete combustion. When ethanol and methanol containing oxygen is blended with gasoline, the combustion of the engine becomes better and therefore CO emission is reduced (Stump et al., 1996; Yasar, 2010). As seen in Fig. 8, the values of CO emission are about 3.654% (3.637%), 3.161% (3.145), 2.842% (2.825), 2.337% (2.306%), 1.851% (1.824%) and 1.275% (1.248%) for E5 (M5), E10 (M10), E20 (M20), E40 (M40), E60 (M60) and E80 (M80) fuels, respectively.

CO_2 is non-toxic but contributes to the greenhouse effect. The CO_2 concentrations at 2000 engine speed with full throttle valve opening using ethanol and methanol gasoline blends were decreased in comparison to gasoline. Because the ethanol and methanol contain less C atoms than gasoline, it gives off lower CO_2 (Knapp, 1998; Celik, 2008). The value of CO_2 emission is about 13.88% for gasoline fuel, while the values of CO_2 are about 13.12% (12.96%), 12.95% (12.78%), 12.25% (12.12%), 11.73% (11.68%), 10.42% (10.39%) and 9.78% (9.57%) with E5 (M5), E10 (M10), E20 (M20), E40 (M40), E60 (M60) and E80 (M80) fuels, respectively.

The HC concentration in the exhaust gas emission at 2000 rpm with full throttle valve opening, for gasoline fuel was 345 ppm, while the HC concentration of E5 (M5), E10 (M10), E20 (M20), E40 (M40), E60 (M60) and E80 (M80) fuels was 341 (304), 301 (297), 282 (223), 265 (234), 273 (261) and 380 (372) ppm, respectively. The HC concentration at 2000 rpm using E5 (M5), E10 (M10), E20 (M20), E40 (M40) and E60 (M60) was decreased by 8.98% (11.88%), 12.75% (13.91%), 18.26% (35.36%), 23.19% (32.17%) and 20.86% (24.35%), respectively, while the HC concentration of E80 (M80) fuels was increased by 10.14% (7.83%), respectively in comparison to gasoline. These results indicate that ethanol and methanol can be treated as a partially oxidized hydrocarbon when they are added to the blended fuel. Therefore, HC emissions decrease to some extent as ethanol/methanol added to gasoline increase. The low ethanol/methanol and high ethanol/methanol content blends reduce the cylinder temperature as the heat of vaporization of ethanol/methanol is higher when compared to gasoline. The lower temperature causes misfire and/or partial burn in the regions near the combustion chamber wall. Therefore, HC emissions increase, and engine power can slightly decrease. This behaviour has been reported by other investigators on various types of engines and conditions (Celik, 2008; Najafi et al., 2009).

It shows that as the percentage of ethanol/methanol in the blends increased, NOx emission was decreased. The NOx concentration in the exhaust gas emission at 2000 rpm with full throttle valve opening, for gasoline fuel was 2247 ppm, while the NOx concentration of E5 (M5), E10 (M10), E20 (M20), E40 (M40), E60 (M60) and E80 (M80) fuels was 1957 (1945), 1841 (1828), 1724 (1574), 1498 (1379), 1366 (1338) and 1223 (1207) ppm, respectively. The NOx concentrations at 2000 rpm using E5 (M5), E10 (M10), E20 (M20), E40 (M40), E60 (M60) and E80 (M80) fuels were decreased by 12.91% (13.44%), 18.07% (18.65%), 23.27% (29.95%), 33.33% (38.63%), 39.21% (40.45%) and 45.57% (53.72%), respectively in comparison to gasoline. Since ethanol/methanol have a higher heat of vaporization relative to that of neat gasoline, the blends temperature at the end of intake stroke decreases and finally causes combustion temperature to decrease. As a result, engine-out NOx emissions decrease (He et al., 2003; Celik, 2008).

The fuel blends containing high ratios of ethanol and methanol had important effects on the reduction exhaust emissions. Experimental results demonstrate that the most suitable fuels were E40 and M20 in terms of HC emission. CO, CO_2 and NOx concentrations of E80 and M80 were the lowest when compared to the other blend fuels.

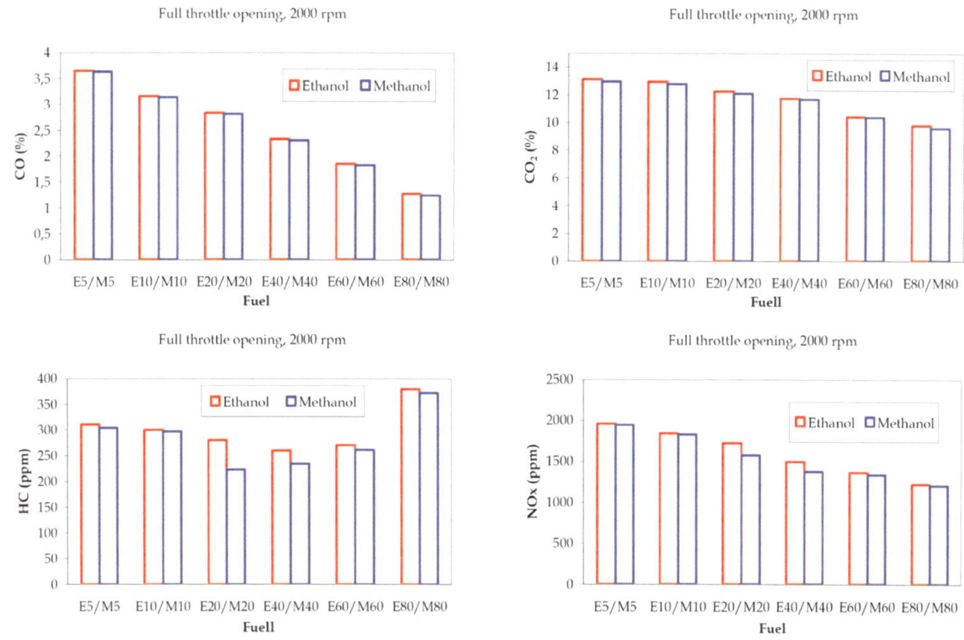

Fig. 8. The effect of various ethanol/methanol gasoline blend fuels on CO, CO_2, HC and NOx emissions.

3. Conclusion

The present chapter demonstrates the influences of ethanol and methanol addition to unleaded gasoline on non-road SI engine performance and emission characteristics. The use of ethanol gasoline blended fuels increase the brake power and brake torque, and decreases the BSFC. Methanol gasoline blended fuels show lower brake power and brake torque and higher BSFC than gasoline. The performance characteristics of methanol gasoline blended fuels are worse than for ethanol due to the influence of the combustion-related properties. The use of fuel blends containing high ratios of ethanol and methanol, at 2000 engine speed with full throttle valve opening, have significant effects on the reduction exhaust emissions.

4. References

Chum, H.L. & Overend R.P. (2001). Biomass and renewable fuels. Fuel Processing Technology, Vol.71, pp. 187-195.

Prasad, S.; Singh, A. & Joshi, H.C. (2007). Ethanol as an alternative fuel from agricultural, industrial and urban residues. Resources Conservation and Recycling, Vol.50, pp. 1-39.

Freudenberger, R. (2009). *Alcohol Fuel*. New Society Publishers, Canada.

Shapouri, H.; Duffield, J.A. & Graboski, M.S. (1995). Estimating the Net Energy Balance of Corn Ethanol: An Economic Research Service Report. U.S. Department of Agriculture, Economic Research Service, Office of Energy and New Uses. Agricultural Economic Report No.721.

(http://www.ethanol-gec.org/corn_eth.htm)
Goldemberg, J.; Coelho, S.T.; Nastari, P.M. & Lucon O. (2004). Ethanol learning curve—the Brazilian experience. Biomass and Bioenergy, Vol.26, pp. 301 - 304.
Kim, S. & Dale B.E. (2004). Global potential bioethanol production from wasted crops and crop residues. Biomass and Bioenergy, Vol.26, No.4, pp. 361-375.
von Blottnitz, H. & Curran M.A. (2007). A review of assessments conducted on bio-ethanol as a transportation fuel from a net energy, greenhouse gas, and environmental life cycle perspective Journal of Cleaner Production, Vol.15, pp. 607-619.
Pimentel, D. (2003). Ethanol Fuels: Energy Balance, Economics, and Environmental Impacts are Negative. Natural Resources Research, Vol.12, No.2, pp. 127-134.
Niven, R.K. (2005). Ethanol in gasoline: environmental impacts and sustainability review article. Renewable and Sustainable Energy Reviews, Vol.9, No.6, pp. 535-555.
Winnington, T.L. & Siddiqui K.M. (1983). Engine performance on gasohol--the Kenyan experience. Automotive Engng, Vol.8, No.3, pp. 12–30.
El-Kassaby, M.M. (1993). Effect of using differential ethanol–gasoline blends at different compression ratio on SI engine. Alexandria Engng J, Vol.32, No3, pp. 135–142.
Abdel-Rahman, A.A. & Osman, M.M. (1997). Experimental investigation on varying the compression ratio of SI engine working under different ethanol-gasoline fuel blends. International Journal of Energy Research, Vol.21, pp. 31-40.
Yacoub, Y.; Bata, R. & Gautam M. (1998). The performance and emission characteristics of C1-C5 alcohol-gasoline blends with matched oxygen content in a single-cylinder spark ignition engine. Proceedings of the Institution of Mechanical Engineers, Vol.212, Part A, pp. 363-379.
Al-Hasan, M. (2002). Effect of ethanol–unleaded gasoline blends on engine performance and exhaust emission. Energy Conversion & Management, Vol.44, No9, pp. 1547–1561.
Wu, C.W.; Chen, R.H.; Pu, J.Y. & Lin, T.H. (2004). The influence of air–fuel ratio on engine performance and pollutant emission of an SI engine using ethanol-gasoline-blended fuels. Atmospheric Environment, Vol.38, pp. 7093–7100.
Yücesu, H.S.; Topgül, T.; Çinar, C. & Okur, M. (2006). Effect of ethanol–gasoline blends on engine performance and exhaust emissions in different compression ratios. Applied Thermal Engineering, Vol.26, pp. 2272–2278.
Liu, S.; Eddy R. Cuty Clemente; Hu, T. & Wei, Y. (2007). Study of spark ignition engine fueled with methanol/gasoline fuel blends. Applied Thermal Engineering, Vol.27, pp. 1904–1910.
Koç, M.; Sekmen, Y.; Topgül, T. & Yücesu, H.S. (2009). The effects of ethanol–unleaded gasoline blends on engine performance and exhaust emissions in a spark-ignition engine. Renewable Energy, Vol.34, pp. 2101–2106.
Yücesu, H.S.; Sozen, A.; Topgül, T. & Arcaklioğlu, E. (2007). Comparative study of mathematical and experimental analysis of spark ignition engine performance used ethanol–gasoline blend fuel. Applied Thermal Engineering, Vol.27, pp.358–368.
Najafi, G.; Ghobadian, B.; Tavakoli, T.; Buttsworth, D.R.; Yusaf, T.F. & Faizollahnejad, M. (2009). Performance and exhaust emissions of a gasoline engine with ethanol blended gasoline fuels using artificial neural network. Applied Energy, Vol.86, pp.630–639
Rajan, S. & Saniee F.F. (1983). Water-ethanol-gasoline blends as spark ignition engine fuels. Fuel, Vol.62, pp. 117-121.

Palmer, F.H. (1986). Vehicle performance of gasoline containing oxygenates. Proceedings of International Conference on Petroleum Based Fuels and Automotive Applications, London.

Bata, R.V. & Roan V.P. (1989). Effects of ethanol and/or methanol in alcohol-gasoline blends on exhaust emission. J Engng Gas Turb Power, Vol.111, No.3, pp. 432-438.

Taylor, A.B.; Moran, D.P.; Bell, A.J.; Hodgson, N.G.; Myburgh, I.S. & Botha, J.J. (1996). Gasoline/Alcohol Blends: Exhaust Emissions, Performance and Burn-Rate in a Multi-Valve Production Engine. SAE Paper, No. 961988, pp. 143-160.

Chao, H.R.; Lin, T.C.; Chao, M.R.; Chang, F.H.; Huang, C.I. & Chen, C.B. (2000). Effect of methanol-containing additive on diesel engine. Journal of Hazardous Materials, Vol.13, No1, pp. 39-54.

Gautam, M.; Martin II, D.W. & Carder, D. (2000). Emissions characteristics of higher alcohol/gasoline blends. Proceedings of the Institution of Mechanical Engineers, Part A: Journal of Power and Energy Vol.214, pp. 165-182.

Hsieh, W.-D.; Chen, R.-H.; Wu, T.-L. & Lin, T-H. (2002). Engine performance and pollutant emission of an SI engine using ethanol-gasoline blended fuels. Atmospheric Environment Vol.36, pp. 403-410.

He, B.-Q.; Wang, J.-X.; Hao, J.-M.; Yan, X.-G. & Xiao, J.-H. (2003). A study on emission characteristics of an EFI engine with ethanol blended gasoline fuels. Atmospheric Environment Vol.37, pp. 949-957.

Yüksel, F. & Yüksel, B. (2004). The use of ethanol-gasoline blend as fuel in a SI engine. Renewable energy Vol.29, pp.1181-1191.

Bayraktar, H. (2005). Experimental and theoretical investigation of using gasoline-ethanol blends in spark-ignition engines. Renewable Energy Vol.30, pp.1733-1747.

Jia, L.-W.; Shen, M.-Q.; Wang, J. & Lin, M.-Q. (2005). Influence of ethanol-gasoline blended fuel on emission characteristics from a four-stroke motorcycle engine. Journal of Hazardous Materials Vol.A123, pp. 29-34.

Ceviz, M.A. & Yüksel, F. (2005). Effects of ethanol-unleaded gasoline blends on cyclic variability and emissions in an SI engine. Applied Thermal Engineering Vol.25, pp. 917-925.

De Caro, P.S.; Mouloungui, Z.; Vaitilingom, G. & Berge, J.Ch. (2001). Interest of combining an additive with diesel-ethanol blends for use in diesel engines. Fuel Vol.80, No.4, pp. 565-574.

Sinor, J.E. & Bailey, B.K. (1993). Current and potential future performance of ethanol fuels. SAE paper 930376, Warrendale, PA.

Bailey, B.K. (1997). Performance of ethanol as a transportation fuel. Handbook on bioethanol: Production and Utilization. Edited by Charles E. Wyman. Applied Energy Technology Series, pp. 37-60.

Reed, T.B. & Lerner, R.M. (1973). Methanol: A versatile fuel for immediate use. Science Vol.182, pp. 1299-1304.

Yamamoto, T. (1972). Environmental pollution and svstematization of chemical techniques - Use of methanol as a fuel. Chemical Economy and Engineering Review, pp. 52-57.

Ceci1, R.R. (1974). Exxon experience with alcohols in motor gasoline. Statement to Subcommittee on Priorities and Economy in Government, Joint Economic Committee.

Fitch, R.E. & Kilgroe, J.D. (1970). Investigation of a substitute fuel to control automotive air pollution. Final Report. National Air Pollution Control Administration.

Adelman, H.G. (1972). Exhaust emissions from a methanol fuelled automobile. SAE Paper No 720693, Society of Automotive Engineers, NY, USA.

Agarwal, A.K. (2007). Biofuels (alcohols and biodiesel) applications as fuels for internal combustion engines. Prog Energy Combust Sci Vol.33, pp. 233–271.

El-Emam, S.H. & Desoky, A.A. (1984). A study on the combustion of alternative fuels in spark-ignition engines. International Journal of Hydrogen Energy Vol.10. No.7/8, pp. 497-504.

Stump, F.D.; Knapp, K.T.; Ray, W.D., Siudak, P. & Snow, R. (1996). Influence of ethanol blended fuels on the emissions from three pre-1985 light-duty passenger vehicles. J Air Waste Mgmt Assoc, Vol.46, pp. 1149–1161.

Yasar, A. (2010). Effects of alcohol-gasoline blends on exhaust and noise emissions in small scaled generators. Metalurgija, Vol.49, No.4, pp.335-338.

Celik, M.B. (2008). Experimental determination of suitable ethanol–gasoline blend rate at high compression ratio for gasoline engine. Applied Thermal Engineering, Vol.28, pp. 396–404.

Knapp, K.T.; Stump, F.D. & Tejada, S.B. (1998). The effect of ethanol fuel on the emissions of vehicles over a wide range of temperatures. J Air Waste Mgmt Assoc, Vol.48, No.7, pp. 646–53.

Part 3

Biomass

8

Use of Charcoal and Partially Pirolysed Biomaterial in Fly Ash to Produce Briquettes: Sugarcane Bagasse

Silvio Rainho Teixeira, Agda Eunice de Souza, Angel Fidel Vilche Peña, Regiane Godoy de Lima and Álvaro Gil Miguel[1]

Universidade Estadual Paulista – UNESP– Presidente Prudente – S.P.

[1]Usina Alto Alegre – Presidente Prudente – S.P.

Brazil

1. Introduction

Currently, many scientists worldwide are studying the recycling or reuse of waste materials to make useful products. For example, silicate-based wastes have been studied for use as raw material in the ceramic industry. Although numerous types of waste have been studied, sugarcane bagasse ash (SCBA) has been overlooked, probably because this type of residue is more common in developing countries that produce sugar and alcohol from sugarcane. Bagasse is a by-product of the cane sugar industry. Brazil is the world's largest producer of alcohol and sugar from sugarcane and has an extensive alternative program for automobile fuel. The new technology of flex-fuel (alcohol/gasoline) vehicles, developed in the 1990s and applied to new cars in 2003 revived the Brazilian PROALCOOL program. Today, several million flex-fuel cars are in use, and more than 85% of the new cars produced in the country run on this system (PróAlcool 2011; History of ethanol fuel in Brazil 2011; Goldemberg 2008).

In the production of sugar and alcohol, sugarcane is ground and the resulting juice is extracted to obtain sugar or used in a fermentation process to produce alcohol. The fibrous matter that remains after sugarcane has been crushed to extract the juice is called bagasse (or "bagaço," in Portuguese). There are many applications and much research effort is directed at exploring the potential of bagasse as a renewable power generation source, for the production of bio-based materials (for example, bioethanol and briquettes or pellets), cattle feed, building materials, etc. Basically, bagasse is a fibrous material containing cellulose (~50%), hemicellulose and lignin as the major (~95%) compounds. There are also waxes, inorganic materials and/or salts inside the plant tissues that will result in the final ash (oxides and carbonates) after firing (Pandey et al. 2000).

Although there is a great deal of research in Brazil on sugarcane bagasse briquetting, showing the many advantages of using briquettes instead of bagasse, factories still burn bagasse. Currently, sugarcane bagasse is used as a primary fuel source in sugar mills. It is burned in a boiler to produce steam which is utilized in the factory's processes and also to power turbines for the production of electrical energy. Today, the cogeneration of electricity by sugar and alcohol mills accounts for ca. 5% of the total electricity produced in Brazil

(BEN 2010; SYS 2001). However, most factories do not participate in this production, which can achieve in the future a rate of about 15% of the market. The combustion yields ashes (bottom and fly ashes) containing high amounts of organic material (~35% by weight of charcoal and bagasse debris) and silica as the major inorganic component. Fly ash contains more charcoal than does the bottom ash. The inorganic fraction can be used by the ceramic industry and the organic one can be used to produce fuel briquettes (Teixeira et al. 2010b).
Fly ash is the solid material that is removed from the gases leaving the boiler to the industry chimneys. In general it consists of a mixture of inorganic materials, charcoal, and partially pyrolyzed material (bagasse debris). Depending on the source and makeup of the material being burned, the components of the fly ash produced vary considerably. Its composition includes substantial amounts of inorganic substances (generally oxides and carbonates) and carbon matter (organic material), which can be detected by loss on ignition or by thermogravimetric analyses. In fly ash from coal, the carbon concentration is low (< 6%), but with biomass for burning, such as sugarcane bagasse, there can be a very high carbon content, as much as 35%. This value can be reduced by improving the efficiency of the firing process or using briquettes instead of burning crude bagasse. Therefore, for this residue to be used as raw material, it is interesting to separate out the charcoal to produce briquettes and to improve the quality of the inorganic part for use in the ceramic industry. Due to the difference in grain size distribution and in density of organic and inorganic fly ash compounds it is possible to separate the two fractions using sieves, hydrocyclones or other processes (Teixeira et al. 2010b, Ahmaruzzaman 2010; Rodríguez-Pedrosa et al. 2010; Zandersons et al. 1999).
The estimated 2009/2010 Brazilian sugarcane harvest was 629 Mton to produce sugar (44.7%) and alcohol (55.3%) (BEN 2010). The volume of ash (fly and bottom) that would be produced from this harvest is approximately 3.2 Mton (1000 kg cane → 250 kg bagasse → 6 kg ash) (Teixeira et al. 2010a). In this chapter, we will show some of the possible utilizations of sugarcane bagasse and will discuss the separation, characterization and utilization of the charcoal in the fly ash of the fired bagasse, to produce briquettes.

2. Energy in Brazil and proalcohol

Brazil has the most renewable energy sources in the industrialized world with 46.8% of its primary energy production from renewable resources such as hydropower plants (13.9%), firewood (10.2%), sugarcane products (18.8%) and others (2.8%). Figure 1 shows the evolution of primary energy production in Brazil over the past 40 years. The non-renewable energy sources (oil, natural gas, coal and uranium) still represent about 53% of the energy produced and consumed in the country. Hydropower plants are responsible for generating over 76% of the electricity in the country. It is worth noting that the global energy matrix is composed of 13% renewable in the case of industrialized countries, falling to 6% among developing nations. The Brazilian energy model has great potential for expansion, which results in a number of opportunities for long-term investment. The estimate of the Ministry of Mines and Energy for the period 2008-2017 indicate public and private contributions of R$ 352 billion (~US$ 207 billion) for the expansion of national energy programs (BG, 2009; BEN, 2010).
The Brazilian domestic energy supply (BEN 2010) shows the participation of each kind of energy source in the country's energy matrix (Fig. 2) in the last seven decades. It is interesting to note the increase in the consumption of oil and electricity, with a

corresponding decrease in consumption of wood, after the World War II until the 1970s. With the oil crisis in the 1970s and 1980s there was a stabilization in the consumption of oil

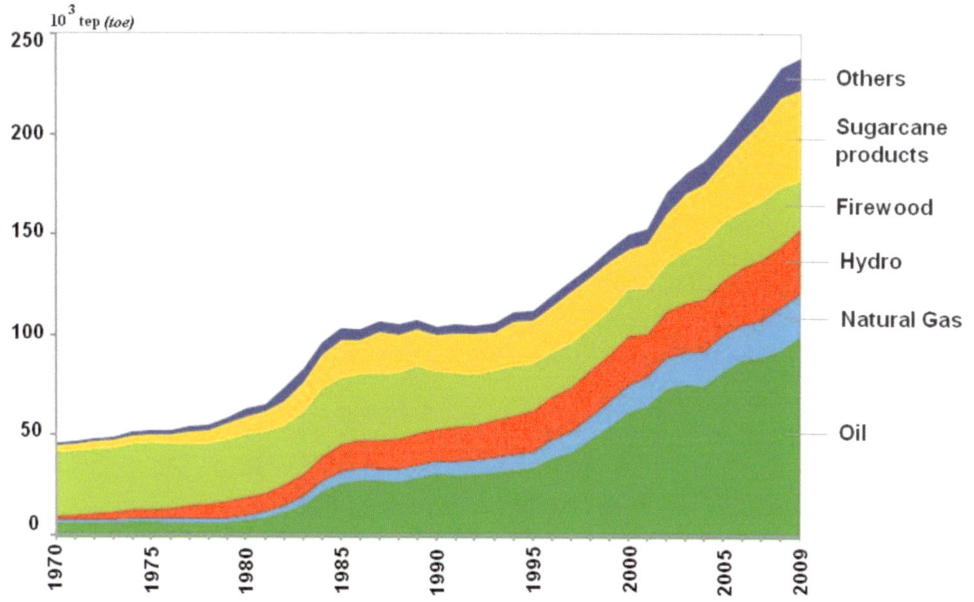

Fig. 1. Brazilian Primary Energy Production (BEN 2010)

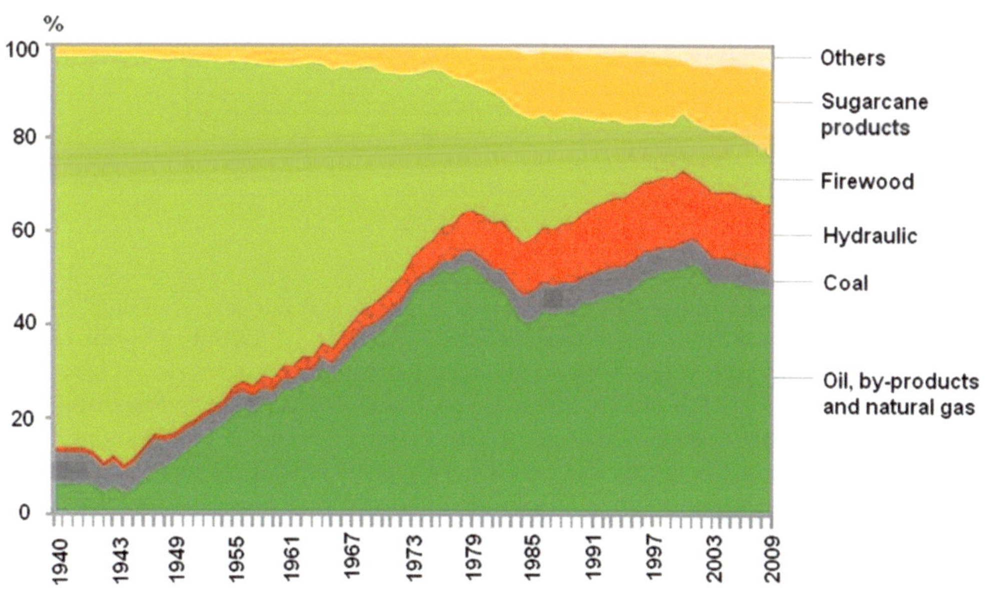

Fig. 2. Brazilian Domestic Energy Supply (BEN 2010)

and increased production of hydraulic power and ethanol, with the Brazilian alcohol program. At the height of this program, between 1975 and 1985, there was a decrease in consumption of oil products including gasoline, which was replaced by alcohol. Table 1 shows the trend in energy production for each sector, every 20 years, for the last 40 years.

The progress in energy production from renewable resources happened mainly in the sugarcane industry. Brazil's sugarcane-based ethanol fuel program allowed the country to become the world's second largest producer of ethanol, and the world's largest exporter. In response to the 1973 oil crisis, the Brazilian government began promoting bioethanol as a fuel. The National Alcohol Program - *Pró-Álcool* (In Portuguese, **Programa Nacional do Álcool**), was a nation-wide program financed by the government to phase out automobile fuels derived from fossil fuels, such as gasoline, in favor of ethanol produced from sugarcane. Several important political and technological developments led Brazil to become the world leader in the sustained use of bioethanol. Government policies and technological advances also allowed the country to achieve a landmark in ethanol consumption, when ethanol retail sales surpassed 50% of the market share of the gasoline-powered vehicle fleet in early 2008.

Year	1970	1990	2009
Oil by-products and natural gas	38,0	43,7	46,6
Coal and by-products	3,6	6,8	4,7
Hydraulic and electricity	5,1	14,1	15,2
Firewood and charcoal	47,6	20,1	10,1
Sugar cane products	5,4	13,4	18,2
Others[1]	0,3	1,9	5,2
Total	100	100	100

Table 1. Domestic Energy Supply (BEN 2010)

The Brazilian National Alcohol Program (known as **Proalcool**) (History of ethanol fuel in Brazil, 2011; PróAlcool, 2011) was established on November 14, 1975 as a result of two serious international crises in the 1970s: in the sugar market and oil market that pushed the average price of a barrel of oil at U.S. $ 2.91 in September 1973 to $ 12.45 in March 1975. At the time, Brazil imported about 80 percent of the crude oil consumed in the country. The program was a success for nearly 15 years. By the end of 1980, there were more than 4 million cars and light trucks running on pure alcohol, which accounted for a third of the country's fleet of motor vehicles. However, at this time, ethanol production and sales of cars running on pure ethanol dropped due to several factors (gasoline prices fell sharply, Brazilian inflation, sugar prices increased sharply, a shortage of ethanol fuel supply, and the

[1] Including others renewable energy and uranium

reduction of ethanol subsidies by the government). In 1990, production of neat ethanol vehicles fell to 10.9% of the total car production as consumers lost confidence in the reliability of ethanol fuel supply, and began selling or converting their cars back to gasoline fuel. Confidence in ethanol-powered vehicles was restored with the introduction in the Brazilian market of flexible-fuel vehicles starting in March 2003 when Volkswagen launched in the Brazilian market the **Gol 1.6** Total Flex, the first commercial flexible fuel vehicle capable of running on any blend of gasoline and ethanol. Some months after others car companies launched their flex-fuel cars in the Brazilian Market. By 2010, there were 12 carmakers in Brazil (Chevrolet, Citroën, Fiat, Ford, Honda, Kia Motors, Mitsubishi, Nissan, Peugeot, Renault, Toyota and Volkswagen) that produced 70 flex fuel models. Flexible-fuel vehicles accounted for 22% of the car sales in 2004, 73% in 2005 and 87.6% in July 2008, with a record 94% set in August 2009. In the year 2010, there were more than 2.6 million cars and light commercial of the flex type produced in Brazil, representing more than 86% of the total manufactured that year (ANFAVEA, 2010). The fleet of **flexible-fuel vehicles in Brazil** is the largest in the world, and since their inception in 2003, a total of almost 12 million cars and light trucks have been produced as of December 2010. The first flex-fuel motorcycle was launched in March 2009, and by December 2010, a total of 515,726 units of the two flexible-fuel motorcycles available in the market had been sold, representing an 18.1% market share of the new motorcycle sales in Brazil in 2010. In addition to the establishment of flexible-fuel cars that use ethanol, the Brazilian government also made it mandatory for gasoline to be blend with ethanol, fluctuating from 1976 until 2010 between 10% and 20-25%, depending on the results of the sugarcane harvest and the levels of ethanol production from sugarcane, resulting in blend variations even within the same year. Brazilian flexible-fuel vehicles are optimized to run on any mix of E20-E25 gasoline and up to 100% hydrous ethanol fuel (E100) (History of ethanol fuel in Brazil, 2011; PróAlcool, 2011; Goldemberg, 2008).

A review article was published in 2008 (Goldemberg, 2008) by a leading authority on energy in Brazil. In this article he describes the rationale for the ethanol program in Brazil, its present 'status' and its perspectives. The environmental benefits of the program, particularly the contribution of ethanol to reducing the emission of greenhouse gases, are discussed, as well as the limitations to its expansion. According to the data presented in 2008, ethanol was used as biofuel to replace about 3% of the gasoline produced from fossil fuels consumed in the world at that time.

Besides the production of ethanol fuel the sugarcane industry co-generates electricity from the burning of bagasse, a byproduct of the production of sugar and alcohol. Today, many alcohol/sugar mills produce electrical energy only for their own use. Most modern plants produce excess energy that is sold to the electricity distributors. Currently, the total contribution from energy co-generation in sugar and alcohol factories to the electricity production is about 5%, and it is estimated that in some years the sector will provide about 15% of the total electricity produced in the country. The conversion of bagasse in energy is given approximately by the relation: 1ton (moist bagasse, 48-52% humidity) $\rightarrow$ 2.2 ton steam (65 kg/cm^2) $\rightarrow$ ~363 kW. At the moment, for the technology available in Brazil, the ideal condition for the sugarcane sector is to use boiler working pressure of 65 kg/cm^2 and temperature of steam of 500°C.

2.1 Biomass and residues: Bagasse

The utilization of organic and agricultural residues for energy is considered to be an important element in any strategy to achieve renewable energy goals and to reduce waste disposal and environmental pollution. This is a global concern that is reflected in the great number of published papers and international conferences, whose main theme is energy production from biomass. There is a wide range of biomass materials that are byproducts, residues (or wastes) from some other process, operation or industry. Many of these have a valuable energy content that can be usefully exploited. Biomass is a natural renewable carbon resource that is large enough to be used as a substitute for fossil fuels.

In many countries, particularly those with the poorest population, it is still common practice to burn organic waste (biomass), such as municipal solid waste, used tires, wood scraps, cardboard, plastics, etc., in garbage dumps or in inappropriate places. This practice produces bad odors and pollutes the air, soil and, in some cases, surface waters. Furthermore, it is meaningless practice considering the waste of energy, environmental pollution and health problems that can result. With increasing world population, demand for more energy and the increase in waste production makes burning even more inappropriate. Nowadays, in many countries, especially developed countries, measures are being adopted that are more interesting from the standpoint of environmental and economic development. In these countries, some solid wastes are collected separately (selective collection) and then go through a sorting process (metals, glass, papers, wood, plastics; composting of organic matter). Ultimately, the remainder is buried or incinerated. Direct incineration of unsorted waste is currently the main route practiced by the waste industry to produce energy (Ryu et al. 2007). The flow diagram (Fig. 3) shows the main processes to convert sugar cane bagasse into energy.

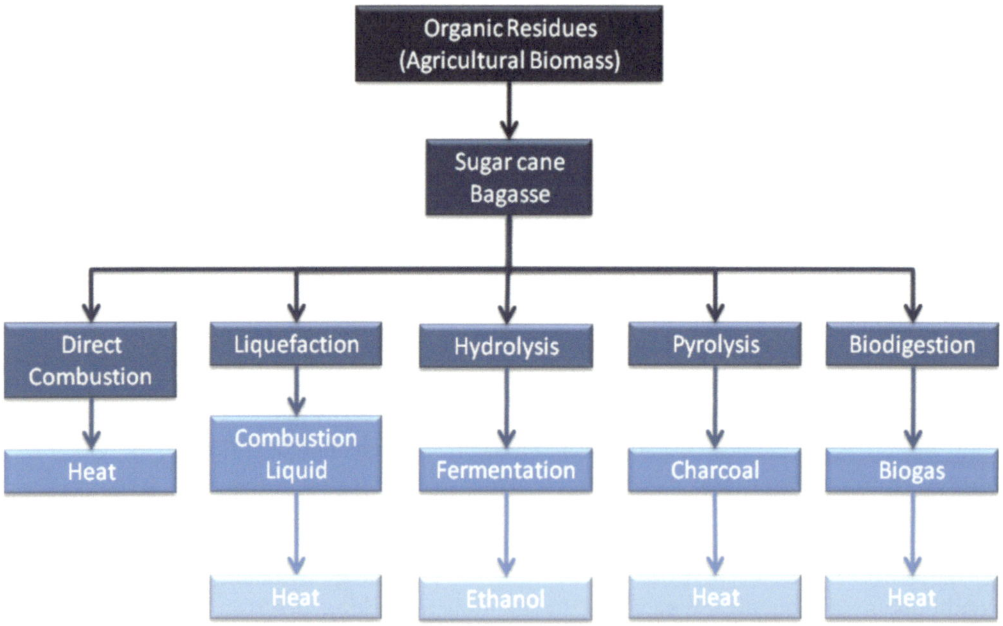

Fig. 3. Schematic diagram of the processes of energy conversion of biomass

Due to its extensive cultivable area, Brazil is one of the largest producers of agricultural products and therefore produces a large volume of biomass residues (around 330 million metric tons/year). Currently, sugarcane bagasse is the most abundant biomass residue generated in Brazil (Felfli et al. 2011).

In the sugar/alcohol industry, sugarcane is crushed to extract the juice used to produce sugar or fermented to produce alcohol. The fibrous matter that remains after the juice extraction is bagasse. The chemical composition of this lignocellulosic waste found in the literature varies widely due to differences in the types of plants, soil where it is grown and analysis conditions (on a washed and dried basis). Considering the ranges found in the literature (Goldemberg et al., 2008; Aguiar, 2010; History of ethanol fuel in Brazil, 2011) its basic composition (in weight %) is: cellulose (41 - 55%), hemicellulose (20 - 27.5%), lignin (18 - 26.3%) and others (~7%).

Sugarcane bagasse can be used either for energy production or for non-energy applications, and currently, there is much research on the uses of sugarcane bagasse (Pandey et al. 2000; Silva et al., 2010). Advances in industrial biotechnology offer potential opportunities for economic utilization of agro-industrial residues (or lignocellulosic residues) such as sugarcane bagasse. For example, it can be used as paper-making material and for cellulose derivatives (methylcellulose, cellulose acetate and microcrystalline cellulose) (Andrade 2001, Covey 2006, Yousef 2005, Pasquini et al., 2005; Vieira et al., 2007; Ilindra and Dhake 2008, Aguiar, 2010; Sun et al., 2004); it can also be utilized as a source of raw material in the chemical industry to produce xylose, xylitol and furfural and their derivatives (Cunha 2005; Lavarack 2002; Aguiar, 2010, Zhang et al., 2007). As it is rich in fermentable sugars (but not directly available), it has aroused the interest of researchers for the production of ethanol with chemical pre-treatments such as enzymatic or hydrolysis (Aguiar, 2010, Dawson and Boopathy, 2008; Botha and Von Blottnitz, 2006). In addition, bagasse can be used in various other ways: as a constituent of animal feed (Hossain et al., 2009; Fontana et al., 1995; Teixeira et al., 2007; Leme et al., 2003), to produce useful materials for the construction industry and furniture manufacturing, such as plywood, particleboard and other engineered wood products (Widyorini et al., 2005; Lu et al., 2006; Maldas and Kokta, 1991; Stael et al., 2001; Tabarsa et al., 2010); and as an adsorbent for environmental remediation to remove oil by-products from contaminated water (Silva 2007).

Nowadays, there are many companies around the world producing paper products (e.g., tableware) made of sugarcane bagasse. These products have many advantages over plastic goods: they are 100% biodegradable, nontoxic, harmless, recyclable, microwaveable and resistant to different temperatures (hot or cold). It can substitute for virgin paper and cellulose in many applications, saving the countless trees used to produce these products.

2.2 Ashes from sugarcane bagasse

For energy production, sugarcane bagasse can be burned as the raw product or in the form of briquettes. Despite the advantages of biomass briquetting (Granada et al. 2002; Yamaji et al. 2010; Felfi et al. 2011) and the huge production of biomass waste in Brazil (Felfi et al. 2011), the briquetting of biomass is insignificant when considering the volume of waste produced annually. Although briquetting results in more energy per unit volume and improves the transport and storage conditions for the sugarcane mills, it is nowadays cheaper to burn the sugarcane bagasse in the raw form at the place where it is produced. Currently, most of the sugarcane bagasse is burned in boilers to produce steam which is utilized in the factory's processes and also to power turbines for the production of electrical energy. The combustion of bagasse yields ashes (bottom and fly ashes) containing high amounts of charcoal (unburned carbon) and silica as the major inorganic component.

Aluminum, calcium, iron and magnesium oxides are the main minor components. The chemical properties of fly ash are associated with the type of material used in the thermal conversion, characteristics of the furnace, temperature of the process and ash extraction system (Rodriguez-Pedrosa et al. 2010; Ahmaruzzaman M. 2010). The loss on ignition (LOI) test and thermal gravimetric measurements show that the organic matter (carbon and biomass unburned) content in the ash from different types of biomass can range from 0 to 80 wt% (Frías et al. 2011; Batra et al. 2011; Rodriguez-Pedrosa et al. 2010; Ahmaruzzaman 2010). In general, the combustion of biomass and organic residues produces more unburned carbon compared to burning coal. The quantity and quality of the ash produced during the combustion process depends on the material burnt and on the technology used. The sugarcane bagasse ash (SCBA) has a high content (about 35 wt%) of carbon and fine particles of bagasse (partial combustion of the biomass) depending on the characteristics of the furnace and the speed of the bagasse combustion process (Teixeira et al. 2010b; Batra et al. 2011; Zandersons et al. 1999). Because of its high concentration of unburned carbon, fly ash can be used in many important environmental applications, such as adsorbent for the removal of various pollutants (Ahmaruzzaman 2009, 2010; Batra et al. 2011; Purnono et al. 2010; Ahmedna et al. 2000; Tai et al. 2009; Wang et al. 2010).

Approximately 65% weight of SCBA is inorganic material. Fly ash contains more charcoal than the bottom ash, and nowadays, in the most modern factories the bottom ash is mixed with the fly ash inside the water channel that comes from the gas washer. The inorganic fraction can be used by the ceramic industry (Teixeira et al., 2008) as non-plastic material to produce bricks and roof tiles, to produce glass-ceramic materials (Teixeira et al., 2010b) or as pozzolan in cementitious materials (Frías et al. 2011; Dutta and Das 2010; Cordeiro et al. 2009).

For each ton of cane processed, between 250 and 300 kg of bagasse are generated, which has a moisture content of 40 to 50%. Considering that about 300 kg of bagasse is produced per ton of sugarcane milled, and considering the amount of sugarcane (630 Mton) harvested in the 2009/2010 season, it can be estimated that 3.2 Mton of ash and about 1 Mton of carbon and unburned biomass were produced in this harvest.

In previous studies (Teixeira et al. 2010a, 2010b), we have demonstrated the feasibility of recovering unburned carbon (soot from bagasse) and bagasse debris from fly ash to produce charcoal briquettes. Although we used in these works a by-product of burning sugarcane bagasse, this can be extended to by-products of combustion processes for any fossil fuel or biomass.

2.2.1 Ashes preparation

Fly ash is the solid material suspended and carried by the exhaust gases, which is collected by electrostatic precipitators, filters or gas washers. Fly ashes are generally highly heterogeneous, consisting of a mixture of inorganic materials, charcoal and, in general, not totally burned material (partially pyrolyzed material). The boiler gas cleaning systems follow two main routes: dry (cyclones and electrostatic separators) and wet (gas washer or scrubbers). The gas-cleaning systems adopted by the sugarcane industry is the wet route (gas washer) because they are simpler to design, less expensive, and provide great efficiency. The scrubber's water treatment consists of waste flocculation and a subsequent rapid sedimentation of flocculated solids. The sludge containing the solid material, extracted in the decanter, is transported to a concentration step in a rotating screen and filter type blanket (suction strainer) where the water with solid waste (ash flay) is then filtered (Fig. 4 and Fig.5). Another option to decrease the moisture content of ash is passing it through an endless thread (Teixeira et al. 2010b). The water returns to the tanks and can be reused, and sludge is transported to deposits to be then spread over the ground. According to the process used and the working conditions (climate), the final ash moisture can vary from 40 to 60% by weight. The amounts of organic and inorganic materials also vary (about 35 and 65 wt%, respectively), as will be shown later in the thermogravimetric analysis.

Fig. 4. System for treatment of water from gas washer to remove the ash using decanters and porous blankets.

Fig. 5. Blanket type filter system (with suction strainer) to remove water and reduce moisture from the ashes

Since SCFA has a high content of charcoal and organic debris, it would be interesting to separate the two fractions (organic and inorganic) for use as raw material suitable for the production of goods useful to society. In briquette production, it is important to separate the combustion by-product with a minimum concentration of the inorganic fraction (useful for the cement and ceramic industries).

There are several methods (for example, cyclonic separation, flocculation and flotation processes) (Rubio et al. 2007, Martignoni et al. 2007) to separate the two fractions but none of them allows complete separation, i.e., there is always a certain amount of ash in the charcoal. Due to the difference in the grain size distribution and in density of organic and inorganic fly ash compounds many different processes can be used to separate them. In this work, we will discuss the separation process using sieves and a hydrocyclone.

2.2.2 Ashe characterization and charcoal separation

Fly ash samples were collected at the solid/water exit of the gas washer at the Alto Alegre mill, nearby Presidente Prudente city in São Paulo State, Brazil. As the charcoal is concentrated in the coarser fraction of the ash, it can be concentrated using sieves. Therefore, we used two different procedures to separate charcoal from ash: (1) ashes were passed through a 1 mm (18 mesh) sieve and (2) ashes were passed through a 0.125 mm (120 mesh) sieve; afterwards, the coarse fraction was passed through a hydrocyclone (Fig. 6). A hydrocyclone (Arterburn, R.A., 2010) is a device that classifies, separates or sorts particles, in a liquid suspension based on the ratio of their centripetal force to fluid resistance. It may be used to separate solid particles of different density (different materials for same grain size) or of different particle size (materials with same density). A hydrocyclone has two exits on its axis: the smaller one on the bottom (for underflow or reject purpose) and the larger one at the top (for overflow or accept purpose). The underflow is generally the denser or thicker fraction, while the overflow is the lighter or more fluid fraction. Internally, centrifugal force is countered by the resistance of the liquid, with the effect that larger or denser particles are transported to the wall for eventual exit at the reject side with a limited amount of liquid, whilst the finer, or less dense particles, remain in the liquid and exit at the overflow side through a tube extending slightly into the body of the cyclone at the center.

In the first procedure, the coarser charcoal fraction (grains > 1mm) that was separated using a 1 mm sieve, was analyzed to determine the calorific value (or heat of combustion) and the final concentration of ash, using a bomb calorimeter, according to Brazilian standards (ABNT-NBR 8633). The results show that this fraction has lower ash content (13.2 wt%) and high calorific value (6,140 kcal/kg). Compared with the literature (Quirino & Brito, 1991), it can be observed that the briquette of sugarcane bagasse generates more energy and lower amount of ash than charcoal briquettes from different sources. As the fraction smaller than 1 mm has a large amount of charcoal, a sieve less than 1 mm (0.125 mm) was used to separate the charcoal in the second procedure.

In the second procedure, the ash was first passed through a 115 mesh (125 mm) sieve. Before going through the hydrocyclone, the three samples, crude and sieved (115 mesh, 125 mm) fractions, were submitted to thermal gravimetric (or thermogravimetric) analysis (TGA) to determine changes in sample weight in relation to change in temperature. This technique determines the amount of ash and charcoal in the samples. The TGA results (Fig. 7) show that the inorganic residue of the ash fired at 800°C is 64.06 wt% for the crude ash, 48.53 wt% for the coarser ash (> 125 mm) and 83.21 wt% for the finer ash (< 125 mm). As observed in previous work (Teixeira et al. 2010b), sieving improves the concentration of charcoal in the samples. The sample moisture measured by TG ranges from 5.7 to 7.2 wt-%. TG data for the finer fraction (< 125 mm) confirms that it holds only 14.86 wt% of charcoal (Teixeira et al. 2010a).

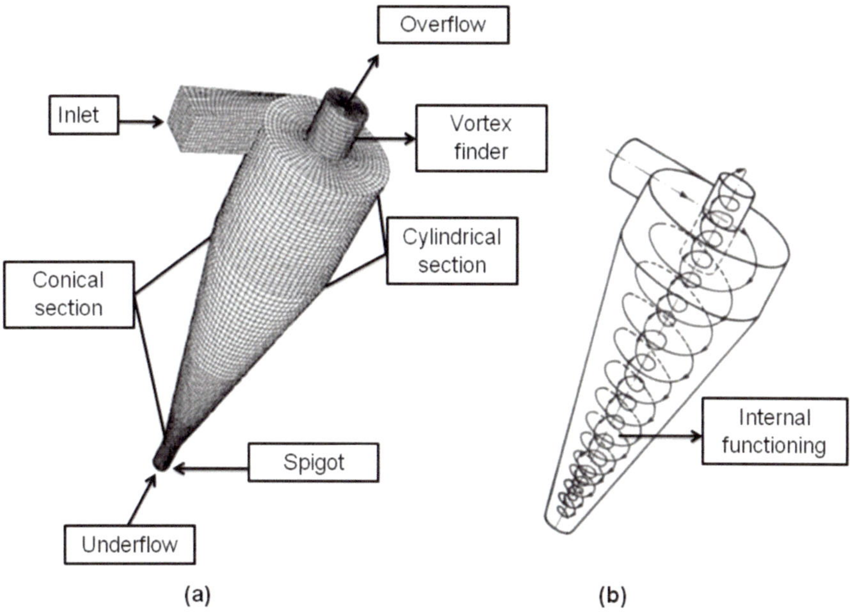

Fig. 6. (a) Illustration of a Hydrocyclone; (b) Internal functioning of the Hydrocyclone

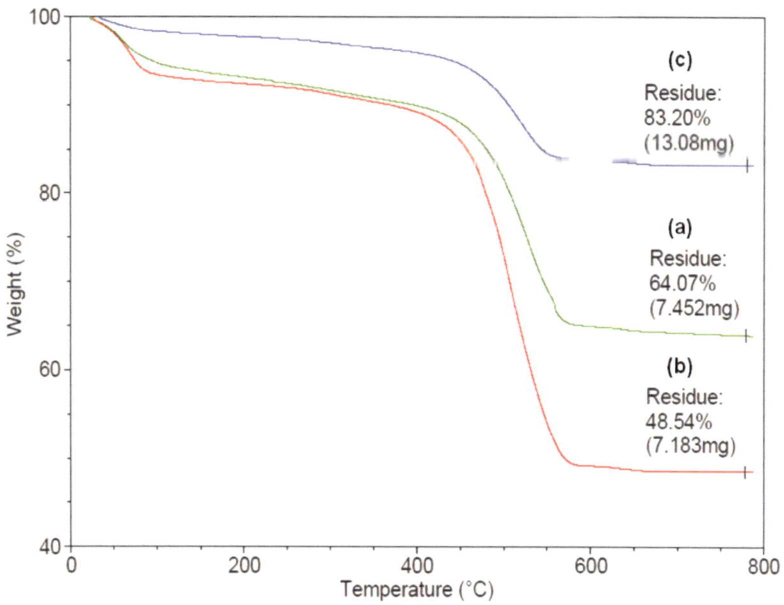

Fig. 7. TGA data of the SCBA: (a) in natura (64.15%), (b) >125 mm (48.60%) and (c) <125 mm (83.21%)

The quality of the sieved coarser charcoal (> 0.125 mm, 120 mesh) was improved, diminishing its fraction of inorganic material using a hydrocyclone (Fig. 8) to separate the fine parts (principally inorganics) and to concentrate the organic particles in the coarser fraction. Although the particles have the same size distribution, there are differences in density between the organic and inorganic particles, and the denser particles will exit preferentially at the bottom (underflow). As the minerals have densities greater than $2g/cm^3$ while the charcoal density is less than $1g/cm^3$, this device is suitable for the selective concentration of charcoal.

The hydrocyclone used has a 250 mm diameter, exit dimension of 40 mm, one ¾ HP (1 HP = 746 W) centrifugal pump (maximum 14 m^3/h) and two 150-liter barrels connected by 40 mm plastic tubes and valves for controlling the feed pressure and liquid inflow; the valves control the amount of time that the overflow (the lighter or finer fraction) passes through the hydrocyclone in a loop circuit. Samples were collected at different times (1, 2, 4, 5 and 20 min) for analysis. This hydrocyclone was used to separate charcoal from SCBA to produce charcoal briquettes.

Fig. 8. Hydrocyclone system (with hydraulic pump) used to separate the charcoal from the SCBA. The charcoal with water goes out of the hydrocyclone through the vortex finder (overflow) and the inorganic components go to the tank (reservoir) through the spigot (underflow)

The X-ray fluorescence (XRF) data show that the ash contains SiO_2 as the major inorganic component (usually over 70% by weight). Among the minor components, Al_2O_3, Fe_2O_3, K_2O, CaO and TiO_2 were the most prominent. The chemical composition of SCBA may be different depending on the type of sugar cane and composition of the soil where it is cultivated. The X-ray diffraction (XRD) data for SCBA shows patterns that suggest that the major crystalline phase is quartz (Teixeira et al. 2008), corresponding to the SiO_2 fraction demonstrated in the chemical analysis (Fig. 9).

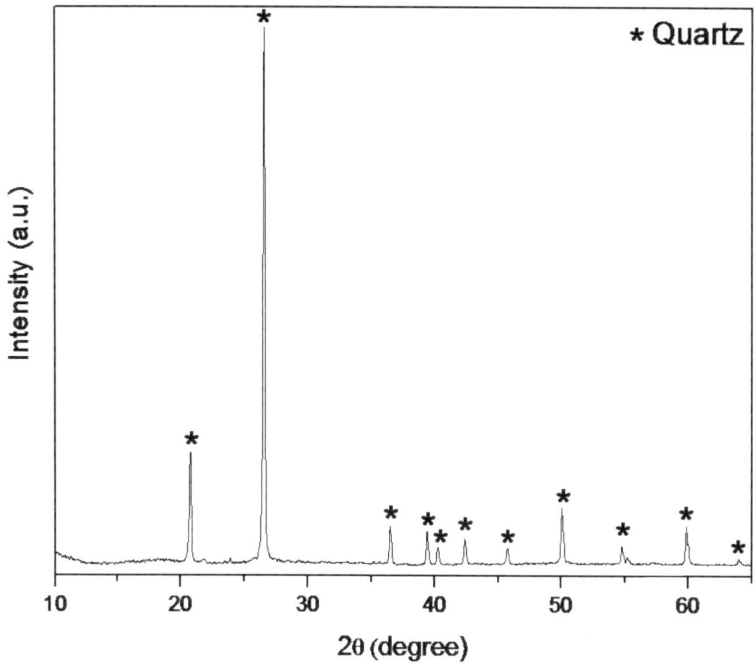

Fig. 9. X-ray diffraction (XRD) pattern of the inorganic fraction of the SCBA showing that it is composed mainly of quartz

2.2.3 Briquetting

Charcoal is a friable material, and therefore, to press it in the form of briquettes, it is necessary to use a binding material to strengthen the briquettes. In this work, cassava (or manioc) starch was used, taking into account that it is widely produced in tropical countries (Charcoal Briquette, 2002; Smith, 2004; Fontes et al., 1991). The binder was prepared in the usual manner, i.e., the starch was added to boiling water until polymer molecules (gum or porridge) were formed (FAO, 1987; Smith, 2004). The briquettes were prepared in two different ways. In the first method, the coal powder was mixed with a gum (8 wt%) to form a homogeneous paste, which was pressed. In the second method, the briquettes were prepared by mixing the moist charcoal powder with starch (8% by weight), and heating at 100°C to form the binder and then pressing it. The cylindrical briquettes (30 mm diameter) (Fig. 10) were pressed using a manual uniaxial hydraulic press and a steel cylindrical mold. An applied compression force of 5 tons was maintained for 1 min for each sample. After some thermal tests, we concluded that it was better to use the second method for preparing the briquettes (Teixeira et al. 2010).

After drying, the briquettes were submitted to various analytical methods and characterization tests to determine their mechanical strength, density, calorific value, moisture and residual ash.

Fig. 10. Charcoal briquettes (ϕ = 30 mm) with different preparation: (a) only moist charcoal, (b) charcol powder mixed with gum (8 wt%) and (c) moist charcol mixed with starch (8 wt%) and heated at 100°C to form the binder

Briquettes prepared with coarse (> 1 mm) charcoal were analyzed and the most important results have been previously published (Teixeira et al. 2010). The mechanical strength value obtained was 7 Mpa, for axial compression of the cylindrical briquette. This value is close to those obtained for briquettes produced using different kinds of wood. The briquette density (measured using caliper and balance) varies between 0.91 to 1.33 g/cm^3. These values are close to those found for wood briquettes manufactured using another type of binder (1.08 g/cm^3) (Fontes et al. 1991). However, this density range is lower than that obtained for briquettes made with eucalyptus charcoal and tar as the binder (1.41 g/cm^3) (Brito & Nucci, 1984). As mentioned before, these briquettes showed a calorific value of 25,700 kJ/kg and a residual quantity of ash of approximately 13 wt%.

The moist charcoal obtained in the second procedure, where the coarse fraction (> 0.125 mm) was passed through the hydrocyclone (for 1, 2, 4, 5 and 20 min), was filtered to remove the excess water and used to prepare the briquettes. The charcoal powder was mixed with starch (8% by weight) and then heated at 100°C (to form the binder). Similarly, the briquettes were pressed uniaxially in a cylindrical shape (30 mm diameter). After air-drying (two days in an open place), density of the briquettes was determined using a caliper and analytical balance. Samples of the briquettes were collected and analyzed using a thermal gravimetric instrument. Table 2 shows the briquette density and percent residue (ash) after TG measurements (up to 800°C) (Teixeira et al. 2010).

Time (min)	01	02	04	05	20
Residue (%)	18.56	18.77	19.99	19.89	16.92
Density (g/cm³)	0.95	0.90	0.75	0.90	0.87

Table 2. Residue and density of the briquettes after (TG) thermal analysis (up to 800°C)

These results indicate that there were no significant changes in briquette density (~0.90 g/cm³) or in the resultant ash (19%) after firing it at 800°C. Therefore, one minute is sufficient time to separate the charcoal from the ash using the hydrocyclone. This period corresponds to the use of three hydrocyclones, similar to the one used, in series.

The briquette calorific value (heat of combustion) measured using a calorimeter bomb was 28,326 ± 920 kJ/kg (6,767 ± 220 kcal/kg), which is approximately the same as for charcoal from wood. The axial compression test shows that briquette resistance was higher than 3,300 N (ϕ = 3cm, 48 kgf/cm²), which is a sufficient value for handling, piling and transporting bags of charcoal.

Figures 11 and 12 show the results of the thermal gravimetric analysis (TGA) and differential scanning calorimetric (DSC) analysis of the charcoal (1 min) and of briquettes prepared by mixing moist charcoal and starch (8 wt-%) and heating at 100°C.

Natural fibers are a complex mix of organic materials (Alvarez et al. 2004). Sugarcane bagasse (SCB) is a natural fiber (lignocellulose) that is not totally pyrolyzed in the boiler of the sugar/alcohol industry. As a result, there is residual natural fiber inside the coarse charcoal particles. Therefore, the thermal decomposition of the charcoal will show the reaction steps of both materials (fiber and charcoal) degrading. Furthermore, cassava starch is added as a binder to the charcoal to raise the mechanical resistance of the briquettes. Lignocellulosic materials are composed of cellulose, hemicellulose and lignin. Therefore, the thermal decomposition of the charcoal shows an overlapping of decomposition peaks of charcoal, fiber and starch. The TGA/DSC curves in nitrogen atmosphere show two decomposition steps at temperatures higher than 200°C. At a temperature of 350 to 400°C the molecular structure of lignocellulosic biomass is degraded (Zandersons et al. 1999). In air (or oxidant atmosphere), besides the material's thermal decomposition, its thermoxidative degradation will also occur. The peaks' positions depend on the heating rate, so the peak near 300°C is attributed to the first peaks representing the thermal decomposition of starch, hemicellulose and cellulose (Alvarez et al. 2004; Suaréz et al. 2010; Guiotoku et al. 2010). The lignin peak is wider and appears between 200°C and 500°C superposing on the other peaks (Alvarez et al. 2004). Its maximum degradation rate occurs above 448°C (Guiotoku et al. 2010) and depends on the kind of lignin. During the thermal degradation of wood components, stable substances are formed around 450 to 530°C and after the completion of the reaction. With regard to charcoal, its combustion in an air atmosphere occurs in this same temperature range. The type of material (or wood) used to produce the charcoal and the pyrolysis conditions influence its temperature of combustion (Suaréz et al. 2010). Suárez A.C. et al. (2010) observed that DSC curves show two exothermic peaks, at ~337°C and 477°C, which overlap with those peaks observed for lignocellulosic materials. These overlapping peaks result in a broad peak (between 400 and 600°C), observed in both samples (Fig. 11 and 12), which is characteristic of the charcoal from SCBA (Fig. 11). Accordingly, the thermal analysis results show that all materials in the briquette composition (charcoal, lignocellulosic material and starch) are degraded in the same temperature ranges. It was observed (Teixeira et al. 2010a) that starch degradation shows three major peaks between 300 and 500°C, and that the transformation of the starch into binder changes only the second degradation peak to a higher temperature (from 470°C to 500°C, approximately). The starch thermal decomposition peak remains at the same temperature at 318°C.

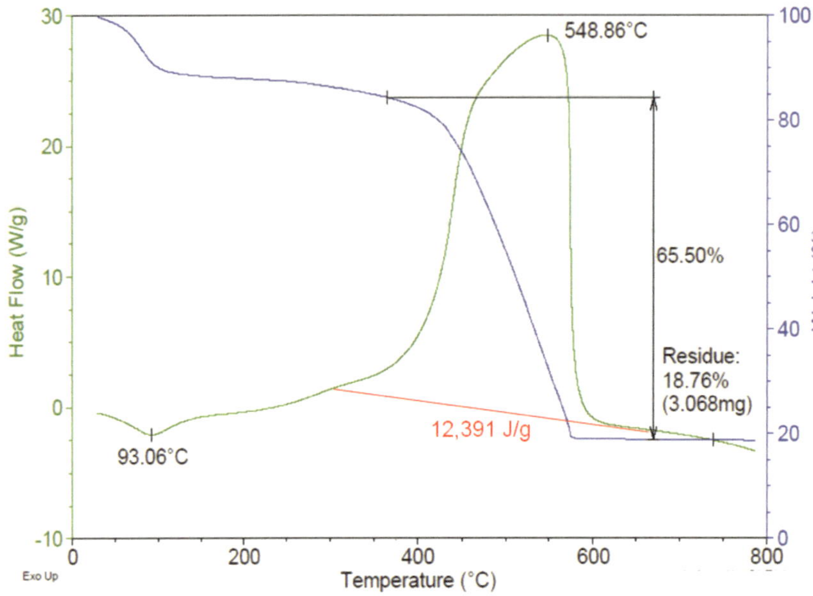

Fig. 11. TGA and DSC data of the charcoal (1 min)

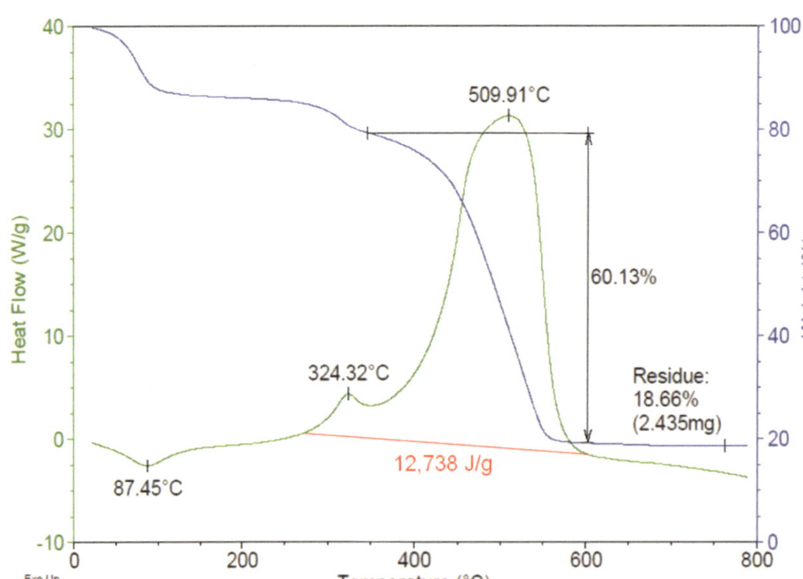

Fig. 12. TGA and DSC data of the charcoal mixed with starch and heated (at 100°C)

3. Conclusion

Brazil is an outstanding position in the production and consumption of charcoal on the international scene, and about 55% of the raw material for the production of charcoal originates from native forests, mainly in areas of the Cerrado. The other 45% comes from homogeneous planted forests of rapidly growing tree species (pine or eucalyptus). The area of planted forests grew on average 20% per annum between 2003 and 2008, with a big drop in 2009 due to an international crisis. Due to its large territory and intense sunlight, the use of plants (carbonized wood) as a source for reducing iron ore (and a source of energy to produce heat) is a valid alternative in countries such as Brazil. The technology used in charcoal production is primitive, mostly we use the same process as a century ago. It is a predominantly artisanal process with low efficiency in which only part of the wood (30-40%) is transformed into charcoal by pyrolysis or a carbonization process. The remainder is released into the atmosphere as gases. The main consumer sectors of charcoal are pig iron, steel and iron-alloy and to a lesser extent, trade and households. Charcoal has many advantages compared to coal. It is renewable, less polluting and virtually free of sulfur/phosphorus, and the technology for its manufacture has been largely consolidated in Brazil. For the forestry economy, the more relevant range of companies with regard to charcoal use includes independent producers of pig iron, which are suppliers of raw material for the steel industry. In 2010, Brazil produced 11.6 million cubic meters of charcoal from planted forests, of which 66.2% were consumed by independent pig iron makers. The trend is for the consumption of native wood to steadily decrease over the years, being replaced with wood from planted forests, and due to the greater control exercised by the inspection agencies and to the increased social pressures on natural resource preservation. (ABRAF 2011)

The use of charcoal in Brazil as a thermal reduction agent in steel production started in Minas Gerais (MG). The lack of coal reserves and abundant availability of native vegetation favored conditions for the use of charcoal. Currently, the consumption of charcoal in Brazil in steelmaking is concentrated in the state of Minas Gerais (MG), the largest consumer. The second steel center is in Carajás, located in the states of Maranhão (MA) and Pará (PA), where the use of charcoal obtained from native vegetation dominates, and in third place at present is the state of Mato Grosso do Sul (MS; Corumbá, Aquidauana and Ribas do Rio Pardo), the youngest pig iron center in the country, surpassing the state of Espírito Santo (ES) (EPAMIG 2010).

In 2009 (BEN 2010), 25,178 x 10^3 tons of wood were used for charcoal production. In this year, firewood and charcoal accounted for approximately 10% of domestic energy supply while sugarcane products made a contribution of 18% (Table 1).

According to the AMS Statistical Yearbook (2008), in 2007 the state of São Paulo (SP) accounted for approximately 3% of domestic consumption of charcoal (1140 x 10^3mdc[2]). Considering that 1mdc is equivalent to an average of 200 kg of charcoal, SP statewide consumption was on the order of 228 x 10^3 tons of charcoal in 2007. Part of this charcoal is imported from other states, mainly Mato Grosso do Sul (MS).

According to CONAB (2009), the state of São Paulo accounts for approximately 58% of the sugarcane (364 Mton) production in Brazil. Considering that 6 kg of ash are produced per

[2]mdc: cubic meter of charcoal

tonne of milled sugarcane and that ash is approximately 33% by mass charcoal and sugarcane debris, then approximately 720,000 tons of charcoal mixed with SCBA were produced in SP State. As shown in the TGA results (Figs. 11 and 12), charcoal has about 19% by mass of ash (residue after firing at 800°C). Therefore the amount of material (charcoal and bagasse debris) that is present in the ash and that may be briquetted is approximately 580,000 tonnes. This value is aproximately twice the amount charcoal produced in SP State.

One difficulty in using this charcoal is the fact that it is produced in industries of different sizes throughout the state. However, if one considers the current value of cubic meters of charcoal, the freight value which is becoming increasingly expensive due to the distance between the sites of production and use, and mainly the environmental benefits, the commercial use of this material should be further assessed in the short term.

4. Acknowledgments

We would like to thank Usina Alto Alegre by the samples and data provided, to FAPESP (2008/04368-4) for funding part of the laboratory equipment and to FUNDUNESP by the financial support to attend conferences. To Dr. A. Leyva for English editing.

5. References

ABNT - Associação Brasileira de Normas Técnicas, NBR 8633 Carvão Vegetal - Determinação do Poder Calorífico (Outubro 1984). In *ABNT Catálogo*, 25.04.2011, Available from:<http://www.abnt.org.br/catalogo>.

ABRAF - Associação Brasileira de Produtos de Florestas Plantadas (Brazilian Association of Forest Plantation Producers), (2011), In *Statistical Yearbook 2011*, 22.05.2011, Available from: <http://www.abraflor.org.br/estatisticas/ABRAF11/ABRAF11-EN.pdf>

Abou-Yousef, H.; El-Sakhawy, M.; Kamel, S. (2005). Multi-stage bagasse pulping by using alkali/Caro's acid treatment. *Industrial Crops and Products An International Journal*, Vol.21 (March 2005), pp. 337-341, ISSN 0926-6690

Aguiar, C.M. (2010). Hidrólise enzimática de resíduos lignucelulósicos utilizando celulases produzidas pelo fungo Aspergillus Níger. *Dissertação de Mestrado*, Universidade Estadual do Oeste do Paraná - UNIOESTE, Toledo, PR, Brazil

Ahmaruzzaman, M. (2009). Role of fly ash in the removal of organic pollutants from wastewater. *Energy & Fuel*, Vol.23, (Febuary 2009), pp. 1494-1511, ISSN 0887-0624

Ahmaruzzaman, M. (2010). A review on the utilization of fly ash. *Progress in Energy and Combustion Science*, Vol.36, (June 2010), pp. 327-363, ISSN 0360-1285

Ahmedna, M.; Marshall, W.E.; Rao, R.M. (2000). Production of granular activated carbons from select agricultural by-products and evaluation of their physical, chemical and absorption properties. *Bioresource Technology*, Vol.71, (January 2000), pp. 113-123, ISSN 0960-8524

Alvarez, V.A. and Vázquez, A. (2004). Thermal degradation of cellulose derivatives/starch blends and sisal fibre biocomposites. *Polymer Degradation and Stability*, Vol.84, (April 2004), pp. 13-21, ISSN 0141-3910

AMS - Associação Mineira de Silvicultura (Abril 2009). In: *Florestas energéticas no Brasil: demanda e disponibilidade.* 20.05.2011, Available from: <http://www.silviminas.com.br/Publicacao/Arquivos/publicacao_585.pdf>

Andrade, A.M.; Duarte, A.P.C.; Belgacem, M.N.; Munaro, E.R. (2001). Produção de papéis artesanais das misturas de aparas com fibras virgens de bambu (*dendrocalamus giganteus*) e de bagaço de cana-de-açúcar (*saccharum officinarum*). *Floresta e Ambiente*, Vol. 8, No.1 (December/January 2001), pp.143-152, ISSN 1985-7775

ANFAVEA - Associação Nacional dos Fabricantes de Veículos Automotores (National Motor Vehicle Manufacturers Association), (January 2011). In *Estatísticas* pp. 4, 20.05.2011, Available from: <http://www.anfavea.com.br/cartas/Carta296.pdf>

Arterburn, R.A. (2010) The Sizing and Selection of Hydrocyclones. 14/03/2011, Available from:<http://www.rockservices.net/83_sizing_select_cyclones.pdf>.

Batra, V.S.; Varghese A.R.; Vashisht, P.; Balakrishnan, M. (2011). Value-added products from unburned carbon in bagasse fly ash. *Asia-Pacific Journal of Chemical Engeneering*, Vol.6, (December 2010), DOI: 10.1002/apj.536 Online Version of Record published before inclusion in an issue, ISSN 1932-2143

BEN: Balanço Energético Nacional 2010, ano 2009 (January 2010). Relatório Final 2010, In *BEN 2010*, 25.04.2011, Available from: <https://ben.epe.gov.br/downloads/Relatorio_Final_BEN_2010.pdf>

BG: Brazilian Government, Sustainable Development Everyone's Commitment and Responsibility (December 2009). Clean Energy Matrix, In *Panorama*, 25.04.2011, Available from: <http://www.cop15brazil.gov.br/pt-BR/?page=panorama/matriz-energetica-limpa>

Botha, T.; Von Blottnitz, H. (2006). A Comparison of the environmental benefits of bagasse derived electricity and fuel ethanol on a life-cycle basis. *Energy Policy*, Vol.34, No.17, (November 2006), pp. 2654-2661, ISSN 0301-4215

Brito, J.A., Nucci, O. (1984). Estudo Tecnológico da Madeira de Pinus spp para a produção de carvão vegetal e briquetagem, In *Instituto de Pesquisas e Estudos Florestais - IPEF, Circular Técnica No. 26, Brasil*, 17.03.2011, Available from: <http://www.ipef.br/publicacoes/ctecnica/nr138.pdf>

Charcoal Briquette: How Products are Made (2002). Gale Cengage, In *Stacey L. Blachford (Ed.)*, 14/03/2011, Available from: <http://www.enotes.com/how-products-encyclopedia/charcoal-briquette>

CONAB - Companhia Nacional de Abastecimento (National Supply Company), (April 2009), In *Safra*, 12.05.2011, Available from: < http://www.conab.gov.br/conabweb/download/safra/1cana_de _acucar.pdf>

Cordeiro, G.C.; Filho, R.D.T.; Fairbairn, E.M.R. (2009). Caracterização de cinza do bagaço de cana-de-açúcar para emprego como pozolana em materiais cimentícios. *Química Nova*, Vol.32, No.1, (January 2009), pp. 82-86, ISSN 0100-4042

Covey, C.; Rainey, T.J.; Shore, D. (2006). The potential for bagasse pulping in Australia. *Appita Journal*, Vol.59, No.1, (January 2006), pp. 17-22, ISSN:1038-6807

Cunha, M.A.A.; Silva, S.S.; Carvalho, W.; Santos, J.C. (2005). Use of PVA-gel immobilized cells: a new strategy for biotechnological production of Xylitol from sugarcane bagasse hidrolysate. Semina: *Ciências Agrárias*, Vol.26, No. 1, (January/March 2005), pp. 61-70, ISSN 1679-0359

Dawson, L.; Boopathy, R. (2008). Cellulosic ethanol production from sugarcane bagasse without enzymatic saccharification. *BioResources*, Vol.3, No.2, (May 2008), pp. 452-460, ISSN 1930-2126

Dutta, S.; Das, A.K. (2010). Analytical perspective on waste management for environmental remediation. *Trends in Analytical Chemistry*, Vol.29, No.7, (July-August 2010), pp. 636-644, ISSN 0165-9936

EPAMIG - Empresa de Pesquisa Agropecuária de Minas Gerais, (2010). A cadeia produtiva do carvão vegetal em Minas Gerais: pontos críticos e potencialidades. *Boletim Técnico*, No. 95, pp. 1-80, ISSN 0101-062X

FAO, Food and Agriculture Organization of the United Nations (1987). Forestry Paper 41, Briquetting of charcoal, In: *Simple Technologies for Charcoal Making (Chapter 11)*, 14/03/2011, Available from: <http://www.fao.org/docrep/X5328E/x5328e0c.htm>

Felfli, F.F.; Mesa P, J.M.; Rocha, J.D.; Filippetto, D.; Luengo, C.A.; Pippo, W.A. (2011). Biomass briquetting and its prespectives in Brazil. *Biomass & Bioenergy*, Vol.35, No.1, (January 2011), pp. 236-242, ISSN 0961-9534

Fontana, J.D.; Ramos, L.P.; Deschamps, F.C. (1995). Pretreated sugar cane bagasse as a model for cattle feeding. *Applied Biochemistry and Biotechnology*, Vol.51-52, No.1, (1995), pp. 105-116, ISSN 0273-2289

Fontes, P.J.P., Quirino, W.F., Brito, J.O. (1991). Aglutinantes para briquetagem de carvão vegetal, In *Instituto de Pesquisas e Estudos Florestais - IPEF, Brasil*, 14/03/2011 Available from: <http://www.funtecg.org.br/arquivos/aglutinantes.pdf>.

Frías, M.; Villar, E.; Savastano, H. (2011). Brazilian sugar cane bagasse ashes from the cogeneration industry as active pozzolans for cement manufacture. *Cement & Concrete Composites*, Vol.33, (April 2011), pp. 490-496, ISSN 0958-9465

Goldemberg, J., Nigro, F.E.B., Coelho, S.T. (2008). O programa de bioenergia do Estado de São Paulo. In *São Paulo: Imprensa Oficial do Estado de São Paulo, Brazil*, 14/03/2011 Available from: <http://www.desenvolvimento.sp.gov.br/noticias//files/livro_bioenergia.pdf>

Goldemberg, J. (2008). The Brazilian biofuels industry. *Biotechnology for Biofuels*, Vol.1, No.6, (May 2008), pp. 1-7, ISSN 1754-6834

Goldemberg, J.; Coelho, S.T.; Nastari, P.M.; Lucon, O. (2004). Ethanol learning curve - the Brazilian experience. *Biomass & Bioenergy*, Vol.26, (May 2004), pp. 301-304, ISSN 0961-9534

Granada, E.; López Gonzáles, L.M.; Míguez, J.L.; Moran, J. (2002). Fuel lignocellulosic briquettes, Die Design and Products Study. *Renewable Energy an International Journal*, Vol.27, No.4, (December 2002), pp. 561-573, ISSN 0960-1481

Guiotoku, M.; Magalhães, W.L.E. (2010). Estudo de estabilidade térmica de materiais lignocelulósicos pré-tratados para produção de etanol. *VII Congresso Brasileiro de Análise Térmica e Calorimetria*, São Pedro, SP, Brasil, April, 25-28, 2010

History of ethanol fuel in Brazil (March 2011). History of ethanol fuel in Brazil, In *Wikipedia, the free encyclopedia*, 25.04.2011, Available from:
< http://en.wikipedia.org/wiki/History_of_ethanol_fuel_in_Brazil>

Hossain, S.; Khalil, M.I.; Alam, M.K.; Khan, M.A.; Alam, N. (2009). Upgrading of animal by solid state fermentation by pleurotus sajor-caju. *European Journal of Applied Science*, Vol.1, No.4, (2009), pp. 53-58, ISSN 2079-2077

Ilindra, A.; Dhake, J.D. (2008). Microcrystalline cellulose from bagasse and rice straw. *Indian Journal of Chemical Technology*, Vol.15, (September 2008), pp. 497-499, ISSN 0971-457X

Lavarack, B.P.; Griffin, G.J.; Rodman, D. (2002). The acid hydrolisis of sugarcane hemicellulose to produce xylose, arabinose, glucose and other products. *Biomass & Bioenergy*, Vol.23, No.5, (November 2001), pp. 367-380, ISSN 0961-9534

Leme, P.R.; Silva, S.L.; Pereira, A.S.C.; Putrino, S.M.; Lanna, D.P.D.; Filho, J.C.M.N. (2003). Utilização de bagaço de cana-de-açúcar em dietas com elevada proporção de concentrados para novilhos Nelore em confinamento. *Revista Brasileira de Zootecnia*, Vol.32, No.6, (Nov/Dec 2003), pp. 1786-1791, ISSN 1806-9290

Lu, J.Z.; Wu, Q.; Negulescu, I.I.; Chen Y. (2006). The influences of fiber feature and polymer melt index on mechanical properties of sugarcane fiber/polymer composites. *Journal of Applied Polymer Science*, Vol.102, No.6, (December 2006), pp. 5607-5619, ISSN 1097-4628

Maldas, D.; Kokta, B.V. (1991). Studies on the preparation and properties of particle boards made from bagasse and PVC: II. Influence of the addition of coupling aAgents. *Bioresource Technology*, Vol.35, No. 3, (February 1991), pp. 251-261, ISSN 0960-8524

Martignoni, W.P.; Bernardo, S.; Quintani, C.L. (2007). Evaluation of cyclone geometry and its influence on performance parameters by computational fluid dynamics (cfd). *Brazilian Journal of Chemical Engineering*. Vol.24, No. 01, (January/March 2007), pp. 83-94, ISSN 0104-6632

Pandey, A.; Soccol, C.R.; Nigam, P.; Socool, V.T. (2000). Biotechnological potential of agro-industrial residues. I: sugarcane bagasse. *Bioresource Technology*, Vol.74, No.1, (August 2000), pp. 69-80, ISSN 0960-8524

Pasquini, D.; Pimenta, M.T.B.; Ferreira, L.H.; Curvelo, A.A.S. (2005). Sugar cane bagasse pulping using supercritical CO_2 associated with co-solvent 1-butanol/water. *The Journal of Supercritical Fluids*, Vol.34, (July 2005), pp. 125-131, ISSN 0896-8446

PróAlcool: Programa Brasileiro de Álcool (January 2011). PróAlcool: Programa Brasileiro de Álcool, In *Biodiesel*, 25.04.2011, Available from: <http://www.biodieselbr.com/proalcool/pro-alcool.htm>

Purnomo, C.W.; Salim, C.; Hidone, H. (2010). Solid acid catalyst preparation from bagasse fly ash based activated carbon. *Proceeding Venice, 2010, Third International Symposium on Energy from Biomass and Waste*, Venice, Italy; November 8-11, 2010

Quirino, W.F.; Brito, J.O. (1991). Características e índice de combustão de briquetes de carvão vegetal. Laboratório de Produtos Florestais - LPF, Série Técnica, No. 13, In *Secretaria do Meio Ambiente*, 13/03/2011, Available from: <www.funtecg.org.br/arquivos/indice.pdf>

Rodrígues-Pedrosa, D.; García-Maraver, A.; Zamorano M.; Díaz, L.F. (2010).Thermal gravimetric study of biomass ash from the combustion of pellets in domestic heating. *Proceeding Venice, 2010, Third International Symposium on Energy from Biomass and Waste*, Venice, Italy; November 8-11, 2010

Ryu, C.; Sharifi, V.N.; Swithenbank, J. (2007). Thermal waste treatment for sustainable energy. *Engineering Sustainability*, Vol.160, (September 2007), pp.133-140, ISSN 1478-4629

Rubio, J.; Carissimi, E.; Rosa, J.J. (2007). Flotation in water and wastewater treatment and reuse: recent trends in Brazil. *Int. J. Environment and Pollution*, Vol. 30, No. 2, (2007) pp.197-212, ISSN 0957-4352

Smith, A. (2004). *Fuel from the fields: a guide to converting agricultural waste into charcoal briquettes*, Massachusetts Institute of Technology - MIT, Massachusetts, USA.

Silva, V.L.M.M.; Gomes, W.C.; Alsina, O.L.S. (2007). Utilização do bagaço de cana de açúcar como biomassa adsorvente na adsorção de poluentes orgânicos. *Revista Eletrônica de Materiais e Processos*, Vol.2, No.1 (2007), pp. 27-32, ISSN 1809-8797

Silva, V.S.; Garcia, C.A.; Silva, C.M. (2010). O destino do bagaço da cana-de-açúcar: um estudo a partir das agroindústrias sucroalcooleiras do Paraná. *Revista em Agronegócios e Meio Ambiente*, Vol.3, No.1, (January/April 2010), pp. 59-76, ISSN 1981-9951

Stael, G.C.; Tavares, M.I.B.; d'Almeida, J.R.M. (2001). Impact behavior of sugarcane bagasse waste-EVA composites. *Polymer Testing*, Vol.20, No.8, (January 2001), pp. 869-872, ISSN 0142-9418

Suaréz A.C.; Tancredi, N.; Pinheiro, P.C.C.; Yoshida, M.I. (2010). Thermal analysis of the combustion of charcoals from Eucalyptus dunnii obtained at different pyrolysis temperatures. *Journal of Thermal Analysis and Calorimetry*, Vol.100, No.3, (March 2010), pp. 1051-1054, ISSN 1388-6150.

Sun, J.X.; Sum, X.F.; Zhao, H.; Sun, R.C. (2004). Isolation and characterization of cellulose from sugarcane bagasse. *Polymer Degradation and Stability*, Vol.84, (February 2004), pp. 331-339, ISSN 0141-3910

SYS, P. A. (September 2001). FIESP/CIESP. In: *Ampliação da oferta de energia através da biomassa*. 26.04.2011, Available from: <www.fiesp.com.br/publicacoes/pdf/ambiente/relatorio_dma.pdf> FIESP/CIESP.

Tabarsa, T.; Ashori, A.; Gholamzadeh, M. (2010). Evaluation of surface roughness and mechanical properties of particleboard panels made from bagasse. *Composites: Part B: Engineering, Vol.42, (July 2011)*, pp. 1330-1335, ISSN 1359-8368

Tai, F.C.; Wei, C.; Chang, S.H.; Chen, W.S. (2010). Raman and X-ray diffraction analysis on unburned carbon powder refines from fly ash. *Joural of Raman Spectroscopy*, Vol.41, (November 2009), pp. 933-937, ISSN 1097-4555

Teixeira, F.A.; Pires, A.V.;Nascimento, P.V.N. (2007). Sugarcane pulp in the feeding of bovine. *REDVET, Revista Eletrônica de Veterinária*, Vol. 8, No.6, (June 2007), pp. 1-9, ISSN 1695-7504 <http://www.veterinaria.org/revistas/redvet/n060607/060708.pdf>

Teixeira S.R.; Peña, A.F.V.; Lima, R.G. (2010a). Use of residues from industrial carbonized sugarcane bagasse to produce charcoal briquette. *Proceeding Venice, 2010, Third International Symposium on Energy from Biomass and Waste*, Venice, Italy; November 8-11, 2010

Teixeira, S.R.; Peña, A.F.V.; Miguel, A.L. (2010b). Briquetting of charcoal from sugar-cane bagasse fly ash (scbfa) as an alternative fuel. *Waste Management*, Vol.30, (February 2010), pp. 804-807, ISSN 0956-053X

Teixeira, S.R.; Souza, A.E.; Santos, G.T.A.; Peña, A.F.V.; Miguel, A.G. (2008). Sugar cane bagasse ash (SCBA) as a potential quartz replacement in red ceramic. *Journal of the American Ceramic Society*, Vol.91, (June 1008), pp. 1883-1887, ISSN 1551-2916

Vieira, R.G.P.; Filho, G.R.; Assumção, R.M.N.; Meireles, C.S.; Vieira, J.G.; Oliveira, G.S. (2007). Syntesis and characterization of methylcellulose from sugar cane bagasse cellulose. *Carbohydrate Polymers*, Vol.67, (July 2007), pp. 182-189, ISSN 0144-8617

Wang, K.-S.; Wei, M.-C.; Peng, T.-H.; Li, H.-C.; Chao, S.-J.; Hsu, T.-F.; Lee, H.-S.; Chang, S.H. (2010). Treatment and toxity evaluation of methylene blue using electrochemical oxidation, fly ash absorption and combined electrochemical oxidation-flay absorption. *Journal of Environmental Management*, Vol.91, (August 2010), pp. 1778-1784, ISSN: 0301-4797

Widyorini, R.; Xu, J.; Umemura, K.; Kawai, S. (2005). Manufacture and properties of binderless particleboard from bagasse I: Effects of raw material type, storage methods, and manufacturing Process. *Journal of Wood Science*, Vol.52, No.6, (December 2005), pp. 648-654, ISSN 1435-0211

Yamaji, F.M.; Chrisostomo, W.; Vendrasco, L.; Flores, W.P. (2010). The use of forest residues for pellets and briquettes production in Brazil. *Proceeding Venice, 2010, Third International Symposium on Energy from Biomass and Waste*, Venice, Italy; November 8-11, 2010

Yousef, H.A.; El-Sakhawy, M.; Kamel, S. (2005). Multi-stage Bagasse pulping by using alkali/Caro's acid treatment. *Industrial Crops and Products*, Vol.21 (May 2005), pp. 337–341, ISSN 0926-6690

Zandersons, J.; Gravitis, J.; Kokorevics, A.; Zhurinsh, A.; Bikovens, O.; Tardenaka, A.; Spince, B. (1999). Studies of the Brazillian sugarcane bagasse carbonisation process and products properties. *Biomass & Bioenergy*, Vol.17, (September 1999), pp. 209-219, ISSN 0961-9534

Zhang, T.; Zhou, Y.; Liu, D.; Petrus, L. (2007). Qualitative analysis of products formed during the acid catalyzed liquefaction of bagasse in ethylene glycol. *Bioresource Technology*, Vol.98, No.7, (May 2007), pp. 1454-1459, ISSN 0960-8524

9

Biomass Waste as a Renewable Source of Biogas Production - Experiments

Adrian Eugen Cioablă and Ioana Ionel
University "Politehnica" from Timisoara
Romania

1. Introduction

Global supply of energy is facing several increasing challenges. Energy consumption is on a moderate increase, especially in rapidly developing countries. The overall size of the world energy market nearly doubled between 1971 and 2003, driven by rapid expansion of energy usage in the developing world, where population and energy activity have grown.

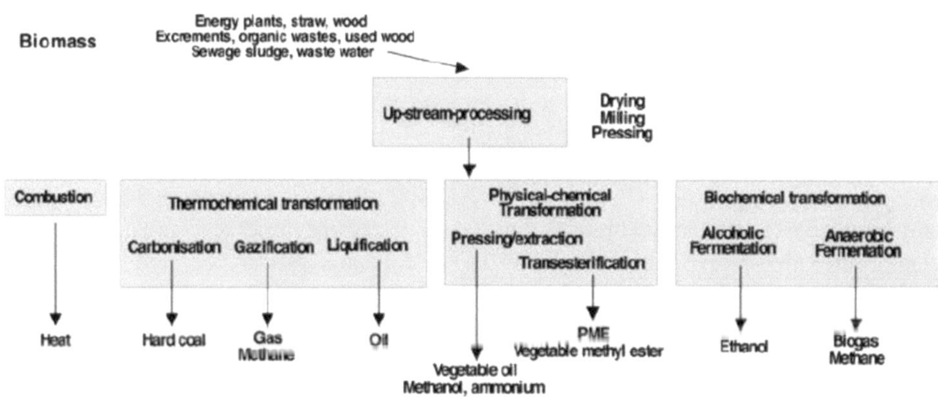

Fig. 1. Applied technologies for transforming biomass in secondary energy sources (Deublein & Steinhauser, 2008)

The International Energy Agency (IEA) has projected an increase in primary energy demand of 1.6 per cent per year until 2030, when the cumulative increase will be equal to half of current demand. At present, fossil fuels - oil, coal, and natural gas - dominate the world energy economy, providing 80 per cent of the world's primary energy supply of 449 EJ/year (Heinimo, 2008).

In Figure 1 the applied technologies for transforming biomass in secondary sources of energy were presented.

The use of biomass has, for millennia, helped human society to fulfill many of its fundamental energy needs, such as for the production of goods, cooking, domestic heating and the transport of people and goods (Petersen, 2008). One of the technologies used for recovering the energy from different types of biomass residues is anaerobic fermentation that has as a primary result the production of biogas.

Anaerobic digestion of energy crops, residues, and wastes is of increasing interest in order to reduce the greenhouse gas emissions and to facilitate a sustainable development of energy supply. Production of biogas provides a versatile carrier of renewable energy, as methane can be used for replacement of fossil fuels in both heat and power generation and as a vehicle fuel (Weiland, 2009).
Organic waste, as a main constituent of solid biomass, has a high potential for biogas generation but the uncontrolled decomposition of waste from agricultural and agro-industrial sources results in large-scale contamination of land, water, and air. All of these cause potentially severe pollution problems and are subject to rigorous environmental regulation in most countries (Busch et al., 2008; Hansen & Cheong, 2007).
Romania, through member countries of the European Union, tries to correlate its activities in the field of green energy production, and different steps are taken in order to accomplish this fact.
According with the Romanian Statistical Register from 2007, the existing agricultural surface decreased reported to the year 2001 from 14852,3 ha to 14731 ha in 2006, while the forest surface increased from 6605,7 ha to 6754,7 ha in 2006.
Related with the European Union's project to achieve an average share of electricity from renewable energy sources of 12% by the year 2010, biomass is expected to provide about 10% of the whole European energy supply, which is equivalent to about 5800 PJ (1 PJ = 1 petajoule = 10^{15} J).
In 2030 it is probable that biomass will be an outstanding solution for individual heating, dominated by pellets in urban areas and by wood chips, wood logs and pellets in rural areas. The boilers and stoves markets will progressively shift from oil based system to biomass based system. Heating oil will progressively disappear because of unaffordable prices. Micro co-generation of heat and power based on biomass will be progressively available in all sizes even at household level and for nearly all part load cases. Most district heating and cooling system will be retrofitted to solar thermal, biomass and geothermal and many new small heats cool and biogas networks will appear (Ionel & Cioablă, 2010).
Related with the existing literature, the field of biogas production and applications is developed at a large scale, the existing studies covering different characteristics of this domain. In connection with the process characteristics, Wu et. al (2009) presented an extensive study related with the pH effect on anaerobic fermentation on sludge material, Li et. al (2011) conducted a review over the solid state anaerobic digestion from organic waste in accordance with methane production, while Busch et. al (2009) presented a new technology for biogas production from solid waste and biomass through means of a double-stage solid-liquid biogas process. Studies were conducted also in the field of biogas production enhancement from solid substrates under different technologies (Yadvika et. al, 2004, Amon et. al, 2007, Karellas et. al, 2010) and in regard to designs of the anaerobic digesters there can be mentioned the contributions of Igoni et. al (2008) or Walla & Schneeberger (2008) in regards to the optimum size of the biogas plants or anaerobic digesters designs for biogas production

2. General presentation of different materials used during the anaerobic fermentation process

One of the directions involved in the research activities regarding biogas production was the usage of different vegetal substrates, having in mind the potential of vegetal biomass in the present context.

In literature there are presented different aspects related with

A. Grains of corn waste

B. Beech dust

C. Linden dust

D. Mix of wheat and corn waste

E. Maize bran

F. Recipe of corn, wheat and sunflower husks

Fig. 2. Materials used in anaerobe fermentation (Cioablă, 2009)

In figures 2.A up to 2.F are presented different sorts of material used in anaerobe fermentation processes.
As it can be seen from the images, the materials that were studied are: beech dust, linden dust, corn waste, mix of wheat and corn waste, degraded maize bran and a recipe of wheat, corn and sunflower husks.
In Table 1 some of the general characteristics of the used materials are presented.
In Table 2 are presented the major elements presented in the analyzed sorts of biomass.
In Table 3 are underlined the minor elements presented in the analyzed biomass.
Related to the study of different substrates of material used to obtain biogas, researches over the influence of municipal wastes in the context of biogas production were conducted.
In figure 3 is presented a general image of municipal residues used in order to produce biogas.
Due to the combined approach of obtaining biogas of different qualities and in various quantities using different sorts of material, the next paragraph is dedicated to the pilot installations built with the goal of obtaining good results in this field of interest.

No	Sample	Total Humidity [%]	Hygroscopic Humidity [%]	Ash content [%]	High calorific value [kJ/kg]	Low calorific value [kJ/kg]
1	Maize bran	10.23	0.16	4,63	17098	15535
2	Recipe of wheat, maize and sunflower husks	11.76	1.04	1,81	16744	15192
3	Grains of corn waste	13.91	1.50	1,88	15933	14488
4	Beech dust	6.43	0.51	0,92	17751	16322
5	Linden dust	8.01	0.23	0,54	17705	16263
6	Mix of wheat and corn waste	10.98	0.88	1,64	16591	15245

Table 1. General characteristics of the biomass sorts (Cioablă, 2009)

No	Elem.	Maize bran [mg/kg]	Cereal recipe [mg/kg]	Grains of corn waste [mg/kg]	Beech dust [mg/ kg]	Linden dust [mg/ kg]	Mix of wheat and corn waste [mg/kg]
1	Mg	1331	764	733	496	289	797
2	Al	71	61	62	143	85	55
3	Si	174	34	26	321	183	-
4	P	5855	2419	1725	267	82	2332
5	S	1165	925	736	109	82	1181
6	Cl	370	388	375	47	44	680
7	K	9697	4359	2955	2035	936	4491
8	Ca	1209	901	382	3605	2808	716
9	Mn	108	14	3	74	34	31
10	Fe	177	117	84	122	77	81
11	Zn	69	25	16	9	7	23

Table 2. Major elements in the composition of analyzed biomass (Cioablă, 2009)

No	Elem.	Maize bran [mg/kg]	Cereal recipe [mg/kg]	Grains of corn waste [mg/kg]	Beech dust [mg/kg]	Linden dust [mg/kg]	Mix of wheat and corn waste [mg/kg]
1	Cr	0.919	8.034	2.225	1.868	3315	0.705
2	Mn	184.127	27.236	10.562	116.715	56.62	59.158
3	Co	1.385	0.866	1.763	1231	1.172	0.6
4	Ni	1.494	4.789	0.776	157	2.138	-
5	Cu	6.053	-	-	-	-	-
6	As	0.366	-	-	2.081	-	-
7	Se	0.833	-	0.18	1.004	0.002	0.289
8	Br	5.472	10.836	11.5	2366	1.632	10.114
9	Sr	7.256	2.819	0.662	12.579	11.824	2.098
10	Cd	4.102	3.905	3.52	3.604	2.457	3.58
11	Sn	-	0.387	0.653	-	-	0.584
12	Hg	-	-	-	-	-	-
13	Pb	6318	8.02	7.737	15.704	7.899	8.291

Table 3. Minor elements in the composition of analyzed biomass (Cioablă, 2009)

Fig. 3. Municipal residues

3. Pilot installations: Role and results

3.1 Small scale installation for the degradation study during the anaerobic fermentation process

The first installation which will be presented is a laboratory scale installation used for the experimental study of the biomass behavior regarding the degradation process during the anaerobic fermentation. In Figure 4 is presented the general schematics of the installation.
As it can be seen in Figure 4, the anaerobic fermentation reactors are made from thermal resistant glass, which is also pressure resistant until 0.5 - 0.6 bar, with a useful volume of about 2 l. In order to guarantee that the gas is properly washed, the reactors are connected to smaller thermal resistant glass vessels filled with 200 ml of water. The formed biogas inside the anaerobic fermentation reactors will pass inside the small glass vessels and will be

washed of the impurities. The process can be inhibited because of the lack of homogenization inside the reactors and because of this the system is equipped with magnets positioned in the lower part of the reactors in order to have a magnetic agitation during the process. The agitation process is made on a daily basis.
Also the heating system is realized with the help of a thermocouple and a temperature controller in order to assure the necessary temperatures during the process. The chosen temperature regime is the mesophilic one (30 - 37 °C).
After passing through the small reactors filled with water, the biogas will be kept inside small gas bags with the potential of analyzing it after the process.

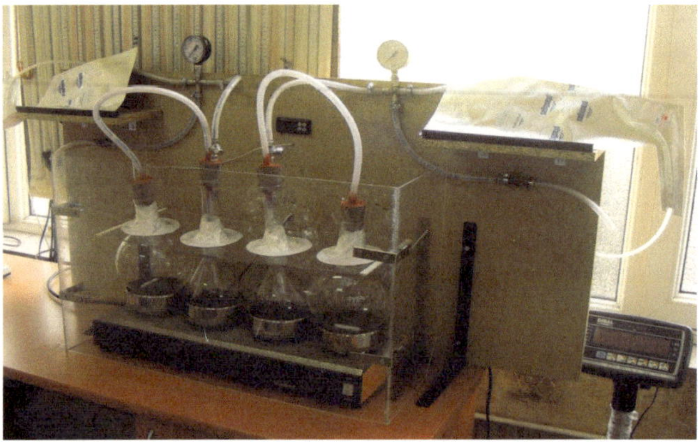

Fig. 4. Components of the small scale installation (Cioablă, 2009)

3.2 Pilot installation for obtaining biomass from vegetal biomass

In order to study at a bigger scale the process of anaerobic fermentation with all its particularities, a pilot installation was built at the Unconventional Energies Laboratory from the Mechanical Engineering Faculty, "Politehnica" University from Timisoara. Its main role was the experimental approach of biogas production from vegetal biomass.
In Figure 5 the general schematics of the pilot installation is presented.
From the biomass deposit, the used material is passed through a mill, and then it's sent to the tank where the preparation of the suspension of biomass is made (1). The biomass suspension is transported with the help of the pump (2) and introduced into the fermentation reactors (3). The correction agent tank for the pH assures, through the control system, the conditions for the process of anaerobic fermentation. The resulted biogas is passed through a filter for retaining the H_2S (5) and after that, through a system used for retaining CO_2 (6), after which takes place the CO_2 desorption and the compression of the CO_2 in the adjacent system (7) and the purified biogas is sent to use (8). The used material is discharged through the means of a gravimetric system (9), and the solid material is retained for being dried using the natural drying. After that it is sent to a compost deposit for being used as a soil fertilizer. When the case, a part of the resulting liquid is neutralized, in the system (10) and sent to the sewerage network, or is transported by the recirculation pump (2) from the suspension preparation tank (1). The fermentation reactors are thermostat heated with the system (11). For the homogenization of the suspension a bubbling system (12) is used, made by polypropylene pipes to avoid the possible corrosion. Also, for depositing small quantities of biogas of the purpose of analyzing, the installation is equipped with a small tank (13) positioned at the top of the reservoirs.

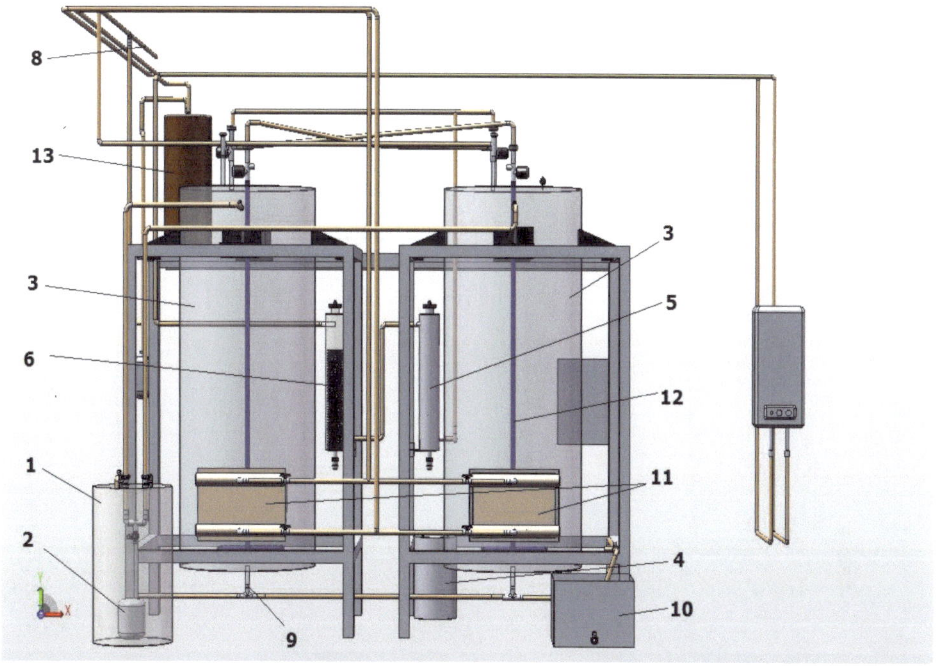

Fig. 5. Principle scheme for the pilot installation used for obtaining biogas from biomass (Cioablă, 2009)

Fig. 6. Front view of the pilot installation (Cioablă, 2009)

From Figure 6 it can be observed the front view of the pilot installation.
In figure 7 is presented the principle scheme for the CO_2 retaining system.

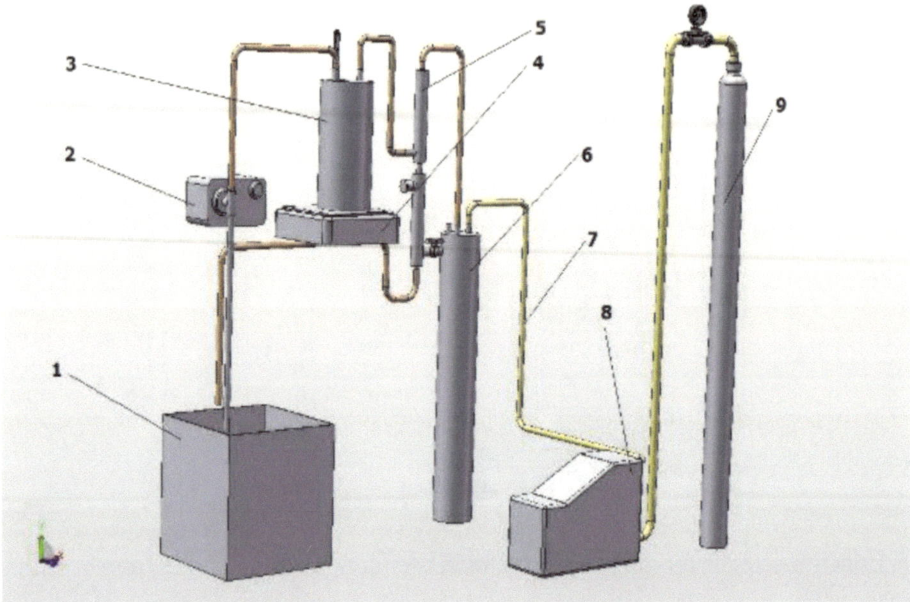

Fig. 7. Principle scheme for the CO_2 retaining system (Cioablă, 2009)

The system is composed from a tank (1) from which the water is passed by the means of a dosing pump (2) - similar with the dosing pumps used for the pH correction -, a stainless steel tank (3) positioned on an heating device (4), a liquid separator (5) and a buffer tank (6) from which the gas is aspirated with the help of a Haug compressor and inserted in a cylinder at a pressure of about 10 - 26 bar. The temperature inside the stainless steel tank can reach values of 50 - 60 °C.

3.3 Pilot installation for obtaining biogas from municipal residues

Relative to the experimental studies involving different substrates used for biogas production, a pilot installation was also built inside an industrial platform, in order to study the potential of Timisoara municipal residues concerning the quality and quantity of biogas.
In Figure 8 it is presented a general view of the pilot installation.
Figure 9 it is presents the general description of the pilot installation.
The biodegradable residues are introduced inside the anaerobic fermentation tank, having in mind that the process parameters (temperature, pressure, pH) are monitored in real time. The used temperature regime is the mesophilic one. The pH is modified when necessary with a lime based suspension in order to bring it to the necessary values.
The produced biogas is passed through a H_2S retaining system and also a CO_2 retaining system into a storage tank. The composition of the produced biogas is determined before and after the retention systems with the help of gas analyzers. The next step is to test the produced biogas in combustion processes in order to establish the applicability to different users.

Fig. 8. Front view of the pilot installation

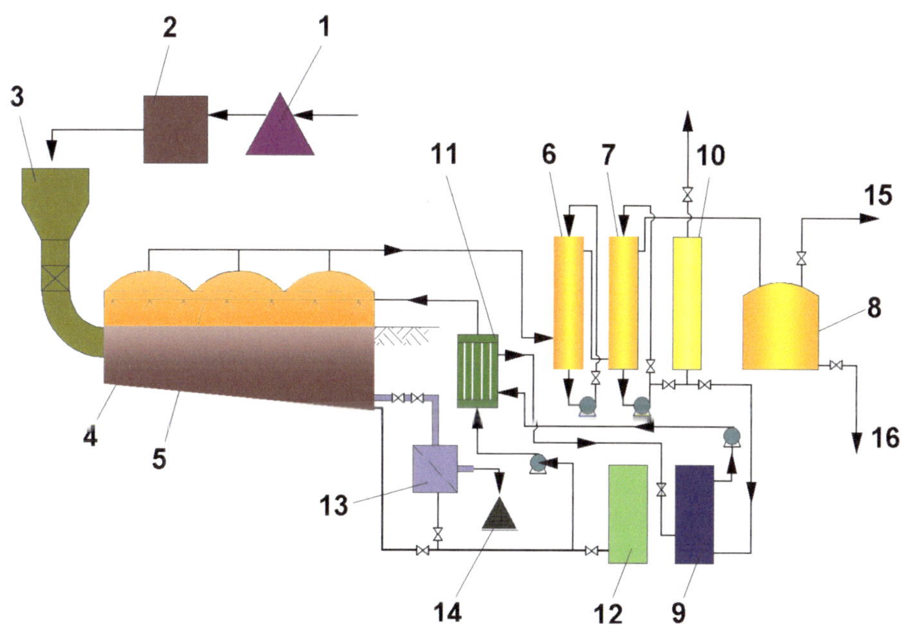

1 - Biodegradable municipal landfill 2 - Grinding waste system 3 - Waste feeding system 4 - Anaerobic fermentation tank 5 - Nozzles equipped system for pH correction 6 - Filter for retaining H_2S from biogas 7 - Filter for retaining CO_2 from biogas 8 - Biogas storage tank 9 - CO_2 desorption tank 10 - CO_2 cooling tank 11 - Heat exchanger 12 - pH correction tank 13 - Filter for fermented residue 14 - Storage for solid residue 15 - Biogas user connection 16 - Condensate removal valve

Fig. 9. General presentation of the pilot installation:

One of the particulate aspects regarding this pilot is the fact that the invention was awarded with the gold medal at the International Exposition IENA 2010, in Nürmberg, Germany and also the gold medal at the "Inventika" Exposition in Bucharest, Romania, 2010.

The principle used has common points with the second pilot installation, which was already presented, but the constructive solution is different and adapted to a much larger scale.

3.4 Results on the pilot installations

This paragraph will underline some of the main results obtained on the presented installations in order to highlight the potential of obtaining biogas from different sorts of material.

3.4.1 Results obtained on the small scale installation

The temperature variation during the anaerobic fermentation process and the degraded aspect of the material at the end of the process will be underlined. The degradation appearance is observed thanks to images taken with the help of a binocular magnifier.
In Figure 10 is presented the temperature variation during the anaerobic fermentation. It can be observed that the average value is around 36°C (mesophilic regime).

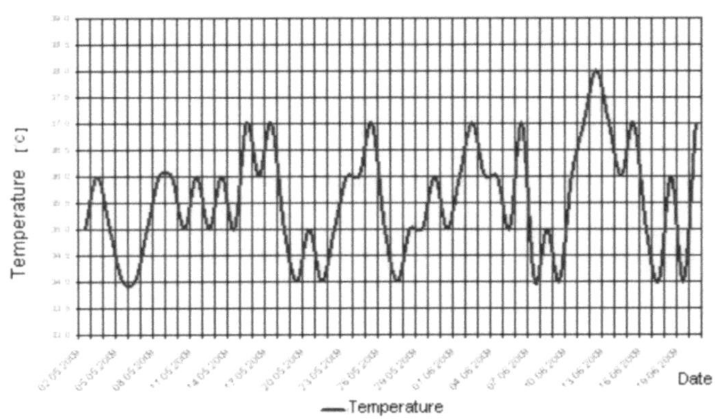

Fig. 10. Temperature variation during anaerobic fermentation (Cioablă, 2009)

From Figure 10 it is underlined the fact that the temperature variation in time is fluctuating in the range of 34 °C - 37 °C, the incorporated system for the temperature control allowing peaks of maximum 37 °C and minimum 34 °C. This temperature domain will have as a main result a biogas production in time starting after a period of 15 - 20 days, during a period of time necessary for the process to develop, taking into consideration the fact that no inoculums were used for process acceleration.
In Figures 11 and 12 it can be seen the general structure for the cereal mix and the bran with the help of a binocular magnifier. This is a step taken in regard to studying the general aspect of the material during anaerobic fermentation from the point of view of the degradation process. It can be observed the fibrous structure of the bran and also it can be partially observed the bacterial hallo inside the suspension. Also, there are signs of degradation related to the general aspect (the lack of color and surrounding bacteria formations on the material surface). Even if this part has a rather imagistic orientation related to the degradation process, the intention was to underline in a certain measure the rather selective behavior of bacteria during the process in regard to the different used substrates.
The second part of this paragraph highlights parts of the obtained results on the pilot installation used in order to obtain biogas from the vegetal biomass. The temperature and

the pH variation from the two reservoirs and production of biogas will be presented for the analyzed batches.

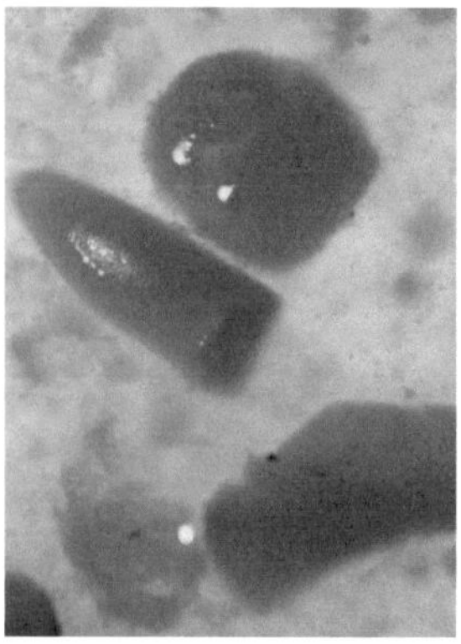

Fig. 11. Detail for the mix (Cioablă, 2009)

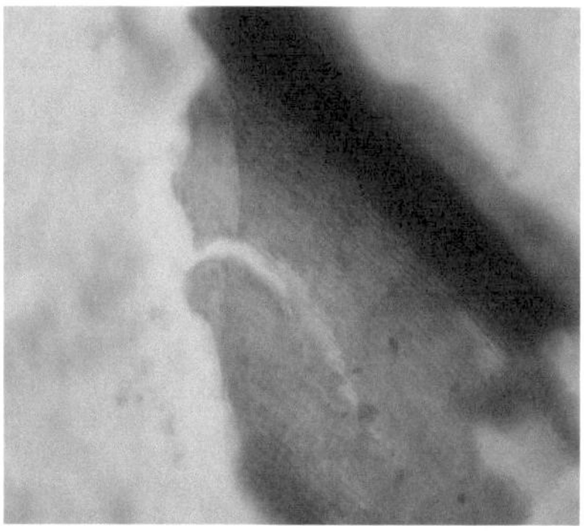

Fig. 12. Structure of the bran (Cioablă, 2009)

3.4.2 Results obtained on the pilot installation

There were made different studies in order to determine the general behavior of substrates with different potentials related to the anaerobic fermentation. This paragraph will underline the trials made on four substrates, from which two are wood residues and two are degraded cereal materials.

The monitored parameters which are presented for all the batches are temperature and pH taking into consideration the fact that the temperature regime chosen for all the batches was the mesophilic one (30 °C - 37 °C).

3.4.2.1 Beech dust batch

$$C/N \text{ ratio} = \frac{47.9}{0.22} = 217 \tag{1}$$

Quantity of solid material introduced in each reservoir = 75 kg
Solid material / liquid volume ratio = 0.037 %

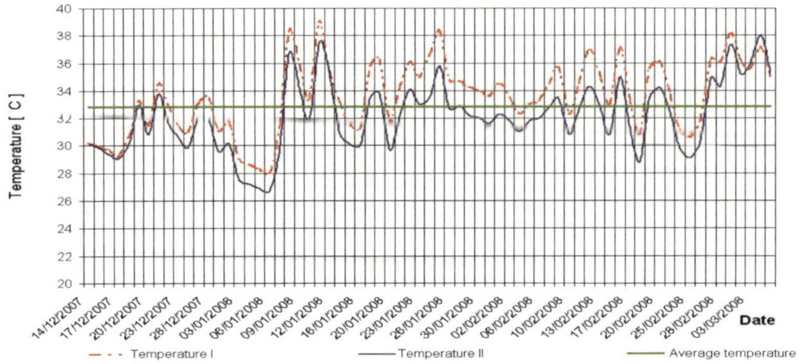

Fig. 13. Temperature variation (Cioablă, 2009)

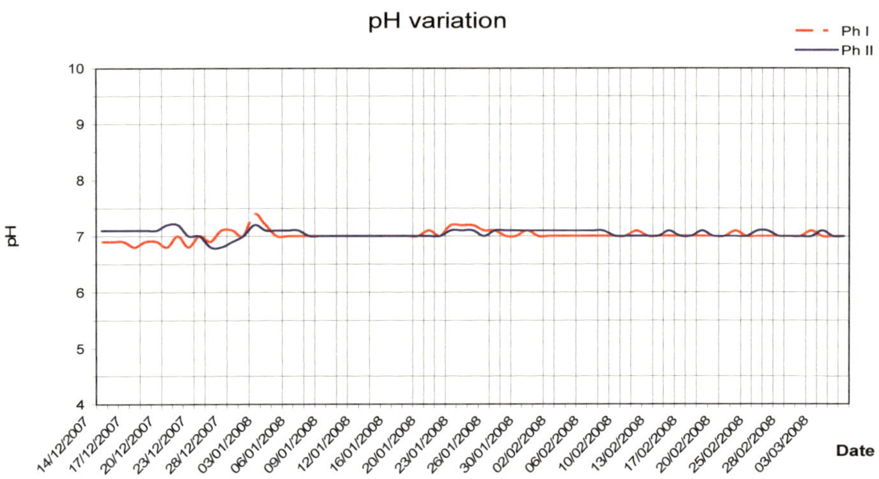

Fig. 14. pH variation (Cioablă, 2009)

From Figures 13 and 14 it can be observed that there is a difference between the variations of temperature for the two reservoirs, inside the established temperature domain, even if the graphic shape is about identical for both reservoirs. The possible explanation of this difference can be related to the general distribution of the material inside the anaerobic fermentation tanks and to the heat transfer inside the tanks as a result of the contact between the heat exchangers and the suspension, even if the general conditions of the process were the same.

The pH values were 6.5 - 7.5, the small differences from one day to another being corrected with the help of the dosing pumps.

The produced biogas quantity was under 0.05 m^3 / day, making this substrate unfit for this type of capitalization. Also the residual material obtained after the process presented small signs of degradation, mainly because of its high lingo - cellulosed content. Even if before the process, the material was subjected to a preliminary acid hydrolysis, using a suspension of water and acetic acid (pH = 3.5) at approximately 55 - 60 °C and normal pressure, this step was not enough for breaking the cellulose chains inside it.

3.4.2.2 Linden dust batch

$$C / N \text{ ratio} = \frac{47.5}{0.42} = 113 \tag{2}$$

Quantity of solid material introduced in each reservoir = 75 kg
Solid material / liquid volume ratio = 0.037 %
The measurements and the monitoring process are taken in similar condition as the first batch. Because the material had also a high content of lingo - cellulose (wood residue), there were taken the same steps preliminary to using the material inside the anaerobic fermentation tanks. Having in mind the conditions related to the temperature which was also determined for the first batch, the pH variation suggests that in the first period of time the general tendency was for the acid domain. In those conditions it was necessary to correct the initial values in order to bring them close to neutral and more suited domain for the anaerobic process to take place.

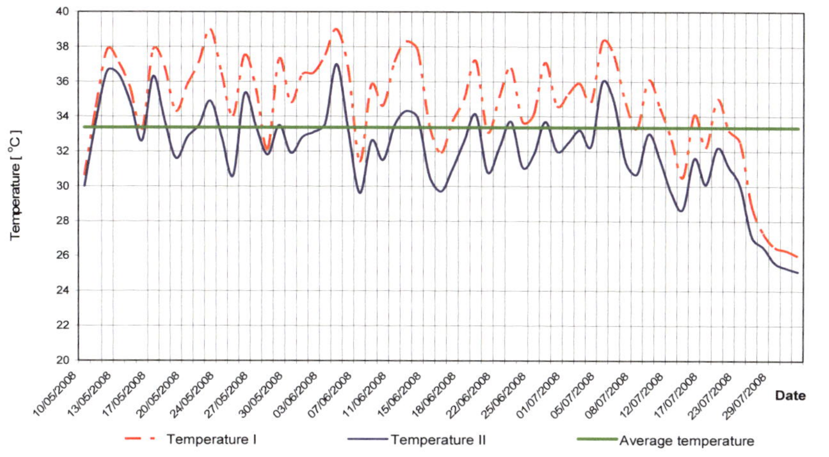

Fig. 15. Temperature variation (Cioablă, 2009)

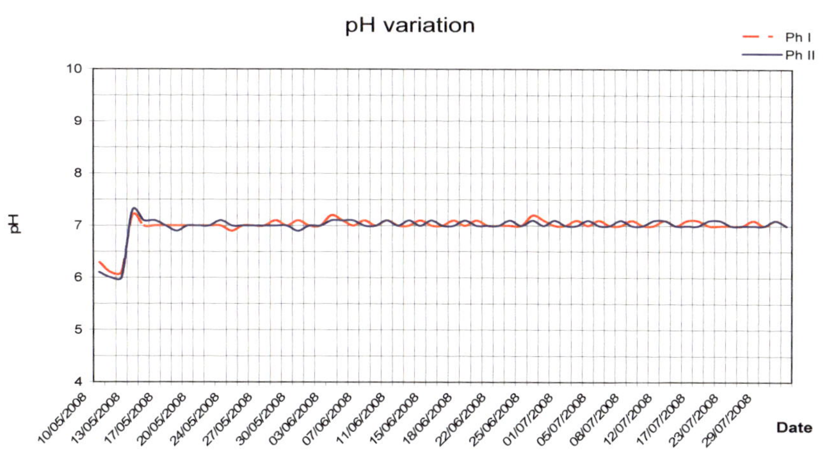

Fig. 16. pH variation (Cioablă, 2009)

The biogas production was also under 0.05 m^3/ day, and the substrate proved to be unfit for this type of capitalization. Like for the first batch, the residual material obtained after the process presented small signs of degradation.

3.4.2.3 Mix of wheat and corn waste

Because the fact that the first studies on wood residues proved to have negative results, next studies were made on agricultural residues in order to determine their potential related with biogas production.

The next batch had in it's composition a mixture of 50% grains of degraded corn and 50 % grains of degraded wheat. The quantity of solid material introduced in each reservoir was about 75 kg.

C / N ration having in mind that in our case there is a mixture of two different types of biomass is calculated with the formula:

$$\frac{C}{N} = \frac{P_1 r_1 + P_2 r_2}{P_1 + P_2} \tag{3}$$

where:

P_1, P_2 – the material quantities introduced in the reservoirs

r_1, r_2 – C / N ration for each type of material

$P_1 = P_{corn\ cobs} = 37.5$ kg

$P_2 = P_{wheat} = 37.5$ kg

$$r_1 = r_{corn\ cobs} = \frac{46.58}{0.47} = 99.1$$

$$r_2 = r_{wheat} = \frac{43.6}{2.28} = 19.1$$

C / N ration for the mix is:

$$C / N = \frac{37.5 \times 99.1 + 37.5 \times 19.1}{37.5 + 37.5} = 59.1$$

The graphics below present the variation of temperature, ph, pressure difference and biogas production, determined in the same conditions like for the first two batches.

From the Figure 17 it can be observed that the average value for the temperature is under 32°C, and the difference between the temperatures of the two reservoirs is about the same as for the first two batches.

The pH variation, presented in Figure 18 shows that the average value is about 7, and this neutral value is, like for the other batches, important because it will not affect the process of anaerobic fermentation.

Even if the pH average value is around 7, in the first period of time (approximately 15 days) the material had an acid pH, characteristic the first stage of the process (acidogenesis). With the help of the dosing pumps, the pH value was corrected and the neutral value was obtained for the rest of the period during the anaerobic fermentation process.

The average value for biogas production was 0.25 m^3 / day, which represented a step forward related with the obtained values from the first two batches.

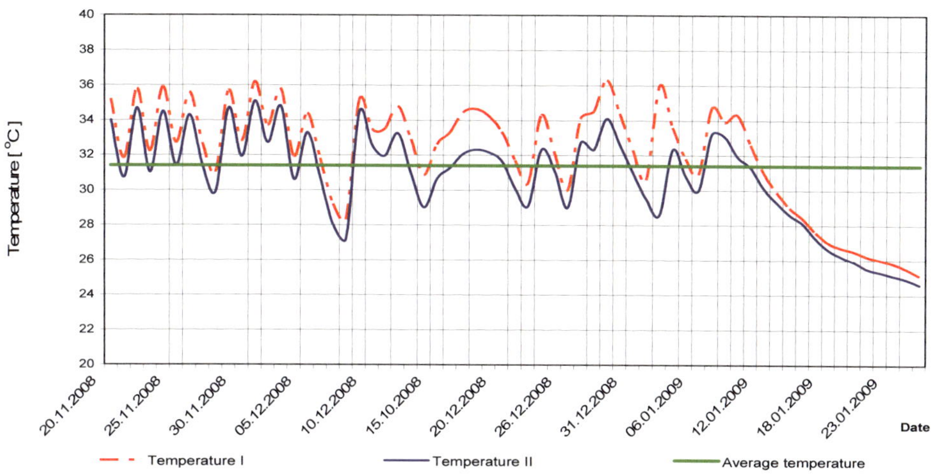

Fig. 17. Temperature variation (Cioablă, 2009)

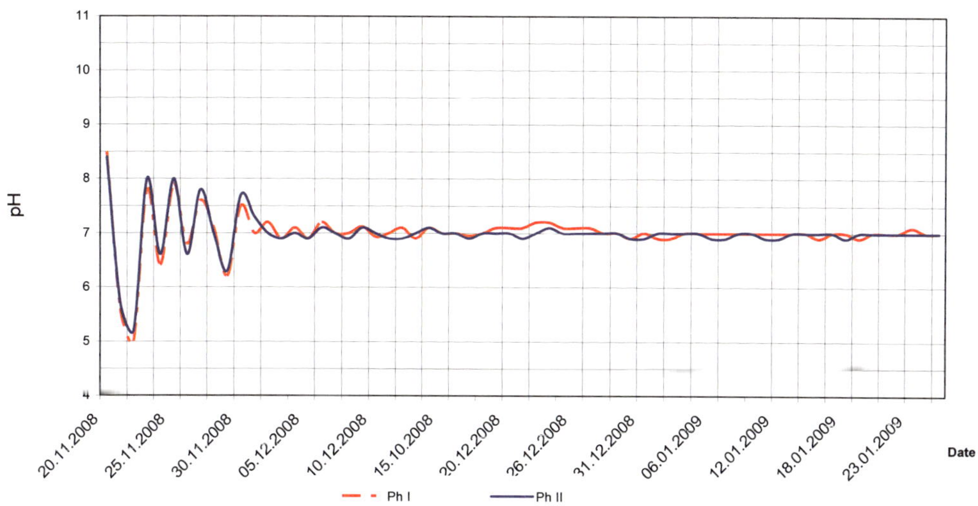

Fig. 18. pH variation (Cioablă, 2009)

3.4.2.4 Corn waste batch

The fourth batch was composed only from corn waste; the total quantity of solid material introduced in each reservoir was 75 kg.
The solid material / liquid volume ratio = 0.037 %

$$C / N \text{ ratio} = \frac{46.58}{0.47} = 99.1 \tag{4}$$

Next graphics present the variation of the main parameters like for the first three batches.

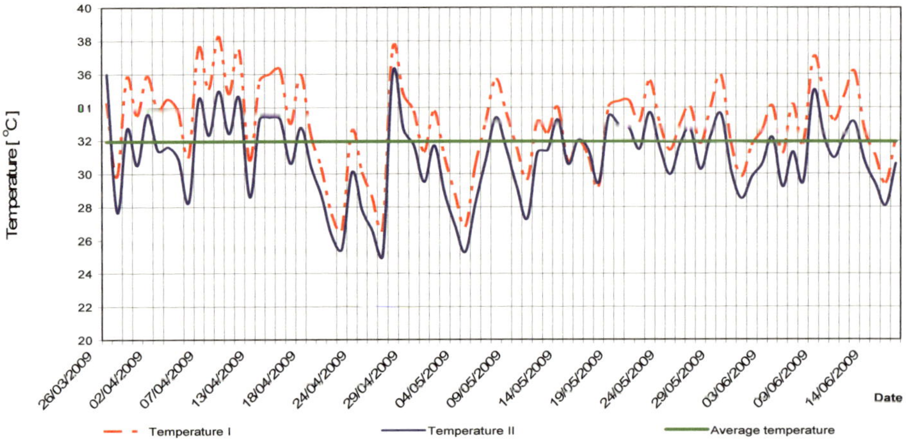

Fig. 19. Temperature variation (Cioablă, 2009)

From Figure 19 it can be observed the same variation between temperatures inside the reservoirs that was noticed for the other batches, and the average value is about 32° C, making it clear that the potential reason of this difference between the reservoirs is connected with the heat transfer between the heat exchanger and the material suspension for each tank..

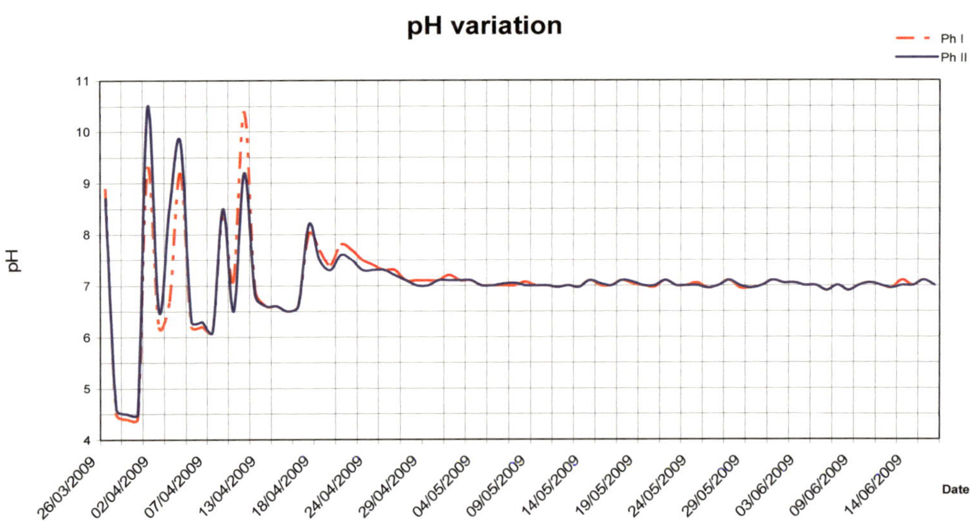

Fig. 20. pH variation (Cioablă, 2009)

The pH values in figure 20 are fluctuating very much in the first part of the process, suggesting that the suspension of material needed to be corrected each day in the first period with the help of dosing pumps. The correction was made to a maximum of 10 during 1 - 2 days, because the decrease in pH until the next day was of 2 - 3 units.

The period of time necessary to bring the pH value to a neutral level was about 25 days, this batch being the hardest to control from this point of view.

Related to the biogas production, this batch had the best results, with an average value of over 0.25 m^3 / day.

3.4.3 Results obtained using municipal residues as substrate

The last part of this paragraph underlines some of the results obtained on the third pilot installation, using municipal residues as substrate.
The measurements were made during a period of 73 days and during this time there were observed and adjusted the pH and temperature for the batch.
In the next figures are presented the variations for the temperature and pH during the batch of material.
The main difference between the two pilot installations is that for the last one presented, the variation for the main parameters is made just for one reservoir, the second one having the buffer role inside the installation.
Also, because the used material is a mix of organic and inorganic residues it is difficult to establish the exact composition of the batch.

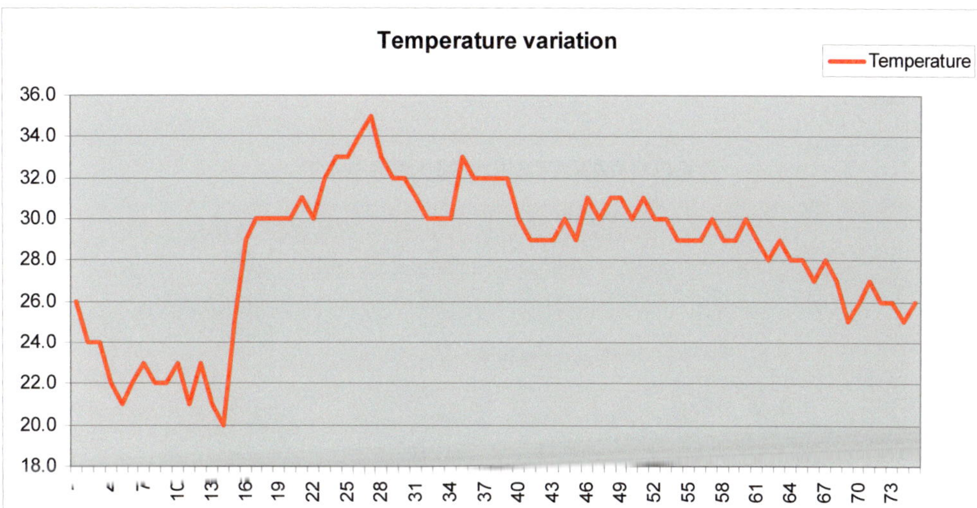

Fig. 21. Temperature variation

The temperature shows a two - domain regime (cryophilic and mesophilic) with variations between 20 °C and 35 °C, meaning a relatively different behavior according to the suspension used in the batch. Because in the first period of time the temperature value was low, the process inside the fermentation tank was developed with a relatively low speed, the production of biogas starting approximately after 25 - 30 days.
The general influence of the temperature over pH is very low, and from Figure 22 it can be observed that the time variations are small; the general values are between 6.4 and 7.2, a domain fit for biogas production in relatively large quantities.
From Figure 23 it can be observed that the municipal residues batch had produced the largest quantity of biogas during the process. The first two batches had the smallest biogas values because of the large content is lingo-cellulose which was very difficult to break during the process, such making wood residues not feet to this technology.

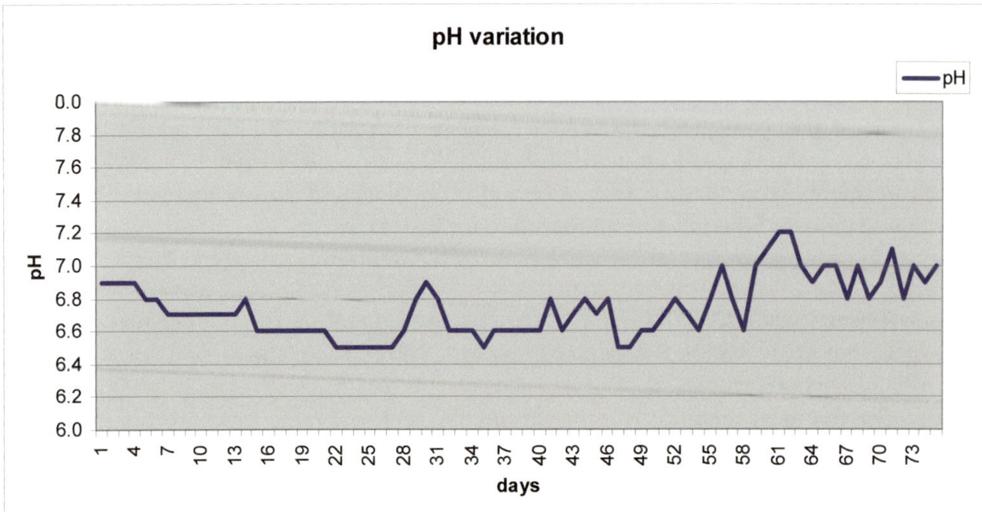

Fig. 22. pH variation

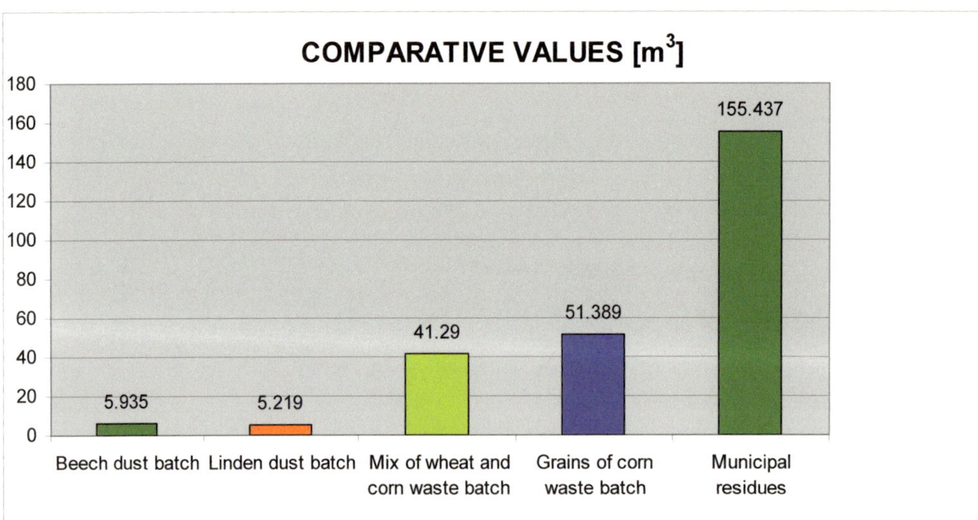

Fig. 23. Comparative values for the batches of used material

The third and fourth batches of material had large quantities of starch, giving them the potential to be used with good results both in quality and quantity related to the biogas production process. Because the fact that the mix of degraded material showed a lower value for biogas production than the corn waste batch, the main idea which can be traced is that wheat is not as suited material as corn for the anaerobic fermentation process.

Also, related with the biogas production form municipal residues there are needed studies in order to determine the exact potential of this kind of substrate for biogas production, especially related with an accurate composition of the used material.

3.5 Possibilities of using the biogas in internal combustion engines

Power generation by using gas fueled engines is widely spread around the world, especially in remote areas with no electric power transport infrastructure. Combined heat and power generation increases overall efficiency and provides environmental as well as economical benefits.

Several operating strategies are investigated, with different air-fuel ratios for a stationary engine. CO_2, CO and NO_x emissions values were calculated and the effect of methane content in the fuel is evaluated. A reduction of both NO_x and CO emissions values is noticed as the biogas methane content decreases, while carbon dioxide emissions are lower for methane rich biogas.

Dedicated lean-burn, turbo-charged spark ignition gas engines can reach electrical efficiency of 45% due to special features including combustion optimization and high mechanical efficiency through reduced losses (Hunt, 2009). Engines like MAN's 5-8 MW 32/40 PGI power generation gas engine with a completely new high-energy ignition system, combine the advantages of diesel engines, high efficiency and power density, with the low emissions of gas engines. Given this operation mode, where a small quantity of gas is injected into a pre-chamber to be ignited on a hot surface, electrical efficiency is as high as 46%, while NO_x emissions are less than 250 mg/Nm^3 at 5% O_2 (Hunt, 2008).

When considering fuel sources, methane is the most environmentally friendly fossil fuel, with minimum CO_2 emissions. While diesel engines deliver high efficiencies up to 50%, nitrous oxide emissions are high and require the use of selective catalytic reduction exhaust gas treatment. Gas engines feature less CO_2 emissions and very low NO_x emissions, under 1 g/kWh without any exhaust gas treatment.

Spark ignition engines are a major source of air pollution due to the exhaust gases that contain nitrous oxides (NO and NO_2 collectively known as NO_x), carbon monoxide (CO), and organic compounds which are unburned or partially burned hydrocarbons (HC) (Ionel et al., 2006).

Nitrous oxide emissions (NO_x) are comprised of nitric oxide (NO) and nitrogen dioxide (NO_2). Out of the two species, NO is predominant with the major source being the oxidation of atmospheric nitrogen. The quantity of NO in the exhaust gas stream is given by the kinetics of formation and dissociation reactions, often called the extended Zeldovich mechanism. Reasonable NO formation rates can be modeled using equation (5) (Heywood, 1988).

$$\frac{d[NO]}{d\tau} = \frac{6 \quad 10^{16}}{T^{0,5}} \quad \frac{1}{e^{69090/T}} \quad [O_2]_e^{0,5} \quad [N_2]_e \qquad (5)$$

where *d[NO]* is the variation of nitric oxide molar concentration measured in mol/cm^3 s, $d\tau$ the variation of time in s, T local temperature in K, $[O_2]_e$ and $[N_2]_e$ are oxygen and nitrogen concentrations at chemical equilibrium measured in mol/cm^3.

A strong dependence between NO formation and temperature is evident through the exponential term. Local temperature during combustion and the expansion stroke can be simulated using a flame let model that predicts a drop in mean effective pressure with decreasing methane content (Figure 24).

Another influence factor is the equivalence ratio or excess air ratio (λ), quantified by O_2 and N_2 molar equilibrium concentrations. This influence becomes important when operating the engine on lean mixtures with certain types of fuels, like power gas. Exhaust gas recirculation indirectly affects NO formation by lowering temperature values during combustion, greatly reducing NO_x emissions even when hydrogen enriched fuel is used.

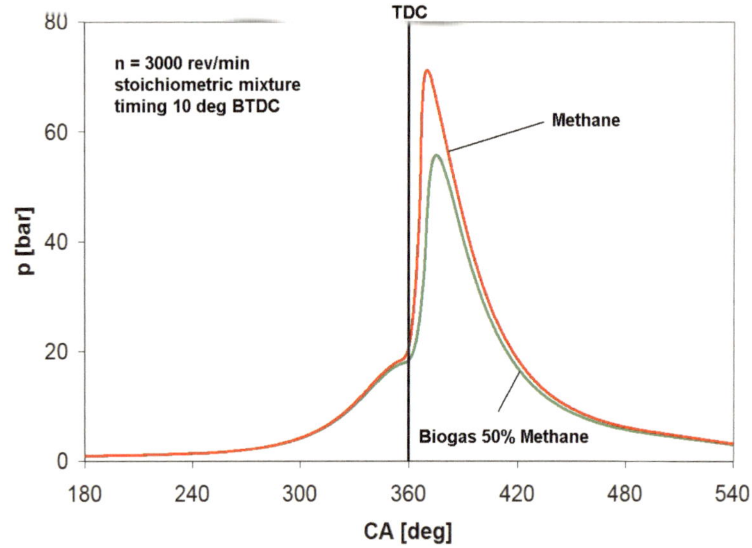

Fig. 24. Simulated pressure traces vs. crank angle rotation for methane and biogas with 50% methane content during compression and expansion strokes (Cioablă, 2009)

Carbon monoxide emission values are kinetically controlled, as the reaction rate of oxidation to CO_2 is slow compared to the CO formation rate. CO concentrations measured for rich mixtures are in good agreement with predicted equilibrium values, while at close to stoichiometric mixtures a correction is necessary. The CO oxidation and formation mechanism can be modeled with specific reaction rates as described by equations (6) and (7) (Papagiannakis et al., 2007).

$$CO + OH \leftrightarrow CO_2 + H, \;\; k_1 = 6{,}76 \quad 10^{10} \quad e^{T_b / 1102} \tag{6}$$

$$CO_2 + O \leftrightarrow CO + O_2, \;\; k_2 = 2{,}51 \quad 10^{11} \quad \frac{1}{e^{22041 / T_b}} \tag{7}$$

where k_1, k_2 are reaction rates measured in m^3/kmol s, and T_b is the burned gas temperature in K.

The presence of CO_2 in biogas has a similar effect on NO_x emissions as exhaust gas recirculation, as temperature values during combustion are lower for biogas with high carbon dioxide content (Figure 25). A significant reduction of NO formation is noticed for the entire excess air ratio range as the biogas is richer in CO_2. Of course, the presence of CO_2 leads to a drop in efficiency because of longer combustion due to a lower flame speed, and an increase of unburned hydrocarbon emissions. The model predicts a peak in NO_x values for slightly lean mixtures (Figure 26), in good agreement with experimental data. Simulated maximum values however, are higher than those actually measured.

CO emissions drop as the methane content decreases (Figure 27), mainly due to lower combustion temperatures and because there is less carbon available for oxidation.
Evidently, less carbon dioxide is present in the exhaust gas stream, as methane content is higher (Figure 28). It must be noted however, that biogas is a CO_2 neutral energy source.
Having in mind the character of this fuel we can determine the CO_2 quantity that can be reduced by not going to be dispersed in the atmospheric air in case of using fossil fuels.

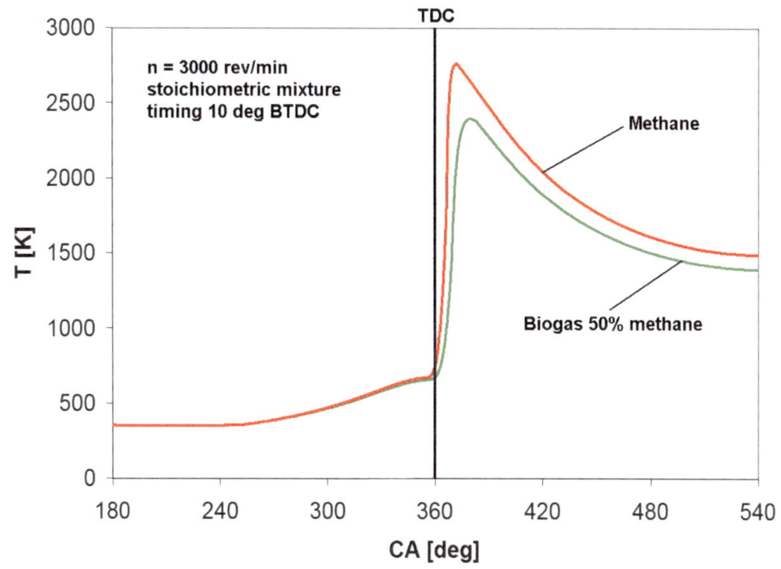

Fig. 25. Simulated temperature values vs. crank angle rotation for methane and biogas with 50% methane content during compression and expansion strokes (Cioablă, 2009)

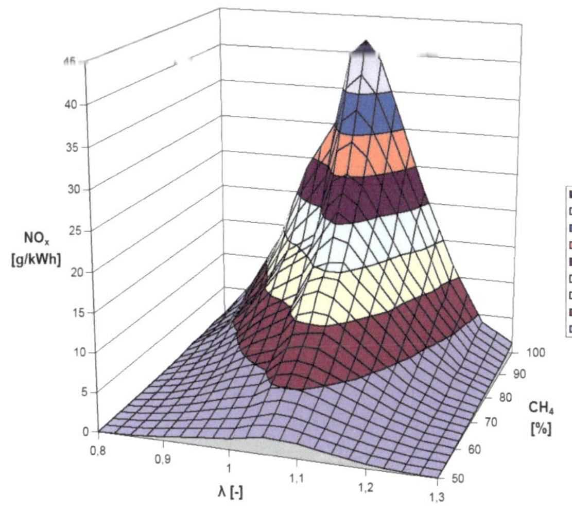

Fig. 26. Calculated NO_x emissions for different biogas quality values, engine speed 3000 rev/min (Cioablă, 2009)

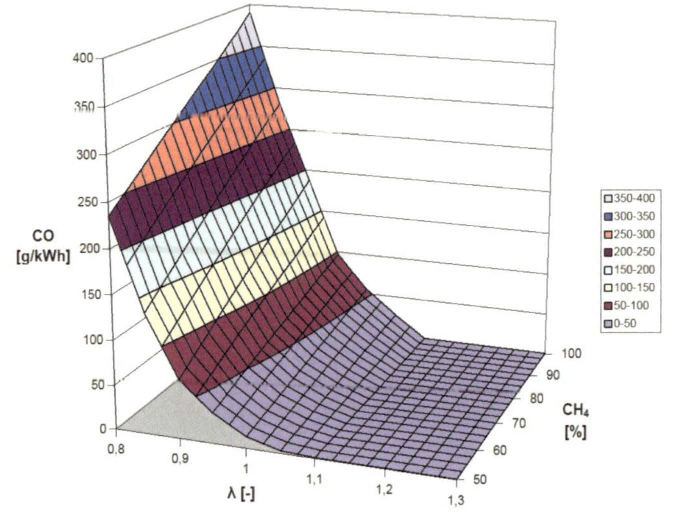

Fig. 27. Calculated CO emissions for different biogas quality values, engine speed 3000 rev/min (Cioablă, 2009)

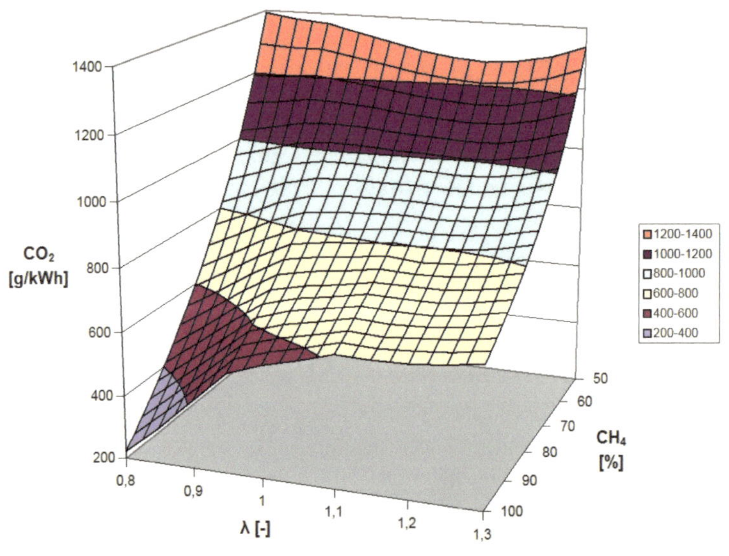

Fig. 28. Calculated CO_2 emissions for different biogas quality values, engine speed 3000 rev/min (Cioablă, 2009)

For low power electrical generators driven by spark-ignition engines in general gasoline, liquefied petroleum gas (LPG) or natural gas are used.
Electric power generated by using a spark ignition engine (MAS) can be calculated with relation (8), where the effective yield (η_e) ranges from ~ 30% under full load operation. Higher values of this efficiency (up to ~ 39% and even higher) can be obtained for MAS supercharged operating with very lean air-fuel mixture at low speed ~ 1500 rpm and have a

high compression ratio. Such engines are limited to operation with natural gas due to high tendency to detonation (Ungureanu & Ionel, 1993).

$$P_e = \eta_g \cdot \eta_e \cdot \dot{m}_{fuel} \cdot H_i \quad (8)$$

where:
P_e - the electrical power [W],
η_g - electric generator efficiency [-]
η_e - actual performance engine [-]
$\dot{m}_{fuel}$ - fuel flow [kg / s]
H_i - fuel calorific value [J / kg].

Internal combustion engine which causes the electric generator requires more maintenance and repairs at regular intervals. Thus, a number of operating hours are necessary to change the oil, the various components of the ignition system, change filters etc.. All these operations effectively reduce the time the engine can be exploited, and the actual operation is reduced compared with the life of unit.
The electricity that can be produced depends on the size of the biogas plant. For small biogas plants, as is the case with the experimental installation on which measurements were made, the power internal combustion engine is on the order of 2.5 kW. For an effective power of the engine (P_e) of 2.2 kW, the actual yield (η_e) is the internal combustion engine of 30%, an effective operating coefficient (η_f) for a yield of 76% and the electric generator (η_g) is 98%. A comparison of CO_2 emissions can be made (Table 4) in case of using gasoline (H_i ~ 43,500 kJ / kg) or natural gas (H_i ~ 50,000 kJ / kg). Obviously the use of biogas in CO_2 emission is practically zero. There is a higher amount of annual emissions of CO_2 in using gasoline compared to the situation when is used natural gas, which is explained by the higher carbon content of gasoline (~ 0.854 per kg of bodyweight / kg_{fuel}) compared with methane (0.75 per kg of bodyweight / kg_{fuel}), this being the main component of natural gas.

Fuel	P_e [kW]	η_e	η_g	η_f	m_{fuel} [kg]	CO_2emission [t/year]
Gasoline	2,2	0,3	0,98	0,76	7131	22329
Natural gas					6203	17058

Table 4. Calculated CO_2 annual emission

Obviously, for an installation which produces electricity and recovers some of the heat evacuated from the engine by using a heat exchanger mounted on the exhaust path, it can be obtained a very high efficiency in the cogeneration system and the potential reduction of CO_2 emission. This is much higher compared to where MAS is used to generate electricity and a central heating is used to get heat. Overall efficiency heat exchangers (η_{sc}) have the maximum values of ~ 70% but this figure varies widely depending on the parameters, which are operated this way for recovering the heat contained in exhaust gases. Another important aspect is the evaluation of the heat to maintain a proper temperature so that the bioreactor provided an environment conducive to development of methanogenic bacteria. Thus, although efficiency of energy production in cogeneration system has high values (around 80%), overall plant efficiency is lower due to the loss of heat needed to maintain optimum temperature in fermentation tanks.

4. Conclusions

Biomass represents an inexhaustible energy resource that can be used partly or wholly for biogas production, both by anaerobic fermentation and other processes (aerobic fermentation, gasification),

Biogas has the potential to become an alternative fuel for the existing fossil fuels related with the field of applications already existing today and developing for the future;

Biogas plants can be of different sizes, thus ensuring the energy needs at both individual (flat, firm) and industry level, with the cost investment depreciation in a reasonable time;

Regionally and globally, stimulated investments are linked to the achievement of plants to produce biogas, a shift in which our country should join the current conditions;

The purity of produced biogas is closely related to the type of biomass used, and a very important element to monitor besides methane and carbon dioxide is the traces of hazardous gases (H_2S);

Washing of produced biogas by spraying water in special filters helps to increase the percentage of methane contained in evacuated, percentage whose influence is reflected in the calculation of calorific value of gas fuel;

Main parameters of influence on the anaerobic fermentation process is the temperature at which the process is done, the pH of slurry, elementary chemical composition of the materials used and the biomass nature;

Reported on the used substrate, it is important to know the mass ratio of solid and liquid used and carbon / nitrogen ratio;

When using wood residues, the main problem is finding a way to break the cellulose chains in order to easily degrade the material during the anaerobic fermentation process with potential good results in terms of biogas quality and quantity;

Connected with the different quantities of produced biogas, the next step of the research involves experimental studies related with using biogas in incineration or co-incineration processes with applications in burners or small engines. The main objective of the research is to determine the composition of the flue gas and the real potential of using the obtained biogas in incineration processes;

Presently, studies are achieved in order to improve the possibilities of obtaining good quality and quantities of biogas from different sources of municipal residues the results representing environmental friendly option for supporting applications in co-incineration and agriculture where the resulted substrates can be used .

5. References

Amon, T. et al. (2007). Biogas production from maize and dairy cattle manure – Influence of biomass composition of the methane yield, *Agriculture Ecosystems and Environment*, 118 (2007), pp. 173 - 182, ISSN 0167 - 8809

Busch, G.; et al. (2008). A New and Sound Technology for Biogas from Solid Waste and Biomass, *Water Air Soil Pollut: Focus* (2009) 9:89–97 DOI 10.1007/s11267-008-9195-5

Cioablă, A. E. (2009). *Theoretical and experimental contributions concerning biogas production from biomass residues*, PhD Thesis, Politehnica Publishing House, Timisoara

Cioablă, A. E.; Ionel, I; Pădurean, I.; Ţenchea, A.; Popescu, F. & Savu, A. (2009). Biogas production from agricultural residues. Test rig and results, *Metalurgia Internaţional*, Vol. XIV, No. 3, pp. 40 - 44, ISSN 1582 - 2214

Deublein, D & Steinhauser, A. (2008). *Biogas from waste and renewable resources, an introduction*, ISBN 978 - 3 - 527 - 31841 - 4

Hansen, C. L. & Cheong, D. Y. (2007). Agricultural waste management in food processing, *Handbook of Farm, Dairy, and Food Machinery*

Heinimo, J. (2008). Views on the international market for energy biomass in 2020: results from a scenario study, *International market for energy biomass*, 547, ISSN 1750 - 7220

Heywood, J. B. (1988) Internal Combustion Engines Fundamentals, *Mc-Graw Hill Series in Mechanical Engineering*

Hunt, J. (January-February 2008). Reciprocating engines continue to evolve. *Cogeneration & On-Site Power Production*, Volume 10, No 1, pp.. 39-47, ISSN 1469-0349

Hunt, J. (March-April 2009). Cleaner, more power and heat, less fuel-today's engines must deliver. *Cogeneration & On-Site Power Production*, Vol. 10, No 2, pp. 23-29, ISSN 1469-0349

Igoni, A.H. et al. (2008), Designs of anaerobic digesters for producing biogas from municipal solid waste, *Applied energy*, 85 (2008), pp. 430 - 438, ISSN 0306 - 2619

Ionel, I.; Ungureanu, C.; et al. (2006). *Thermo energy and environment*, Ed. 2ª, Ed. Politehnica Publishing House, ISBN: (10) 973-625-387-2, Timisoara

Ionel, I. & Cioabla, A. E. (2010). Biogas production based on agricultural residues. From history to results and perspectives, *WSEAS Transactions on Environment and Development*, Issue 8, Vol. 6, pp. 591 - 603, ISSN 1790 - 5079

Karellas et al. (2010), Development of an investment decision tool for biogas production from agricultural waste, *Renewable and sustainable energy reviews,* 14 (2010), pp. 1273 - 1282, ISSN 0364 - 0321

Li, I. et al. (2011), Solid state anaerobic digestion for methane production from organic waste, *Renewable and sustainable energy reviews,* 15 (2011), pp. 821 - 826, ISSN 0364-0321

Papagiannakis, R. G.; Rakopoulos, C. D.; Hountala, D. T. & Giakoumis, E. G. (2007). Study of the performance and exhaust emissions of a spark-ignited engine operating on syngas fuel, *International Journal of Alternative Propulsion*, Vol. 1, Nr. 2/3, ISSN 1741-8011

Petersen, J. E. (2008). Energy production with agricultural biomass: environmental implications and analytical challenges, *European Review of Agricultural Economics*, Vol 35 (3), ISSN o165 - 1587

Romanian statistical register (2007)

Ungureanu, C. & Ionel, I. (1993). *Steam generators. Thermal calculation*, Politehnica Publishing House, Timisoara,

Walla, C. & Schneeberger, W.(2008), The optimal size for biogas plants, *Biomass and bioenergy,* 32 (2008), pp. 551- 557, ISSN 0961 - 9534

Weiland, P. (2009). Biogas production: current state and perspectives, *Appl Microbiol Biotechnol*, DOI 10.1007/s00253-009-2246-7

Wu, H. et al. (2009), The effect of pH on anaerobic fermentation of primary sludge at room temperature, *Journal of hazardous materials*, 172 (2009), pp. 196 - 201, ISSN 0304 - 3894

Yadvika et al. (2004), Enhancement of biogas production from solid substrates using different techniques - a review, *Bioresource technology,* 95 (2004), pp. 1 - 10, ISSN 0960 - 8524

10

Influential Parameters on Biomethane Generation in Anaerobic Wastewater Treatment Plants

Meisam Tabatabaei[1,2], Alawi Sulaiman[3], Ali M. Nikbakht[4], Norjan Yusof[5] and Ghasem Najafpour[6]
[1]*Microbial Biotechnology and Biosafety Department, Agricultural Biotechnology Research Institute of Iran (ABRII), Karaj*
[2]*Department of Bioprocess Technology, Faculty of Biotechnology and Biomolecular Sciences, Universiti Putra Malaysia, Selangor*
[3]*Faculty of Plantation and Agrotechnology, Universiti Teknologi MARA, Shah Alam*
[4]*Department of Farm Machinery Engineering Faculty of Agriculture, Urmia University, Urmia*
[5]*Department of Biology, Faculty of Science and Mathematics Universiti Pendidikan Sultan Idris , Tanjong Malim, Perak*
[6]*Biotechnology Research Lab., Faculty of Chemical Engineering Noshirvani University of Technology, Babol*
[1,4,6]*Iran*
[2,3,5]*Malaysia*

1. Introduction

The nature of wastewater and its complexities are clear to all research scientists. The impact of industrial wastes on our environments is based on their characteristics and chemical composition as well as their degradability and treatment through biological processes. Application of biogas as a source of energy for heating and household usage has been promoted as an alternative solution to energy demands (Metcalf & Eddy, 2003).

In anaerobic fermentation process, generally four groups of microorganisms are sequentially active leading to the biodegradation of organic matters. In the process of anaerobic decomposition, organic compounds are hydrolyzed in the primary stage, as a result, long chain of aliphatic and aromatic organic compounds break down to small size volatile organic compounds and volatile fatty acids. Hydrolytic microorganisms are naturally capable of degrading biomaterials and organic compounds in wastewater. As a matter of fact, the hydrolysis takes place at the initial stage and the hydrolytic enzymes create a suitable environment for acid forming bacteria while intermediate metabolites are synthesized. In other word, large molecules; long chain natural polymers such as polysaccharides and proteins break down to their soluble and insoluble monomers. The monomers are then converted into fatty acids with small amount of hydrogen. The most frequently detected organic acids are acetic, propionic and butyric acids with small quantities of valeric acid produced. This stage is called acetogenesis (acid formation). The bacteria, producing acetic acid is called acetogenic bacteria and their optimal pH value is in the range of 5-6. Therefore, the main products of acetogens are a number of organic acids, hydrogen and carbon dioxide (Najafpour et al. 2008). In the acidification of organic wastes the pH of media drops to acidic condition. In the next stage of digestion, consortia of microbial populations utilize organic acids and hydrogen, methane and carbon dioxide are formed. In fact, anaerobic process decomposes all organic pollutants of wastewater to methane and carbon dioxide.

Anaerobic processes are normally operated at a pH near 7 which is in favor of methane formers, responsible for the final conversion of organic acids into methane. The final stage of anaerobic digestion is called methanogenesis. The product of acid forming phase, acetic acid mainly is converted into methane and carbon dioxide. The bacteria utilizing acetic acid, is called acetophilic bacteria. Alkalinity and hydrogen formers may seriously act on the bases of commensalisms which exhibit quite satisfactory coexistences with the acid formers. Energy is liberated through catabolic reactions while organic compounds are decomposed. Microorganisms grow on wastewater, using soluble organic compounds as sole source of energy for treatment and the useful end-product such as methane can be used as energy sources to generate electricity and heat. Moreover, it has been frequently reported that anaerobic treating of various wastewaters has achieved considerably high percent removal of chemical oxygen demand (COD)(>90%) (Nandy et al. 1998; Metcalf & Eddy 2003; Najafpour et al. 2005; Najafpour et al. 2006; Khademi et al. 2009; Zinatizadeh et al. 2009; Alrawi et al. 2010).

The purpose of the present chapter is to investigate the influential parameters on biomethane generation and efficient treatment of various wastewaters such as granulation, lipids in particular long chain fatty acids, organic loading rate, mixing, sludge recycling, temperature, alkalinity and pH, ammonia, and trace metal ions. Moreover, the applications of modern modeling techniques in providing support to the plant operation and the decision-making process are also discussed in detail.

2. Factors affecting biomethane generation

2.1 Granulation

The retention of a considerable amount of active biomass as happens in a number of reactor such as up-flow anaerobic blanket reactor (UASB)(Table 1), anaerobic baffled reactor (ABR) and upflow anaerobic sludge fixed film reactor (UASFF) is known as granulation. These anaerobic granules harbor several metabolic groups of microorganisms, including hydrolytic, fermentative, syntrophic, and methanogenic microorganisms, involved in the anaerobic degradation of complex organic compounds (Satoh et al., 2007). Granules characteristically have a spherical form with a diameter from 0.14 to 5 mm (Schmidt and Ahring, 1996). Granules have been reported to be in different colors i.e. black, gray and brown reflecting the different stages in their life cycle (Liu, 2003). The spatial distributions of important phylogenetic groups in the granules have been determined by fluorescent *in situ* hybridization (*FISH*) (Santegoeds et al., 1999; Ariesyady et al., 2007; Satoh et al., 2007). In situ hybridization results showed that the outer layer of the granules was dominated by *Bacteria* whereas the inner layer was dominated by *Archaea* affiliated with *Methanosaeta* (>70%) (Santegoeds et al., 1999; Sekiguchi et al., 1999). However, the center of the granules was composed of dead or resting cells, or both, which were used as a support for active archaeal and bacterial cells near the surface (Yuko Saiki et al., 2002; Díaz et al., 2006).

For start-up of reactors in which granules are present such as UASB, the development of stable granules is the key factor for successful operation (Sekiguchi et al., 2001). Several factors are known to have prominent effects on the granulation process including hydrophobicity of constituents in sludge such as surface of cells (Daffonchio et al., 1995; Schmidt and Ahring, 1996), the presence of extracellular polymers (ECP) (Schmidt and Ahring, 1996; Veiga et al., 1997), and the composition of different layers and microorganisms (Wiegant and de Man, 1986; Schmidt and Ahring, 1996; Sekiguchi et al., 2001; Yuko Saiki et al., 2002). Among these, the microbial constituents are thought to be the decisive factor in many cases.

As CO_2 and CH_4, the main biogases produced during biomethanation (Tabatabaei et al, 2010) are poorly water-soluble, granular sludge with a very hydrophobic surface obviously tend to float (Daffonchio et al., 1995). The latter could adversely affect biomass retention in self-immobilized cell systems, such as UASB reactors (Daffonchio et al., 1995). Therefore, the formation of consortia with more hydrophilic cells such as *Chloroflexi* in the outer layer is of great importance. In addition, granules containing methanogens deep within them as well as those with a *Bacteria* layer on the surface tend to float, therefore, well-settled granules are considered to have methanogens that develop near the granule surface so that the gases generated during methane fermentation are readily released (Yuko Saiki et al., 2002).

The extracellular polymers in the granular sludge vary between 0.6 and 20% of the volatile suspended solids (VSS) and consists mainly of protein and polysaccharides (Schmidt and Ahring, 1996). These extracellular polymers are important for the structure and maintenance of granules (Veiga et al., 1997). *Methanosarcina* spp. and more importantly *Methanosaeta* spp. have been identified as essential aceticlastic methanogens for the initial granulation and development of granular sludge under mesophilic conditions (Schmidt & Ahring, 1996). They are considered of significant importance for making cores of sludge granules by constructing structures similar to a spider's web (Sekiguchi et al., 2001).

In contrast, in granules under thermophilic conditions, the long-filament type of *Methanosaeta* cells are replaced by dispersed type of *Methanosaeta* cells (Uemura & Harada, 1995; Syutsubo et al., 1997). Therefore, in thermophilic granules, the settleability of the granules is linked with the predomination of another group of very thin filamentous microorganisms instead of *Methanosaeta* (Sekiguchi et al., 2001). This type of microbe which is affiliated with a Clone Cluster, the Green Non-Sulfur Bacteria, Subdivision I, entirely covers the thermophilic granules, forming a web-like outside layer on granules (Uemura & Harada, 1993; Uemura & Harada, 1995; Sekiguchi et al., 1999; Sekiguchi et al., 2001). A combination of volatile fatty-acid mixtures and an addition of sucrose or glucose to the influent wastewater seemed to be essential for the optimum growth of these microbes and consequently the formation of granular sludge with good settleability (van Lier et al., 1994; van Lier et al., 1996).

Several studies have shown that the granulation process in different reactors such as anaerobic sequencing batch reactors (ASBRs) (Wirtz & Dague, 1996; El-Mamouni et al. 1997), ABR (Uyanik et al., 2002) and UASB reactors (Tiwari et al., 2004; Tiwari et al., 2005) can be influenced by conditioning the sludge with polymers. A study conducted by Uyanik et al. showed that a polymer-amended ABR reactor contained sludge that had a greater density of anaerobic bacteria and larger and denser granules than the control reactor, indicating that polymer addition possibly contributed to the retention of active biomass and resisted washout within the ABR (Uyanik et al., 2002). Moreover, it was found that the time required to form granules in a polymer-amended ASBR by using powdered activated carbon was reduced by approximately 75% compared to a control reactor (Wirtz & Dague, 1996). In fact, freely moving polymeric chains may form a bridge between cells, and this would facilitate the formation of initial microbial nuclei, which is the first step towards granulation (Liu et al., 2003) This is especially essential for UASB reactors which in them the long start-up period required for the development of anaerobic granules seriously limits the application of this technology. Therefore, it appears that polymers (synthetic and natural) can assist anaerobic bacteria to aggregate together and then form granules faster.

Abbreviation	Definition	Abbreviation	Definition
ABR	anaerobic baffled reactor	MAPE	mean absolute percentage errors
AFB	anaerobic fluidized bed	MLP	multilayer perceptrons
AF	anaerobic filters	MOO	multi-objective optimization
ANFIS	adaptive network based fuzzy system	MSE	mean squared error
ANN	artificial neural networks	OLR	organic loading rate
ASBRs	anaerobic sequencing batch reactors	ORP	oxidation-reduction potential
BOD	biological oxygen demand	PCA	principal component analysis
COD	chemical oxygen demand	POME	palm oil mill effluent
CSTR	completely stirred tank reactor	SBR	sequencing batch reactor
EGSB	expanded granular sludge blanket	STP	standard temperature and pressure
EPA	Environmental Protection Agency	TAN	total ammonia nitrogen
FISH	fluorescent in situ hybridization	TVFA	Total volatile fatty acids
Ga	genetic algorithm	UASB	up-flow anaerobic blanket reactor
GM-ANN	Grey Model ANN	UASFF	upflow anaerobic sludge fixed film reactor
HLR	hydraulic loading rate	VFA	Volatile Fatty Acids
HRT	hydraulic retention time	VSS	volatile suspended solids
KSOFM	Kohonen self-organizing feature maps	WTP	wastewater treatment plant
LCFA	long chain fatty acids		

Table 1. List of the abbreviations and definitions

2.2 Lipids in particular long-chain fatty acids (LCFAs)

Lipids, largely in the form of neutral fats, are commonly present in domestic sewage and industrial effluents such as food-processing wastewater (for instance dairy wastewater), wool-scouring wastewater, slaughterhouse wastewater and edible oil-processing effluents (Sousa et al., 2009). Lipid content and composition of industrial wastewaters is considerably variable. Lipid concentration of 0.2-1.3 g/l was measured in wastewaters from a sunflower oil mill (Saatci et al., 2003). Relatively higher levels of lipids were detected in wastewaters from a dairy industry (1.5–4.6 g/L) (Mendes et al., 2006). Becker et al. and Beccari et al. reported the presence of the minimum of 10 g/l and 16 g/l of lipids in olive oil-processing and wool-scouring effluents, respectively (Becker et al., 1999; Beccari et al., 2002). Furthermore, wastewaters significantly differ in their LCFA composition but in general palmitic and oleic acids are the most abundant saturated and unsaturated LCFA, respectively (Sousa et al., 2009).

Lipids are in general, glycerol bonded to LCFA, alcohols, and other groups by an ester or ether linkage. Triacylglycerides (neutral fats), are the most abundant family of lipids and are hydrolyzed by extracellular lipases to glycerol and LCFA. Glycerol is further degraded via acidogenesis while LCFA are broken down to acetate, H_2 and CO_2 through the β-oxidation process (syntrophic acetogenesis) (Stryer, 1995).

The application of anaerobic digestion of fat-rich wastewaters was hindered by problems related to LCFA adsorption to the biomass, with consequent sludge flotation and washout (Rinzema et al., 1993; Hwu et al., 1998a; Hwu et al., 1998b) and inhibitory or toxic effects of LCFA to different groups of microorganisms such as acetoclastic methanogens (Koster & Cramer, 1987; Rinzema et al., 1994; Angelidaki & Ahring, 1995; Lalman & Bagley, 2001) and to a much less extent to hydrogenotrophic methanogens (Hanaki et al., 1981; Lalman & Bagley, 2001). Besides the potential metabolic inhibition, LCFA accumulation onto the sludge can create a physical barrier or damage the cell wall, with consequent limitations in the transport of substrates and products (Pereira et al., 2005). Hence, treatment of fat-rich effluents in order to reduce the fat concentration before the anaerobic digestion was advised (Perle et al., 1995). However, it consequently leads to loss of their energetic potential as theoretically at standard temperature and pressure (STP), 1 g of oleate (unsaturated LCFA, C18:1) results in approximately 3 times higher methane production than 1 g of glucose (Cavaleiro et al., 2008) marking them attractive sources for methane production (Kim et al., 2004; Pereira et al., 2004). This potential is however limited due to the above mentioned operational problems.

Several technologies have been proposed for the anaerobic treatment of oily effluents, namely UASB reactors (Kim et al., 2004; Jeganathan et al., 2006), expanded granular sludge bed (EGSB) reactors (Hwu et al., 1998; Pereira et al., 2002a), and anaerobic filters (AF) (Pereira et al., 2002a). However, as explained earlier, due to fat-related operational problems, that efficient treatment of LCFA-rich wastewater can only be accomplished, if a correct equilibrium between accumulation and degradation is assured (Pereira et al., 2004; Cavaleiro et al., 2008). Different approaches has been practiced to achieve this equilibrium such as cycles of continuous LCFA feeding, followed by batch degradation of the accumulated substrate (Pereira et al., 2004; Pereira et al., 2005) and continuous operation after a step feeding start-up (Cavaleiro et al., 2006). Nielsen and Ahring also showed that the addition of oleate pulses to thermophilic reactors treating mixtures of cattle and pig manure had a promising effect on the overall process (Nielsen & Ahring, 2006). A study by Pereira et al. confirmed LCFA adsorption during continuous oleate feeding in anaerobic bioreactors (Pereira et al., 2004) which is a necessary state for LCFA degradation (Hwu et al., 1998b). They proposed a value of about 1000 mg COD-LCFA/g biomass for the optimal specific adsorbed-LCFA content that led to the maximal degradation rate. Long-term operation at higher LCFA-loading rates resulted in LCFA accumulation in bioreactors, with consequent limitations in the transport of substrate to the biomass and decreased removal efficiency (Pereira et al., 2004).

In addition, the organic suspended solids such as lipids and proteins present in seafood processing wastewaters are considered as a serious drawback for the use of high rate anaerobic reactors, such as UASB or AF, in which case the implementation of a previous hydrolysis acidification step would be required (Guerrero et al., 1999). Therefore for this type of effluents, two-phase systems are recommended, since these compounds could be degraded to Volatile Fatty Acids (VFA) in the first reactor by hydrolytic and acidogenic bacteria and finally converted into methane in the second reactor (Guerrero et al., 1999). Tagawa et al. encountered severe operational problems and unsatisfactory COD removal efficiency (60 to 70%) during the anaerobic treatment of a food processing wastewater containing high strength of lipid and protein in a thermophilic multi-staged UASB reactor (Tagawa et al., 2002). They reported the formation of a severe scum and insolubilized substance within the UASB sludge bed due to the presence of high strength of lipid and protein along with high cations concentration including Mg and Ca in the raw wastewater. This resulted in hindering the contact efficiency between substrate and sludge. Moreover, the replacement of active microbial granules in the sludge bed with the insolubilized protein and lipid led to reduced methanogenic activity (Tagawa et al., 2002). Enzymatic pretreatment (lipases) was reported as a very promising alternative for treating wastewaters having high-fat contents (Cammarota & Freire, 2006; Mendes et al., 2006; Leal et al., 2006). Mendes et al. reported higher biogas formation due to lipids liquefaction and bioavailability for anaerobic microorganisms when a dairy wastewater (lipids, 1.5 to 4.7 g/l; free fatty acids, 5 to 69 mg/L) was pretreated with a low-cost lipase for 12 h (Mendes et al., 2006). In addition, high COD and color removal was achieved.

Cammarota et al. obtained promising results when evaluated the efficacy of an enzymatic hydrolysis stage in dairy industry wastewater prior to the anaerobic biological treatment using *Penicillium restrictum* lipases (Cammarota et al., 2001). The authors obtained COD and lipid removal efficiencies of 91% and 82% when the reactor was operated with low loads of lipids (203 mg lipid/L). However, the removal efficiencies diminished significantly when the reactor received wastewater containing high lipid levels (868 mg/L), with COD and lipids removal efficiencies climbing down 50% and 40% ,respectively, due to the toxic effects of LCFAs on the anaerobic consortia. When wastewater pre-treated with 0.1% (w/v) of fermented babassu cake containing *P. restrictum* lipases was fed, the average COD and lipid removal efficiencies rose to their initial values of 92% and 89%, respectively, within approximately 15 days (Cammarota et al., 2001). These findings support the idea of applying hydrolytic enzymes as coadjutants in the anaerobic treatment of fat-rich wastewaters (Cammarota et al., 2001; Leal et al., 2002). Taken all together, pretreatment of lipid- and protein-rich food industrial wastewaters is a perquisite to their successful anaerobic treatment.

Concerning the microbiological aspects of fat-rich wastewaters, about 14 species have been described with the ability to grow on fatty acids in syntrophy with methanogens, all belonging to the families *Syntrophomonadaceae* and *Syntrophaceae* in the subclass of the *Deltaproteobacteria* (Jackson et al., 1999; Sousa et al., 2009). Sousa et al. applied molecular approaches to investigate the microbial diversity of anaerobic sludge after extended contact with long LCFA and found the predominance of the members of the phylum *Firmicutes* (87%), among which members of the *Clostridiaceae* and *Syntrophomonadaceae* families represented 69% (Sousa et al., 2007). Several studies investigating the diversity of methanogenic archaea in up-flow bioreactors fed with oleate or palmitate found a predominance of hydrogenotrophic methanogens belonging to the genera *Methanobacterium* (Pereira et al., 2002b; Sousa et al., 2007). Similarly, in a study conducted by Shigematsu et al. on a chemostats fed with a mixture of oleate (C18:1) and palmitate (C16:0), hydrogenotrophic methanogens (genus *Methanospirillum*) were found predominant (Shigematsu et al., 2007). Acetoclastic archaea mainly *Methanosaeta* and *Methanosarcina* species (Pereira et al., 2002b; Shigematsu et al., 2007; Sousa et al., 2007; Tabatabaei et al., 2009) have an important role in the overall efficient conversion of LCFA to methane, because methane derived from acetate accounts for about 70% of the total theoretical methane potential of LCFA (Sousa et al., 2009). Pereira et al. explained the decrease of methanogenic activity after cell contact with LCFA temporarily mainly because of mass transfer limitations (Pereira et al., 2005). However, in different study, Pereira et al. (2004) reported increased specific methanogenic activity after LCFA adsorption to biomass followed by efficient conversion of biomass to methane (Pereira et al., 2004). Therefore, it could be concluded that a possible acclimation of methanogens takes place after contact with LCFA and consequently the adverse effects of LCFA on anaerobic sludge functionality are not irreversible and that anaerobic sludge, under the appropriate conditions reaching the mentioned equilibrium is able to efficiently mineralize them (Pereira et al., 2003). Sousa et al. reported the possibility of the domination of sulfate-reducing hydrogen and acetate consumers (*Desulfovibrio-*, *Desulfomicrobium-* and *Desulforhabdus*-related species) during the anaerobic treatment of sulphate-LCFA-rich wastewaters (Sousa et al., 2009).

2.3 Organic loading rate (OLR)

The OLR variation can be derived from either variation in influent COD or variation in flow rate with constant COD. An increase in OLR beyond the optimum level are followed by a decrease in the main process parameters such as COD removal, specific methane production. In addition, high amount of suspended solids "known as biomass wash-out" are observed in the effluent, indicating that the reactor suffered a process imbalance and that biomass accumulated in the reactor (Converti et al., 1993; Rizzi et al., 2006; Fezzani & BenCheikh, 2007; Rincón et al., 2008). This could be ascribed to an increase in the concentrations of the VFAs with a consequent decrease in pH (Tiwari et al., 2006) or to escalated levels of inhibitory or toxic compounds such as phenols, LCFAs, lignin and etc. Therefore, there is a maximal operational value for this parameter. For instance, Rizzi et al. reported a decrease in COD removal and specific methane production when OLR was increased from 10 to 15 g COD/l/day. With the OLR increase to 20 g COD/l/day the biomass excess started to wash out, followed by deterioration of the reactor performance (Rizzi et al., 2006). In a different study, stable reactor performance was observed when the OLR increased from 1.5 to 9.2 g COD/l/day with the maximum methane production rate achieved for an OLR of 9.2 g COD/l/day. However, a significant decrease in the pH value (from 7.5 to 5.3) was observed when OLR was further raised to 11.0 g COD/l/day. In addition, the increase in the effluent CODs with increased OLR was paralleled to a sharp increase in the effluent total volatile fatty acids (TVFA, g acetic acid/L) by about 400% (Rincón et al., 2008). This indicates that, at higher OLR, the effluent total COD and mainly soluble COD is largely composed of the unused volatile acids produced in the reactor due to the inhibition of methanogenesis.

Methanobacteriaceae and *Methanosaeta* were found the main methanogens in a laboratory-scale up-flow anaerobic digester treating olive mill wastewater (Rizzi et al., 2006). However, the authors also reported an interesting population shift by OLR variation. At lower OLR i.e.

6 g chemical oxygen demand (COD/l/day, hydrogenotrophic *Methanobacterium* predominated in the reactor but the number of cells/g sludge showed a 1000 fold decrease from 10^{11} to 10^8 when the OLR was increased to 10 g COD/l/day. In contrast, phylotypes belonging to the acetoclastic *Methanosaeta* were not affected by OLR variation and at 10 g COD/l/day, dominated in the biofilm (10^9 cells/g sludge) (Rizzi et al., 2006). Olive oil wastewater is characterized by high levels of inhibitory/toxic compounds such as tannins, and lipids. As a result, increased OLR leads to higher concentration of these substances and a consequent inhibition of methanogenic cells. However, acetoclastic *Methanosaeta* due to its high affinity for acetate is capable of occupying the deepest and thus more protected niches in the granule or biofilm with low concentrations of substrate (acetate) (Gonzales-Gil et al., 2001). Phylotypes belonging to the genus *Methanosaeta* were also dominant independent of different OLRs in other anaerobic digesters (Rincón et al., 2006; Rincón et al., 2008). So with the above findings, we may suggest that these acetoclastic methanogens i.e. *Methanosaeta* be the least susceptible methanogens to the elevated concentrations of toxic compounds when bioreactors are operated at high OLRs. In a different study, Kalyuzhnyi et al. investigated the microbial ecology of granules in UASB-reactor fed by synthetic wastewater under various OLRs. The authors showed that the predominant microbial biomass was *Methanosaeta*. However, increasing the OLR led to a substantial increase of *Methanosarcina* in the granules (Kalyuzhnyi et al., 1996). The increase of *Methanosarcina* in the studied synthetic wastewater (toxin-free) due to increasing OLR is explained by the low affinity of these methanogens for acetate in comparison with *Methanosaeta*. Hence, by increasing OLR and consequent VFAs concentration, *Methanosarcina* is favored. These findings would have been more interesting if they had included a group of inhibitory/toxic compound/s in the synthetic medium.

As reviewed earlier, under mesophlic conditions *Methanosaeta* plays a significant role in making cores of sludge granules (Schmidt & Ahring, 1996; Sekiguchi et al., 2001) and thus their ratio seems to control the speed of granulation (Rincón et al., 2006). Higher OLRs result in consequent higher concentrations of substrate i.e. acetate in the reactor. Morvai et al. attempted to investigate the influence of organic load ranging from 0.5 to 3.0g/1 on granular sludge development in an acetate-fed system (Morvai et al., 1990). They argued that in the range of feed acetate levels examined, higher concentrations of feed (acetate) caused faster granulation of the sludge bed and, presumably, of the microbial population, and resulted in better sludge structure and improved sludge settleability (Morvai et al., 1990). Low OLR has been reported to cause acute mass transfer limitation leading to disintegration of the larger granules (Ahn et al., 2002). The disintegration begins at the core of the granules due to substrate limitation with a consequent loss of granule's strength and stability (Kosaric et al., 1990). However, this was not in agreement with the studies conducted by Teer et al. and Tiwari et al. which low OLR (<1.5 kg COD/m^3/d) did not lead to disintegration of the granules in UASB reactors (Teer et al., 2000; Tiwari et al., 2005). This could be ascribed to the different experimental settings and wastewaters used in these studies. Teer et al. attempted to treat a high iron bearing wastewater in a UASB reactor. Evidence shows that the presence of divalent and trivalent cations ions, such as Fe^{2+}, and Fe^{3+}, helps bind negatively charged cells together to form microbial nuclei that promote further granulation (Teo et al., 2000). On the other hand, Tiwari et al. tried to enhance the granulation process by using natural ionic polymer additives. These may thus reduced the effect of low OLR (i.e. substrate limitation) on the granules and delayed the disintegration (Tiwari et al., 2004).

Converti et al. reported that COD removal rate, the COD specific removal rate (r_s) and methane production rate were not suppressed by increasing OLR when treating wine wastewater and sewage mixture (Converti et al., 1990). That indicated that no inhibition factor related to the organic content of the effluent was present in both wine wastewater and sewage mixture studied. This was further supported by the cell mass concentration varied very little with increasing the OLR. However, as completely noticed by the authors, even at the absence of inhibitory/toxic compounds, in the initial part, the removal rate increased with the OLR, following a first order kinetic. In the second part, instead, the removal rate, tended to a constant maximum value, following a zero order kinetic. Afterwards, the removal efficiencies as well as the methane production yield gradually decreased with increasing influent COD due to increasing the OLR, which evidently showed a substrate inhibition occurrence (Converti et al., 1990). This supports the idea that even at the absence of the inhibitory/toxic compounds in the wastewater, increasing influent COD by the means of increasing OLR could lead to substrate inhibition and consequent reduced removal efficiencies. Converti et al also described the observed dependence of the removal rate on the OLR by an empirical equation similar to Monod's model to compare the degradability of different effluents (see Equation 1)(Converti et al., 1990):

$$r_s = r_{s(max)} OLR / (k + OLR) \quad (1)$$

where $r_{s(max)}$ [kg COD/(kg_{VSS} d)] is the maximum value of r_s, and k a constant which physically is expressed in units of OLR and an increase of k indicates increased treatment ability of the studied effluent.
Taken all into consideration, the desired OLR is the function of the favorable effect of OLR on stimulating the growth of methanogens in the bioreactor by providing them with higher substrate concentrations, its reverse effect on elevating the concentration of inhibitory/toxic compounds and the buffering capacity of methanogenic community. In the other words, the maximal operational value of OLR (OLR_{max}) is translated into the highest methane production (indicating the highest conversion efficiency of the system) that the buffering capacity of methanogenic community is still capable of compensating for elevated concentrations of inhibitory/toxic compounds.

2.4 Mixing

There is only a limited number of studies found specifically focused on the effects of mixing on the treatment efficiency and biogas production using various types of organic wastes including cow manure, palm oil mill effluent (POME), animal waste, municipal solid waste, primary sludge and fruit and vegetable wastes (Stafford, 1982; Stroot et al., 2001; Karim et al., 2005a; Karim et al., 2005b; Kaparaju et al., 2007; Sulaiman et al., 2009). Adequate mixing is very important in order to achieve successful anaerobic treatment of organic rich wastewater. In another word, it enhances the anaerobic process rate by preventing stratification of substrate, preventing the formation of surface crust, ensuring the remaining of solid particles in suspension, transferring heat throughout the digester, reducing particle size during the digestion process and releasing the biogas from the digester content (Kaparaju et al., 2008; Sulaiman et al., 2009).

Prior to 1950s, anaerobic digesters treating sewage sludge were not equipped with mechanical mixing and thus caused the formation of scum layer at the surface (Fannin, 1987). To overcome this problem, mixing was employed to disrupt scum formation and enhance contact between microorganisms and substrates. It has been reported that the acetate-forming bacteria and methane-forming bacteria are required to be in close contact to achieve continuous degradation of organic materials (Gerardi, 2003). In addition to the mentioned advantages, mixing also helps to eliminate thermal stratification inside the digesters, maintain digester sludge chemical and physical uniformity, rapid dispersion of metabolic products and toxic materials and prevent deposition of grit (Gerardi, 2003).

In modern anaerobic digesters, mixing could be achieved in various ways such as gas injection, mechanical stirring and mechanical pumping as presented in Table 2 (Tchobanoglous & Burton, 1991). In more advanced applications where higher temperature is needed, digester is also equipped with a heater used to heat up the digester content or feed as illustrated in Figure 1 (Gerardi, 2003). Generally for large scale applications, agitator or mixer system is commonly used to mix substrate homogenously inside the bioreactor and to provide a good contact between microorganisms and the substrate. All previously published articles on the effects of mixing on the stability and methane production reported similar observations which led to a consistent conclusion that vigorous mixing that is turbulent flow in nature is unsuitable for microorganisms growth and consequently results in an unsatisfactory methane production (Stafford, 1982; Stroot et al., 2001; Karim et al., 2005a; Karim et al., 2005b; Kaparaju et al., 2007; Sulaiman et al., 2009). This is basically due to the effect of high shear force on separating the hydrolytic bacteria from their substrate (Stafford, 1982). This is supported by another study on the effect of shear force on both aerobic and anaerobic sludge (Sheng et al., 2008). They found out that the anaerobic sludge or flocs (with shear sensitivity of 0.088) was less stable compared to the aerobic sludge flocs (shear sensitivity of 0.032). Through further investigation, Sheng et al. also discovered that the adhesion enthalpy based on the shear experiment (ΔH_G) for anaerobic flocs was larger (-919 l/s) than that of the aerobic flocs (-1600 l/s). This reflects that the deflocculating process is less energy demanding for anaerobic flocs or in other word, the flocs in the anaerobic sludge is easier to be broken than the aerobic flocs.

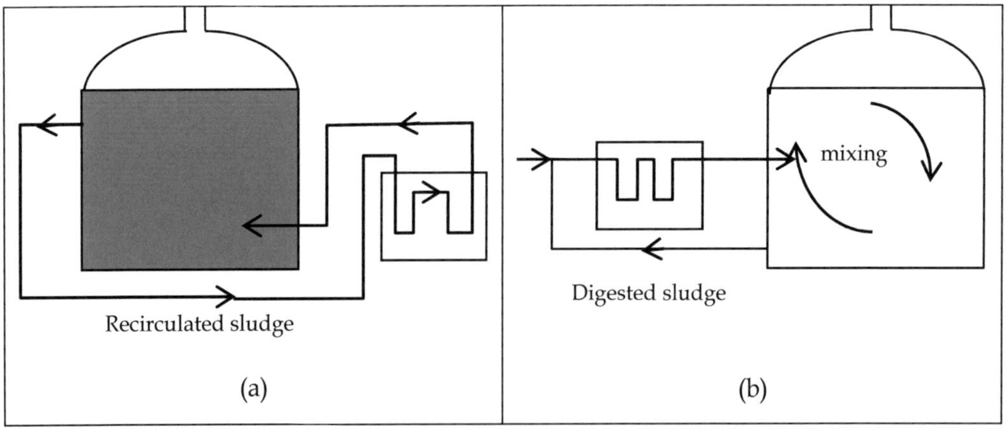

Fig. 1. The mixing methods that incorporate heating with limited mixing achieved through recycling sludge through a heat exchanger (a) and simultaneous mixing and heating of the digested sludge inside the bioreactor through sludge recycling system. (Gerardi, 2003).

No	Type of Devices	Description	Type of digesters the device suits	Schematic Figure	References
1	Unconfined gas injection system	Gas is collected at the top, compressed, recycled and discharged at the bottom of the digester. The rising gas inside the digester will mix the content and the sludge.	Suitable for digesters with fixed, floating or gas holder covers such as continuous stirred tank reactor (CSTR).	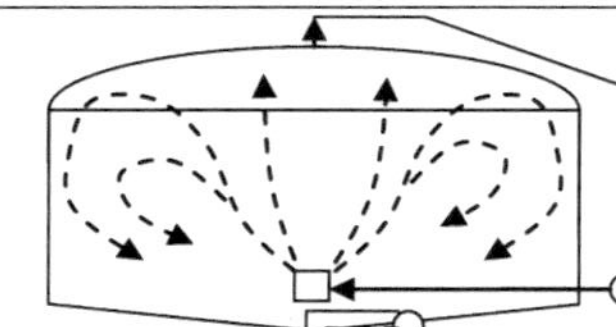	Tchobanoglous and Burton, (1991)
2	Confined gas injection system	Gas in collected at the top, compressed and returned into through confined tubes.	Suitable for fixed, floating or gas holder covers such as continuous stirred tank reactor (CSTR).	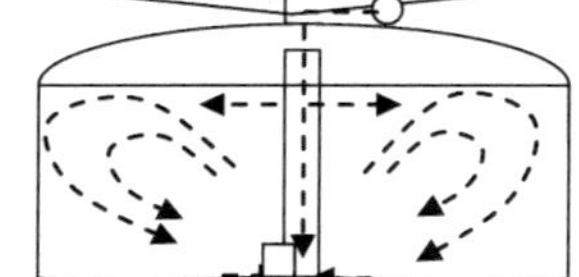	Tchobanoglous and Burton, (1991)
3	Mechanical stirring system	Mixer is used to displace the sludge and mix the digester contents. The mixer usually has one cover-mounted mixer at the top of the digester.	Suitable for fixed or floating covers such as continuous stirred tank reactor (CSTR), sequencing batch reactor (SBR).	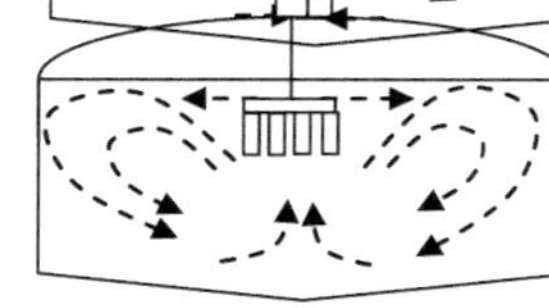	Tchobanoglous and Burton, (1991); Song et al (2004). Fannin and Biljetina (1987)
4	Mechanical pumping system	Turbine is installed which usually has one cover-mounted motor at the top and the turbine impeller is located at different sludge depths. Mixing pump may also be used for effluent recycling and mixing.	Suitable for fixed or floating covers such as continuous stirred tank reactor (with semi-continuous mixing system)	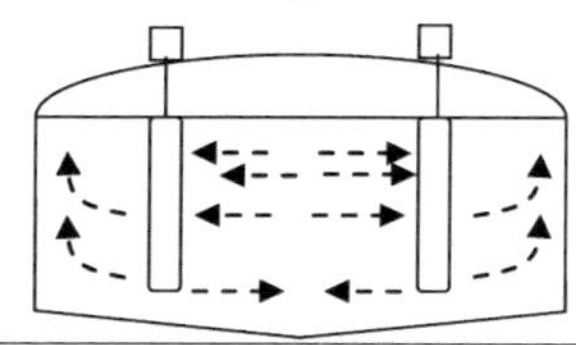	Sulaiman et al., (2009); Visser et al., (1993);

Table 2. Various mixing devices used in high rate anaerobic digesters systems

In addition to high turbulent mixing, continuous mixing could also reduce the performance of the biogas production as reported by Stroot et al. (2001) and Sulaiman et al. (2009). The former studied mixing organic fraction of municipal solid waste co-digested with primary sludge and waste activated sludge and found the continuous and vigorous mixing inhibitory when the system was operated at a high organic loading rate due to the disruption of syntrophic relationships and spatial juxtaposition of the microorganisms responsible for organic degradation and methane production (McMahon et al., 2001; Stroot et al., 2001). Sulaiman et al. investigated anaerobic treatment of POME for biogas production using a pilot scale bioreactor under constant mixing with turbulent flow regime condition and discovered that the methanogenesis process was totally inhibited just after 13 days of operation. The biogas was not produced at all and the VFA content was recorded higher than its inhibitory limit (i.e 1000 mg/L). The inhibition has severely affected the whole digester content and it took more than 3 months to recover the process through addition of sodium hydroxide to bring up the alkalinity and pH to neutral (Sulaiman et al., 2001).

In terms of power requirement for digesters mixing system, there is no specific value enforced, however US Environmental Protection Agency (EPA) recommended the power input of 0.20-0.30 HP/1000 cu ft (5.26-7.91 W/m^3) of the digester content (Karim et al., 2005a). In general, the mixing power requirement is very much related to the type of wastewater in terms of total solid contents as confirmed byvarious studies (Karim *et al.*, 2005a; Karim *et al.*, 2005b). For the dilute cow manure of 5% total solids, the influence of mixing was insignificant in terms of biogas productivity for the mixed and unmixed digesters (Karim et al., 2005a). In their study, the biogas productivity and methane yield recorded for the mixed and unmixed digesters were 0.84-0.94 L/L/d and 0.26-0.31 L CH_4/ g volatile solid loaded, respectively.

On the other hand, Sulaiman et al. reported that natural mixing could occur in the digester due to the biogas rising (Sulaiman et al., 2009). They also highlighted that a satisfactory level of methane production could be achieved at 1.0 $m^3/m^3/d$ just by allowing natural mixing without employing any additional mixing devices. Although this value (1.0 $m^3/m^3/d$) was lower than that achieved by the minimal mixing (1.1 – 1.4 $m^3/m^3/d$), but was still higher than the methane production record of only 0.84-0.94 $m^3/m^3/d$ reported by Karim et al. (2005 a-b). The low yield obtained in their study could be attributed to the lower solids content recorded which led to an easier biogas rising due to less resistance available inside the digester. In summary, there is no specific measurement of power requirement for optimum methane production in any particular anaerobic system and it is very much related to the type of wastewater itself whether it is diluted or concentrated. However as a general rule, minimal mixing is the best mode of mixing as confirmed by many studies (Karim et al., 2005a; Karim et al., 2005b; Karaparju et al., 2007; Sulaiman et al., 2009).

2.5 Sludge recycling

As for any anaerobic treatment system designs, it is always desirable to maintain optimum levels of microorganisms inside the digester in order to meet an appropriate ratio of food-to-microorganism (F/M) which results in an efficient bioconversion process. For a complete-mix system Tchobanoglous et al. suggested F/M ratio of 0.2-0.6 $IbBOD_5/IbMLVSS/d$ and slightly lower at 0.2-0.4 $IbBOD_5/IbMLVSS/d$ for a conventional system (Tchobanoglous et al., 1991). In order to maintain high levels of biomass inside the digester, several strategies have been adopted by many researchers. Among those, sludge recycling (Visser et al., 1993; Setiadi et al., 1996; Bae et al., 1998; Faisal & Unno, 2001; Najafpour et al., 2006; Sulaiman et al., 2009; Busu et al., 2010), is probably the simplest strategy to increase the biomass concentration in the digester.

Busu et al. conducted a simple experiment on anaerobic treatment of POME and the sludge was recycled at 6 m^3/d with different feeding strategies. Both feeding strategies (every three and six hours) resulted in a satisfactory COD removal efficiency of higher than 90% but the latter led to a more stable process with the total VFA concentration recorded below 500 mg/l and VFA:Alk ratio of less than 0.3 at the maximum OLR of 6.0 $kgCOD/m^3/d$ applied (Busu et al., 2010). Prior to their study and by using the same digester, Sulaiman et al. proved that by increasing the sludge recycling rate from 6 m^3/d to 12 m^3/d and to the maximum of 18 m^3/d, the anaerobic treatment process could be proceeded to the maximum organic loading rate of 10 $kgCOD/m^3/d$ with biogas and methane productivity of 1.5 m^3 $m^3/m^3/d$ and 0.9 $m^3/m^3/d$, respectively (Sulaiman et al., 2009). Thus based on these two studies on POME, it could be concluded that by increasing the sludge recycling rate during treatment at higher OLRs, the treatment process maintained stable with high COD removal efficiency and satisfactory methane production. Using the same substrate but different digester designs, Setiadi et al., Faisal and Unno and Najafpour et al. revealed that sludge recycling strategy is compulsory to ensure system stability by maintaining pH higher than 6.8 without alkalinity supplementation, eliminating high organic over loading and supplying alkalinity by blending the fresh feed and recycled stream (Setiadi et al., 1996; Faisal & Unno, 2001; Najafpour et al., 2006). Setiadi et al conducted a simple study on the effects of sludge recycling on the baffled system stability and found out that a recycle system of more than 15 times of the fresh feed was required in order to maintain the system's pH higher than 6.8 without alkalinity supplementation and the implementation of the recycle system was an effective means to reduce alkalinity requirements (Setiadi et al., 1996). Najafpour et al. used an upflow anaerobic sludge-fixed film bioreactor and applied a very high recycling ratio of 11.25 of the fresh feed in order to eliminate high organic overloading and to supply alkalinity by blending the fresh feed with the recycle stream characterized by low COD and high alkalinity (Najafpour et al., 2006). As a result, no alkali additional was required to buffer the system. In a different study, Faisal and Unno using modified anaerobic baffled reactor showed that a recycling ratio of 30 was required in order to stabilize the modified anaerobic baffled reactor operation (Faisal & Unno, 2001). On the same concept, but using solid waste substrate, Bae et al. concluded that the continuous addition of active methanogenic population into the anaerobic digester by sludge recycling was effective for rapid and significant methane gas recovery (Bae et al, 1998). Visser et al. also reported that effluent recirculation (of factor 6.7) was necessary in order to improve mixing and to ensure sufficient buffer capacity of extreme pH situation (pH 2-4) (Visser et al., 1993).

2.6 Temperature

It is interesting to note that anaerobic digestion in the natural environments occurred in a wide range of temperatures between 4°C (lake sediment) to 60°C (thermophilic digestion process); however, for the industrial practices, the temperature range is limited to 20-55°C (Fannin, 1987). In the natural environments, the optimum temperature for the growth of methane forming archaea is 5-25°C for psychrophilic, 30-35°C for mesophilic, 50-60°C for thermophilic and >65°C for hyprethermophilic (Tchobanoglous & Burton, 1996). It is generally understood that higher temperature could produce higher rate of reaction and thus promoting higher application of organic loading rate (OLR) without affecting the organic removal efficiency (Desai et al., 1994; Chae et al., 2007, Choorit & Wisarnwan, 2007; Poh & Chong, 2009;). Using palm oil mill effluent as the substrate, Choorit and Wisarnwan, demonstrated that when the digester was operated at thermophilic temperature (55°C), showed higher OLR application than the that of mesophilic (17.01 against 12.25 gCOD /l/d) and the methane productivity was also higher (4.66 against 3.73 l/l/d) (Choorit & Wisarnwan, 2007). A similarly study by Chae et al. indicated that the higher temperature of 35°C led to the highest methane yield as compared to 30°C and 25°C although the methane contents only changed slightly (Chae et al., 2007). Using cheese whey, poultry waste and cattle dung as substrates, Desai et al. showed that when the temperature was increased from 20 to 40 and finally to 60 °C, the gas production and methane percentage in biogas increased as well (Desai et al., 1994). This could be explained by the following reaction rate equation (Tchobanoglous & Burton, 1996). As clearly indicated, as the temperature is raised, the power fraction of (T-20) would be increased as well (as a result of higher enzymatic activity) which could result in higher reaction rate; r_T (see Equation 2).

$$r_T = r_{20}\theta^{(T-20)} \quad (2)$$

Where r_T = reaction rate at T °C, r_{20} = reaction rate at 20 °C, θ = temperature-activity coefficient and T = temperature , °C.

Although the thermophilic anaerobic process could increase the rate of reaction, the yield of methane that could be achieved over the specified organic amount is the same regardless of the mesophilic or thermophilic conditions. That value is 0.25 kg CH_4/kgCOD removed or 0.35 m^3 CH4/kgCOD removed (at standard temperature and pressure i.e 0°C, 1 atm.) which is derived by balancing the following equation (see Equation 3):

$$CH_4 + 2O_2 \longrightarrow CO_2 + 2H_2O \quad (3)$$

On the other hand, although thermophilic condition could result in higher application of organic loading rates and better destruction of pathogens, at the same time it is more sensitive to toxicants and temperature control is more difficult (Gerardi, 2003; Choorit & Wisarnwan, 2007). Furthermore, biomass washout that could lead to volatile fatty acids accumulation and methnogenesis inhibition could also occur if the thermophilic temperature could not be controlled (Poh and Chong, 2009). As a result, in tropical regions, mesophilic temperatures are the preferred choice for anaerobic treatment (Yacob et al., 2005, Sulaiman et al., 2009).

2.7 Alkalinity and pH

As far as the anaerobic digestion process is concerned, it is more appropriate to discuss alkalinity and pH together because these parameters are related to each other and very promising to ensure a suitable environment for successful methanogenesis process. Alkalinity is produced in the wastewaters as results of the hydroxides and carbonates of calcium, magnesium, sodium, potassium or ammonia and may also include borates, silicates and phosphates (Tchobanoglous & Burton, 1991). The alkalinity plays an important pH controlling role in the anaerobic treatment process by buffering the acidity derived from the acidogenesis process (Gerardi, 2003; Fannin, 1987).

Methane producing archaea or methanogens are known to be strongly affected by pH (Poh & Chong, 2009) and could only survive on a very narrow range of pH as listed in Table 3 (Gerardi, 2003). As such, the methanogenic activity will be severely affected once the optimum pH range is not met. Steinhaus et al. studied the optimum growth conditions of *Methanosaeta concilii* using a portable anaerobic microtank (Steinhaus et al., 2007). They reported an optimum pH level of 7.6 revealing that even little variations on both sides of the optimum pH suppressed the growth of the methanogens. Several studies have also reported

Genus	pH range
Methanosphaera	6.8
Methanothermus	6.5
Methanogenium	7.0
Methanolacinia	6.6-7.2
Methanomicrobium	7.0-7.5
Methanosprillium	7.0-7.5
Methanococcoides	6.5-7.5
Methanohalobium	6.5-6.8
Methanolobus	6.5-6.8
Methanothrix	7.1-7.8
Methanosaeta	7.6

Table 3. The optimum pH range for selected methanogens (Gerardi, 2003; Steinhaus et al. 2007)

reactor failure or underperformance simply due to pH reduction caused by accumulation of high volatile fatty acids in the anaerobic treatment system (Visser et al., 1993; Fabián & Gordon, 1999; Poh & Chong, 2009). In a study using synthetic wastewater in the thermophilic temperature, Visser et al. found that at the pH of above 8.0, the methanogenesis was strongly inhibited and the value recorded for acetotrophic methanogenic test was zero (Visser et al., 1993). When investigating the role of pH in anaerobic degradation test, Fabián and Gordon, found out that the acidification led to the low performance of the anaerobic degradation, however the biodegradation was significantly increased once the waste was saturated with water and the pH was adjusted to above 6.5 (Fabián & Gordon, 1999).

There are various types of chemicals that could be introduced into anaerobic digesters for alkalinity supplementation as summarized in Table 4. However on the other hand, over-supplementing these chemical into the system could affect the system in many ways including pH over range, development of vacuum conditions inside the digester, chemical precipitation inside the digester, excess release of ammonia gas, Na^+ and K^+. The last could potentially cause toxicity to microorganisms, lead to foaming problems and increase the oxidation-reduction potential (ORP).

No	Name	Buffering Cation
1	Sodium bicarbonate	Na^+
2	Potassium bicarbonate	K^+
3	Sodium carbonate (soda ash)	Na^+
4	Potassium carbonate	K^+
5	Calcium carbonate (lime)	Ca^{2+}
6	Calcium hydroxide (quick lime)	Ca^{2+}
7	Anhydrous ammonia (gas)	NH^{4+}
8	Sodium nitrate	Na^+

Table 4. Chemicals used for alkalinity supplementation (Fannin, 1987; Gerardi, 2003)

2.8 Ammonia

There are various type of waste/wastewater containing high organic nitrogen such as municipal solid waste, young landfill leachate (Yusof et al., 2009), wastewater from seafood processing factories (Guerrero et al., 1997; Gebauer, 2004), animal waste (Garcia & Angenent, 2009) and waste activated sludge (Appels et al., 2008). Therefore, as a results of degradation of the organic nitrogen fraction, high concentration of total ammonium (ammonium (NH_4^+) plus free ammonia (NH_3) in these waste/wastewater is common (Jokela & Rintala, 2003). Many studies have shown that free ammonia and not ammonium is responsible to inhibit the methanogenic activity during the anaerobic digestion (Sawayama et al., 2004; Sossa et al., 2004; Calli et al., 2005a; Garcia & Angenent, 2009). In a solution, the ammonia exists in a pH dependent equilibrium between ammonium ion (NH_4^+) and unionized ammonia or free ammonia (NH_3) (see Equation 4).

$$Organic\ Nitrogen \rightarrow NH_4^+ + OH^- \leftrightarrow NH_3 + H_2O \tag{4}$$

The free ammonia concentration is influenced by total ammonium concentration (NH_4^+ + NH_3), pH and temperature as described by equations 5 and 6 (Anthonisen et al., 1976).

$$FA = {}^{17}\!/_{14} \times \frac{TAN \times 10^{pH}}{\left({}^{K_b}\!/_{K_w}\right) + 10^{pH}} \tag{5}$$

Where, FA (mg/L) is the free ammonia concentration, TAN (mg/L), the total ammonia nitrogen, K_b, the ionization constant of the ammonia equilibrium equation and K_w, the ionization constant of water. The ratio of the ionization constant of the ammonia equilibrium equation and the ionization constant (K_b/K_w) of water is dependant on the temperature (see Equation 6) (Anthonisen et al., 1976).

$${}^{K_b}\!/_{K_w} = e^{\left({}^{6344}\!/_{(273+°C)}\right)} \tag{6}$$

Therefore, the free ammonia inhibition on methanogens is controlled mainly by total ammonium concentration, pH and temperature. Ammonium ion is an essential nutrient for methanogens growth, however excess concentrations of free ammonia can inhibit these microorganisms. Earlier investigation observed that methanogens inhibition strongly occurred at a wide range of total ammonia nitrogen (TAN) concentrations; 1500-3000 mg/L at pH above 7.4 (Hashimoto, 1986); 3300-4100 mg/L at pH 8.1 and temperature 55°C (Hansen et al., 1998); 6000 mg/L at temperature of 35°C (Calli et al., 2005a) and 5250 mg/L at temperature of 35°C (Garcia & Angenent, 2009). The remarkable differences reported in ammonia inhibitive concentration could be ascribed to variations in the TAN concentrations as well as the operating parameters such as pH and temperature (Sung and Liu, 2003). Besides, the acclimatization of the seed sludge to the elevated concentrations of ammonium also has been reported to contribute to the success of methanogenesis performance (Hashimoto, 1986; Calli et al., 2005a). The adaptation of the methanogens to high concentrations of ammonium was confirmed by detecting resistant methanogens present in the seed sludge with a high concentration of ammonium (Sprott & Patel, 1986). During the adaptation to high ammonium concentrations, it was found that the acetate utilizing methanogens i.e. *Methanosarcina* and *Methanosaeta,* were more sensitive as compared to hydrogen utilizing methanogens (Angelidaki & Ahring, 1993; Sawayama et al., 2004). The *Methanosarcina* sp. was more severely affected by high ammonium concentration than the hydrogen utilizing methanogens (Sawayama et al., 2004). In contrast, a study by Calli et al. an abundant number of *Methanosarcina*-like methanogens at elevated ammonium concentration (6000 mg/l TAN) was observed (Calli et al., 2005b). However, both studies showed that *Methanosaeta*-related species vanished during the adaptation period.

Methanogenesis is more sensitive to ammonia when the pH value is increased. By increasing the pH, greater fraction of the total ammonium will be in the form of free ammonia (NH_3) (eq.1) (Anthonisen et al., 1976) which is believed to inhibit the process and consequently results in low methanogenic activity. The pH of anaerobic digestion reactors treating high organic nitrogen wastes such as poultry leachate and livestock waste were observed high in a range of 7.5-9.0 (Angelidaki & Ahring, 1993; Gangagni et al., 2008). This resulted in 10 times higher free ammonia concentration than the tolerable level and led to the reduction in biogas and methane productions. On the other hand, instability of anaerobic treatment process always results in high VFA concentrations (Strik et al., 2006; Sulaiman et al., 2009). Hence, the low pH encountered will facilitate the reduction of free ammonia concentrations. The relation between free ammonia, VFA and pH is known as inhibited steady state, a condition where the anaerobic process is stable but with low methane yield (Angelidaki & Ahring, 1993).

The anaerobic treatment process has been long known to be influenced by temperature. An increase in temperature will increase the anaerobic degradation rate of the organic substances. It has been shown that the methanogenic activity was higher at higher mesophilic temperature (Masse & Masse, 2001). It was postulated that effects of higher kinetic rate at higher temperature in mesophilic range overwhelmed the free ammonia inhibition (Garcia & Angenent, 2009). However, high ammonia concentrations inhibit methanogenesis more seriously under thermophilic as compared to mesophilic condition (Angelidaki & Ahring, 1993). It was demonstrated that the degradation of swine manure at thermophilic temperature of 55°C was feasible even at high ammonia concentration (6000 mg/l). However, inhibition on methanogenesis was severe which resulted in low methane yield (Hansen et al., 1998). Therefore, although there is a kinetic advantage when operating anaerobic treatment at high temperatures, the thermophilic process is more sensitive to the environmental factors including ammonia toxicity (Sung & Liu, 2003). Higher free ammonia fraction to the TAN has been observed in thermophilic range due to the temperature dependant constant which led to the methanogensis inhibition (El-Mashad et al., 2004; Strik et al., 2006).

2.9 Heavy metals inhibition

Heavy metals are present in various types of waste/wastewater including industrial wastewater, landfill leachate and sludge (Appels et al., 2008; Colussi et al., 2009; Yusof et al., 2009). Although many metals are required in trace amounts to provide sufficient growth to methanogens, the methanogenic activity in anaerobic reactors is strongly affected by excess amounts of heavy metals (Altas, 2009; Colussi et al., 2009). The toxic effects of metals to the biological process is particularly due to the inhibition of enzymes activity as a result of metals binding to the SH group of the enzyme (Nies, 1999). The inhibitory concentrations of four heavy metals on methane-producing granular sludge that caused 50% reduction in cumulative methane production was found to be 7.5 mg/L of Zn, 27 mg/L of Cr, 35 mg/L of Ni and 36 mg/L of Cd with an order of Zn>Cr>Ni≈Cd (Altas, 2009). Whereas a different study revealed that 50% reduction in methane production occurred at 6.4 mg/L of Cu (II), 4.4 mg/L of Cd(II) and 18.0 mg/L of Cr(VI) with an order of Cd(II)>Cu(II)>Cr(VI) in anaerobic digestion of cattail with rumen culture (Yue et al., 2007). The differences reported in the metals inhibitory concentration might be due to the several factors including variation in sludge characteristics, chemical form of heavy metals and microbial resistance to metals (Nies, 1999; Yue et al., 2007; Altas, 2009). Various heavy metals presence in wastewater also showed synergistic effects during anaerobic treatment process. For instance, the presence of chromium in the sludge results in higher toxicity of copper (Colussi et al., 2009).

3. Modeling anaerobic treatment process: New trends

It is well known that the functional and operational state of a bioreactor is subject to wide fluctuations due to process disturbances such as loading rates and pH changes. This has gained considerable concerns within the last years to overcome the alterations and consequently guarantee the safe operation of anaerobic digesters. The attention toward such issues gets more crucial when high rate bioreactors are utilized. On the other hand, an exact picture of what happens in bioreactors has been so much sophisticated and as a result, massive simplifications have been assumed to predict the necessary outputs of the process (Lin, 1991). Mathematical modeling of wastewater treatment processes plays an outstanding role as a tool capable of providing diagnostics that will give support to the plant operation and the decision-making process.

3.1 Artificial neural networks

Many analytical models, mostly kinetic models, have been developed to describe the anaerobic treatment in bioreactors, however, they are not routinely used for control and on-line applications. The reason lies in their complexity and insolvable parameters. Furthermore, the kinetic models are highly affected by the environmental conditions rendering them to be too generalized for other substrates or environments (Tay & Zhang, 2000).

Artificial neural networks (ANN), as powerful modeling methods have found a great deal of interest among the researchers in the last two decades (Feng et al., 2007; Zeng et al., 2003). They can solve a wide range of problems and differential equations, particularly when the conventional approaches fail (Ghobadian et al., 2009). The ability of ANNs in the modeling is traced back to their training mechanism which is inspired by biological neurons (Fig 2 a). Fig 2b presents a three-layer network. The data processing can extend over multiple layers and the final error criteria such as mean squared error (MSE) are calculated at the output layer which measures the accuracy of modeling (Hertz et al., 1991).

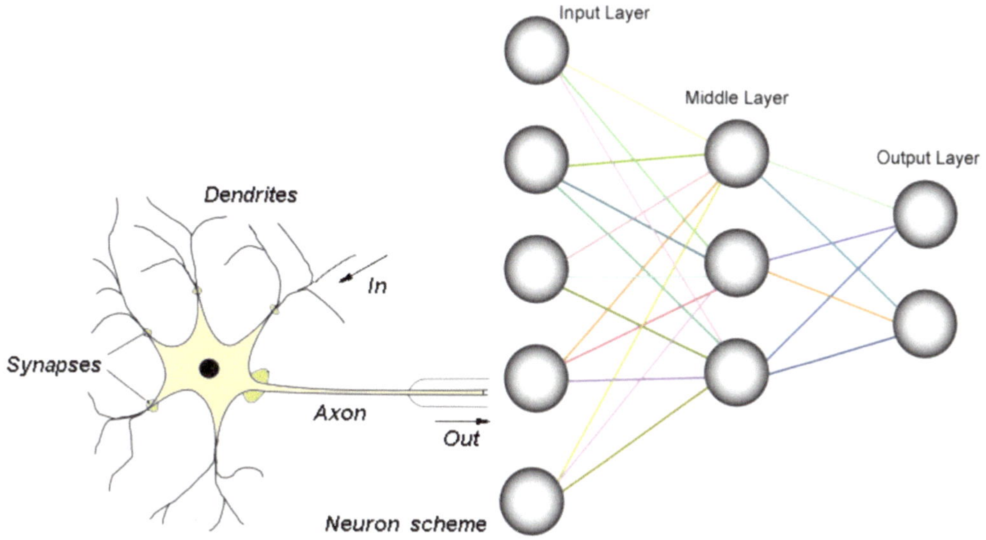

Fig. 2. a) a schematic picture of biological neurons, b) representation of biological neurons as a feed-forward three-layer network.

A well trained network learns from the pre-seen experimental dataset (training data) and generalizes this learning beyond to the unseen data which is called 'prediction' (Haykin, 1994). Furthermore, ANNs are able to model non-linear behaviors and complex processes. This is highly important in anaerobic treatment and bio-processes due to the special hydrodynamics and non-linear nature of the anaerobic digestion (Holubar et al., 2002). Based on the mechanism of human nervous systems, the ANN models can be classified into three major groups; 1) feedforward network, 2) recurrent network and 3) unsupervised network (Fig 3). The most usual model is feedforward network which has frequently been used in anaerobic treatment studies (Zhu et al., 1998; Gontarski et al., 2000; Strik et al., 2005; Ozkaya et al., 2007; Parthiban et al., 2007). Although, recurrent and unsupervised networks are powerful models developed to map the variables of anaerobic processes non-linearly, a few studies have applied them in on-line or control applications (Kecman, 2001). This could be ascribed to the simplicity, accuracy and swiftness that feedforward networks offer (Kalogirou, 2000, Kiani Deh Kiani et al., 2009).

ANN together with other intelligent methodologies could be a promising alternative to the conventional techniques. Of the premier studies of this field, was the investigation carried out by Zhu et al., on modeling the wastewater process by time-delay neural networks in order to predict the quality attributes of the process (Zhu et al., 1998). A few years later, Gontarski et al. also proved the successful application of neural networks in the simulation of industrial anaerobic treatment plant in Brazil. They showed that the liquid flow rate and pH of the inlet stream were the major variables in controlling the plant and the neural network presented desirable results in minimizing the plant fluctuations (Gontarski et al.,

2000). In a different study, a hybrid technique providing principal component analysis (PCA) together with neural networks was used for optimal control of a wastewater treatment process (Choi & Park, 2001). The application of PCA in that case emerged as a novel idea at the time, since the input dataset could be reduced in order to solve the overfitting problem of the model. Zeng et al. employed back-propagation multilayer perceptrons (MLP) networks to model the nonlinear relationships between the removal rates of pollutants and the chemical dosages. Using this technique, the system could be adapted and operated in a variety of conditions showing a more flexible performance (Zeng et al., 2003).

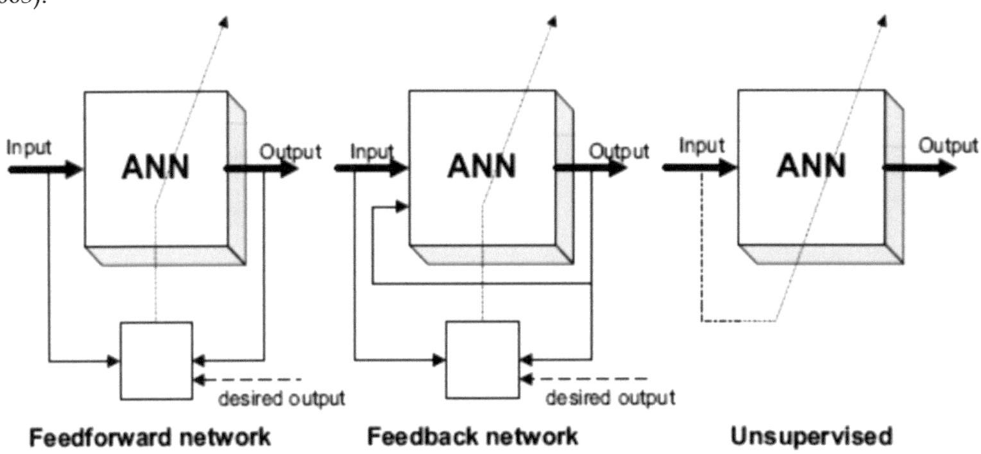

Fig. 3. Three general models of neural networks used in engineering analyses (Timothy Hong et al., 2003)

A few studies have reported the utilization of unsupervised learning algorithms. Hong et al. applied Kohonen Self-Organizing Feature Maps (KSOFM) neural networks to analyze the process data obtained from a municipal wastewater treatment plant (WTP) (Timothy Hong et al., 2003). KSOFM differ from feedforward networks in a way that they provide a clustering methodology which leads to data reduction (Kasabov, 1998) and also project the data nonlinearity onto a lower-dimensional display. The latter creates an abstraction of various features of input signals in the absence of any supervisor signals (Kasabov, 1998). As the authors reported, the KSOFM was computationally efficient, accurate, and reliable for the analysis of WTP and since the analysis and diagnosis of WTPs is a difficult task, the developed technique yields a great deal of significance (Timothy Hong et al., 2003). There are also some other studies which have utilized unsupervised networks for modeling the wastewater treatment process (Hong & Bhamidimarri, 2003; Garcia and Gonzalez, 2004; Cinar, 2005).

Tay and Zhang developed a fast predicting neural fuzzy model to predict the response of high-rate anaerobic systems to different system disturbances 1 h in advance (Tay & Zhang, 2000). Three laboratory scale systems including an anaerobic fluidized bed reactor (AFB), an anaerobic filter (AF), and an up-flow anaerobic sludge blanket (UASB) reactor were utilized. The reactors underwent two disturbing shocks, organic loading rate and hydraulic loading rate. The adaptive network based fuzzy system (ANFIS) used a database of system performance and implemented these data to predict the response of the anaerobic wastewater treatment system in the presence of OLR, hydraulic loading rate (HLR) and alkalinity loading shocks. The adaptability of the neural fuzzy modeling used was proven to

be acceptable in different operation conditions and therefore, it was suggested to be of high potential in real time control (Tay & Zhang, 2000).
Moreover, advanced controlling of anaerobic digestion has been also achieved using hierarchical neural networks (Holubar et al., 2002). Holubar et al. developed several configurations of feedforward back-propagation neural networks to predict gas composition, methane production rate, volatile fatty acid concentration, pH, redox potential, volatile suspended solids and chemical oxygen demand of feed and effluent of a CSTR under pulse like disturbances of OLRs. The correlation coefficient between the measured data and calculated values were found to be bigger than 0.8 for all the predicted variables which was a promising result by which the controlling of CSTR could be accomplished.
On the other hand, hydrogen sulfide (H_2S) and ammonia (NH_3) as the main gaseous trace compounds in anaerobic digestion process were predicted in a CSTR digester by using ANN (Strik et al., 2005). The obtained coefficients of determination (R^2) were 0.91 and 0.83 for hydrogen sulfide and ammonia, respectively. A similar research was conducted on an AFB reactor for starch wastewater using MLP and the effects of OLR, hydraulic retention time (HRT) and efficiency of the reactor on its steady-state performance were modeled using ANN (Parthiban et al., 2007). The MSE of the network performance was found the desirable value of only 0.0146. The findings of this research were of a great value since effluent COD and pH, alkalinity, VFA and biogas production were predicted by two input parameters: influent pH and OLR.
Neural networks have shown a more extensive application predicting methane fraction in biogas produced by field-scale land-fill bioreactors (Ozkaya et al., 2007). The methane production was modeled based on inputs such as pH, alkalinity, COD, sulfate, conductivity, chloride and substrate temperature. As a result, predicting hourly methane production, control achievement, optimization of energy conversion and construction time could be achieved. Pai et al. employed Grey Model ANN (GM-ANN) to predict suspended solids (SS) and COD of hospital wastewater treatment reactor effluents (Pai et al., 2007). Results showed that GM-ANN could predict the hospital wastewater variations with the same accuracy as ANN did. Besides, while ANN needed a large quantity of data, GM-ANN could handle smaller quantities.
In another study, a real-time monitoring of wastewater treatment process was achieved with the aid of multivariate statistical methods and ANN (Luccarini et al., 2010). In that study, Luccarini et al. installed some probes for the acquisition of signals such as pH, ORP and dissolved oxygen in a SBR and manipulated these data in an ANN model. The research aimed to verify the treatment process based on the continuous signals obtained on-line from the reactor (Luccarini et al., 2010). The signals from the plant were transmitted by telecommunication facilities and the data were reduced using PCA method and then analyzed. Therefore, by remote monitoring of a small-scale reactor ANN could approximate the performance criteria of the plant on-line. This novel work opened a new window for the remote controlling of reactors and benefiting ANN potentials to model such facilities.

3.2 Genetic algorithms

Genetic algorithms (Ga) fall in a class of stochastic search strategies modeled after evolutionary mechanisms based upon evolutionary principles of natural selection, mutation, and survival of the fittest (Sivanandam and Deepa, 2008). They have become a very popular strategy to optimize non-linear systems with a large number of variables. Gas are very different from most optimization methodologies comprising well-defined algorithms. The GA approach is to generate a large number of potential solutions and "evolve" a solution to the problem. One of the big keys to a successful genetic algorithm is in the development of a good "fitness function". The fitness function is how each potential solution is evaluated by the algorithm, and is in essence, how the problem to be solved by the algorithm is defined (Melanie, 1999). Optimization of wastewater treatment variables using GA models was reported by Cho et al. (2004). In their study, GA was integrated with a mathematical management model to reduce the treatment costs at a Korean plant and the application of GA was promising as concluded by the authors (Cho et al., 2004).

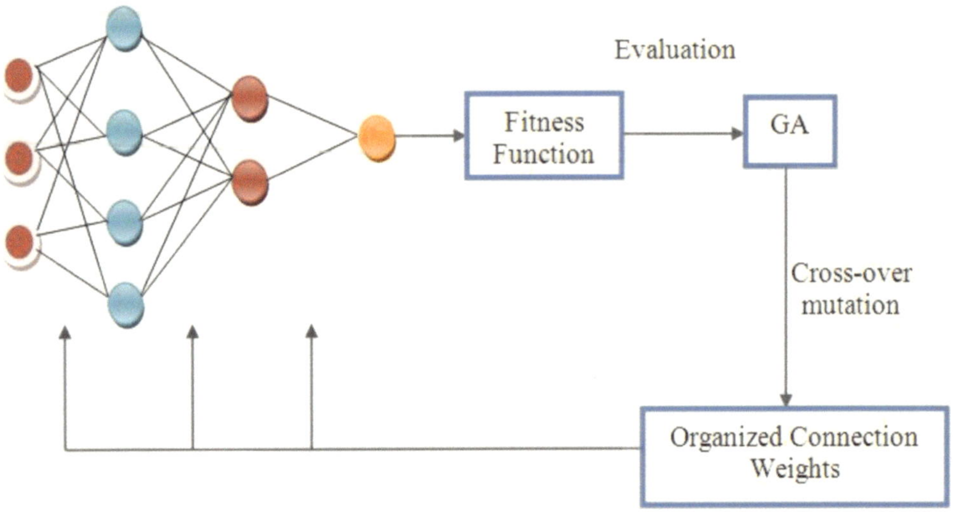

Fig. 4. Overview of the GA-ANN model

There have been several hybrid systems developed that combined neural networks with Gas in various ways (Haupt & Haupt, 2004). These generally fall into three categories: (1) using a GA to determine the structure of a neural network (Kecman, 2001), (2) using a GA to calculate the weights of a neural network as depicted in Fig 4, and (3) using a GA to both determine the structure and the weights of a neural network (Sivanandam & Deepa, 2008). Yang Mu and Yu developed a hybrid model for simulation of biological hydrogen production in a UASB reactor was reported (Yang Mu & Yu, 2007). OLR, HRT, and influent bicarbonate alkalinity were fed to a model of ANN combined with a GA. H_2 concentration, H_2 production rate, H_2 yield, effluent total organic carbon, and effluent aqueous products including acetate, propionate, butyrate, valerate and caporate were determined as outputs of the model. Simulations showed that the model could describe the daily variations of the studies UASB reactor by predicting the steady-state performance of reactor regarding various HRTs and substrate concentrations. Hybrid models have also been studied on the biodegradation process of phenol in a FBR. The authors suggested that the feedforward ANNs trained by a real-coded GA acted desirably for the simulation of biodegradation process in the FBR (Venu Vinod et al., 2009). In a different study, Iqbal and Guira developed an optimization procedure for an activated sludge reactor. They built a multi-objective optimization (MOO) to take into consideration the maximized influent flow rate of wastewater and minimized exit effluent biological oxygen demand (BOD) (Iqbal & Guriaa, 2009). The correct utilization of this method would be a necessity in the treatment process modeling and optimization as the non-linear and complex behavior dominating the bioreactor performance requires.

3.3 Fuzzy systems

Fuzzy logic, a novel technique introduced in 1965 by Lotfi Zadeh, is a mathematical tool for dealing with uncertainty and relative importance of precision. It is a convenient way to map an input space to an appropriate output space (Kasabov, 1998). The basis for fuzzy logic is similar to that for human communication or natural language. This statement underpins the whole concept of the technique. The theory of fuzzy logic is based upon the notion of relative graded membership and so are the functions of cognitive processes. The utility of fuzzy sets lies in their ability to model uncertain or ambiguous data so often encountered in real life, esp. biological issues (Sivanandam et al., 2007). At present, numerous applications of fuzzy logic in control and automation exist and many devices, plants and industrial facilities are instrumented with fuzzy control systems (Kandel & Langholz, 1993). However, designing fuzzy logic controls would require a series of prerequisites, including the determination of the input and output variables, the parameters of membership functions, and the fuzzy control rules (Chen et al., 2003). In spite of the extensive application in several areas of industry and research, fuzzy logic control has not been so popular in anaerobic digestion processes. Moreover, the need for a proper tool for screening out the essential control rules based on the experimental knowledge about the plant operation seems beyond the question. In order to overcome this challenge, Chen et al. conducted a three-stage study using fuzzy-neural hybrid controller for industrial wastewater treatment (Chen et al., 2003). The first stage was identifying the state function of wastewater treatment system followed by searching for multi-objective control strategies, and finally, tuning fuzzy control rule base. The results of the simulations proved that the hybrid fuzzy control approach effectively achieved the required real-time control objectives and was shown as an efficient and cost-effective tool to deal with the unexpected uncertainties in the wastewater treatment process (Chen et al., 2003). The authors also stated that their control architecture could be generalized to other physical, chemical and biological waste treatment systems.

In a different investigation, fuzzy-neural control system was utilized to real-time control and supervise the submerged biofilm wastewater treatment reactor (Mingzhi et al., 2009a). The research aimed to maximize cost efficiency of the treatment operation and address the problem of controlling air flow rate. The results obtained indicated that using fuzzy logic combined with ANN models led to better outputs and less MSEs compared to pure ANN. As a general conclusion, the authors reported that the fuzzy-neural control system used performed desirably when dealing with unexpected uncertainties in the small-scale bioreactor (Mingzhi et al., 2009a).

Mingzhi et al. developed a fuzzy neural network to model the nonlinear relationships between the removal rate of pollutants and their chemical dosages in a paper mill wastewater treatment plant. The objective of their research was to adapt the system to a variety of operating conditions and also to achieve a more flexible performance. The developed model reached a reasonable prediction of the COD and BOD in a high efficient reactor (Mingzhi et al., 2009b). A similar study was conducted by Pai et al. which employed fuzzy systems in combination with neural networks to predict SS and COD in the effluents of a hospital wastewater treatment. Regarding the maximum coefficient of correlation (R) and the minimum mean absolute percentage errors (MAPE) of the predictions, the developed model showed a satisfactory performance in comparison with the pure ANN models. Therefore, the model developed could be recommended in order to optimize design considerations of the treatment process (Pai et al., 2009).

Model	Application Domain	Advantages	Drawbacks	Reported applications in wastewater treatment
ANN	Modeling-regression analysis, classification, clustering,	Models non-linear data, Adaptive nature, robustness, parallel nature (continues to work when an element fails)	Needs training, high processing time for large networks, subject to over-fitting and under-fitting	modeling the wastewater process by time-delay neural networks (Zhu et al., 1998), optimal control of a wastewater treatment process integrated with PCA (Choi and Park, 2001), Kohonen Self-Organizing Feature Maps (KSOFM) to analyze the process data of municipal wastewater treatment plant (Timothy Hong et al., 2003), Unsupervised networks for modeling the wastewater treatment process (Garcia and Gonzalez, 2004; Hong and Bhamidimarri, 2003; Cinar, 2005), Grey Model ANN (GM-ANN) to predict suspended solids (SS) and COD of hospital wastewater treatment reactor effluents (Pai, 2007), on-line monitoring of a reactor (Luccarini, 2010)
FUZZY	Control systems, automation, modeling, integration with ANN for classification	Not constrained to the crisp logic, benefits from linguistic nature	Hard to implement in machines and practical systems	Fuzzy-neural hybrid controller for industrial wastewater treatment (Chen, 2003), control and supervise the submerged biofilm wastewater treatment reactor (Mingzhi et al., 2009), modeling the nonlinear relationships between the removal rate of pollutants and their chemical dosages in a paper mill wastewater treatment plant (Mingzhi et al., 2009).
GA	Optimization, control engineering, integration with ANN	Mimics the process of natural evolution, desirable for problem domains that have a complex fitness landscape	requires expensive fitness function evaluations for complex high dimensional, multimodal problems, subject to converge toward local optima,	GA-ANN for simulation of biological hydrogen production in a UASB reactor (Yang Mu and Yu, 2007), GA-ANN for biodegradation process of phenol in a fluidized bed reactor (FBR). (Venu Vinod et al., 2009), multi-objective optimization (MOO) for optimization procedure for an activated sludge reactor (Iqbal and Guriaa, 2009)

Table 5. Summary of introduced models with examples and aspects to be considered

Table 5 summarizes the aforementioned models together with the advantages and drawbacks which might be considered for selection in applied projects and utilization in industrial scales. As observed the most extensively used model is ANN. The reason is already discussed in this chapter. However, the application domain of Fuzzy logic and Genetic algorithms provide a high potential for integration of these models with ANN. Fields such as optimization and classification can be highly enhanced when integrated models are applied.

4. Conclusion

This chapter reviewed and discussed various influential parameters on biomethane generation during anaerobic treatment of wastewaters i.e. granulation, OLR, LCFA, mixing, sludge recycling, temperature, alkalinity and pH, ammonia and heavy metals inhibitions. By fully understanding these factors, anaerobic plant operators could enhance the process performance for improved biomethane generation and COD removal. Besides, the successful operation of anaerobic digesters also involves efficient process control component as the anaerobic treatment process relies on the complex biochemical reactions involved. This has led to the development and consistent improving of the anaerobic mathematical modeling techniques to further enhance the operation. In this chapter, the applications of modern modeling techniques such as ANN, GA and Fuzzy system as capable tools for optimizing the plant operation and assisting in the decision-making process were also discussed in detail.

5. References

Ahn, Y.H., Song, Y.J., Lee, Y.J., Park, S., 2002. Physicochemical characterization of UASB sludge with different size distributions. Environ. Technol. 23, 889–897.

Alrawi, R., Ahmad, A., Norli, I., AK, M., 2010. Methane Production during Start-Up Phase of Mesophilic Semi-Continues Suspended Growth Anaerobic Digester. Int. J. Chem. React. Eng. 8(8), 89.

Altas, L., 2009. Inhibitory effect of heavy metals on methane producing anaerobic granular sludge. J. Hazard. Mater. 162, 1551-1556.

Angelidaki, I., Ahring, B.K., 1993. Thermophilic anaerobic digestion of livestock waste: effect of ammonia. Appl. Microbiol. Biot. 38, 560-564.

Angelidaki, I., Ahring, B.K., 1995. Establishment and characterization of an anaerobic thermophilic (55 °C) enrichment culture degrading long-chain fatty acids. Appl. Environ. Microbiol. 61, 2442–2445.

Anthonisen, A.C., Loehr, B.C., Prakasam, T.B.S., Srinath, E.G., 1976. Inhibition of nitrification by ammonia and nitrous acid. J. Water. Pollut. Control. Fed. 48(5), 835-852.

Appels, L., Baeyens, J., Degreve, J., Dewil, R., 2008. Principles and potential of the anaerobic digestion of waste-activated sludge. Prog. Energ. Combust. 34, 755–781.

Ariesyady, H.D., Ito, T., Okabe, S., 2007. Functional bacterial and archaeal community structures of major trophic groups in a full-scale anaerobic sludge digester. Water Res. 41(7), 1554-1568.

Bae, J.H., Cho, K.W., Lee, S.J., Bum, B.S., Yoon, B.H., 1998. Effects of leachate recycle and anaerobic digester sludge recycle on the methane production from solid wastes. Water Sci. Technol. 38, 159–168.

Beccari, M., Carucci, G., Lanz, A.M., Majone, M., Papini, M.P., 2002. Removal of molecular weight fractions of COD and phenolic compounds in an integrated treatment of olive oil mill effluents. Biodegradation 13, 401–410.

Becker, P., Koster, D., Popov, M.N., Markossian, S., Antranikian, G., Markl, H., 1999. The biodegradation of olive oil and the treatment of lipid-rich wool scouring wastewater under aerobic thermophilic conditions. Water Res. 33, 653–660.

Busu, Z., Sulaiman, A., Hassan, M.A., Shirai, Y., Abd-Aziz, S., Yacob, S., Wakisaka, M., 2010. Improved anaerobic treatment of palm oil mill effluent is a semi-commercial closed digester tank with sludge recycling and appropriate feeding strategy. Pertanika J. Trop. Agric. Sci. 33(1), 27–37.

Calli, B., Mertoglu, B., Inanc, B., Yenigun, O., 2005a. Effects of high free ammonia concentrations on the performances of anaerobic bioreactors. Process Biochem. 40, 1285-1292.

Calli, B., Mertoglu, B., Inanc, I., Yenigun, O., 2005b. Methanogenic diversity in anaerobic bioreactors under extremely high ammonia levels. Enzyme. Microb. Tech. 37, 448-455.

Cammarota, M.C., Teixeira, G.A., Freire, D.M.G., 2001. Enzymatic pre-hydrolysis and anaerobic degradation of wastewaters with high oil contents. Biotechnol. Lett. 23, 1591–1595.

Cammarota, M.C., Freire, D.M.G., 2006. A review on hydrolytic enzymes in the treatment of wastewater with high oil and grease content. Bioresour. Technol. 97, 2195–2210.

Cavaleiro, A.J., Pereira, M.A., Alves, M., 2008. Enhancement of methane production from long chain fatty acid based effluents. Bioresour. Technol. 99, 4086–4095.

Cavaleiro, A.J., Salvador, A.F., Alves, J.I., Alves, M.M., 2009. Continuous high rate anaerobic treatment of oleic acid based wastewater is possible after a step feeding start-up. Environ. Sci. Technol. 43, 2931–2936.

Chae, K.J., Jang, A., Yim, S.K., Kim, I.S., 2007. The effect of digestion temperature and temperature shock on the biogas yields from the mesophilic anaerobic digestion of swine manure. Bioresour. Technol. 99(I), 1–6.

Chen, W.C., Chang, N.B., Chen, J.C., 2003. Rough set-based hybrid fuzzy-neural controller design for industrial wastewater treatment. Water Res. 37, 95–107.

Cho, J.H., Sung, K.S., Ha, S.R., 2004. river water quality management model for optimising regional wastewater treatment using a genetic algorithm. J. Environ. Manage. 73, 229–242.

Choi, D.J., Park, H., 2001. A hybrid artificial neural network as a software sensor for optimal control of a wastewater treatment process. Water. Res. 35(16), 3959–3967.

Choorit, W., Wisarnwan, P., 2007. Effect of temperature on the anaerobic digestion of palm oil mill effluent. Electron. J . Biotechnol. 10(3), 376–385.

Cinar, O., 2005. New tool for evaluation of performance of wastewater treatment plant: Artificial neural network. Process Biochem, 40, 2980–2984.

Colussi, I., Cortesi, A., L., Vedova, D.V., Gallo, V., Robles, F.K.C., 2009. Start-up procedures and analysis of heavy metals inhibition on methanogenic activity in EGSB reactor. Bioresour. Technol. 100, 6290-6294.

Converti, A., Zilli, M., Del Borghi, M. and Ferraiolo, G., 1990. The fluidized bed reactor in the anaerobic treatment of wine wastewater. Bioproc. Eng. 5, 49–55.

Converti, A., del Borghi, M., Ferraiolo, G., 1993. Influence of organic loading rate on the anaerobic treatment of high strength semisynthetic wastewaters in a biological fluidized bed. Chem. Eng. J. 52(1), 21-28.

Daffonchio, D.D., Thaveesri, J., Verstraete, W., 1995. Contact angle measurement and cell hydrophobicity of granular sludge from upflow anaerobic sludge bed reactors. Appl. Environ. Microbiol. 61, 3676–3680.

Desai, M., Patel, V., Madamwar, D., 1994. Effect of temperature and retention time on biomethanation of cheese whey-poultry waste-cattle dung. Environ. Pollut. 83, 311–315.

Díaz, E.E., Stams, A.J.M., Amils, R., Sanz, J.L., 2006. Phenotypic properties and microbial diversity of methanogenic granules from a full-scale upflow anaerobic sludge bed reactor treating brewery wastewater. Appl. Environ. Microbiol. 72, 4942-4949.

El-Mamouni, R., Leduc, R., Guiot, S.R., 1997. Influence of synthetic and natural polymers on the anaerobic granulation process. Water Sci. Technol. 38(8–9), 341–347.

El-Mashad, H.M., Zeeman, G., van Loon, W.K.P., Bot, G.P.A., Lettinga, G., 2004. Effect of temperature and temperature fluctuation on thermophilic anaerobic digestion of cattle manure. Bioresour. Technol. 95, 191-201.

Fabián, R.M., Gourdon, R., 1999. Effect of baling on the behavior of domestic wastes: laboratory study on the role of pH in biodegradation. Bioresour. Technol. 69, 15–22.

Faisal, M., Unno, H., 2001. Kinetic analysis of palm oil mill wastewater treatment by a modified anaerobic baffled reactor. Biochem. Eng. J. 9, 25–31.

Fannin, K.F., 1987. Start-up, Operation, Stability and Control. In Anaerobic Digestion of Biomass, Eds., Chynoweth, D.P. and R. Isaacson., Elsevier Applied Science Publishers Ltd, pp: 171-196.

Fannin, K.F., Biljetina, R., 1987. Reactor designs. In: Chynoweth DP, Isaacson R, editors. Anaerobic digestion of biomass. England: Elsevier Science, pp. 141–66.

Feng, J., Jost, J., Qian, M. (Eds.)., 2007. Networks: from biology to theory. Springer-Verlag London Limited, pp. 313.

Fezzani, B., BenCheikh, R., 2007. Anaerobic co-digestion of olive mill waste water with olive mill solid waste in a tubular digester at a mesophilic temperature. Bioresour. Technol. 98, 769–774.

Gangagni, R.A., Reddy, T.S.K., Prakash, S.S, Vanajakshi, J., Joseph, J., Jetty, Reddy, A.R., Sarma, P.N., 2008. Biomethanation of poultry litter leachate in UASB reactor coupled with ammonia stripper for enhancement of overall performance. Bioresour. Technol. 99, 8679-8684.

Garcia, H.L., Gonzalez, I.M., 2004. Self-organizing map and clustering for wastewater treatment monitoring. Eng. Appl. Artif. Intel. 17, 215–225.

Garcia, M.L., Angenent, L.T., 2009. Interaction between temperature and ammonia in mesophilic digesters for animal waste treatment. Water Res. 43, 2373-2382.

Gebauer, R., 2004. Mesophilic anaerobic treatment of sludge from saline fish farm effluents with biogas production. Bioprocess Technol. 93, 155-167.
Gerardi, M.H., 2003. The microbiology of anaerobic digesters. John Wiley and Sons Inc., New Jersey, pp. 91-118.
Ghobadian, B., Rahimi, H., Nikbakht, A.M., Najafi, G., Yusaf, T.F., 2009. Diesel engine performance and exhaust emission analysis using waste cooking biodiesel fuel with an artificial neural network. Renew. Energ. 34 (4), 976–982.
Gontarski, C.A., Rodrigues, P.R., Mori, M., Prenem, L.F., 2000. Simulation of an industrial wastewater treatment plant using artificial neural networks. Comput. Chem. Eng. 24, 1719-1723.
Gonzales-Gil, G., Lens, P.N.L., Van Aelst, A., Van As, H., Versprille, A.I., Lettinga, G., 2001. Cluster structure of anaerobic aggregates of an expanded granular sludge bed reactor. Appl. Environ. Microbiol. 67, 3683-3692.
Guerrero, L., Omil, F., Mendez, R., Lema, J.M., 1997. Treatment of saline wastewater from fish meal factories in an anaerobic filter under extreme ammonia concentrations. Bioresour. Technol. 61, 69-78.
Guerrero, L., Omil, F., Mendez, R., Lema, J.M., 1999. Anaerobic hydrolysis and acidogenesis of wastewaters from food industries with high content of organic solids and protein. Water Res. 33, 3281-3290.
Hanaki, K., Nagase, M., Matsuo, T., 1981. Mechanism of inhibition caused by long-chain fatty acids in anaerobic digestion process. Biotechnol. Bioeng. 23, 1591–1610.
Hansen, K.H., Angelidaki, I., Ahring, B.K., 1998. Anaerobic digestion of swine manure: inhibition by ammonia. Water Res. 32(1), 5-12.
Harper, S.R., Suidan M.T., 1991. Anaerobic treatment kinetics: discussers' report. Water. Sci. Tech. 24(8), 61-78.
Hashimoto, A.G., 1986. Ammonia inhibition of methanogenesis from cattle waste. Agr. Wastes. 17, 241-261.
Haupt, R.L. and Haupt, S.E., 2004. Practical genetic algorithms. A John Wiley & Sons, Inc., publication, pp. 253.
Haykin, S., 1994. Neural Networks: A Comprehensive Foundation, Mac-millan, New York.
Hertz, J., Krogh, A., Palmer, R.G., 1991. Introduction to the Theory of Neural Computation. Addison-Wesley Publishing Company, Redwood City, NJ.
Holubar, P., Zani, L., Hager, M., Froschl, W., Radak, Z., Braun, R. 2002. Advanced controlling of anaerobic digestion by means of hierarchical neural networks. Water. Res. 36, 2582–2588.
Hong, Y.S., Bhamidimarri, R., 2003. Evolutionary self-organising modelling of a municipal wastewater treatment plant. Water Res. 37, 1199–1212.
Hwu, C-S., van Lier, J.B., Lettinga, G., 1998a. Physicochemical and biological performance of expanded granular sludge bed reactors treating long-chain fatty acids. Process Biochem. 33, 75–81.
Hwu, C-S, Tseng, S.K., Yuan, C.Y., Kulik, Z., Lettinga, G., 1998b. Biosorption of long-chain fatty acids in UASB treatment process. Water Res. 32, 1571–1579.

Iqbal, J., Guriaa,, C., 2009. Optimization of an operating domestic wastewater treatment plant using elitist non-dominated sorting genetic algorithm. Chem. Eng. Res. Des. 87, 1481–1496.

Jackson, B.E., Bhupathiraju, V.K., Tanner, R.S., Woese, C.R., McInerney, M.J., 1999. Syntrophus aciditrophicus sp. nov., a new anaerobic bacterium that degrades fatty acids and benzoate in syntrophic association with hydrogen-using microorganisms. Arch. Microbiol. 171, 107–114.

Jeganathan, J., Nakhla, G., Bassi, A., 2006. Long-term performance of high-rate anaerobic reactors for the treatment of oily wastewater. Environ. Sci. Technol. 40, 6466–6472.

Jokela, J.P.Y., Rintala, J.A., 2003. Anaerobic solubilisation of nitrogen from municipal solid waste (MSW). Rev. Environ. Sci. Biotechnol. 2, 67-77.

Kalogirou SA., 2000. Application of artificial neural-networks for energy systems. Appl. Energ. 67, 17–35.

Kalyuzhnyi, S.V., Sklyar, V.I., Davlyatshina, M.A., Parshina, S.N., Simankova, M.V., Kostrikina, N.A., Nozhevnikova, A.N., 1996. Organic removal and microbiological features of UASB-reactor under various organic loading rates. Bioresour. Technol. 55,47–54

Kandel A., Langholz, G., 1993. Fuzzy Control Systems. CRC Press, CRC Press LLC, pp. 656.

Kaparaju, P., Buendia, I., Ellegaard, L., Angelidakia I., 2007. Effects of mixing on methane production during thermophilic anaerobic digestion of manure: Lab scale and pilot scale studies, Bioresour. Technol. 99, 4919–4928.

Karim, K., Hoffmann, R., Klasson, K.T., Al-Dahhan, M.H., 2005a. Anaerobic digestion of animal waste: Waste strength versus impact of mixing. Bioresour. Technol. 96, 1771-1781.

Karim, K., Hoffmann, R., Klasson, K.T., Al-Dahhan, M.H., 2005b. Anaerobic digestion of animal waste: Effect of mode of mixing. Water Res. 39, 3597–3606.

Kasabov NK., 1998. Foundations of neural networks, fuzzy systems, and knowledge engineering. Massachusetts Institute of Technology, pp. 550.

Kecman, V., 2001. Lerning and soft computing. Massachusetts Institute of Technology, pp. 361.

Khademi, M., Najafpour, G., Nia, B., Zinatizadeh, A., Kalantary, R., 2009. Biological Treatment of Antibiotic Plant Effluent in an UASFF Bioreactor. World Appl. Sci. J. 5, 1-8.

Kiani Deh Kiani, M., Ghobadian, B., Tavakoli, T., Nikbakht, A.M., Najafi, G., 2009. Application of artificial neural networks for the prediction of performance and exhaust emissions in SI engine using ethanol- gasoline blends. Energy. 35:65-69.

Kim, S.H., Han, S.K., Shin, H.S., 2004. Two-phase anaerobic treatment system for fat containing wastewater. J. Chem. Technol. Biotechnol. 79, 63–71.

Kosaric, N., Blaszczyk, R., Orphan, L., 1990. Factors influencing formation and maintenance of granules in upflow anaerobic sludge blanket reactors (UASBR). Water Sci. Technol. 22, 275–282.

Koster, I.W., Cramer, A., 1987. Inhibition of methanogenesis from acetate in granular sludge by long chain fatty acids. Appl. Environ. Microbiol. 53, 403–409

Lalman, J.A., Bagley, D.M., 2001. Anaerobic degradation and methanogenic inhibitory effects of oleic and stearic acids. Water Res. 35, 2975–2983.

Leal, M.C.M.R., Cammarota, M.C., Freire, D.M.G., Sant'Anna Jr., G.L., 2002. Hydrolytic enzymes as coadjuvants in the anaerobic treatment of dairy wastewaters. Braz. J. Chem. Eng. 19(2), 175–180.

Leal, M.C.M.R., Freire, D.M.G., Cammarota, M.C., Sant'Anna Jr., G.L., 2006. Effect of enzymatic hydrolysis on anaerobic treatment of dairy wastewater. Process Biochem. 41, 1173–1178.

Lee, M.W., Hong, S.H., Choi, H.C., Kim, J.H., Lee, D.S., Park, J.M., 2008. Real-time remote monitoring of small-scaled biological wastewater treatment plants by a multivariate statistical process control and neural network-based software sensors. Process Biochem. 43, 1107–1113.

Lin, L., 1991. An anaerobic treatment process model: Development and calibration. PhD dissertation, Michigan Technological University.

Liu, Y., Xu, H.L., Yang, S.F. and Tay, J.H., 2003. The mechanisms and models for anaerobic granulation in upflow anaerobic sludge blanket reactor. Water Res. 37, 661–673. Melanie, M., 1999. An Introduction to Genetic Algorithms. The MIT Press, pp. 158.

Luccarini, L., Bragadin, G.L., Colombini, G., Mancini, M., Mello, P., Montali, M., Sottara, D., 2010. Formal verification of wastewater treatment processes using events detected from continuous signals by means of artificial neural networks. Case study: SBR plant. Environ. Modell. Softw. 25, 648–660.

Masse, D.I., Masse, L., 2001. The effect of temperature on slaughterhouse wastewater treatment in anaerobic sequencing batch reactors. Bioresour. Technol. 76, 91-98.

McMahon, K.D., Stroot, P.G., Mackie, R.I., Raskin, L., 2001. Anaerobic codigestion of municipal solid waste and biosolids under various mixing conditions-II: Microbial population dynamics. Water Res. 35, 1817–1827.

Mendes, A.A., Pereira, E.B., de Castro, H.F., 2006. Effect of the enzymatic hydrolysis pretreatment of lipids-rich wastewater on the anaerobic biodigestion. Biochem. Eng. J. 32, 185–190.

Metcalf, E., Eddy, H., 2003. Wastewater engineering: treatment and reuse. 3rd ed. McGraw-Hill, Inc., New York.

Mingzhi, H., Jinquan, W., Yongwen, M., Yan, W., Weijiang, L., Xiaofei, S., 2009. Control rules of aeration in a submerged biofilm wastewater treatment process using fuzzy neural networks. Expert. Syst. Appl. 36, 10428–10437.

Mingzhi, H., Yongwen, M., Jinquan, W., Yan, W., 2009. Simulation of a paper mill wastewater treatment using a fuzzy neural network. Expert. Syst. Appl. 36, 5064–5070.

Morvai, L., Miháltz, P., Czakó, L., P'terfy, M., Holló, J., 1990. The influence of organic load on granular sludge development in an acetate-fed system. Appl. Microbiol. Biotechnol. 33, 463-468.

Najafpour G.D., Zinatizadeh A.A.L., Mohamed A.R., Hasnain Isa M., Nasrollahzadeh H. 2006. High-rate anaerobic digestion of palm oil mill effluent in an up-flow anaerobic sludge-fixed film bioreactor. Process biochem. 41, 370–379.

Najafpour, G., Yieng, H.A., Younesi, H., Zinatizadeh, A., 2005. Effect of organic loading on performance of rotating biological contactors using Palm Oil Mill effluents. Process Biochem. 40(8), 2879-2884.

Najafpour, G.D., Zinatizadeh, A.A.L., Mohamed, A.R., Hasnain Isa, M., Nasrollahzadeh, H., 2006. High-rate anaerobic digestion of palm oil mill effluent in an upflow anaerobic sludge-fixed film bioreactor. Process Biochem. 41(2), 370-379.

Najafpour, G., Hashemiyeh, B., Asadi, M., Ghasemi, M., 2008. Biological Treatment of Dairy Wastewater in an Upflow Anaerobic Sludge-Fixed Film Bioreactor. American-Eurasian J. Agric. Environ. Sci. 4(2), 251-257.

Nandy, T., Kaul, S., Szpyrkowicz, L., 1998. Treatment of herbal pharmaceutical wastewater with energy recovery. Int. J. Environ. Stud. 54(2), 83-105.

Nielsen, H.B., Ahring, B.K., 2006. Responses of the biogas process to pulses of oleate in reactors treating mixtures of cattle and pig manure. Biotechnol. Bioeng. 95, 96–105.

Nies, D.H., 1999. Microbial heavy-metal resistance. Appl. Microbiol. Biot. 51, 730-750.

Ozkaya, B., Demir, A., Bilgili., M.S., 2007. Neural network prediction model for the methane fraction in biogas from field-scale landfill bioreactors. Environ Modell Softw. 22, 815-822.

Pai, T.Y., Tsai, Y.P., Loa, H.M., Tsai, C.H., Lin, C.Y., 2007. Grey and neural network prediction of suspended solids and chemical oxygen demand in hospital wastewater treatment plant effluent. Comput. Chem. Eng. 31, 1272–1281.

Pai, T.Y., Wan, T.J., Hsu, S.T., Chang, T.C., Tsai, Y.P., Lin, C.Y., Su, H.C., Yu, L.F., 2009. Using fuzzy inference system to improve neural network for predicting hospital wastewater treatment plant effluent. Comput. Chem. Eng. 33, 1272–1278.

Parthiban, P., Iyer, P., Sekaran, G., 2007. Anaerobic tapered fluidized bed reactor for starch wastewater treatment and modeling using multilayer perceptron neural network. J. Environ. Sci. 19, 1416–1423.

Pereira, M.A., Mota, M., Alves, M.M., 2002. Operation of an anaerobic filter and an EGSB reactor for the treatment of an oleic acid-based effluent: influence of inoculum quality. Process Biochem. 37, 1025–1031.

Pereira, M.A., Roest, K., Stams, A.J.M., Mota, M., Alves, M., Akkermans, A.D.L., 2002. Molecular monitoring of microbial diversity in expanded granular sludge bed (EGSB) reactors treating oleic acid. FEMS Microbiol. Ecol. 41, 95–103.

Pereira, M.A., Cavaleiro, A.J., Mota, M., Alves, M.M., 2003. Accumulation of long chain fatty acids onto anaerobic sludge under steady state and shock loading conditions: effect on acetogenic and methanogenic activity. Water Sci. Technol. 48, 33–40.

Pereira, M.A., Sousa, D.Z., Mota, M., Alves, M.M., 2004. Mineralization of LCFA associated with anaerobic sludge: kinetics, enhancement of methanogenic activity, and effect of VFA. Biotechnol. Bioeng. 88, 502–511.

Pereira, M.A., Pires, O.C., Mota, M., Alves, M.M., 2005. Anaerobic biodegradation of oleic and palmitic acids: evidence of mass transfer limitations caused by long chain fatty acid accumulation onto the anaerobic sludge. Biotechnol. Bioeng. 92, 15–23.

Perle, M., Kimchie, S., Shelef, G., 1995. Some biochemical aspects of the anaerobic degradation of dairy wastewater. Water Res. 29, 1549–1554.

Poh, P.E., Chong, M.F., 2009. Development of anaerobic digestion methods for palm oil mill effluent (POME) treatment. Bioresour. Technol. 100, 1–9.

Rincón, B., Raposo, F., Borja, R., González, J.M., Portillo, M.C., Saiz-Jiménez, C., 2006. Performance and microbial communities of a continuous stirred tank anaerobic reactor treating two-phases olive mill solid wastes at low organic loading rates. J. Biotechnol. 121, 534–543.

Rincón, B., Borja, R., González, J.M., Portillo, M.C., Saiz-Jiménez, C.S., 2008. Influence of organic loading rate and hydraulic retention time on the performance, stability and microbial communities of one-stage anaerobic digestion of two-phase olive mill solid residue. Biochem. Eng. J. 40, 253–261.

Rinzema, A., Alphenaar, A., Lettinga, G., 1993. Anaerobic digestion of long-chain fatty acids in UASB and expanded granular sludge bed reactors. Process Biochem. 28, 527–537.

Rinzema, A., Boone, M., van Knippenberg, K., Lettinga, G., 1994. Bactericidal effect of long chain fatty acids in anaerobic digestion. Water Environ. Res. 66, 40–49.

Rizzi, A., Zucchi, M., Borin, S., Marzorati, M., Sorlini, C., Daffonchio, D., 2006. Response of methanogen populations to organic load increase during anaerobic digestion of olive mill wastewater. J. Chem. Technol. Biot. 81(9), 1556-1562.

Saatci, Y., Arslan, E.I., Konar, V., 2003. Removal of total lipids and fatty acids from sunflower oil factory effluent by UASB reactor. Bioresour. Technol. 87, 269–272.

Saiki, Y., Iwabuchi, C., Katami, A., Kitagawa, Y., 2002. Microbial analyses by fluorescence in situ hybridization of well-settled granular sludge in brewery wastewater treatment plants. J. Biosci. Bioeng. 93(6), 601-606.

Santegoeds, C.M., Damgaard, L.R., Hesselink, G., Zopfi, J., Lens, P., Muyzer, G., de Beer., D., 1999. Distribution of sulfate-reducing and methanogenic bacteria in anaerobic aggregates determined by microsensor and molecular analyses. Appl. Environ. Microbiol. 65, 4618-4629.

Satoh, H., Miura, Y., Tsushima, I., Okabe, S. 2007. Layered structure of bacterial and archaeal communities and their in situ activities in anaerobic granules. Appl. Environ. Microbiol. 73, 7300–7307.

Sawayama, S., Tada, C., Tshukahara, K., Yagishita, T., 2004. Effect of ammonium addition on methanogenic community in a fluidized bed anaerobic digestion. J. Biosci. Bioeng. 97, 65-70.

Schmidt, J.E., Ahring, B.K., 1996. Granular sludge formation in upflow anaerobic sludge blanket (UASB) reactors. Biotech. Bioeng. 49, 229–246.

Sekiguchi, Y., Kamagata, Y., Nakamura, K., Ohashi, A., Harada, H., 1999. Fluorescence in situ hybridization using 16S rRNA-targeted oligonucleotides reveals localization of methanogens and selected uncultured bacteria in mesophilic and thermophilic sludge granules. Appl. Environ. Microbiol. 65, 1280-1288.

Sekiguchi, Y., Takahashi, H., Kamagata, Y., Ohashi, A., Harada, H., 2001. In situ detection, isolation, and physiological properties of a thin filamentous microorganism abundant in methanogenic granular sludges: a novel isolate affiliated with a clone cluster, the green non-sulfur bacteria, subdivision I. Appl. Environ. Microbiol. 67, 5740-5749.

Setiadi, T., Husaini, Djajadiningrat, A., 1996. Palm oil mill effluent treatment by anaerobic baffled reactors: Recycle effects and biokinetic parameters. Water Sci. Technol. 34, 59–66

Sheng, G.P., Yu, H.Q., Li, X.Y., 2008. Stability of sludge flocs under shear conditions. Biochem. Eng. J. 38, 302–308.

Shigematsu, T., Tang, Y., Mizuno, Y., Kawaguchi, H., Morimura, S., Kida, K., 2006. Microbial diversity of mesophilic methanogenic consortium that can degrade long-chain fatty acids in chemostat cultivation. J. Biosci. Bioeng. 102, 535–544.

Sivanandam, S.N., Sumathi, S., Deepa, S.N., 2007. Introduction to Fuzzy Logic using MATLAB. Springer-Verlag Berlin Heidelberg, pp. 430.

Sivanandam, S.N. and Deepa, S.N., 2008. Introduction to Genetic Algorithms. Springer Berlin Heidelberg New York. Pp: 442.

Song, Y.C., Kwon, S.J., Woo, J.H., 2004. Mesophilic and thermophilic temperature co-phase anaerobic digestion compared with single-stage mesophilic and thermophilic digestion of sewage sludge. Water Res. 38,1653–1662.

Sossa, K., Alarcon, M., Aspe, E., Urrutia, H., 2004. Effect of ammonia on the methanogenic activity of methylaminotrophic methane producing *Archaea* enriched biofilm. Anaerobe. 10, 13-18.

Sousa, D.Z., Pereira, M.A., Smidt, H., Stams, A.J.M., Alves, M.M., 2007. Molecular assessment of complex microbial communities degrading long chain fatty acids in methanogenic bioreactors. FEMS Microbiol. Ecol. 60, 252–265.

Sousa, D.Z., Alves, J.I., Alves, M.M., Smidt, H., Stams, A.J.M., 2009. Effect of sulfate on methanogenic communities that degrade unsaturated and saturated long-chain fatty acids (LCFA). Environ. Microbiol. 11, 68–80.

Sousa, D.Z., Smidt, H., Alves, M.M., Stams, A.J., 2009. Ecophysiology of syntrophic communities that degrade saturated and unsaturated long-chain fatty acids. FEMS Microbiol. Ecol. 68(3), 257-272.

Sprott, G.D., Patel,G.B., 1986. Ammonia toxicity in pure cultures of methanogenic bacteria. Syst. Appl. Microbiol. 7, 358-363.

Stafford, D.A., 1982. The effects of mixing and volatile fatty acid concentrations on anaerobic digester performance. Biomass, 2, 43–55.

Steinhaus, B., Garcia, M.L., Shen, A.Q., Angenent L.T., 2007. A Portable Anaerobic Microtank Reveals Optimum Growth Conditions for the Methanogen Methanosaeta concilii. Appl. Environ. Microbiol. 73, 1653–1658.

Strik, D., Domnanovich, A.D., Zani, L., Braun, R., Holubar, P., 2005. Prediction of trace compounds in biogas from anaerobic digestion using the MATLAB Neural Network Toolbox. Environ. Modell. Softw. 20, 803–810.

Strik, D.P.B.T.B., Domnanovich, A.M., Holubar, P., 2006. A pH-based control of ammonia in biogas during anaerobic digestion of artificial pig manure and maize silage. Process Biochem. 41, 1235-1238.

Stroot, P.G., McMahon, K.D., Mackie, R.I., Raskin, L., 2001. Anaerobic codigestion of municipal solid waste and biosolids under various mixing conditions-I.Digester performance. Water Res. 35, 1804–1816.

Stryer, L., 1995. Fatty acid metabolism, in: Stryer, L., (Eds.), Biochemistry. W.H. Freeman and Co., New York, pp. 603–628.

Sulaiman, A., Hassan, M.A., Shirai, Y., Abd-Aziz, S., Tabatabaei, M., Busu, Z., Yacob, S., 2009. The effect of mixing on methane production in a semi-commercial closed digester tank treating palm oil mill effluent. Aust. J. Basic Appl. Sci. 3(3), 1577–1583.

Sulaiman, A., Tabatabaei, M., Hassan, M.A., Shirai, Y., 2009. The effect of higher sludge recycling rate on anaerobic treatment of palm oil mill effluent in a semi-commercial closed digester for renewable energy. Am. J. Biochem. Biotechnol. 5(1), 1–6.

Sung, S., Liu, T., 2003. Ammonia inhibition on thermophilic anaerobic digestion. Chemosphere. 53, 43-52.

Syutsubo, K., Harada, H., Ohashi, A., Suzuki, H., 1997. An effective start-up of thermophilic UASB reactor by seeding mesophilically-grown granular sludge. Water Sci. Technol. 36, 391–398.

Tabatabaei, M., Zakaria, M.R., Abdul Rahim, R., Wright, A.D.G., Shirai, Y, Abdullah, N., Sakai, K., Ikeno, S., Mori, M., Kazunori, N., Sulaiman, A., Hassan, M.A., 2009. PCRbased DGGE and FISH analysis of methanogens in anaerobic closed digester tank treating palm oil mill effluent (POME). Electron. J. Biotech. 12(3) [available from Internet: http://www.ejbiotechnology.cl/content/vol12/issue3/ full/4/index .html].

Tabatabaei, M., Abdul Rahim, R., Wright, A.D.G., Shirai, Y., Abdullah, N., Sulaiman, A., Sakai, K., Hassan, M.A., 2010. Importance of the methanogenic archaea populations in anaerobic wastewater. Process Biochem. 45, 1214–1225.

Tagawa, T., Takahashi, H., Sekiguchi, Y., Ohashi, A., Harada, H., 2002. Pilot-plant study on anaerobic treatment of a lipid- and protein-rich food industrial wastewater by a thermophilic multi-staged UASB reactor. Water Sci. Technol. 45, 225-230.

Tay, J.H., Zhang, X., 2000. A fast predicting neural fuzzy model for high-rate anaerobic wastewater treatment systems. Water. Res. 34(11), 2849-2860.

Tchobanoglous, G., Burton, F.L., 1991. Wastewater engineering treatment disposal and reuse. McGraw-Hill International Edition, Singapore, pp. 84–827.

Teer, J.E., Leak, D.J., Dudeney, A.W.L., Nayaranan, A., Stuckey, D.C., 2000. Changes in pre-formed granule composition and structure in sulphidogenic UASB treating a synthetic ferric oxalate wastewater. Environ. Technol. 21, 1325–1335.

Teo, K.C., Xu, H.L., Tay, J.H., 2000. Molecular mechanism of granulation: II. Proton translocating activity. J. Environ. Eng. 126, 411–418.

Timothy Hong, Y.S., Rosen, M.R., Bhamidimarri, R., 2003. Analysis of a municipal wastewater treatment plant using a neural network-based pattern analysis. Water Res. 37, 1608–1618.

Tiwari, M.K., Guha, S., Harendranath, C.S., 2004. Enhanced granulation in UASB reactor treating low-strength wastewater by natural polymers. Water Sci. Technol. 50, 235–240.

Tiwari, M.K., Guha, S., Harendranath, C.S., Tripathi, S., 2005. Enhanced granulation by natural ionic polymer additives in UASB reactor treating low-strength wastewater. Water Res. 39, 3801–3810

Tiwari, M.K., Guha, S., Harendranath, C.S., Tripathi, S., 2006. Influence of extrinsic factors on granulation in UASB reactor. Appl. Microbiol. Biotechnol. 71(2), 145–154.

Uemura, S., Harada, H., 1993. Microbial characteristics of methanogenic sludge consortia developed in thermophilic UASB reactors. Appl. Microbiol. Biotechnol. 39, 654–660.

Uemura, S., Harada, H., 1995. Inorganic composition and microbial characteristics of methanogenic granular sludge grown in a thermophilic upflow anaerobic sludge blanket reactor. Appl. Microbiol. Biotechnol. 43, 358–364.

Uyanik, S., Sallis, P.J., Anderson, G.K., 2002. The effect of polymer addition on granulation in an anaerobic baffled reactor (ABR). Part II: compartmentalization of bacterial populations. Water Res. 36, 944–955.

van Lier, J.B., Boersma, F., Debets, M.M.W.H., Lettinga, G., 1994. High-rate thermophilic anaerobic wastewater treatment in compartmentalized upflow reactors. Water Sci. Technol. 30, 251–261.

van Lier, J.B., 1996. Limitations of thermophilic anaerobic wastewater treatment and the consequences for process design. Antonie van Leeuwenhoek. 69,1–14.

Veiga, M.C., Jain, M.K., Wu, W.M., Hollingsworth, R.I., Zeikus, J.G., 1997. Composition and role of extracellular polymers in methanogenic granules. Appl. Environ. Microbiol. 63, 403–407.

Venu Vinod, A., Arun Kumar, K., Venkat Reddy, G., 2009. Simulation of biodegradation process in a fluidized bed bioreactor using genetic algorithm trained feed-forward neural network. Biochem. Eng. J. 46, 12–20.

Visser, A., Gao, Y., Letingga, G., 1993. Effects of pH on methanogenesis and sulphate reduction in thermophilic (55°C) UASB reactors. Bioresour. Technol. 44, 113–121.

Wiegant, W.M., de Man A.W.A., 1986. Granulation of biomass in thermophilic upflow anaerobic sludge blanket reactors treating acidified wastewaters. Biotechnol. Bioeng. 28, 718–727.

Wirtz, R.A., Dague, R.R., 1996. Enhancement of Granulation and Start-up in the Anaerobic Sequencing Batch Reactor. Water Environ. Res. 68, 883–892.

Yacob, S., Shirai, Y. Hassan, M.A., Wakisaka, M., Subash, S., 2005. Start-up operation of semi-commercial closed anaerobic digester for palm oil mill effluent treatment. Process Biochem. 41, 962–964.

Yang Mu, Y., Yu, H.Q., 2007. Simulation of biological hydrogen production in a UASB reactor using neural network and genetic algorithm. Int. J. Hydrogen Energ. 32, 3308–3314.

Yue, Z.B., Yu, H.Q., Wang, Z.L., 2007. Anaerobic digestion of cattail with rumen culture in the presence of heavy metals. Bioresour. Technol. 98, 781-786.

Yusof, N., Haraguchi, A., Hassan, M.A., Othman, M.R., Wakisaka, M., Shirai, Y., 2009. Measuring organic carbon, nutrients and heavy metals in rivers receiving leachate from controlled and uncontrolled municipal solid waste (MSW) landfills. Waste. Manage. 29 (10), 2666-2680.

Zeng, G.M., Qin, X.S., He, L., Huang, G.H., Liu, H.L., Lin, Y.P., 2003. A neural network predictive control system for paper mill wastewater treatment. Eng. Appl. Artif. Intel. 16: 121–129.

Zhu, J., Zurcher, J., Rao, M., Meng, M.Q.H., 1998. An on-line wastewater quality predication system based on a time-delay neural network. Eng. Appl. Artif. Intel. 11, 747-758.

Zinatizadeh, A.A.L., Younesi, H., Bonakdari, H., Pirsaheb, M., Pazouki, M., Najafpour, G.D., Hasnain Isa, M., 2009. Effects of process factors on biological activity of granular sludge grown in an UASFF bioreactor. Renew. Energ. 34(5), 1245-1251.

11

Alternative Fuels in Cement Manufacturing

Moses P.M. Chinyama
University of Malawi – The Polytechnic
Malawi

1. Introduction

Fossil fuels such as coal, petroleum and natural gas provide most of the energy needs of the world today. Coal and natural gas are used in their natural forms, but petroleum and other fossil fuels such as shale and bituminous sands require distillation and refinement to give usable fuels. These fuels exist in any of the following forms: solid, liquid and gas. The finite nature of global fossil fuel resources, high prices and most importantly, their damaging effect on the environment underscore the need to develop alternative fuels[1] for many industrial systems that rely on fossil fuels. Increased use of renewable and alternative fuels can extend fossil fuel supplies and help resolve air pollution problems associated with the use of conventional fuels.

This chapter reviews in detail some of the main alternative fuels used in cement production. It focuses on types of alternative fuels used, the environmental and socio-economic benefits of using alternative fuels, challenges associated with switching from conventional to alternative fuels, combustion characteristics of the alternative fuels concerned, and their effect on cement production and quality. The aim of this chapter is to provide empirical evaluation of alternative fuels. It offers an invaluable source of information for cement manufacturers that are interested in using alternative fuels. Researchers and students would also find this information valuable for their professional and academic development.

Cement is considered one of the most important building materials around the world. Cement production is an energy-intensive process consuming thermal energy of the order of 3.3 GJ/tonne of clinker produced. Electrical energy consumption is about 90 - 120 kWh/tonne of cement (Giddings, et al, 2000; European Commission [EC] 2001). Historically, the primary fuel used in cement industry is coal. A wide range of other fuels such as gas, oil, liquid waste materials, solid waste materials and petroleum coke have all been successfully used as sources of energy for firing cement-making kilns, either on their own or in various combinations.

The cement manufacturing industry is also under increasing pressure to reduce emissions. Cement manufacturing releases a lot of emissions such as carbon dioxide (CO_2) and nitrogen oxide (NO_x). It is estimated that 5 percent of global carbon dioxide emissions originate from cement production (Hendriks, et al, 1998). The use of alternative fuels in cement manufacturing, therefore do not only afford considerable energy cost reduction, but they also have significant ecological benefits of conserving non-renewable resources, the

[1] Alternative fuels here refer to fuels that can be used instead of conventional fuels such as coal, oil and natural gas.

reduction of waste disposal requirements and reduction of emissions. Use of low-grade alternative fuels in some kiln systems reduces NO_x emissions due to reburn reactions. There is an increased net global reduction in CO_2 emissions when waste is combusted in the cement kiln systems as opposed to dedicated incinerators.

1.1 Overview of alternative fuels

Alternative fuels and alternative sources of energy usually fall under eight broad headings: biofuels; natural gas; waste-derived fuels; wind energy; hydroelectric power; solar energy; hydrogen; and nuclear energy. Alternative fuels discussed in this chapter are predominantly agricultural biomass, non-agricultural biomass (e.g. animal waste and by-products), chemical and hazardous waste, and petroleum-based fuels.

Biofuels are from organic origin (plants or animals based) including organic waste, residues from agriculture and energy crops, meat and bone-meal, methane from animal excrement or as a result of bacterial action, ethanol and biodiesel from plant materials, as well as the organic part of waste.

Solid biofuels (generally called biomass) include plant tissues such as wood, charcoal and yarns; farm wastes such as coffee husks, straw, sugarcane leaves, sugarcane bagasse, rapeseed stems, palm nut shells, rice husks, etc.; and non-agricultural biomass such as animal fat, dung, meats and bones; and household or industrial biological degradable wastes. These materials are primarily composed of carbon-based organic matter, which releases energy when it reacts or combusts with oxygen (Seboka et al., 2009).

Solid biofuels should be distinguished from solid fossil fuels which are of biological origin but which are non-renewable. Similarly, liquid biofuels should be distinguished from fossil liquid fuels which are also of biological origin but which are non-renewable. Liquid biofuels are transport fuels, primarily biodiesel and ethanol. Another form of biofuel is biogas. Biogas is the product of organic material decomposition, composed mainly of methane and carbon dioxide.

Candidate materials for the hazardous waste fuel/waste derived fuels are too many to list. They include almost every residue from industrial or commercial painting operations from spent solvents to paint solids including all of the wash solvents and pot cleaners, metal cleaning fluids, machining lubricants, coolants, cutting fluids, electronic industry solvents (chlorinated/fluorocarbon solvents), oils, resins and many more. The list of candidate materials for use as alternative waste fuels continues to expand. Regulatory pressures, economic considerations, shrinking traditional solid waste disposal capabilities, and a host of similar factors are reflected in the constant change of the candidate waste fuel universe (Gabbard, 1990).

2. Alternative fuel options for the cement industry

Coal is the primary fuel burned in cement kilns, however, the use alternative fuels in cement kilns is now common and increasing. The range of alternative fuels is extremely wide. They are usually available as gas, liquid and solid as shown in Table 1.

Before proceeding to a consideration of some of these fuels and their properties it is necessary to consider briefly the cement production process.

2.1 Cement production process

Cement is considered one of the most important building materials around the world. In 1995 the world production of cement was about 1420 million tonnes (Cembureau, 1997).

Cement production is an energy-intensive process consuming thermal energy of the order of 3.3 GJ/tonne of clinker produced, which accounts for 30 – 40 percent of production costs (Giddings et al., 2000; EC, 2001). Worldwide, coal is the predominant fuel burned in cement kilns. Cement production consumes approximately 120 kg of coal per tonne of cement. In the European Union about 25 million tonnes of coal is required annually by the Cembureau[2] members to service the demand of cement in Europe. In 2005, the global cement industry consumed about 9 exajoules (EJ) of fuels and electricity for cement production (IEA 2007, as cited in Murray & Price, 2008).

Category	Fuels
Gaseous fuels	Refinery waste gas, landfill gas, pyrolysis gas, natural gas
Liquid fuels	Tar, chemical wastes, distillation residues, waste solvents, used oils, wax suspensions, petrochemical waste, asphalt slurry, paint waste, oil sludge
Solid fuels	Petroleum coke (petcoke), paper waste, rubber residues, pulp sludge, sewage sludge, used tyres, battery cases, plastics residues, wood waste, domestic refuse, rice husks, refuse derived fuel, nut shells, oil-bearing soils, diapers, etc.

Table 1. Alternative fuel options for the cement industry

Cement production involves the heating, calcining and sintering of blended and ground raw materials, typically limestone ($CaCO_3$) and other materials containing calcium, silicon oxides, aluminium and iron oxides to form clinker. Clinker production takes place at material temperatures of about 1450°C in either rotary or shaft kilns. Carbon dioxide is released during the production of clinker. Specifically, CO_2 is released as a by-product during calcination, which occurs in the upper, cooler end of the kiln, or a precalciner, at temperatures of 600-900°C, and results in the conversion of carbonates to oxides. Most modern cement kiln systems[3] have a special combustion chamber called a 'precalciner' as part of the preheating tower, as shown in Fig. 1. The limestone (calcium carbonate) decomposition process known as 'calcination' ($CaCO_3 \rightarrow CaO + CO_2$) is virtually completed (approximately 95 percent) in the precalciner if 50 – 60 percent of the total fuel required for clinker production is added to this chamber (Taylor, 1990).
The clinker is then removed from the kiln to cool, ground to a fine powder, and mixed with a small fraction (about five percent) of gypsum to create the most common form of cement known as Portland cement.

2.2 Benefits of using alternative fuels in cement production

Cement producers worldwide are striving to lower their production costs. One effective method of achieving this end is the use of alternative fuels. Use of low-grade alternative fuels such as waste coal, tyres, sewage sludge, and biomass fuels (such as wood products, agricultural wastes, etc.) in precalciners is a viable option because combustion in a

---

[2] CEMBUREAU – The European Cement Association.

[3] The kiln system comprises a tower of pre-heater cyclones, precalciner and the rotary kiln.

precalciner vessel takes place at a lower temperature. In precalciners where kiln exhaust gases pass through, the NO_x emissions are much reduced due to reburn reactions. There is an increased net global reduction in CO_2 emissions when waste is combusted in the cement kiln systems as opposed to dedicated incinerators, resulting in reduction in the CO_2 penalties. Since alternative fuels are often deemed cheaper than conventional fossil fuels, the possibility of a competitive edge is generated.

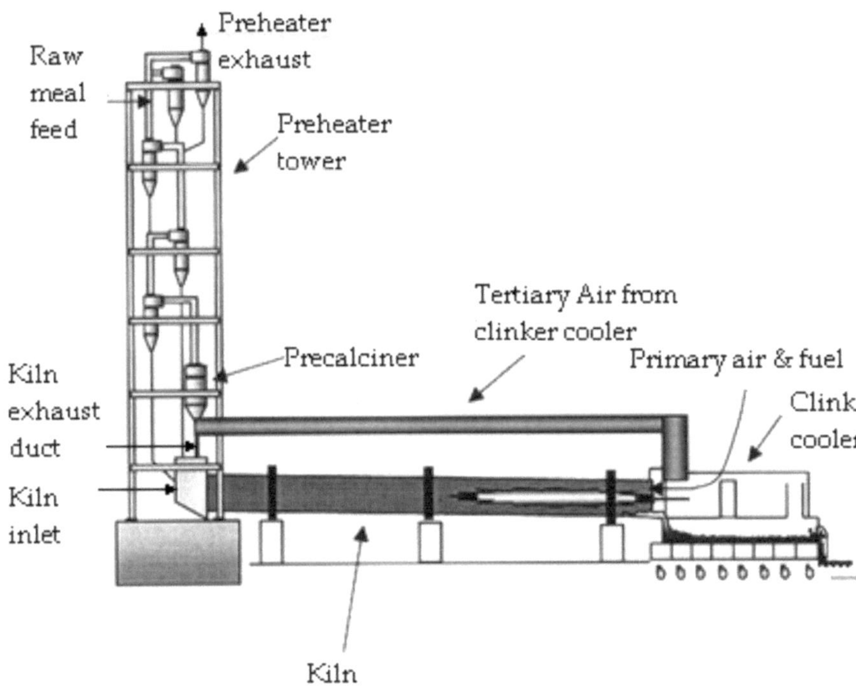

Fig. 1. The kiln system

The use of alternative fuels in cement manufacture is also ecologically beneficial, for two reasons: the conservation of non-renewable resources, and the reduction of waste disposal requirements. The use of alternative fuels in European cement kilns saves fossil fuels equivalent to 2.5 million tonnes of coal per year (Cembureau, 1999). The proportion of alternative fuels used in cement kiln systems between 1990 and 1998 in some European countries are as follows in order of importance: France 52.4 percent; Switzerland 25 percent; Great Britain 20 percent; Belgium 18 percent; Germany 15 percent; Czech Republic 9.7 percent, Italy 4.1 percent; Sweden 2 percent; Poland 1.4 percent; Portugal 1.3 percent and Spain 1 percent (Mokrzycki et al., 2003).

The process of clinker production in kiln systems creates favourable conditions for use of alternative fuels. These include: high temperatures, long residence times, an oxidising atmosphere, alkaline environment, ash retention in clinker, and high thermal inertia. These conditions ensure that the fuel's organic part is destroyed and the inorganic part, including heavy metals is trapped and combined in the product.

The wastes used as alternative fuels in cement kilns would alternatively either have been landfilled or destroyed in dedicated incinerators with additional emissions as a consequence. Their use in cement kilns replaces fossil fuels and maximises the recovery of energy. Employing alternative fuels in cement plants is an important element of a sound waste management policy. This practice promotes a vigorous and thriving materials recovery and recycling industry (Cembureau, 1999).

2.3 Key considerations for co-processing alternative fuels

The potential benefits of burning alternative fuels at cement plants are numerous. However, the contrary is possible, where poor planning results in projects where cement kilns have higher emissions, or where alternative fuels are not put to their highest value use. Five guiding principles outlined by the German development agency, GTZ, and Holcim Group Support Ltd., reproduced in Table 2, provide a comprehensive yet concise summary of the key considerations for co-incineration project planners and stakeholders (GTZ and Holcim, 2006 as cited in Murray & Price, 2008).

Principle	Description
Co-processing respects the waste hierarchy	- Waste should be used in cement kilns if and only if there are not more ecologically and economically better ways of recovery. - Co-processing should be considered an integrated part of waste management. - Co-processing is in line with international environmental agreements, Basel and Stockholm Conventions.
Additional emissions and negative impacts on human health must be avoided	- Negative effects of pollution on the environment and human health must be prevented or kept at a minimum. - Air emissions from cement kilns burning alternative fuels cannot be statistically higher than those of cement kilns burning traditional fuels.
The quality of the cement must remain unchanged	- The product (clinker, cement, concrete) must not be used as a sink for heavy metals. - The product must not have any negative impacts on the environment (e.g., leaching). - The quality of the product must allow for end-of-life recovery.
Companies that co-process must be qualified	- Have good environmental and safety compliance records. - Have personnel, processes, and systems in place committed to protecting the environment, health, and safety. - Assure compliance with all laws and regulations. - Be capable of controlling inputs to the production process. - Maintain good relations with public and other actors in local, national and international waste management schemes.
Implementation of co-processing must consider national circumstances	- Country specific requirements must be reflected in regulations. - Stepwise implementation allows for build-up of necessary management and handling capacity. - Co-processing should be accompanied with other changes in waste management processes in the country.

Table 2. Guiding principles for co-processing alternative fuels in cement kilns

2.4 Challenges of using alternative fuels in cement production

Alternative fuels used in cement manufacturing have different characteristics compared to the conventional fuels. Switching from conventional fuels to alternatives fuels presents several challenges that must be addressed in order to achieve successful application. Poor heat distribution, unstable precalciner operation, blockages in the preheater cyclones, build-ups in the kiln riser ducts, higher SO_2, NO_x, and CO emissions, and dusty kilns are some of the major challenges (Roy, 2002; F.L. Smidth & Co., 2000).

The cement industry, like other industrial sectors, is strictly regulated by the national and international legislation as well as internal regulatory procedures regarding environmental protection, health and safety, and quality of products. Strict regulations are applied and plants are operated on the basis of permits from national authorities. Emissions are regularly checked by the authorities. Special approval from relevant authorities is therefore required to burn alternative fuels in many countries on account of potential environmental hazards (Hewlett, 2004).

The operation of cement kiln system is not only affected by the chemical composition of the main components of the raw meal but also the combustion and consequently the fuel used. The type of fuel used can introduce some material components which can interfere with the chemistry of the cement materials as well as affect the operation of the system. The use of a type of fuel is therefore subject to the constraints imposed by any deleterious effect on cement quality, refractory life, gas and material flow or potential emissions to the atmosphere (Bye, 1999).

In most kiln systems the fuel ash is incorporated into the clinker thereby changing the compound composition of the product. The main constituents of fuel ash are silica and alumina compounds which combine with the raw materials to become part of the clinker. The composition of fuel ash tend to limit the level of replacement of more conventional fuels, for instance rice husks have been used to replace 5 – 7 percent of traditional fuels since the ash contains 78 – 90 percent silica. Fuel ash with high content silica can on the other hand provide a very satisfactory means of increasing the silica modulus of the clinker, thus making it possible to reduce the amount of ground sand incorporated into the feedstock (Hewlett, 2004).

Approximately 95 percent of clinker consists of oxides of CaO, SiO_2, Al_2O_3, and Fe_2O_3 and the remainder consists of the so-called minor constituents. In cement manufacturing care is taken to avoid constituents which, even when present in small amounts (< 1percent), may have adverse effect upon the performance of the product and/or the production process. The most important of these are probably the oxides of potassium and sodium (K_2O and Na_2O) commonly known as alkalis. High levels of alkalis in cement can, in the presence of moisture, give rise to reactions with certain types of aggregates to produce a gel which expands and gives rise to cracking in concretes and mortars.

The alkali metals Na_2O and K_2O have a very strong affinity for SO_3 and where there is sufficient sulphate present in the clinker, the alkalis are normally present as compounds of sulphates such as K_2SO_4, Na_2SO_4, aphthilalite $Na_2SO_4{\cdot}K_2SO_4$ and langbeinite $2CaSO_4{\cdot}K_2SO_4$ (Hewlett, 2004; Newman, et al., 2003). Higher levels of alkali sulphates in cements affect the reactivity of the cement, thus leading to possible setting problems (Hewlett, 2004). From kiln operational point of view, it is desirable that as much as possible the alkalis (and sulphates) get discharged from the system with the clinker. If this does not take place, the presence of these alkalis (and sulphates) can have an extremely disruptive effect upon production especially in kiln systems with high efficient heat exchangers such as the cyclones.

It is reported that in kiln systems equipped with high efficient pre-heaters, volatilized recirculation loads of 150 - 200 percent (of the total input) for K_2O, 100 – 150 percent for Na_2O and 350 - 400 percent for alkali sulphate exist because volatilized alkalis, chlorine and some heavy metal condense on the raw meal grains in the suspension preheater. With the raw meal they return to the kiln where they are volatilized again thereby increasing the recirculation load (Hewlett, 2004; Ghosh, 1991; Taylor, 1990). Some of the alkalis volatilized in the high temperature part of the system (kiln) condense in the cooler parts, causing build-up and blockages in the heat-exchange systems. The sticky deposits attract dust and bind it together to form build-up, which in an extreme case can completely throttle the flow of gas and/or cement solid materials. There are mainly two ways of maintaining the alkalis at a required level, firstly through the careful selection of materials and secondly by bleeding high-alkali dusts from the kiln system (Hewlett, 2004; Taylor, 1990).

If chloride is introduced into the system either through the raw meal or the fuel, the melting point of the sulphates is reduced and sulphate–spurrite ($C_2S{\cdot}CaSO_4$), which is stable within the temperature range 900–1200°C can also be formed. Sulphate–spurrite is considered to be associated with the formation of rings in cement kilns. It is recommended that chloride content must be kept low to avoid formation of kiln rings and preheater deposits (Taylor, 1990). Contents below 0.02 percent are preferred, though higher ones can be acceptable if a sufficient proportion of the kiln gases is bypassed or in less energy-efficient (e.g. wet process) plants. Ring formation is known to increase with an increase in the amount of excess sulphur over that which combines with alkalis (Hewlett, 2004).

Since sulphur is introduced into the system through the fuel and also with the raw materials, the sulphur content of the fuel can become an important factor in kiln system operation. It is however, important to distinguish between the sulphur in the raw meal that enters the kiln system in the form of sulphates (such as calcium sulphates) and that which enters as sulphides (such as pyrite, marcasite and organic sulphides). The latter can oxidize through an exothermic reaction at 400 – 600°C in sections of the system (for example cyclones) where there is less calcium oxide available. Consequently, the SO_2 released is emitted and treated. On the other hand, the calcium sulphates present do not decompose until 900 – 1000°C. This gives the oxides of sulphur an opportunity to react with the alkalis which have been volatilized and also with CaO that has already formed thereby increasing the chance of alkalis and sulphate being removed from the kiln system in the clinker. This is why it is generally possible to use fuels with high sulphur content in the cement industry without significant harmful consequences to the environment (Ghosh, 1991). As already mentioned, if significant amounts of the low melting point mixtures of calcium and alkali sulphates form in and around the preheater sections can lead to blockages.

Some of the waste materials used as alternative fuels in cement kiln systems such as polyvinylchloride (PVC), chlorinated hydrocarbons, sewage sludge, and meat and bone meal can increase the amount of chlorine (Cl) introduced into the system (Saint-Jean et al, 2005). Fuels containing high (> 0.7 percent Cl) can adversely affect the performance of some types of electrostatic precipitators on wet process kilns. It is also reported that in kilns with cyclone pre-heaters, only about 20 percent of the chlorine input is retained in the clinker, with the result that a recirculation chloride load of some 400 – 500 percent develops in the kiln/preheater system (Hewlett, 2004) . Clogging may occur in the cyclone pre-warmer if chlorine content of fuel is more than 0.2 – 0.5 percent (Werther et al., 1997). Chlorine content is also known to affect the quality of the product. High Cl content increases the corrosion of reinforcement in concrete. From quality point of view most standards for Portland cements

restrict the amount of chloride present to 0.10 percent of the raw meal feedstock. However, in preheater kiln systems operational problems normally manifest themselves long before this quality point is reached (Hewlett, 2004).
The effect of other trace elements such as fluorine, barium, chromium, lead, manganese, thallium, titanium, vanadium and zinc on quality of cement range from very small to negligible. However, it is important to note the exceptions of fluorine and zinc from this list. There are indications that the use of fluorine as a mineraliser[4] may give rise to the build-up of excessive coating in the kiln and that this may be due to the formation of additional spurrite. CaF_2 acts both as a mineraliser and as a flux in promoting the formation of alite[5] (Taylor, 1990; Newman et al., 2003). Small amounts of zinc (0.01 – 0.2 percent) have been reported to increase the reactivity of C_3A[6] and in consequence lead to possible setting time problems. However, the presence of up to 0.5 percent of ZnO does not appear to have a profound effect upon other hydraulic properties (Hewlett, 2004). The zinc content in tyres is, from cement quality point of view, the main constraint in the use of scrap tyre as a fuel. Type ash contains about 20 percent Zn (Al-Akhras et al., 2002).
The incomplete combustion, poor heat distribution and unstable precalciner operation are problems associated with switching from conventional fuel to alternative fuels (Roy, 2002). The arrangement of combustion in such a manner as to create a reducing condition in some zones of the precalciner is useful for the diminution of NO_x emissions. On the other hand, it is important to note that low combustion efficiency at the precalciner stage can create reducing zones in deposited material at the kiln inlet, significantly increasing the volatilization rate of sulphur (Ghosh, 1991). Desmidt, 1987, observed a 78 percent volatilization of SO_3 at 93 percent combustion and 0.2 percent CO at kiln inlet, and a 42 percent volatilization of SO_3 at 98percent combustion and 0.06 percent CO at kiln inlet. Incomplete combustion also gives rise in the carbon content in the product. This is undesirable for the following reasons. First, high carbon content accelerates corrosion of steel in concretes. Secondly, the carbon absorbs water reducing the quantity available for hydration reaction. Thirdly, the alkalinity of the cement is affected. Finally, high carbon content ash darkens the concrete reducing its aesthetic appeal and leading to inaccurate prejudgements of the concrete quality (Ha et al., 2005; Freeman et al., 1997; CIF, 2000).

2.5 Main alternative fuels used in the cement industry

Well-established technology, on the one hand, allows the rotary kiln of any cement plant to be fired with low-volatile fuels such as petcoke, low-volatile bituminous coal, and anthracite, without problem (Nielsen et al., 1986). On the other hand, high volatile-low calorific value alternative fuels have limited use in the kiln primary firing system due to their relatively low combustion temperatures. They are utilised more in the precalciner firing than in the kiln unless their calorific value exceeds about 16.8 MJ/kg (Hochdahl, 1986). Experience has shown that it is difficult to obtain complete combustion of low-volatile fuels in precalciners. The use of low-volatile fuels in precalciners, often requires design and operational modifications of the precalciner, or specially designed precalciners (Roy, 2002; Nielsen & Hundebol, 1986).

---

[4] A mineraliser is an agent that promotes the formation of a particular solid phase by affecting the equilibria through incorporation in one or more of the solid phases (Taylor1990).
[5] Alite is a cement phase $2CaO.SiO_2$ or C_2S.
[6] C_3A is another cement phase $3CaO.Al_2O_3$.

Switching from conventional fuels to alternatives fuels therefore presents several challenges that must be addressed. Poor heat distribution, unstable precalciner operation, blockages in the preheater cyclones, build-ups in the riser ducts, higher SO_2, NO_x, and CO emissions, and formation of rings in kilns are some of the major challenges (Roy, 1986; Pipilikaki et al., 2005; F.L Smidth, 2000).
The quality, type and quantity of fuel exert a profound effect on the quality of clinker produced. Consequently, selection of the proper type of fuel is vital for optimum efficiency. Waste derived fuels have different characteristics compared to conventional fuels such as coal and to be able to use them in cement manufacturing, knowledge of their composition is important. The energy and ash content as well as the moisture and volatile contents are all important limiting factors. A comparison of coal with some of the most common alternative fuels used in cement kiln systems is presented in Table 3. Ultimately, cost and availability of the alternative fuels remain the main influencing factors for their choice.

2.5.1 Petroleum coke

Petroleum coke (petcoke) is the solid residue that remains after extraction of all valuable liquid and gaseous components from crude oil. The world production of petcoke grew by 50 percent from 1997 to 1998. It reached 50 million tonnes in 1999 and was expected to reach 100 million tonnes by 2010 (International Energy Agency [IEA], 2001). Petcoke is a low-volatile fuel whose volatile content range is typically 5 - 15 percent, depending on the coking process (Roy, 2002). There are three processes of coking: delayed, fluid and flexi-coking with delayed coking producing over 90 percent of total production (IEA, 2001). Petcoke is composed mainly of carbon and it also contains high levels of sulphur and heavy metals such as vanadium and nickel (IEA, 2001; Bryers, 1995). The fixed carbon varies from 80 - 92 percent (Bryers, 1995). Irrespective of the coking process, petcoke has higher calorific value than coal, typically LHV of about 32.5-35 MJ/kg (Commandre & Salvador, 2005). The use of petcoke as fuel presents several challenges due to its high sulphur content, poor ignition and burnout characteristics because of its low volatile content. Low-volatile fuels are generally fired in an arch-type furnace to induce ignition and ensure flame stability (Bryers, 1995).
Petcoke is widely used in cement kilns worldwide. However, owing to the challenges associated with its burning, it is not possible to fire 100 percent petcoke in the kiln and precalciner in many existing cement kiln systems without co-firing with a high-volatile fuel or special design considerations (IEA, 2001; Roy, 2002; Nielsen et al., 1986; Tiggesbäumker & Beckum 1986). New plants specifically designed to enable 100 percent petcoke firing are coming on the market while the many classical precalciners are retrofitted to enable petcoke firing. One traditional solution to using petcoke is to grind the coke to a much finer residue than standard coal, up to 0.5 - 1 percent retained on 90 μm (Roy, 2002; Bryers, 1995). The burning rate of an individual char particle depends primarily on its particle size, the amount of oxygen present in the local atmosphere and the local temperature.
In precalciner application where temperature is lower than in kilns, besides the particle size, considerable retention time is required to complete the combustion. In precalciners designed for coal firing where the gas retention time is less than 3 s, often petcoke is introduced directly into the tertiary air where oxygen is highest before mixing with kiln exhaust gases. In some of new precalciner designs the gas retention time is increased significantly to about 7 s by injecting petcoke in a long loop duct before joining the main calciner (Roy, 2002). The

combustion of petcoke in a relatively raw meal free hot-zone in the precalciner away from the walls is an important aspect in new precalciner designs and retrofits to achieve high burning rate and avoid build-ups.

The sulphur content of petcoke is several times higher than that of coal. It is therefore expected that combustion of petcoke will lead to higher emissions of SO_2. SO_2 emissions in pulverized fuel firings as mentioned earlier, normally correlate strongly with the sulphur content of the fuel and generally almost all the sulphur in the fuel is released as SO_2 (Commandre, 2005; Werther & Ogada, 1997; Spliethoff & Hein, 1998).

Petcoke has low ash content which easily fuses in the cement clinker. Table 4 shows typical ash analyses of various types of petcokes. Vanadium pentoxide (V_2O_5) is by far the highest except in a type of coke where iron oxide was highest. Molten V_2O_5 (melting temperature 675-690°C) is known to dissolve most refractory and metal oxides which could expose virgin metal surfaces to oxidising environment (Bryers, 1995; Hewlett, 2004). Vanadium, if present in clinker, tends to be found in the alite, the crystal size of which increases with additions of up to 0.4 percent (Hewlett, 2004). A 0.2 percent addition is reported to lead to a 10 percent reduction in 28-day strength of cement. However, due to low ash content of petcoke such high contents of vanadium in cement are unlikely.

	Coal	Petcoke, (Kääntee et al., 2004)	Predried Sewage sludge (Winter et al., 1997)	Meat and bone-meal (Kääntee et al., 2004)	Tyre
Proximate analysis (wt. percent)					
			As received (wt. percent)		
Moisture (wt. percent)	9.20	1.50	7.00	8.09	0.6
Ash (db)	8.85	0.90	26.70	28.30	19.1
Volatiles (daf)	36.22	11.80	60.60	56.41	56.6
C - fixed	45.63	85.80	5.70	7.20	23.7
Ultimate analysis (wt. percent daf)			As received (wt. percent)		
C	68.6	89.50	37.20	42.10	71.85
H	3.05	3.08	4.29	5.83	6.07
N	1.3	1.71	4.18	7.52	0.20
S	0.49	4.00	0.53	0.38	1.06
O	8.51[7]	1.11	20.10	15.30	1.12
Ash	7.0	0.90	26.70	28.30	19.1
LHV (MJ/kg)	27.89	33.7	14.8	16.2	31
Density (kg/m^3)	1300	2023	1140	720	1179

Table 3. Analyses of some common fuels used in cement manufacturing

[7] By difference

Composition (wt. percent)	Delayed coke	Short coke	Fluid coke	Flexicoke
SiO_2	10.1	13.8	23.6	1.6
Al_2O_3	6.9	5.9	9.4	0.5
TiO_2	0.2	0.3	0.4	0.1
Fe_2O_3	5.3	4.5	31.6	2.5
CaO	2.2	3.6	8.9	2.4
MgO	0.3	0.6	0.4	0.2
Na_2O	1.8	0.4	0.1	0.3
K_2O	0.3	0.3	1.2	0.3
SO_3	0.8	1.6	2.0	3.0
NiO	12.0	10.2	2.9	11.4
V_2O_5	58.2	57.0	19.7	74.5

Table 4. Typical ash analysis of various types of cokes (Bryers, 1995)

2.5.2 Sewage sludge

A large amount of sewage sludge is produced worldwide. Sludge is formed during wastewater treatment. Wastewater is a combination of the liquid- or water-carried wastes removed from residential, institutional, commercial and industrial establishments (Werther & Ogada, 1997). Landfill, dumping in the sea and use in agriculture as organic fertiliser and soil conditioner are the main conventional methods of disposal. There are however, economical and ecological constraints to these methods. The increasing costs of land for landfill coupled with increasing stringent environmental standards are making landfill a less attractive option. The use of sewage sludge in agriculture poses human health and environmental risks. Uncontrolled addition of sludge to the agricultural land may increase the concentration of heavy metals in farmland. These factors are making the thermal utilisation of sewage sludge an attractive means of its management.

The thermal utilisation of sewage sludge is deemed feasible when its secondary environmental impacts are minimised. The most common sewage sludge disposal alternative is to incinerate it and deposit the ash in controlled landfill. Incineration accounts for 21 percent of the sludge produced in Denmark, 20 percent in France, 15 percent in Belgium and 14 percent in Germany (Hall & Dalimier, 1994). In the USA and Japan, 25 and 55 percent of the sludge produced, respectively, is incinerated (Werther & Ogada, 1997). Incineration ash of municipal solid waste accounts for a great portion of the matter in landfills. A total annual incineration of municipal waste of 26 million tonnes was estimated in the EU in 1997 (Kikuchi, 2001).

The formation of poisonous solid and gaseous by-products during sludge incineration is, however, noted to be a source of public concern. These include the release of heavy metals and the emission of substances such as NO_x, N_2O, SO_2, HCl, HF and C_xH_y (Ogada &Werther, 1996). SO_2 emissions in pulverized fuel fired systems normally correlate strongly with the sulphur content of the fuel and generally almost all the sulphur in the fuel is released as SO_2 (Spliethoff & Hein, 1998; Werther & Ogada, 1997). The sulphur content of the sludge is comparable with that of coal.

Other various processes have been proposed for the thermal utilisation of sewage sludge, including the co-briquetting with coal, the co-combustion with coal (Åmand et al., 2004; Folgueras et al., 2003; Lopes et al., 2003); the use of sewage sludge pyrolysis volatiles as a

reburn fuel in the air-staged combustion of coal (Boocock et al., 1992a, 1992b; Konar et al., 1994); incineration/combustion of sewage sludge alone (Sänger et al., 2001; Arai et al., 1989) as well as co-combustion with other fuels (Åmand & Leckner, 2004).
In countries like Japan, USA, Denmark, Netherlands, Switzerland and Belgium sewage sludge is used in cement production. In cement production, sludge is usually co-fired with coal in predried form. Predried sludge is easier to store, transport and feed (Werther & Ogada, 1997). The sewage sludge for co-combustion is dried, pulverised and pneumatically fed to the burners. Either the sludge is preblended with coal and fed together, or the two fuels can be fed separately if multi-fuel burners are used. The environmental concerns associated with sewage incineration are significantly reduced when sewage sludge is used as fuel in cement kilns. The organic part is destroyed and the inorganic part, including heavy metals, is trapped and combined in the product (CEMBUREAU, 1997).
Combustion of sewage sludge is expected to lead to higher emissions of SO_2. In cement production this might not affect the SO_2 emissions significantly, since about 60 to 80 percent of the sulphur is captured by the calcium oxide in the kiln system (Manning et al., 2003; Cement Industry Federation [CIF], 2000). However, as discussed earlier on, in cement kilns sulphur is known to cause hard build-ups due to formation of sulphate compounds. The higher nitrogen content of the sewage sludge does not translate into a proportionate increase in NO_x emissions in precalciners. This is due to lower combustion temperatures, well below 1200°C, that suppresses thermal NO_x formation. The in-line precalciner in particular combines the merits of both the air staging and fuel staging technologies. In this arrangement, the fuel fired in the precalciner is used in reburn reactions.
Sewage sludge ash, however, has a high content of SiO_2, Al_2O_3 and Fe_2O_3 which could affect the quality of cement if excess amounts of sewage sludge are used. Table 5 shows the comparison of the composition of sewage sludge ashes with those of cements. A chlorine content of the sludge of more than 0.2 - 0.5 percent may cause clogging in the cyclone preheaters. To keep the levels of the oxides within limit (Werther et al., 1997) suggested the maximum sewage sludge feed rate to be no more than 5 percent of the clinker production capacity of the cement plant unless the sludge is conditioned and stabilised by lime, normally 0.3 – 0.5 kg CaO/kg dry sludge.

	Cement (Bye, 1999)	Sewage sludge ash (Werther & Ogada, 1997)
CaO	63 -67	9 - 22
SiO_2	19 - 23	30 – 49
Al_2O_3	3 - 7	8 - 15
Fe_2O_3	1.5 - 4.5	5 - 23
MgO	0.5 - 2.5	1 - 2
K_2O	0.1– 1.2	
Na_2O	0.07 - 0.4	
SO_3	2.5 - 3.5	

Table 5. Comparison of the compositions of sludge ash with those of cements

Sewage sludge has significantly higher contents of nitrogen, volatile matter and ash, and very low fixed carbon than typical coals. Up to 80 percent of the sludge carbon is volatile carbon and sludge combustion is characterised by the gas-phase combustion of volatiles (Werther & Ogada, 1997). Published calorific values (HHV) of sludge typically range between 8 - 17 J/kg. The cause of this variation in HHVs is attributed to the treatment of producing the sludge and the heterogeneous nature of sludge (Vesilind & Ramsey, 1996).

The main products of sludge pyrolysis are gas (volatiles); char and oil; the quantities of which depend on factors such as pyrolysis temperature. Sewage sludge releases volatiles over a wide range temperature 250 - 850°C (Stolarek & Ledakowicz, 2001; Inguanzo et al., 2002). The percentage of the gaseous component increases whereas the amounts of oil and char decrease with increasing temperature (Rumphorst & Ringel, 1994; Inguanzo et al., 2002). The composition of pyrolysis gas from predried sludges depends upon on the type of sludge. However, in general the main gaseous components are CO, CO_2, H_2, O_2, N_2 and C_xH_y. CO and H_2 are reported to increase whereas CO_2 and C_xH_y decrease with increasing temperature (Inguanzo et al, 2002; Ogada & Werther, 1996).

2.5.3 Used tyres

Scrap motor tyres have been utilised as a supplementary energy source in Japan, Europe and USA since the 1970s (Gray, 1996) and represent a rapidly growing application in most developed countries where scrap tyres are an environmental nuisance. About 290 million tyres were discarded in the USA in 2003, and nearly 45 percent of these scrap tyres were used as tyre derived fuel (TDF). Approximately 58 percent of the TDF was used in the cement industry (Rubber Manufacturers Association [RMA], 2003). In the EU over 2.5 million tonnes of tyres are produced per year and almost 40 percent of these tyres are thrown away untreated (Diez et al., 2004).

Although the environmental acceptability of the use of tyres as fuel in kiln systems is dependent on individual plant performance, extensive environmental data has been generated for a variety of kiln configurations and fuel displacement. In general, the different test results have shown that TDF has no adverse effect upon the emissions; that is to say, the use of TDF has not caused a facility to exceed its operating limits (Gray, 1996; Environmental Agency, 1998; Blumethal, 1992a, 1992b). In comparison with coal, particulates, SO_x, NO_x and HCl emissions generally decline or remain constant with TDF use. Organic emissions, dioxins and furans are also observed to decline while changes in heavy metal concentrations are nominal (Gray, 1996; Scrap Tyre Management Council [STMC], 1992).

Table 6 shows an elemental ash analysis of tyres in comparison with coal. The use of TDF in cement kiln systems is technically sound as the rubber is destroyed and the inorganic part, including heavy metals, is trapped and combined in the product (CEMBUREAU, 1997).

An additional advantage of TDF use is its steel portion. The steel can substitute, in part, for the iron requirement in the raw meal recipe. The content of iron oxide in Portland cement is 1.5 - 4.5 percent on weight basis. On the other hand, although zinc oxide acts as a flux as well as mineraliser, it is known to have detrimental effect on the quality of cement if it is in excess. It strongly retards the setting time and strength of the cement if the total zinc content of all fuels exceeds 4000 parts per million (PPM) (STMC, 1992; Olmo et al., 2001; Murat & Sorrentino, 1996). Therefore, other than the problems of incomplete combustion, the zinc oxide content in tyres tend to limit its displacement of conventional fuels in cement production (STMC, 1992).

Combustion of whole tyres requires long residence times to obtain complete conversion. In some cement installations, tyres are fired whole, mostly in the rotary kiln. More commonly, they are shredded in a slashing process, producing tyre chunks or chips, and co-fired with coal in the precalciner. They cannot, however, be finely comminuted economically. At best they can be shredded to pieces (chips) of about 25 mm, typically, their size ranges from 25 - 100 mm.

Element (oxide)	Coal	TDF with wire	TDF without wire
Aluminium	20.7	1.93	13.11
Calcium	3.3	0.56	3.8
Iron	18.89	0.35	2.37
Magnesium	0.79	0.10	0.68
Phosphorous	0.62	0.10	0.68
Potassium	2.06	0.14	0.95
Titanium	0.82	0.14	0.95
Silicone	47.98	5.16	35.05
Sodium	0.48	0.13	0.88
Sulphur	4.33	0.99	6.72
Zinc	0.02	5.14	34.81
Metal		85.26	
Total	100	100	100

Table 6. Elemental ash analysis of tyres in comparison with coal (Gray, 1996)

The majority of precalciners are basically entrained flow combustion vessels. Due to the relatively short residence time in the precalciners, circa 2 to 4s, firing tyre chips often results in incomplete combustion. Some of the tyre chips drop directly into the kiln back end or into the tertiary air duct, in an in-line precalciner arrangement before they are fully devolatilised. At the kiln back end there is very little oxygen in the kiln gas for the combustion of the tyre chips. Smaller chips and fragments of devolatilised chips levitate much more easily and are carried over before their combustion is complete in the precalciner. This means that a considerable fraction of tyre chips may also pass to the rotary kiln as carbon particles mixed with the calcined meal. Besides the under utilisation of the fuel energy, an increase in carbon content in ordinary Portland cement accelerates corrosion of the steel reinforcing in concretes, the alkalinity of the cements is affected and the cement loses its characteristic colour (Kääntee et al., 2004; Winter et al., 1997).

The problems of using scrap tyre as alternative fuel emanate from the lack of understanding of their devolatilisation and combustion behaviour. Tyres, like most of the alternative fuels suffer from insufficient characterisation. Tyres are a hydrocarbon-based material derived from oil, natural rubber and gas. Some inorganic materials, as shown in Table 7, are added to enhance reactions or performance properties. As such tyres are very non-homogeneous and exhibit major property variations. The non-homogeneity arises because of the cord re-enforcing materials and steel beading used in their construction. The property variations arise due to applications dependence, extreme size and geometric differences as shown in the differences between passenger car and truck tyres in Table 8. The property variations could also arise due to variable degree of used tyre wear and country dependent construction. Table 8 presents comparisons of analyses of different tyres, and petcoke and bituminous coal. In addition to these variations there are also as received shredded chip size and shape distributions variations.

The thermal degradation of tyres is known to produce a wide variety of products in the liquid (oil) and gas phases in addition to the residual char. The main gases produced during the pyrolysis of tyres are: CO_2, CO, H_2, CH_4, C_2H_6, C_3H_6, C_3H_8, and C_4H_6, with lower concentrations of other hydrocarbon gases (Williams et al., 2001; Conesa et al., 1997).

Composition	Approximate percent
Rubber (natural & synthetic)	38
Fillers (carbon black, zinc oxide, etc.)	30
Reinforcing materials (fibres, steel cords, bead wire)	16
Plasticizers (oils & resins)	10
Chemicals	5
Miscellaneous	1

Table 7. Typical tyre composition (Bhowmick et al., 1994)

	Tyres					Conventional fuels	
	Passenger car (Chen et al., 2001)	Truck (Chen et al., 2001)	Samples used by Williams et al., (2001)	Samples used by Giddings et al., (2001)		Petcoke, (Giugliano et al., 1999)	Bituminous Coal (Kääntee et al., 2004)
Ultimate Analysis, (wt. percent)							
C	81.16	85.19	85.90	78.35		85-90	70.6
H	7.22	7.42	8.0	6.62		5-10	4.3
N	0.47	0.31	0.4	0.22		1.5-2	1.2
S	1.64	1.52	1.0	1.15		3-5	1.3
O	2.07	1.72	2.3	1.22			11.8
Ash	7.44	3.84	2.4	11.74			11.1
Proximate Analysis (wt. percent)							
Fixed Carbon			30.3	23.7		84-88	53.0
Volatile matter			66.5	56.6		11-13	35.9
Ash			2.4	10.8		0.3-0.5	11.1
Moisture			0.8	0.6		1-3	3.0
Wire				8.3			
Heating Value: High Heat Value, (HHV, MJ/kg); and Low Heat Value (LHV, MJ/kg)							
HHV: Wire free			40.0	34.8			28.4
With Wire				31.9			
LHV	32.8	31.6				34.5	27.4

Table 8. Comparative analysis between tyres and conventional fuels

The devolatilisation and combustion burnout times of tyre chips remain sources of controversy. Devolatilisation of an average tyre chip held in a precalciner at temperatures of between 1050 - 1150°C, at a gas speed of 22 m/s, was reported nearly complete in 30 seconds (Giuliamo et al., 1999). Complete devolatilisation and total burnout times of tyre

chips held at about 900°C in a refractory-lined furnace as long as 2 minutes and 20 minutes respectively have been recorded (Giddings et al., 2002).
When tyre chips are inserted into the furnace, they burn with a vigorous flame after ignition. The ignition time is clearly temperature dependent. At 900°C an average time to ignition of 3 s was observed (Chinyama et al., 2007). The vigorous flame is evidence of intense devolatilisation of the tyre chip (Giddings et al., 2002, Atal & Levendis, 1995). Upon burnout of the volatiles, a low-intensity flame follows and burns to extinction - characterising the burning of char. Tyre char originates from the reinforcing carbon black used as fillers in tyre production. Tyre char also contains almost all the inorganic compounds present in tyres (Helleur et al., 2001).

2.5.4 Meat and bone meal

Meat and bone meal (MBM) is produced in rendering plants where animal offal and bones are mixed, crushed and cooked. Tallow is extracted during the cooking process, and the remaining material is then dried and crushed. Feeding MBM to cattle, sheep or other animals was banned within the EU in 1994 and disposal to landfill is not an option since this does not destroy any potential bovine spongiform encephalopathy (BSE) pathogens (Gulyurtlu et al., 2005). This change in legislation increased the interest in using MBM as fuel to ensure that any living organism is thermally destroyed totally and its energy potential is utilised.
Co-incineration in cement kiln systems is the most common way for MBM destruction (Deydier et al., 2005; Conesa et al., 2005). Compared with coal, MBM has lower fixed carbon and high ash content and chlorine. Most of the chloride in MBM is present as common salt (NaCl) (Conesa et al., 2005). Proximate and ultimate analyses of typical MBM samples and coal are presented in Table 3. The sulphur content of MBM is slightly lower than coal. However, on the basis of calorific value, the sulphur content of MBM could be about same as coal. MBM has calorific value (HHV) of approximately 14 -17 (MJ/kg), this is about half that of coal (Deydier et al., 2005; Gulyurtlu et al., 2005). The high content of calcium in MBM could be an advantage as it could act to retain most of SO_2 formed during MBM combustion. Using MBM in cement production further reduces the possibility of increase in SO_2 emissions.
In France where about 850,000 tonnes of MBM are produced per year, about 45 percent is burnt in cement plants. The remaining 55 percent is usually stored waiting for further destruction or valorisation (Deydier et al., 2005). Apart from use in cement plants, in other countries, for example England, dedicated MBM incinerators are used. The feeding rates of MBM in cement kilns vary from country to country, in Spain the limit is 15 percent of the energy needed in the kilns (Conesa et al., 2005). However, the limit in the feed rate of MBM is due to the effects of chlorides. Chlorides readily volatilise in the burning zone of the kiln and condense in the heat exchangers to combine with alkalis and sulphates to form low melting point mixtures. This leads to build-up and blockages in preheater units. Their effect upon the operation of kilns with cyclone preheaters and gate preheaters is so serious that for the former it is normal practice to limit the total amount of chloride introduced into the process to a maximum of 0.015 percent of the raw meal feed (Hewlett, 2004).
As the nitrogen content in the MBM is about 7-8 times higher than that in coal, it could be expected that NO_x emissions would increase with an increase of MBM in co-combustion. However, the fuel-N conversion to both NO_x and N_2O was observed to decrease with increasing MBM content in coal-MBM blend (Gulyurtlu et al, 2005). A 20 percent (wt.) MBM addition to the fuel gave rise to a reduction in the NO_x concentration of about 25 percent compared to the combustion of coal alone, although the N input was almost double. This

was attributed to a significant part of the fuel-N being released as NH_3 during devolatilisation which reduces NO_x to N_2 through the known $DeNO_x$ mechanism (Wenli et al., 1990). The minimisation of the emission of nitrogen compounds in MBM co-combustion was also observed by Goeran et al., (2002).

As MBM ash mainly arises from bone combustion, it contains a high amount of phosphate (56.3 percent) and calcium (30.7 percent), the two major constituents of bone. It also has significant levels of sodium (2.7 percent), potassium (2.5 percent) and magnesium (0.8 percent) (Deydier et al., 2005; Gulyurtlu et al., 2005). Whereas the high content of calcium in MBM is an advantage in cement, high levels of phosphate, sodium, potassium and magnesium can have harmful effects on the production process and/ or cement quality. Phosphate is a compound of phosphorus. The normal range of P_2O_5 contents in Portland cement clinker are from 0.03 to 0.22 percent. When higher amounts of P_2O_5 are present, the dicalcium silicate (C_2S)[8] is stabilised to an extent that the conversion to tricalcium silicate (C_3S) is inhibited. When the amount of P_2O_5 present exceeds 1 percent, it has been reported that 10 percent of C_3S is lost for each additional 1 percent of P_2O_5 (Hewlett, 2004; Taylor, 1990). Potassium and sodium are alkalis and in cement, high alkali levels can, in the presence of moisture give rise to reactions with certain types of aggregates to produce a gel which expands resulting in cracking in concretes and mortars. Where there is sufficient sulphate present in the clinker, the alkalis are normally present as sulphates. Higher alkali levels in cements (over $\approx$ 0.8 percent $(Na_2O)_e$)[9] when present as alkali sulphates have the effect of increasing the early strength ($\approx$ 10 percent) of cements at the expense of their 28 day strength (Hewlett, 2004; Taylor, 1990). The presence of alkalis (and sulphates) also causes blockages in preheater units. Excessive amounts of magnesia (MgO) (usually considered to be over 5 percent of the clinker as a whole), can crystallise out from the flux as a periclase[10], the presence of which has been associated with long term unsoundness[11] (Boynton, 1980; Hewlett, 2004).

Use of MBM in cement production can therefore be limited by the constituents of the ash produced.

2.5.5 Agricultural biomass

Biomass and biomass residues, if sourced in an environmentally and socially sustainable fashion, represent a vast - and largely untapped - renewable energy source. Crop and agro-industrial residues have low bulk and energy density, and for these reasons cannot be transported far from production sites without some form of processing. Residues from large commercial farms and agro-industries can be converted to relatively high-quality and high-energy density fuels for use in the domestic, commercial and industrial sectors through a number of physical, biological and thermo-chemical conversion processes (Seboka et al., 2009). The use of agricultural biomass residues in cement manufacturing is less common in industrialized countries and appears to be concentrated in more rural developing regions such as India, Thailand, and Malaysia. The type of biomass utilized by cement plants is highly variable, and is based on the crops that are locally grown. For example rice husk, corn stover, hazelnut shells, coconut husks, coffee pods, and palm nut shells are among the many varieties of biomass currently being burned in cement kilns (Murray & Price, 2008).

---

[8] C_2S and C_3S are cement clinker phases.

[9] $(Na_2O)_e$ is sodium oxide equivalent given as 0.658 (K_2O) + percent Na_2O

[10] Periclase - MgO becomes sintered into a dense, stable form.

[11] A cement is said to be unsound if the hydration of a hardened paste of it is eventually accompanied by excessive expansion, causing cracking and reduction in strength.

Biomass fuels are considered carbon neutral because the carbon released during combustion is taken out of the atmosphere by the species during the growth phase (Intergovernmental Panel on Climate Change [IPCC] 2006). Because the growth of biomass and its usage as fuel occurs on a very short time-scale, the entire cycle is said to have zero net impact on atmospheric carbon emissions. An important caveat to this assumption is that growing biomass and transporting it to the point of use requires inputs like fuel and fertilizer that contribute to the carbon footprint of biomass. When biomass is grown specifically for fuel, the upstream GHGs that are typically attributed to the biomass are those associated with fertilizer, collection, and transportation to the facility. When biomass residues are used, fertilizer is only considered part of the carbon footprint if residues that would normally stay in the fields to enrich the soil are collected (Murray & Price, 2008).

In addition to serving as an offset for non-renewable fuel demand, the use of biomass residues has the added benefit of reducing a cement kiln's nitrogen oxide (NO_x) emissions. Empirical evidence suggests that the reductions in NO_x are due to the fact that most of the nitrogen (N) in biomass is released as ammonia (NH_3) which acts as a reducing agent with NO_x to form nitrogen (N_2) (McIlveen-Wright 2007 as cited by Murray & Price, 2008). Interestingly, there does not seem to be a strong relationship between the N content in the biomass and the subsequent NO_x emissions reductions (McIlveen-Wright 2007 as cited by Murray & Price, 2008).

There is a wide range in the calorific values reported in the literature for agricultural biomass categorically, as well as for individual types. The range in lower heating values[12] (LHV) of agricultural biomass is from 9.2 - 19.4 GJ/dry ton. The quantity of agricultural biomass residues that are necessary to replace one tonne of coal depends on the fuel's energy value and water content. As a rule of thumb, a 20 percent substitution rate of agricultural biomass residues for fossil fuel (on a thermal energy basis[13]) is quite feasible in cement kilns without the need for major capital investment (Seboka et al., 2009; Demirbas 2003 as cited by Murray & Price, 2008).

Major challenges of using agricultural biomass residues include the relatively low calorific value which can cause flame instability, and availability since most of the agricultural residues are seasonal (not available all year round). The flame instability problems could be overcome with lower substitution rates and ability to adjust air flow and flame shape. Collection and storage of residues during the months of availability or alternatively, sourcing different residues at different times of the year could overcome the availability problem. Another challenge is that biomass is prone to change with time, thus care must be taken to use the material before it begins to breakdown. Importantly, new biomass should be rotated into the bottom of storage facilities such that the oldest material is injected into the kiln first. Related to biomass conveyance, the flow behaviour of different materials is quite variable, therefore, cement kiln operators must choose the method for injecting fuel into the kiln that will facilitate a constant and appropriate heat value.

As discussed in section 2.3, the presence of halogens (e.g., chlorine) found in biomass such as wheat straw and rice husks may be a concern for slagging and corrosion in the kiln;

[12] The energy content of fuels can be reported in terms of the lower heating value (LHV) or the higher heating value (HHV), alternatively referred to as net and gross calorific value, respectively. The LHV assumes that the latent heat of vaporization of water in the material is not recovered, whereas the HHV includes the heat of condensation of water.

[13] Biomass can replace up to 20 percent of the total energy demand. Substitution rates on a mass basis are relative to the heat content of the alternative fuel in comparison to coal.

however studies have shown that co-firing biomass with sulphur containing fuels (such as coal) prevents the formation of alkaline and chlorine compounds on the furnaces (Demirbas 2003; McIlveen-Wright 2007 as cited in Murray & Price, 2008). However, ash deposits may decrease heat transfer in the kiln.

Biomass can be used in cement plants through two major modes, namely direct combustion and transformation into producer gas. Direct combustion of biomass in pre-heaters / pre-calciners and in the kiln by part-replacing the fossil fuel used in raising the temperature of the raw meal. This can happen in two ways: first, by mixing crushed and pulverized biomass with coal or petcoke for use in the kiln, and secondly, by direct feeding of biomass in solid lump form (such as pellets and briquettes) into the rotary kiln and / or pre-heater/pre-calciner combustion chamber. The biomass can also be transformed into producer gas (also known as 'synthesis gas' or 'syngas') and co-firing it in the kilns using a gas burner (Seboka et al., 2009).

3. Conclusions

This chapter presents the current fuel alternatives to fuel of fossil origin for cement manufacturing. The chapter introduces different potential alternative fuels that can be used in the cement manufacturing industry and how these fuels are to be considered in order to avoid negative effects on the final product. The type of fuel used in cement production is subject to the constraints imposed by any deleterious effect on cement quality, refractory life and emissions released to the atmosphere. The benefits of using alternative fuels are highlighted, showing that good planning is needed before the alternative fuel to be used is chosen. The chapter has included detailed study of the main alternative fuels used in the cement industry including petcoke, sewage sludge, used tyres, meat and bone meal, and agricultural biomass.

4. References

Al-Akhras, N.M, & Smadi, M.M. (2004). Properties of tyre rubber ash mortar. *Cement & Concrete Composites*, 26: 821 - 826.

Åmand, L.E, & Leckner, B. (2004). Metal emissions from co-combustion of sewage sludge and coal/wood in fluidized bed. *Fuel* 83: 1803 - 1821.

Arai, N, Hasatani, M, & Nakai, Y. (1989). In furnace reduction of NOx during the parallel-flow moving-bed combustion of surplus activated sludge char by an active use of self-evolved ammonia. *Fuel*. 68: 591 - 595.

Atal, A., and Levendis, Y.A. (1995). Comparison of the Combustion Behaviour of Pulverized Waste Tyres and Coal. *Fuel*; Vol. 74, No. 11: 1570 - 1581.

Bhowmick, A.K., Hall, M.M, & Benarey, H.A. (1994). *Rubber Products Manufacturing Technology*. Marcel Dekker, Inc. New York.

Blumenthal, M.H. (1992a). *A report on An Emission Study on the Use of Scrap Tyres in Cement Kilns.* Scrap Tyre Management Council, Washington, DC.

Blumenthal, M.H. (1992b). The Use of Scrap Tyres in the US Cement Industry. *World Cement.*

Boocock, D.G.B., Konar, S.K., Leung, A, & Ly, L.D. (1992a). Fuels and chemicals from sewage sludge: 1. The solvent extraction and composition of a lipid from a raw sewage sludge. *Fuel*. 71: 1283 - 1289.

Boocock, D.G.B., Konar, S.K., Mackay, A, Cheung, P.T.C, & Liu, J.N. (1992b). Fuels and chemicals from sewage sludge: 2. The production alkenes by the pyrolysis of triglycerides over activated alumina. *Fuel*. 71: 1291 - 1297.

Boynton, R. S. (1980). *Chemistry and Technology of Lime and Limestone.* (2nd edition). John Wiley & Sons. Inc., New York.
Bryers, R.W. (1995). Utilisation of petroleum coke and petroleum coke/coal blends as a means of steam raising. *Fuel Processing Technology.* 44: 121 - 141.
Bye, G.C. (1999). *Portland Cement.* (2nd edition). Thomas Telford Publishing, London E14 4JD.
CEMBUREAU. (1997), *Alternative Fuels in Cement Manufacturing: Technical and Environmental Review.* Brussels, The European Cement Association: 24.
CEMBUREAU. (1999), *Environmental Benefits of Using Alternative Fuels in Cement Production.* Brussels, The European Cement Association: 25.
Cement Industry Federation (CIF). (2000). *Applied Energy, 74:* 101-111.
Chen, J.H., Chen, K.S., and Tong, L.Y. (2001). On the Pyrolysis of Scrap Automotive Tires. *Journal of Hazardous Materials;* B84: 43 - 55.
Chinyama, M. P. M, and Lockwood, F. C. (2007). Devolatilisation behaviour of shredded tyre chips in combusting environment. *Journal of the Energy Institute,* Vol. 80, no. 3: 162 - 167.
Cement Industry Federation (CIF). (2002). Australian Cement Industry.
Commandre, J.M, & Salvador, S. (2005). Lack of correlation between the properties of petroleum coke and its behaviour during combustion. *Fuel Processing Technology,* 86, pp. 795 - 808.
Conesa, J.A., Font, R, & Fullana, A. (2005). Kinetic Model for the combustion of tyre wastes. *Chemosphere.* 59: 85 - 90.
Conesa, J.A., Font, R, & Marcilla, A. (1997). Mass Spectrometry Validation of Kinetic Model for the Thermal Decomposition of Tyre Wastes. *Journal of Analytical and Applied Pyrolysis;* 43, pp. 83 - 96.
Desmidt, F, 1987. *World Cement,* 18: 404.
Deydier, E., Guilet, R., Sarda, S, & Sharrock, P. (2005). Physical and chemical characteristics of crude meat and bone meal combustion residue: "waste or raw material?" *Journal of Hazardous Materials* B121: 141 - 148.
Diez, C., Martinez, O., Calvo, L.F, & Cara, J. (2004). Pyrolysis of tyres. Influence of the final temperature of the process on emissions and the calorific value of the products recovered. *Waste Management.* 24: 463 - 469.
Environmental Agency (Bristol) (1998). Tyres in the Environment.
European Commission (EC) (2001). Integarted Pollution Prevention and Control. Reference Document on Best Available Techniques in the Cement and Lime Manufacturing Industries.
F.L Smidth & Co. (2000). Dry process kiln systems, technical brochure.
Freeman, E., Gao, Y., Hurt, R, & Suuberg, E. (1997). Interactions of Carbon-Containing Fly Ash with Commericial Air - Entraining Admixtures for Concrete. *Fuel;* Vol. 76, No. 8: 761 - 765.
Gabbard, W. D, & Gossman, D. (1990). Hazardous waste fuels and the cement kilns - The Incineration Alternative. *ASTM Standardization News, September 1990,* 10th March 2011, http://gcisolutions.com/HWF&CKS.htm
Ghosh, S.N. (1991). *Cement and Concrete Science & Technology.* (Vol. 1, Part 1), ABI Books Private Limited.
Giugliano, M., Cernuschi, S., Ghezzi, U, & Grosso, M. (1999). Experimental Evaluation of Waste Tires Utilization in Cement Kilns. *J. Air & Waste Manage. Assoc.* 49: 1405 -1414.
Giddings, D., Eastwick, C.N., Pickering, S.J, & Simmons, K. (2000). Computational fluid dynamics applied to a cement precalciner. *Proc. Instn. Mech. Engrs. Vol. 214 Part A.*

Giddings, D., Eastwick, C.N., Pickering, S.J, & Simmons, K. (2002). Combustion and aerodynamic behaviour of car tyre chips in a cement works precalciner. *Journal of the Institute of Energy*. 75: 91 - 99.

Goeran, O., Wang, W., Ye, Z., Bjerle, I, & Anderson, A. (2002). Repressing NO_x and N_2O emissions in fluidized bed biomass combustor. *Energy Fuels*. 16 (4: 915 - 919.

Gray, T.A. (1996). *Tyre derived fuel: an environmentally friendly resource*. Texas Natural Resource Seminar.

Gulyurtlu, I., Boavida, D., Abelha, P., Lopes, M.H, & Cabrita, I. (2005). Co-combustion of coal and meat and bonemeal. *Fuel*. 84: 2137 - 2148.

Ha, T., Muralidharan, S., Bae, J., Ha, Y., Lee, H., Park, K.W. & Kim, D. (2005). Effect of Unburnt Carbon on the Corrosion Performance of Fly Ash Cement Mortar. *Construction and Building Materials*; 19: 509–515.

Hall, J.E, & Dalimier, F. (1994). Waste management - sewage sludge: survey of sludge production, treatment, quality and disposal in the EC. EC Reference No: B4-3040/014156/92, Report N: 3646.

Helleur, R., Popovic, N., Ikura, M. & Stanciulescu. (2001). Characterisation and potential applications of pyrolytic char from ablative pyrolysis of used tyres. *Journal of Analytical and Applied Pyrolysis*. 58-59: 813-824.

Hendriks, C.A., Worrell, E., de Jager, D., Block, K., & Riemer, P. (1998). Reduction of Greenhouse Gases from the Cement Industry. IEA GHG Paper presented at GHGT-4 Interlaken, Switzerland.

Hewlett, P.C. (2004). *LEA's Chemistry of Cement and Concrete*. (Fouth Edition). Elsevier Butterworth-Heinemann.

Hochdahl, O. (1986). Fuels and heat economy. *Zement-Kall-Gips*, No. 4: 90 - 96.

Inguanzo, M., Dominguez, A., Menendez, J.A., Blanco, C.G, & Pis, J.J. (2002). On the pyrolysis of sewage sludge: the influence of pyrolysis conditions on solid, liquid and gas fractions. *Journal of Analytical and Applied Pyrolyisis*. 63: 209 - 222.

International Energy Agency (IEA) (2001). Coal Research. *The use of petroleum coke in coal-fired plant.*

Intergovernmental Panel on Climate Change (IPCC) (2006). Guidelines for National Greenhouse Gas Inventories, Vol. 5 Waste.

Kääntee, U., Zevenhoven, R., Backman, R, & Hupa, M. (2004). Cement Manufacturing Using Alternative Fuels and the Advantages of Process Modelling. *Fuel Processing Technology*; 85: 293 - 301.

Kikuchi, R. (2001). Recycling of municipal solid waste for cement production: pilot scale test for transforming incineration ash of solid waste into cement clinker. *Resources, Conservation & Recycling*. 31: 137-147.

Konar, S.K., Boocock, D.G.B., Mao, V, & Liu, J.N. (1994). Fuels and chemicals from sewage sludge: 3. Hydrocarbon liquids from the catalytic pyrolysis of sewage sludge lipids over activated alumina. *Fuel*. 73: 642 - 646.

Manning, R., Cooper, S, and Macfadyen, J. (2003). Petcoke firing in lime recovery kilns becomes option as energy cost rise. *Pulp & Paper*, Vol. 77, No. 12.

Mokrzycki, E, Uliasz-Bochenczyk, A, & Sarna, M. (2003). Use of Alternative Fuels in the Polish Cement Industry. *Applied Energy*, 74: 101 - 111.

Murray, A, & Price, L. (2008). Use of Alternative Fuels in Cement Manufacture: Analysis of Fuel Characteristics and Feasibility for Use in the Chinese Cement Sector, *Ernest Orlando Lawrence Berkeley National Laboratory*, 10 March 2011, http://ies.lbl.gov/iespubs/LBNL-525E.pdf

Murat, M, & Sorrentino, F. (1996). Effects of large additions of Cd, Pb, Cr, Zn to cement raw meal on composition and properties of clinker and the cement. *Cement and Conrete Research*. 26 (3): 377 - 385.

Newman, J, & Choo, B.S. (2003). Advanced Concrete technology; Constituent Materials. Elsevier Butterworth Heinemann.

Nielsen, P.B, & Hundebol, S. (1986). Precalcing with low-volatile solid fuels. *Zement-Kall-Gips*, No. 4: 99 - 100.

Ogada, T, & Werther, J. (1996). Combustion characteristics of wet sludge in a fluidized bed: Release and combustion of the volatiles. *Fuel*, Vol. 75. No. 5: 617 - 626.

Olmo, I.F., Chacon, E, & Irabien, A. (2001). Influence of lead, zinc, iron (III) and chromium (III) oxides on the setting time and strength of Portland cement. *Cement and Concrete Research*. 31: 1213 - 1219.

Pipilikaki, P., Katsioti, M., Papageorgiou, D., Fragoulis, D, & Chaniotakis, E. (2005). Use of tyre derived fuel in clinker burning. *Cement & Concrete Composites*. 27: 843-847.

Rubber Manufacturers Association (RMA). (2003). *U.S Scrap tyre Markets 2002.*

Roy, G.R. (2002). Petcoke combustion characteristics. *World Cement.*

Rumphorst, M.P, & Ringel, H.D. (1994). Pyrolysis of sewage sludge and use of pyrolysis coke. *Journal of Analytical Applied Pyrolysis*. 28: 137- 155.

Saint-Jean, S.M., Jons, E, Lundgaad, N, & Hansen, S. (2005). Chlorellestadite in the preheater system of cement kilns as an indicator of HCl formation. *Cement and Concrete Research*, 35: 431 - 437.

Sänger, M, Werther, J, & Ogada, T. (2001). NO_x and N_2O emission characteristics from fluidized bed combustion of semi-dried municipal sewage sludge. *Fuel*. 80: 167 - 177.

Seboka, Y. Getahun, M.A & Haile-Meskel, Y. (2009). Biomass Energy for Cement Production: Opportunities In Ethiopia. CDM Capacity Development in Eastern and Southern Africa. United Nations Development Programme.

Spliethoff, H, & Hein, K.R.G. (1998). Effects of combustion of biomass on emissions in pulverized fuel furnaces. *Fuel Processing Technology*, 54: 189 - 205.

Scrap Tyre Management Council (STMC) (1992). The Use of Scrap Tyres in Cement Rotary Kilns.

Stolarek, P, & Ledakowicz, S. (2001). Thermal processing of sewage sludge by drying, pyrolysis, gasification and combustion. *Water Science and Technology*, Vol. 44, No. 5: 333-339.

Taylor, H.F.W. (1990). *Cement Chemistry*. Academic Press, London.

Tiggesbäumker, P, & Beckum, K.M. (1986). Investigations on combustion of solid fuels in calciners. *Zemet-Kall-Gips*, No. 4, pp. 104 - 106.

Vesilind, P.A, & Ramsey, T.B. (1996). Effect of drying temperature on the fuel value of wastewater sludge. *Waste Management & Research*. 14, pp. 189-196.

Wenli, D., Dam-Johansen, K, & Ostergaard, K. (1990). Widening the temperature range of thermal DeNOx process. An experimental investigation. *Proceedings of the 23rd International Symposium on Combustion. Pittsburg, PA: The Combustion Institute*, pp. 297 - 303.

Werther, J, & Ogada, T. (1997). Sewage sludge combustion. *Progress in Energy and Combustion Science*, 25: 55 - 116.

Williams, P.T., Cunliffe, A.M, & Brindle, A.J. (2001). Enhanced Pyrolysis Processing of Scrap Tyres. *Journal of the Institute of Energy*; 74: 100 - 112.

Winter, F., Prah, M.E, & Hofbauer, H. (1997). Temperature in a fuel particle burning in a fluidized bed: The effects of drying, devolatilisation and char combustion. *Combustion and Flame*, 108: 302 - 314.

12

Waste Biomass as Alternative Bio-Fuel - Co-Firing versus Direct Combustion

Gavrilă Trif-Tordai and Ioana Ionel
"Politehnica" University of Timisoara
Romania

1. Introduction

Efficient and environmental friendly energy production technologies are strongly requested in the developing countries due to rapid economic growth. Especially for the southeastern countries, biomass resources are abundant. Therefore, the developing countries need the utilization technologies of biomass to produce energy. This study contributes to constructing ecological energy production systems in the developing countries due to enhanced possibilities of biomass utilization. Use of biomass, which is considered to produce no net CO_2 emissions in its life cycle, can reduce the effective CO_2 emissions of a coal-fired power generation system, when co-fired with coal, but may also reduce system efficiency. In this context of fossil fuels depletion and the desire to reduce greenhouse gases emissions (GHG), the use of biomass in energy purposes needs more attention, in respect to practical applications.

As it is known, the term biomass means total amount of living matter that is included in all living creatures on our planet: plants, animals, and micro and macro organisms. As origin, biomass is a storage form for solar energy.

In general, two main originated groups of biomass sources exist: vegetal and animal. Vegetal biomass often called phytomass, holds the lead in capturing solar energy through the process of photosynthesis and provides almost exclusively the energy required for all other forms of life from planet, including animal biomass.

Obviously phytomass is used mainly to feed animals and people; another important part is raw material for industry and only scrap (waste) from these activities is capitalized in energy purpose. The main possible sources of biomass, convertible into fuel, are listed in Table 1.

Co-firing is regarded as a great opportunity for replacing coal used for power generation with renewable fuels with low costs and a direct repercussion in the decrease of greenhouse gas emissions. During the last decades researches has provided very diverse solutions for co-firing biomass in coal power stations with a limited impact in efficiency, operation and lifespan (Ionel et all., 2009).

In fact, the difficulties in harvesting (collecting), transport, primary processing and eventual phytomass storage before conversion to energy - are often some of the hardest problems to solve because of the large territorial dispersion and quite low energy density (Ionel et all., 2007). The data presented in Table 2 shows that compared with fuel oil, the amount of biomass which through combustion releases the same amount of heat is about 10-17 times higher for sawdust and about 18-23 times higher for straw bales.

Wastes	Residues
1. Ranch: - manure. 2. Food industry: - sugar factory (grain, molasses). 3. Milk industry. 4. Wood industry: - sawdust, wood chips, small pieces of wood - wood from demolished buildings.	1. From agricultural and vegetable production: - straw (straw cereals), - cobs and corn ear, - spindle (potatoes, tomatoes, pepper), - sugar beet. 2. From forestry: - bark, kindling, rind. 3. From wine and fruit growing docks: sprout, branch.

Table 1. Possible sources of biomass convertible into fuel

Phytomass	Net calorific value H_i [MJ/kg]	Density σ [kg/m³]	Energy density σ_E [MJ/m³]	Energy density relative to fuel oil $(\sigma_E)_R$
Fuel oil	39.39	854	33.64	1.00
Pit coal	20.05	830	16.64	2.02
Brown coal	9.85	720	7.09	4.74
Small pieces of wood	18.5-18.8	297-380	5.55-7.05	4.77-6.06
Sawdust	17.8-18.0	110-170	1.96-3.06	10.99-17.16
Wood chips	18.2-18.6	90-153	1.64-2.85	11.80-20.51
Bark	18.0-19.2	205-320	3.69-6.10	5.51-9.12
Straw bale	17.1-17.5	85-104	1.45-1.82	18.48-23.20
Hemp stalk	16.9-17.1	140-170	2.37-2.91	11.56-14.19
Cereal grains	16.9-17.0	730-760	12.34-12.92	2.60-2.73
Wood pellets	18.0	650	11.7	2.875

Table 2. Energy characteristics for biomass

Romania possesses a large potential of biomass, and is expected to make considerable efforts to make benefit of the European knowledge and dissemination of recent technologies in order to use the potential energy accumulated by the biomass. As Figure 1 indicates, in addition to existing large renewable energy sources (RES) assured especially by the hydraulic energy system, that is quite developed, the next future necessitates development in areas such as biomass, wind and solar energy exploitation (Hansson et all., 2009).

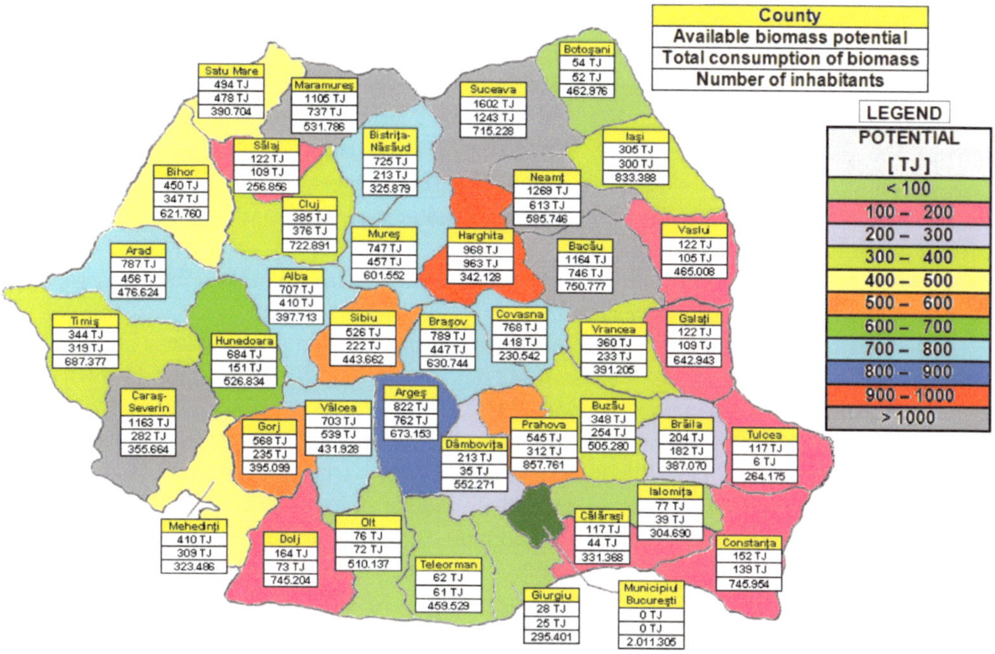

Fig. 1. Biomass distribution in Romania (Ţânţăreanu, 2009)

2. Current techniques for energy conversion

In Table 3 are mentioned the main energy conversion techniques for the biomass. Currently, the world's attention is directed particularly to the possibilities of producing alcohol, vegetable oil and biogas, in a word, bio-fuels used mainly in internal combustion engines (Ghergheleş, 2007).

Process	Product
a) Thermal	
- direct combustion	Thermal energy, ash and flue gases (CO_2, H_2O, N_2)
- pyrolysis	Charcoal, oil, fuel gases (CO, CH_4, H_2)
- gasification	Fuel gases (CO, H_2) and inert gases (N_2, CO_2)
- gasification+catalytic synthesis	Methanol, hydrocarbons
b) Biological	
- fermentation with yeasts	Ethanol, CO_2
- anaerobic decomposition	CH_4, CO_2
c) Mechanical	
- extraction by pressing	Oil
d) Chemical	
- extraction with solvents	Oil

Table 3. Techniques for converting biomass

Direct combustion of biomass for heating, hot water production and steam generation is performed in specific facilities, provided with stoves, hot water boilers or steam generators.

Conversion efficiency is around 30 % for stoves, 65 % for hot water boilers and can increase to 90 % for modern facilities with intensified combustion in circulating fluidized bed.

Pyrolysis is a complex process of destructive distillation of phytomass without air, which occurs at temperatures between 200 and 1100 °C. Depending on the used technique, in general, the resulted products are: 35 % charcoal, 35 % liquid distillate and 30 % fuel gases (CO, CH_4 and H_2). Charcoal is used as activated carbon for gas purification, liquid distillate can be used directly to produce fuel or raw chemical materials; combustible gases can be burned in the furnace of power plants.

Gasification allows the thermal conversion of biomass in a combustible gas mixture. Mainly, two gasification techniques are known:

- **Gasification in air flow**, which leads to the formation of a gas mixture (CO, H_2, CH_4, N_2, CO_2), with high nitrogen content (42-45 %) and lower energy value (4-8 MJ/m^3_N) used as fuel in small combustion plants;
- **Gasification in oxygen flow**, which leads to the formation of synthesis gas with a much higher energy value (10-20 MJ/m^3_N). This can be burned to produce heat used for different purposes or can be used to produce methanol for internal combustion engines.

Direct liquefaction is used to convert biomass into liquid bio-fuels through catalytic hydrogenation of wood. The result is a mixture of liquids, used as fuel or as raw material for chemisation. Direct liquefaction can be applied to cellulosic or lingo-cellulosic wastes. Carbon monoxide and water vapor are used for hydrogenation of biomass in the presence of a catalyst (sodium carbonate) resulting oils and gases with high methane content.

Anaerobic decomposition (fermentation) is a process for conversion of various organic substances under the action of enzymes produced by microorganisms in the absence of oxygen in the air. Anaerobic fermentation processes take place in three stages - liquefaction, formation of acids and methane production - as a result of simultaneous action of three communities of microorganisms that appear and develop during those stages, resulting in biogas and sludge formation. Sludge can be used as organic fertilizer for plants.

Specific biogas production and composition vary according to the nature and composition of raw materials, decomposition degree of organic substances, fermentation temperature, fermentation period of organic material and other factors. The biogas produced in this way usually contains 50-65 % CH_4, 25-45 % CO_2 and in very small proportion CO, H_2, N_2, O_2 and H_2S. The biogas with 60 % CH_4 and 40 % CO_2 has the net calorific value H_i = 21.5 MJ/m^3_N and ignites at a temperature of 600-750 °C if mixed with air within 6-12 %.

Remarkable progresses in the use of waste biomass for biogas production have been reported in many countries (India, China, USA etc.). In Romania, concerns in this direction date over 30 years, but without a coordinated approach in results implementation from scientific researches.

3. Co-firing pilot in bubbling fluidized bed

Co-firing of biomass and coal has been demonstrated successfully in several coal-fired boilers in Europe and the United States. The results have been promising - boiler efficiencies have not suffered considerably. However, when converting existing coal-fired boiler for biomass/coal co-firing, the save share of biomass depends on biomass and coal properties as well as boiler design, and must be determined case by case.

A fluidized bed is a bed of solid particles suspended or fluidized by forcing air through the bed. The high heat transfer and mixing encourage complete combustion. When the air velocity is increased above the minimum fluidization velocity, air flows through the bed as bubbles. This type of bed is called bubbling fluidized bed (BFB). When the air velocity is increased, the particles are carried higher up in the reactor. With a circulating fluidized bed (CFB) it is no longer possible to distinguish between the bed and freeboard area. A large fraction of the particles rise up from the bed and are circulated with the help of a cyclone back to the bed. The circulating bed material can be used for temperature control in the boiler. Fluidization velocity of a bubbling fluidized bed boiler is typically 1-3.5 m/s, whereas in a circulating fluidized bed boiler it is 3-6 m/s (European Commission, 2003).

Fluidized bed boilers can be designed to burn almost any solid, semi-solid or liquid fuel as long as the calorific value is sufficient to heat the fuel, drive off the moisture and preheat the combustion air. They achieve high fuel-to-steam efficiency, typically over 90 %, even with challenging, low-grade fuels. Because of the high amount of hot circulating bed material, it is possible to burn moist, heterogeneous fuels with low calorific value. With high moisture content fuels, a support fuel can be used. More than 90 % of the bed is sand or ash and the rest is fuel.

At the moment, fluidized bed combustors are the best combustion systems thanks to their ability to burn a wide assortment of fuels and still keep emissions low. The choice between BFB and CFB technology has been largely linked to the choice of fuels. As a simpler and cheaper technology, BFB has been favoured in plants fuelled exclusively with biomass or similar low-grade fuels containing highly volatile substances. The new enhanced CFB designs can be a competitive alternative even in smaller biomass fired plants. Fluidized bed combustion systems are suitable for large-scale applications exceeding 30 MW_{th} in size (European Commission, 2005).

The co-firing facility, presented in Figure 2, comprises several main parts and is based on original design (Romanian patent no. 121362, 2007). The main characteristics of the co-firing facility are presented in Table 4.

The bubbling bed combustion furnace has a parallelepiped shape (1 x 0.5 x 1 m) and it's made from 3 mm heat-resisting steel with truncated pyramid base. At the bottom is mounted the air distributor, divided in compartments for injection of the fluidization air and main combustion air through 37 nozzles, and an appropriate air feeding system including a Cole-Parmer regenerative blower and all necessary adaptors. The furnace is equipped with two fuel bunkers each having a helical conveyer coupled to variable speed motors which allow flow adjustments. For start up, on one side of the furnace is connected a Riello gas burner working with natural gas or biogas. Exterior insulation applied to furnace, ash cooler and convective case is made with several layers of ceramic fiber and one layer of mineral wool. For safety, the furnace is fitted with copper explosion valves.

The heat transfer subassembly components are mainly formed by the convective case and one heat exchanger. The convective case has a parallelepiped shape and it's made from W4841 heat-resisting steel. The case is split in two by a stainless steel vertical wall, attached to the top. The heat exchanger is made from 8 cold-drawn steel pipes, 4 in each side, connected to inlet and outlet manifolds. The ash is collected at the bottom in a pyramidal bunker.

The flue gases de-dusting system components are formed by a cyclone dust separator-cylindrical body mounted in vertical position with a tangential upside connection, flow measuring sockets, powder/flue gas sampling, thermocouples and manometers.

The flue gases cleaning subassembly is formed by a scrubbing tower, neutralization reactor and the demister. The scrubbing tower and the neutralization reactor for gaseous pollutants have a cylindrical body mounted in vertical position with a cone shaped basis. The parallelepiped shaped demister is equipped with vertical baffles for retention of excessive humidity in cleaned flue gases. Both scrubber and reactor are equipped with the necessary equipments for hydraulic circuit of washing liquids and a 100 mm layer of Rasching rings to increase the residence time and to provide a large surface area for interaction between washing liquid and flue gases.

In order to control the processes, the experimental facility is equipped with measuring instruments, devices for temperatures, pressures, water and gas flows and visors. All facility components are mounted on a steel frame with overall dimensions: L = 6 m, W = 1.3 m and H = 2.6 m.

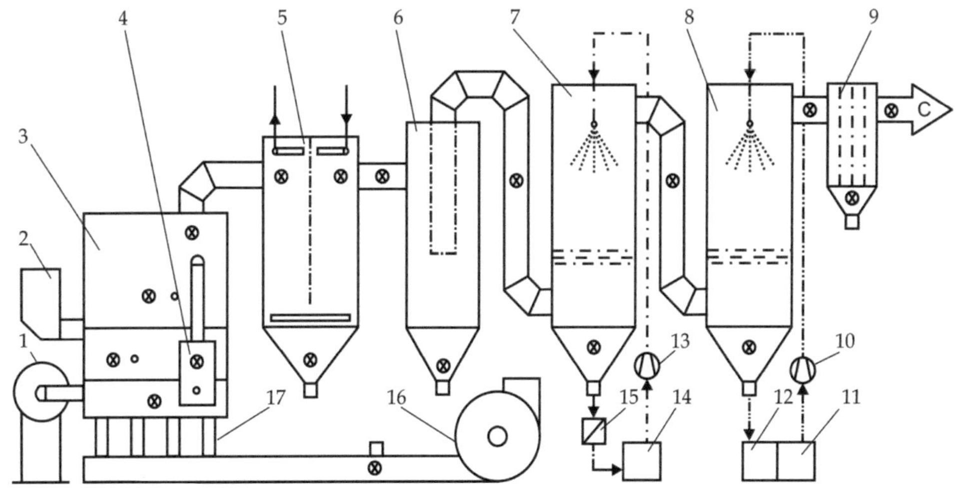

1-Start-up burner, 2-Fuel bunkers, 3-Bubling fluidized bed furnace (BFB), 4-Ash cooler, 5-Convective case, 6-Dust separator-cyclone, 7-Scrubbing tower, 8-Neutralization reactor, 9-Demister, 10, 13-Reagents circulation pumps, 11, 12, 14-Containers, 15-Filter, 16-Air feeding system, 17-Air distributor, C-Chimney

Fig. 2. Design of the co-firing facility in fluidized bed (Trif-Tordai, 2008)

Characteristics	Value
Thermal energy output	45 - 90 kW_{th}
Electrical power consumption	2 - 4 kWh_{el}
Water flow (in heat transfer system)	2 - 4 m^3/h
Combustion / fluidization air flow	max. 270 m^3/h
Compressed air flow (for washing pumps)	0.5 - 1 m^3/h
Coal mass flow	25 - 50 kg/h
Biomass mass flow	15 - 30 kg/h
Washing liquid flow	0.2 - 0.6 m^3/h
Resulted ash mass flow	10 - 20 kg/h

Table 4. Main characteristics of the co-firing facility

4. Co-firing fuels and potential emissions

According to the promising biomass potential available for heat and electricity production, based in Romania on the stock of wood and agriculture, researches have been developed, in order to utilize different qualities of available biomass as second fuel in the co-firing process, characterized by low price compared to coal, for a comparative energy offer (Romanian Ministry of Environment and Forests, 2007).

Figure 3 presents four different fuels used in co-firing facility: waste biomass (corncob "Cc", respectively sawdust "Sd") and coal (pit coal "Pc" from the Jiu Valley basin and lignite "L" from Oltenia basin).

After drying and milling, the secondary fuels were mixed externally with the basic fuel, before being fed to the combustion system. All other methods - as separate feeding - determined lack of stability and non-homogenous temperature levels, and the risks of thermal strength were raised, and were abandoned.

Table 5 gives the elementary analysis for in use coal and biomass qualities. Notable is the S content, as well the lower humidity of the pit coal in comparison to the used biomass. Also the N content in biomass is sensible reduced in comparison to that of the pit coal.

Fig. 3. Types of coal and biomass used as fuels for co-firing facility

Characteristics	Symbol	IS unit	Pit coal "Pc"	Lignite "L"	Sawdust "Sd"	Corncob "Cc"
Carbon	C^i	%	58.84	23.48	35.97	43.62
Hydrogen	H^i	%	2.24	2.24	4.60	4.64
Oxygen	O^i	%	10.64	11.35	28.96	21.11
Nitrogen	N^i	%	2.26	0.59	0.35	0.44
Sulfur	S^i	%	1.80	0.85	0.01	0.01
Humidity	W_t^i	%	8.00	43.29	30.00	29.87
Ash	A^i	%	16.22	18.20	0.12	0.31
NHV	H_i^i	kJ/kg	21089	8035	13023	16516

Table 5. Elementary proximate analysis for the used fuels, in reference to humid state

Figure 4 presents eight different mixtures, between waste biomass and coal, used as fuels in co-firing facility. The visual analysis indicates disperse homogeneous, quite raw aspect and various grinding.

a) 15 % by mass corncob with 85 % pit coal mixture

b) 15 % by mass sawdust with 85 % pit coal mixture

c) 15 % by mass corncob with 85 % lignite mixture

d) 15 % by mass sawdust with 85 % lignite mixture

e) 30 % by mass corncob with 70 % pit coal mixture

f) 30 % by mass sawdust with 70 % pit coal mixture

g) 30 % by mass corncob with 70 % lignite mixture

h) 30 % by mass sawdust with 70 % lignite mixture

Fig. 4. Eight different mixtures, between biomass and coal, used as fuels in co-firing facility

Based on the chemical composition of biomass compared to other solid fuels in a Van Krevelen diagram, the biomass fuels are high in the O/C - and H/C - ratios compared to peat and coals (Figure 5). These high ratios are responsible for the biomass fuels being more volatile than coals and peat, fact that determines a better stability in the ignition process. The O/C - ratios is responsible for the fuels lower heating values.

Depending on the fuel composition, the design of the combustion chamber and the operation technology of the system, biomass combustion can lead to emissions of CO, HC, (VOC, UHC), PAH, tar, soot, particles, NOx, N_2O, HCl, SO_2, salts, PCDD/F and heavy metals (Pb, Zn, Cd and others).
Two main groups of pollutants from the combustion of the fuel combination (coal with waste biomass) are expected: unburnt pollutants such as CO, HC, PAH and soot, and oxidized pollutants such as NOx, SO_2 and CO_2.

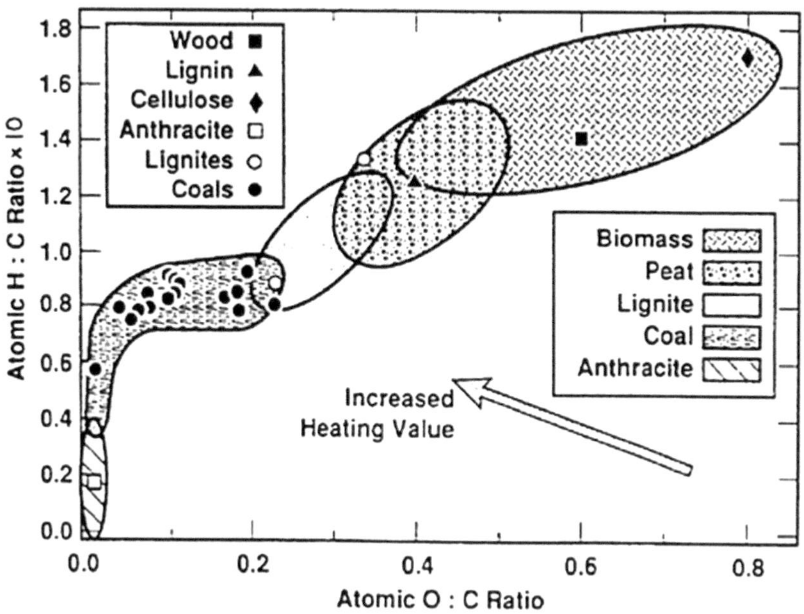

Fig. 5. Van Krevelen diagram for various solid fuels (Van Loo & Koppejan, 2007)

Further, additional pollutants can be emitted if biomass contains Cl, metals etc.
The emissions from biomass combustion are distinguished as:

- Emissions which are mainly influenced by the quality of the combustion process and operation of the convective system (unburnt pollutants which can be avoided by complete combustion: CO, HC, PAH etc.).
- Emissions which mainly originate from the fuel properties (emissions which are formed from elements found in the waste and are not to the same extend dependent on the combustion process: NOx from N, HCl from Cl etc.).
- Nitrogen compounds include NO (nitrogen oxide), NO_2 (nitrogen dioxide) usually summed up as NOx (nitrogen oxides) and N_2O (nitrous oxide). While NO, formation and emissions have been widely investigated for many years, emissions of N_2O have been in focus in the last years, due to its contribution to the greenhouse effect. Formation of NOx in combustion systems involves three main paths:
- Formation of thermal NOx, which requires sufficiently high temperatures for dissociation of the atmospheric diatomic species N_2 and O_2;
- Formation of fuel NOx, originating in the fuel bound N;
- Formation of prompt NOx, involving fuel-bound hydrocarbon radicals and atmospheric N_2, forming HCN as the most important intermediate species.

Nitrogen oxides emissions from small-scale waste biomass regular and fluidized bed co-firing originate mainly from the fuel bound nitrogen, thermal NOx emissions are less important.
However, some discussions on the role of thermal NOx, as well as prompt NOx, have been noticed in the literature regarding the contribution in fluidized bed systems, due to incomplete mixing giving possibilities of fuel rich zones and high temperature zones. For large-scale suspension combustion and co-firing, all mechanisms must be considered.
Co-firing was the selected technology of the present research, and one measured experimental emissions of NO, NO_2, SO_2 and CO from combustion of mixtures of coal and biomass in bubbling fluidized bed, in order to complete the information with peculiar aspects regarding Romanian bio-fuels and fuel.
At the resource part of biomass supply chain, the following area will have key importance in relation to increase the availability of bio-energy resources (Center for Promotion of Clean and Efficient Energy in Romania [ENERO], 2009):

- Increased crop yields, plant breeding, introduction of new energy crops - this trend is a result of improved land management, improvement of conventional and introduction of new plant breeding techniques, selection of improved varieties and hybrids, increasingly mechanized irrigation, cultivation and harvesting and improved post-harvest handling and storage methods, more intensive inputs of fertilizers and agro-chemicals. Future land availability for energy crops will depend on the rate of further crop yield improvements. Nutrient recycling and the success and acceptance of genetically modified species will also have an impact on land use and crop yield, as will future water availability, its use and the impacts of climate change.
- Improved harvesting, logistics and pre-treatment of biomass including biochemical conversion process - reduction of supply chain costs could significantly improve economy of bio-energy projects. Larger plants can benefit from economies of scale, but this advantage could be offset by increased transport distances needed to obtain the required volume of biomass. Most often, both small and large scale biomass supply solutions need new types of machinery, new logistics and a new management approach in order to be economical. The improved pre-treatment processes include physical, chemical and biological (enzymatic and other) pre-treatment and conversion processes, biomass fractionation and separation technologies, utilization of residual solids and liquids etc.

5. Experimental results

The co-firing tests have been achieved at a ratio of 15-30 % by mass of biomass, the rest being coal.
The temperatures and pressures have been recorded during tests with a data acquisition system, on line, in several important points. All values were in the range of expected relevance: in the furnace 800-1200 °C, in the convective part 300-1200 °C, in the cyclone 150-300 °C, in the scrubber 90-150 °C, and in the neutralization reactor 70-90 °C. Figure 6 and 7 shows the temperature variation in furnace and along flue gases path when co-firing 15 % by mass corncob with 85 % pit coal and respectively for co-firing 30 % by mass corncob with 85 % pit coal.
Flue gas composition was measured with Testo 350XL gas analyzers mounted after the cyclone; combined measurement uncertainty for Testo 350XL gas analyzer is 1.19 %

according to SR EN 50379-1:2004 standard. For dust concentration a Strohlein STE4 device was mounted before cyclone; combined measurement uncertainty for Strohlein STE4 particulate sampling train is 0.03 % according to SR EN 9096:2003 standard.

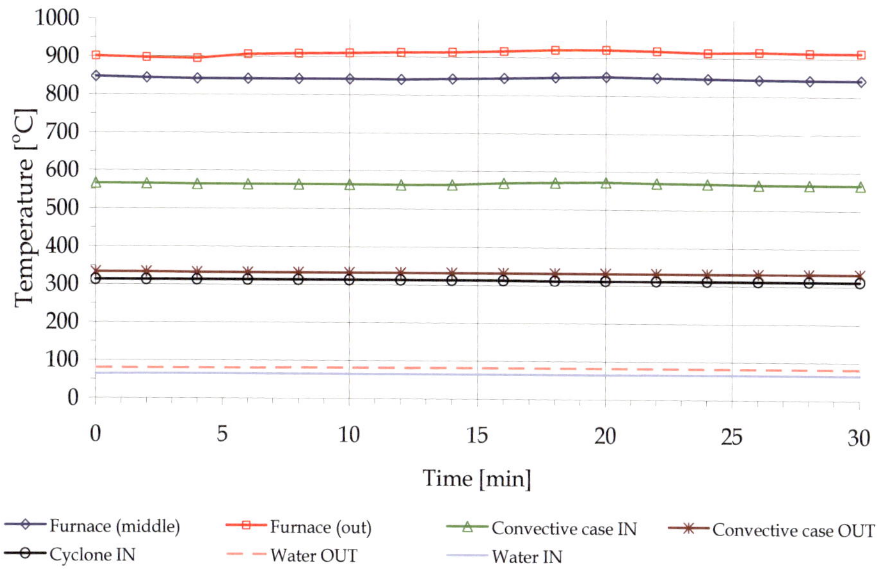

Fig. 6. Temperature variation in furnace and along flue gases path when co-firing 15 % corncob with 85 % pit coal mixture (% by mass)

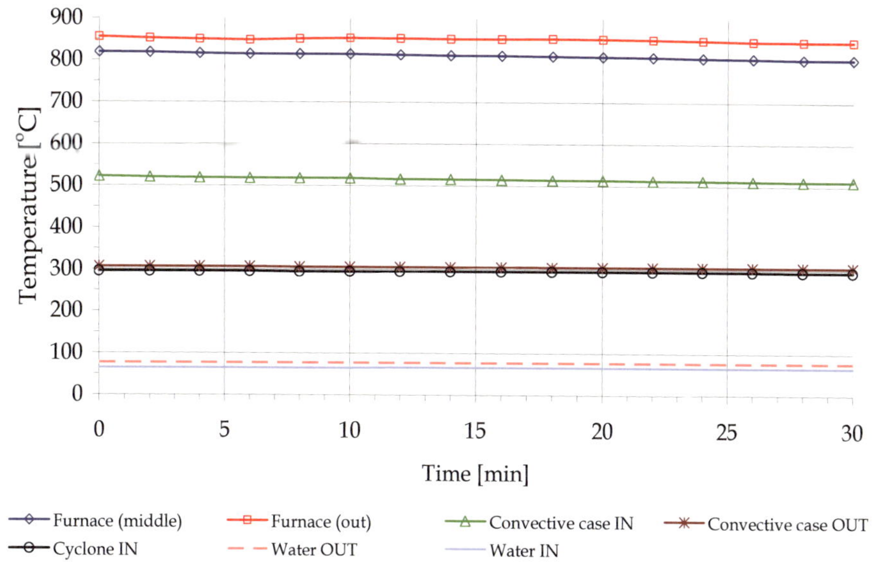

Fig. 7. Temperature variation in furnace and along flue gases path when co-firing 30 % corncob with 70 % pit coal mixture (% by mass)

Main results representing average values obtained after achieving a steady state, in several points along the flue gases lay out, are given in Figures 8 - 12. The pit coal (Pc) and lignite (L) co-firing ratio with sawdust (Sd), corncob (Cc) are indicated. For basic comparison, the experiment with no biomass addition was used. Thus, as reference value, one considered the concentration of a species in stack without the biomass mixture. All figures were experimentally determined after the application of the proposed co-firing process.
The results regarding NO_x emissions from co-firing are comparable to those resulting from burning coal alone, as unique fuel. Nitrogen content of biomass is lower comparative to coal, fact that supposes to reduce the formation of NO_x. Nevertheless, the formation of thermal or proximate NO_x is directly related to the operation techniques, as well, mainly the range of temperature levels covered and oxygen content in reaction zones. Thus one may conclude, that the N from the biomass and also the thermal mechanism of the NO_x formation are not activated as expected, due to the fluidized system combustion that limits the temperature levels, and influences the residence time and the oxygen content, as well.
The higher the biomass support, the less SO_2 concentration in the flue gases is resulting. The explanation consists of the reduced S content of the used biomass sorts. The achieved desulphurization efficiency accomplished only by the biomass addition (sawdust and respectively corncob), is between 15 and 31 %, compared to the reference, when no biomass was added.

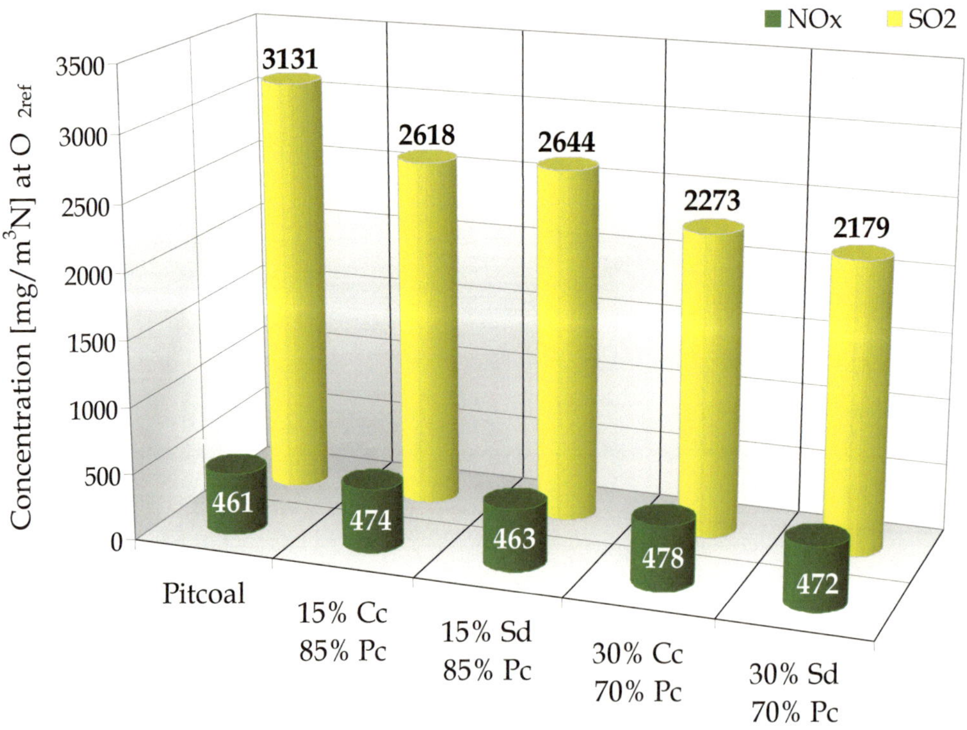

Fig. 8. Average concentration of SO_2 and NO_x in the case of pit coal - biomass co-firing, at different mass participation of biomass in the mixture (% by mass); combined measurement uncertainty is 1.19 %

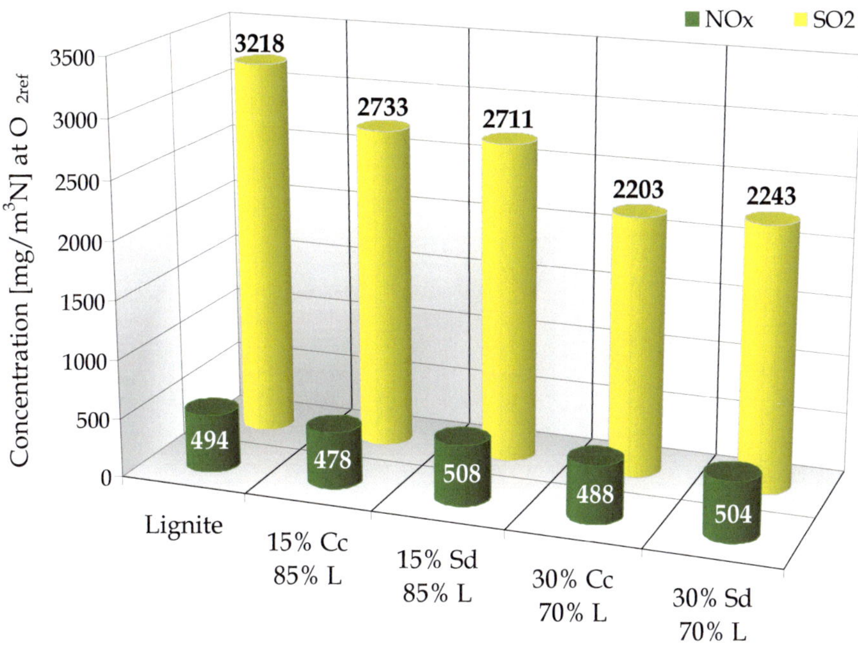

Fig. 9. Average concentration of SO_2 and NO_x in the case of lignite - biomass co-firing, at different mass participation of biomass in the mixture (% by mass); combined measurement uncertainty is 1.19 %

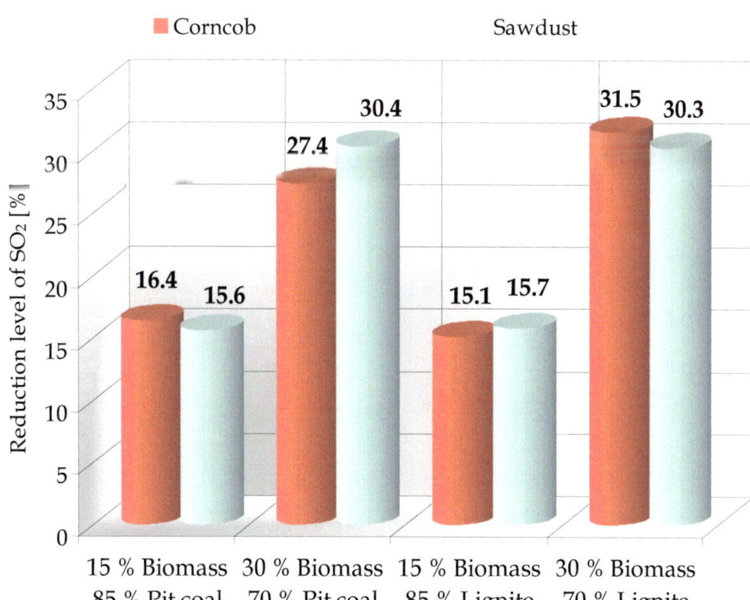

Fig. 10. Desulphurization rate resulted from tests with biomass - coal co-firing (participation of biomass in the mixture is by mass)

Analyzing the particle concentration in the exhaust flue gases, one notes that the co-firing determines a reduction of the particles amount, explicable by the better combustion conditions, due to the higher volatile content of the biomass, which supports the stability of the ignition and combustion process. With the increasing of biomass mixture ratio, the particles in the flue gases are reduced.

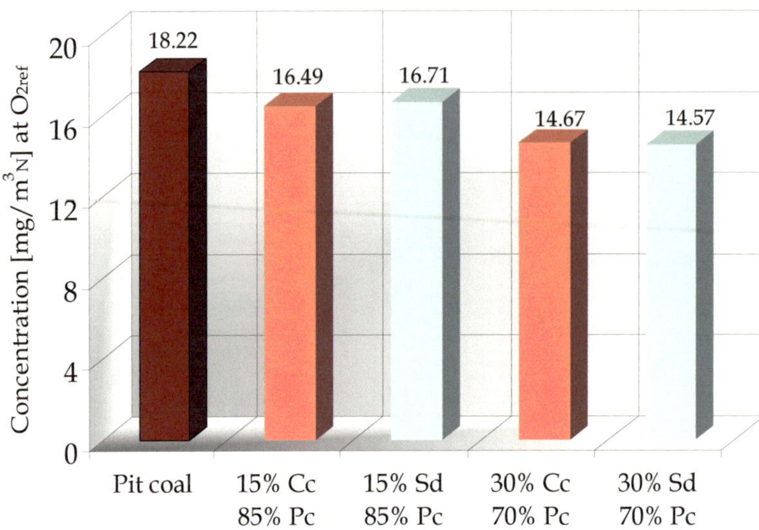

Fig. 11. Mass concentration of dust in the flue gases resulted from co-firing of corncob, respectively sawdust, with pit coal (% by mass); combined measurement uncertainty is 0.03 %

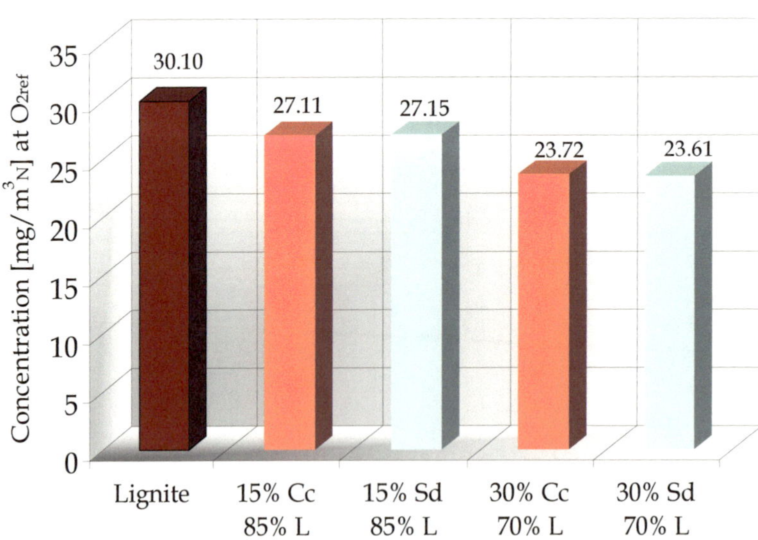

Fig. 12. Mass concentration of dust in the flue gases resulted from co-firing of corncob, respectively sawdust, with lignite (% by mass); combined measurement uncertainty is 0.03 %

Figures 13-20 review the stability of the process, as the measured values are quite constant and the variations are negligible, versus a mean value. Increasing the share of biomass (from 15 to 30 % by mass) was found to lead to lower concentrations of SO_2, NO and NO_2 at the bed exit except CO. This is considered to be due to introduction of higher volatile matter with increasing biomass share.

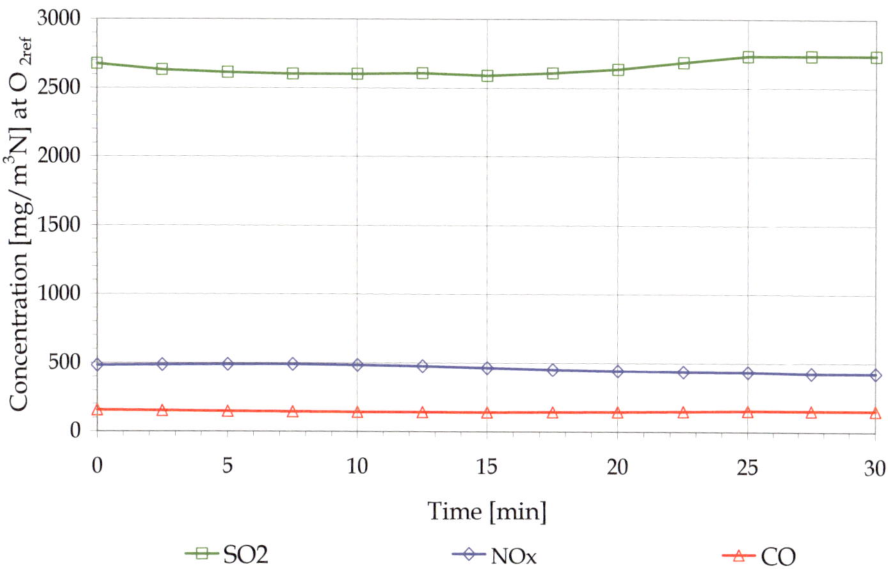

Fig. 13. Time dependence of SO_2, NO_x and CO concentration when co-firing 15 % by mass sawdust with pit 85 % coal mixture

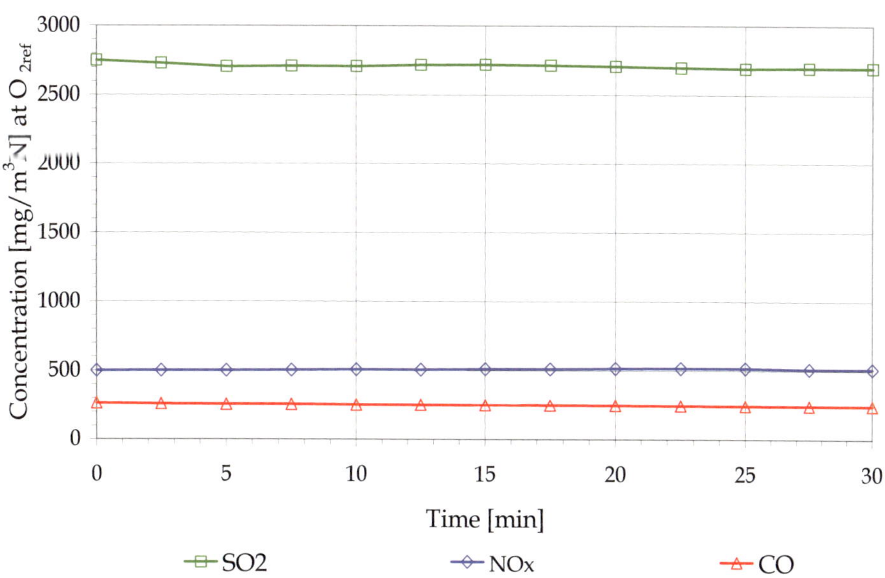

Fig. 14. Time dependence of SO_2, NOx and CO concentration when co-firing 15 % by mass sawdust with 85 % lignite mixture

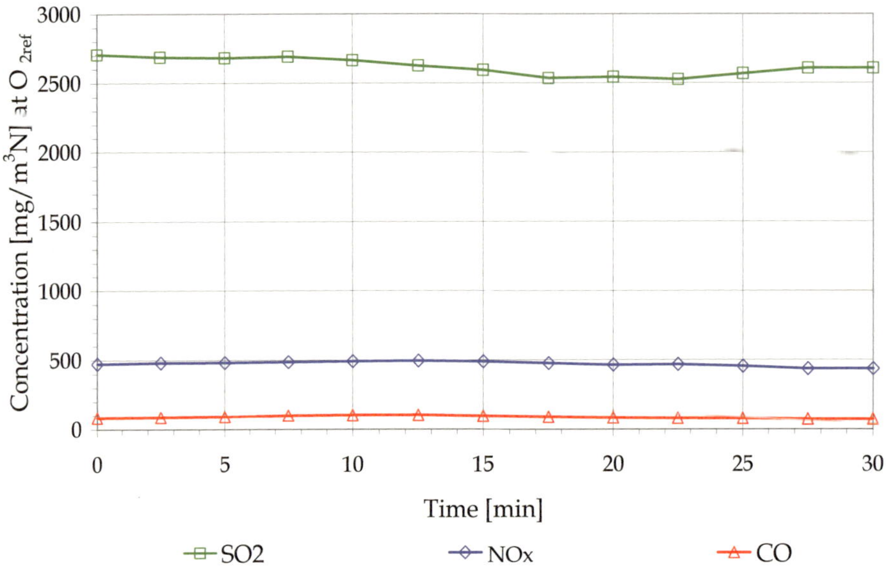

Fig. 15. Time dependence of SO_2, NO_x and CO concentration when co-firing 15 % by mass corncob with 85 % pit coal mixture

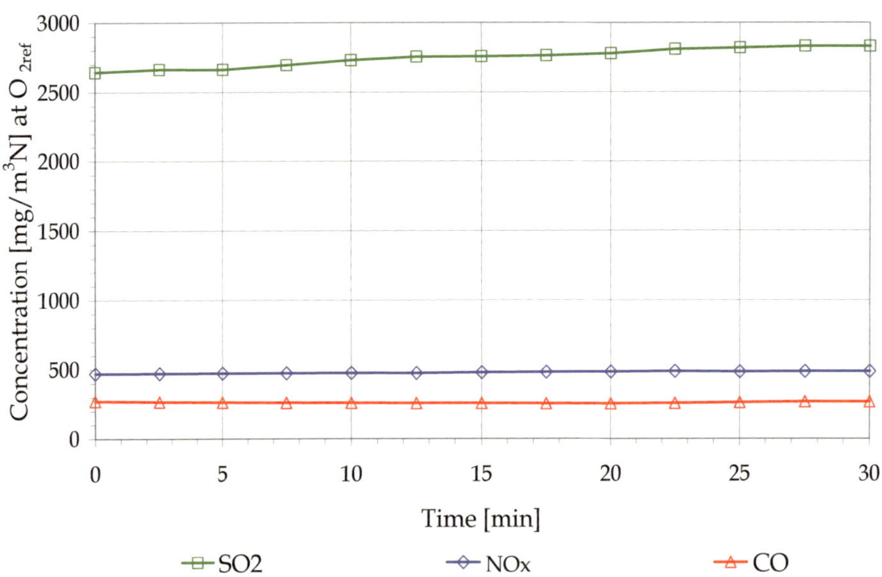

Fig. 16. Time dependence of the SO_2, NO_x and CO concentration when co-firing 15 % by mass corncob with 85 % lignite mixture

The measured NO_x, SO_2 and CO concentrations at stack are considered reasonable constant. Concerning the unburned C (CO), the amounts are considerable (between 100 and 350 mg/m^3_N at O_2 reference) and thus, the efficiency due to unburnt matter of the global co-firing process is reduced, in comparison to the basic case (only coal). This phenomenon suggests that the biomass addition can enhance the ignition of coal, since volatile matter in

biomass is easily evolved even at relatively low temperature, and because lack of oxygen, the CO is generated more intensively as if no biomass is added.
In general, when the volatile matter content is low the reactivity is low too and the combustion of this fuel is more difficult. The coal has, in some cases, a lower reactivity or comparative, in other, even higher than the biomass species. The presence of unburned particles in the ash is higher or less, resulting, in majority of cases, a decrease of the combustion efficiency. The high reactivity of the biomass species compared to that of coal char, results in a rapid burnout of the biomass particles in case such particles would have survived the passage through the combustion chamber to burn in the cyclone together with the combustible gases.
If the share of biomass is less than 5 % by mass, it was been demonstrated that there are no relevant CO increased values in the flue gases (Van Loo & Koppejan, 2007).
The advantage of the co-firing is for sure the benefit concerning the CO_2 emission, knowing that the CO_2 exhaust from fossil fuel origin is reduced by the same amount as the biomass ratio. Despite the fact that normally not the total CO_2 resulted from the biomass share might be considered green, neutral, as according to the lack of relevant differences - in terms of experimental results - concerning the two used biomass qualities and ratio attest the fact that fuel flexibility is possible. Varying qualities and quantities of fuels can be partially compensated by adjusting the co-firing rate.

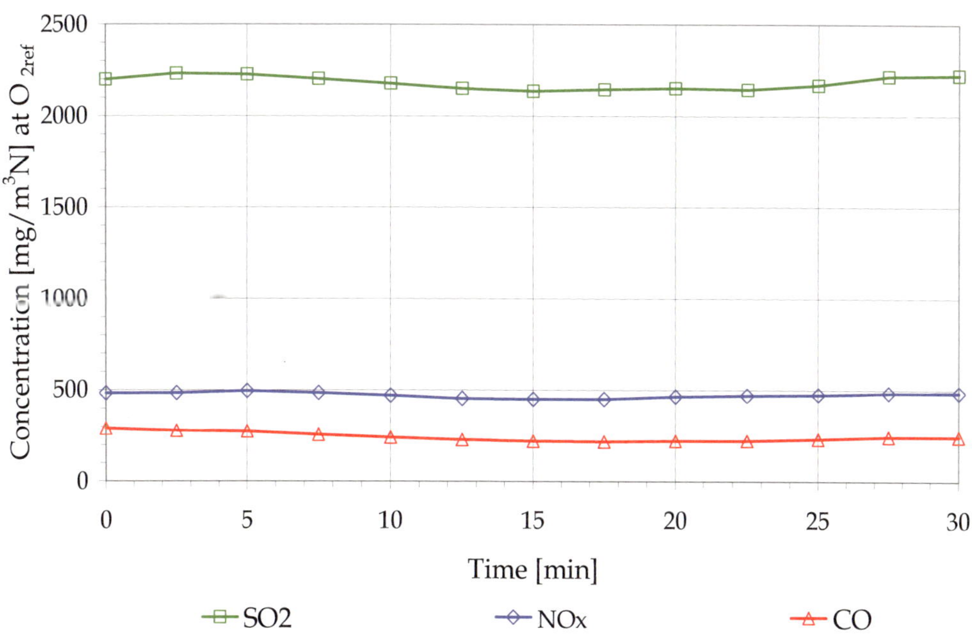

Fig. 17. Time dependence of SO_2, NO_x and CO concentration when co-firing 30 % by mass sawdust with 70 % pit coal mixture

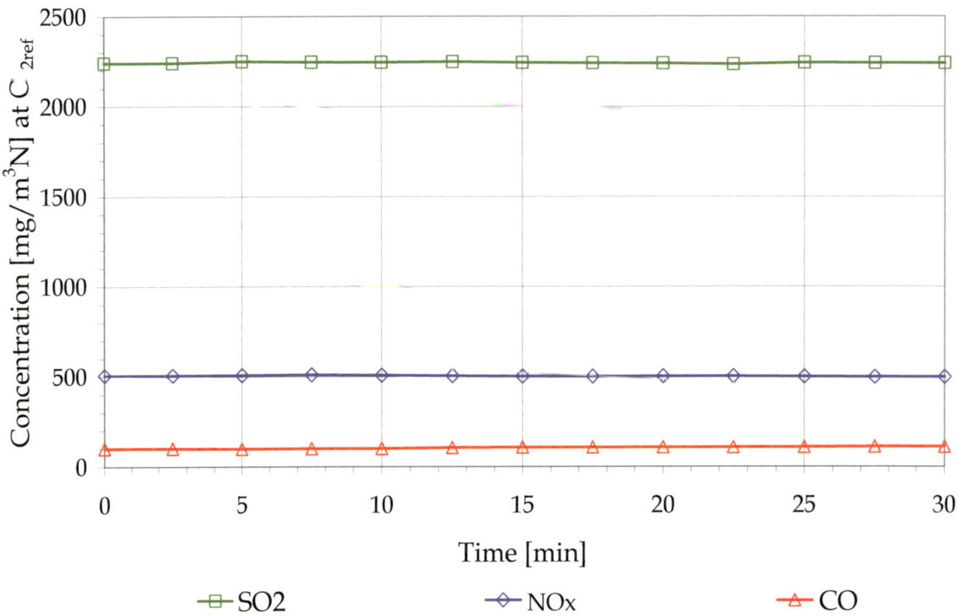

Fig. 18. Time dependence of SO_2, NO_x and CO concentration when co-firing 30 % by mass sawdust with 70 % lignite mixture

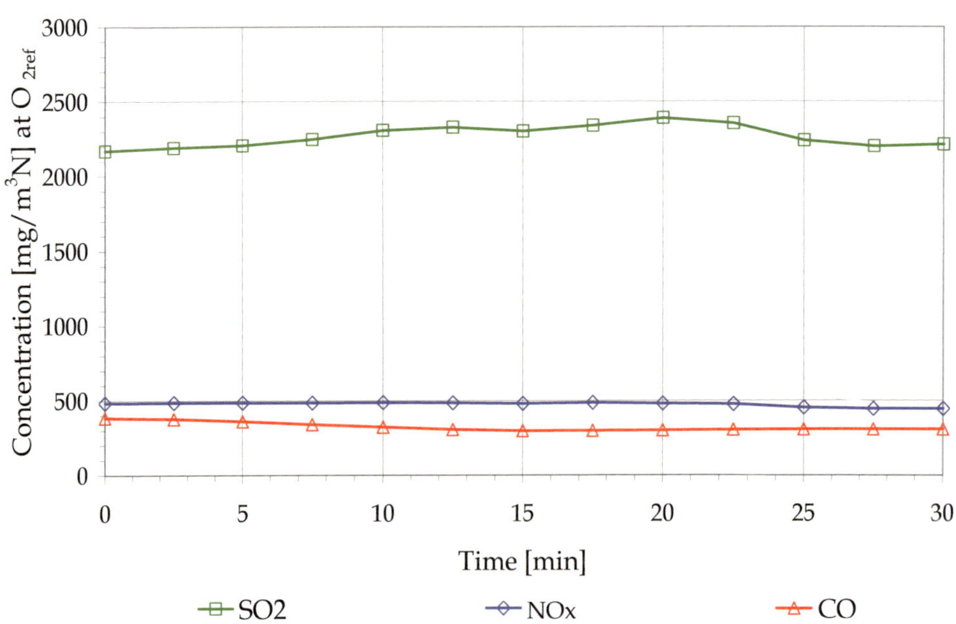

Fig. 19. Time dependence of the SO_2, NOx and CO concentration when co-firing 30 % by mass corncob with 70 % pit coal mixture

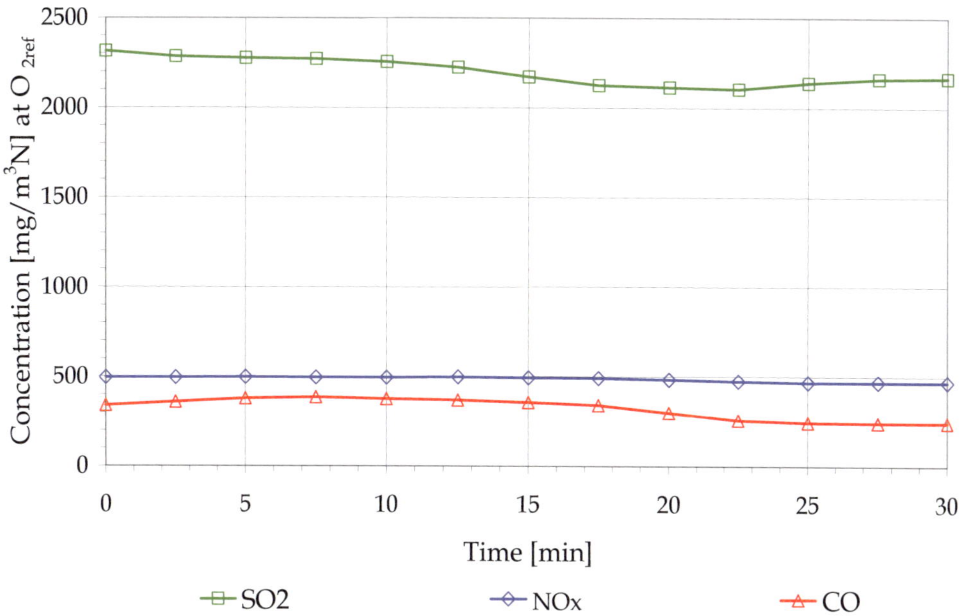

Fig. 20. Time dependence of the SO_2, NOx and CO concentration when co-firing 30 % by mass corncob with 70 % lignite mixture

6. Conclusions

As specific comparative result, one indicates that between the two test series accomplished on co-firing of biomass species (sawdust and corncob) in addition to coal no major relevant differences resulted, despite the difference of heat value, determining higher temperatures, with all resulting benefits and disadvantages. Both species of waste biomass are appropriate to be used in co-firing with coal, in slightly retrofitted existing facilities, normally used for one fuel. The support of other fossil fuel is not needed. In order to improve the efficiency of the co-firing process, one suggests to continue the tests for reducing the unburned carbon (CO) values in the flue gases, that is the main reason for a lower thermal efficiency of the process, in comparison to the classic (fossil fuel) case. Also the CO amount is over the limits permitted for a clean combustion and stack concentration.

Important relevant experimental conclusions, compared to the unblended combustion tests, are noticed:

- The fuel cost under the co-firing circumstances is, under the specific conditions from Romania, lower as in comparison to alone fossil fuel utilization;
- Reduction of the SO_2 concentration in the flue gases occurs, in accordance to the biomass ratio; there are theories that the composition of the biomass might act also in influencing this process, as well;
- Using fluidized bed technology, not notable NOx enhanced emissions in case of the co-firing was attested, due to the higher heating value of the biomass in comparison to the coal.
- No special deposit problems have been recorded; one reason might be the special outfit of the furnace, according to the design of the fluidized bed combustor.

- A reduction of the thermal efficiency (due to a higher unburned C in the flue gases) is caused by the presence of biomass in the blended fuel;
- The particle concentration in the flue gases is reduced in the case of the co-firing, in accordance to the biomass content, the explanation being also connected to the reduced ash content in biomass;
- In order to generate a total CO_2 lean global process, one suggests adding finally, a technology of capturing the CO_2 by absorption (through scrubbing with monomethanol-amine MEA). The CO_2 emission might be also reduced and controlled by paying the price for the supplementary technology.

As further plans one will study the corrosion aspects, knowing that they are related to alkali metals and chlorine and might turn into a major problem. In addition, the utilization of the residues (ash) is term of further and peculiar analysis and research.

The experimental results demonstrate that the co-firing technology, for Romania, represents a progress, and is worth to be applied in industrial environment also, taking into consideration the potential of local biomass availability. The technology is stable, is expected to be applied with a large share of waste biomass, and by applying flue gas cleaning technologies, the limits for the pollutants' concentration in stack might be respected.

When proper choices of biomass, coal, boiler design, and boiler operation are made, traditional pollutants (SO_x, NO_x, etc.) and net greenhouse gas (CO_2, CH_4, etc.) emissions decrease. Ancillary benefits include a reduced dependency on imported fuels and decreased demand for disposal of residues. These advantages can be realized in the very near future with relatively low technical risks. Therefore, commercialization of co-firing technologies offers among the best short- and long-term solution to greenhouse gas emissions from power generation. Since co-firing is not an option in all areas, a robust biomass utilization strategy requires development of alternative technologies as well. The effectiveness of the co-firing option, combined with its low cost and low technical risk, should place it high on a priority list of institutions considering an array of greenhouse gas options (Baxter & Koppejan, 2005). However, there are many issues that, if not carefully managed, could compromise the boiler or downstream processes. Results to date indicate that these are all manageable but they require careful consideration of fuels, boiler operating conditions, and boiler design.

It is anticipated that the biomass co-firing activities in Europe will continue to expand over the next few years. It is clear, however, that the limits to the expansion of co-firing will depend on the attitude of the specific governments and the limitations applied to the eligibility of co-firing under the Renewable Obligation Order. At present, co-firing is perceived by some governments as a relatively short-term approach to the production of electricity from renewable sources, which will help to create a biomass fuel supply infrastructure, and in particular will help to develop the cultivation and supply of indigenous energy crop materials. In a broader context, and taking a longer term view, co-firing clearly provides the most efficient and lowest risk means for generating electricity from biomass, which in Europe, at least, is a relatively scarce and expensive energy source. In those countries where utilization of coal will continue to play a significant role in the fuel mix for power generation, co-firing has a key role to play in the meeting the challenges represented by global warming, and the imperative to develop means of progressively reducing fossil fuel utilization (Livingston, 2005).

The results of this paper confirm that biomass is the only renewable energy source that can replace fossil fuels directly. By blending solid biomass fuels with corresponding fossil fuels different applications (from small scale to large scale) are possible. Thus co-utilization of biomass fuels with fossil fuels is a quick and relatively reliable way to reduce greenhouse gas emissions and preserve natural resources. It is therefore a sustainable, interim mechanism for meeting commitments to the Kyoto Protocol.

Co-firing of solid biomass with fossil fuels (mainly coal) represents an attractive option for the large-scale power plant's infrastructure and high electrical conversion efficiency. The technology is available and is based on many different types of biomass. Compared to most other forms of bio-energy, co-utilization of biomass in modern, highly efficient CHP (combined heat and power) and power plants is commercially available, represent technically, and environmentally an attractive solution for all countries that aim to increase the proportion of renewable electricity in the general energy balance. However, as all other renewable energy sources, this too is dependent on policy instruments.

Secure supply and an offer of mixture in the potential biomass share, also including short rotation coppices are main requirements for further implementation, especially under the conditions, where lack of organization in biomass flow is attested, despite best climate and soil conditions exist. Combustion of solid fossil fuel with biomass should represent both a retrofit option, but also a very good solution from the start of the project. The bio-energy creates jobs and secures energetically the region/country, being also an alternative for developing local skills, and economic growth.

7. References

Baxter, L. & Koppejan, J. (2005). Biomass-coal co-combustion: opportunity for affordable renewable energy, Fuel, Vol. 84, Issue 10, July 2005, pp. 1295-1302, ISSN: 0016-2361

Center for Promotion of Clean and Efficient Energy in Romania (2009). Market development Biomass Romania. Survey: Current status & possibilities market chains, Ref. no. BIORM 08002, Developed by ENERO Romania, September 2009, Bucharest

European Commission (2005). EUR 21350 - Biomass – Green energy for Europe, Office for Official Publications of the European Communities, ISBN 92-894-8466-7, Luxembourg

European Commission (2003). Biomass co-firing – an efficient way to reduce greenhouse gas emission, European Bioenergy Networks, ALTENER programme and the Finnish Ministry of Trade and Industry, Finland

Ghergheleş, V. (2006). Future energy, renewable sources, Mediamira publishing house, ISBN 973-713-146-0, Cluj-Napoca

Hansson, J.; Berndes, G.; Johnsson, F. & Kjärstad, J. (2009). Co-firing biomass with coal for electricity generation - An assessment of the potential in EU 27. Energy Policy, Vol. 37, Issue 4, April 2009, pp. 1444-1455, ISSN: 0301-4215

Ionel, I.; Popescu, Fr.; Lontis, N.; Trif-Tordai, G. & Russ W. (2009). Co-combustion of fossil fuel with bio-fuel in small cogeneration systems between necessity and achievements, Proceeding of 11th WSEAS International Conference on Sustainability in Science Engineering, ISBN 978-960-474-080-2, Timisoara, Romania, 27-29 May 2009

Ionel, I.; Ungureanu, C.; Trif-Tordai, G.; Tenchea, A.; Silaghi, Diana. & Constantin, C. (2007). Management of biomass energy recovery, "Politehnica" publishing house Timisoara, ISBN 978-973-625-541-0, Timisoara

Livingston, W. R. (2005). A review of the recent experience in Britain with the co-firing of biomass with coal in large pulverized coal-fired boilers, IEA ExCo55 Workshop on Biomass Co-firing, Copenhagen, Denmark, 25 May 2005

Romanian patent no. 121362/30.07.2007 (2007). Procedure and facility for the combined combustion of agricultural or municipal waste with coal and flue gas cleaning technologies, Owners: SC SAVPROD SA & "Politehnica" University of Timisoara

Romanian Ministry of Environment and Forests (2007). www.mmediu.ro, Romania

Trif-Tordai G. (2008). Researches on biomass with coal co-firing-PhD Thesis, "Politehnica" publishing house Timisoara, ISBN 978-973-625-713-1, Timisoara

Țânțăreanu, C. (2009). Biomass potential in Romania. The CEUBIOM project approach. In: ENERO-Centre for promotion of clean and efficient energy in Romania, 29 January 2009, Available from www.enero.ro

Van Loo, S. & Koppejan, J. (2007). The handbook of Biomass Combustion & Co-firing, Earthscan publishing house, ISBN 978-1-84407-249-1, London

Part 4

Hydrogen

13

Nano-Porous Silicon-Based Mini Hydrogen Fuel Cells

Tayyar Dzhafarov[1] and Sureyya Aydin Yuksel[2]
[1]*Institute of Physics, Azerbaijan National Academy of Sciences, Baku*
[2]*Department of Physics, Yildiz Technical University, 34210 Esenler/Istanbul*
[1]*Azerbaijan*
[2]*Turkey*

1. Introduction

Hydrogen is an important chemical material that is utilized in a large scale in synthetic chemical industries in modern society. On the other hand, technologies for utilizing hydrogen as clean source of energy are considered to assume an important position in order to overcome problems of lack of energy and environment in future. Hence, fuel cells that store hydrogen and operate using it as fuel are developed (Hoffmann, 2002). Such fuel cell is a battery which is actuated by gas. Energy obtained upon a reaction of hydrogen and oxygen is directly converted into electric energy. Fuel cells offer a significant advantage over traditional combusting-based thermal energy conversion, so they provide efficiencies of electrical power supply in the range of 35 to 55%, causing very low level of pollutant emission. They can be used in a wide variety of applications from miniaturized portable power to stationary power stations (Mench, 2008; O'Hayre et al., 2009; Vialstich et al., 2003). Nowadays the rise in portable electronics requires energy sources compatible with the environmental constrains. Portable fuel cells are expected to come first and open the market for other applications. The potential of portable fuel cells includes portable electronics devices (portable computers, cameras, electronic games, phones, audio players, televisions etc.), weather stations, signal units, medical devices etc. Proton exchange membrane (PEM)-type fuel cell is attractive and alternative option for production of clean energy for portable and transportation applications (Hockaday, 2000; Hahn et al., 2007). It consists of an organic polymer membrane as proton conductor (Nafion®) sandwiched between two the platinum-based catalysts. Fuel cell operates under clear hydrogen exposition at 100-300°C. An external reformer is required to convert fuel such as methanol or natural gas to hydrogen. However, the major hurdles for commercialization of system with PEM fuel cell consist of generation and direct supply of pure hydrogen gas. Recently, direct methanol fuel cell (DMFC), based on the PEM-type cell appropriate for micro-scale electronics power has been developed (Varga, 2007; Ohashi et al.,2010; Kamarudin et al., 2009). But, direct methanol fuel cells, in its turn, have disadvantages, related with high temperature operation and formation of the CO pollutant gas. Moreover, using the organic polymer membrane in PEM-type direct-fueled cells and the high-cost platinum catalysts mitigates durability and increases a cost of fuel cells. Degradation of structural characteristics of polymer membrane material due to

diffusion and electro-corrosion processes during electrochemical energy conversion in fuel cells becomes the major challenge in fuel cell basic research. Now, the fabrication technology of most portable electronic devices is based on standard micro-fabrication technique, but PEM-type fuel cells with polymer membrane are not readily integrated with this technique. On this view, it would be desirable to develop an inorganic proton conducting membrane for fuel cell with technology compatible with standard micro-fabrication techniques.
In order to get over the abovementioned disadvantages, we proposed to fabricate novel Schottky-type fuel cell for micro power generation, based on nano (micro)-porous silicon, operating at room temperature under direct exposition of hydrogen-containing gas and liquid compounds (Dzhafarov & Can, 2000; Dzhafarov et al., 2002, 2004, 2005, 2008, 2010, 2011). Fabrication of porous silicon is compatible with micro-fabrication technology. Moreover, high porosity of porous silicon and very large surface-to-volume ratio (up to 10^3 $m^2.cm^{-3}$) ensures the high proton conductivity comparable with that of polymer membrane Nafion®. Additionally, such direct-fueled fuel cell can operate at room temperature under exposition hydrogen-containing compounds (including the sea-water) without formation of polluting gases.
The review of our and other published researches on fabrication and characterization of nanoporous silicon-based direct fuelled hydrogen fuel cells has been presented below. Moreover, the literature search concerning different types of hydrogen fuel cells is briefly given.

2. Literature search

Hydrogen is the most attractive and ultimate candidate for a future fuel and an energy carrier. Hydrogen is recognized as the environmentally desirably clean fuel of the futute since it can be used directly in different types of hydrogen fuel cells. In hydrogen fuel cells the electrical energy will be derived from the reaction of hydrogen and oxygen gases within the fuel cell to make water.
In the water electrilizer, electric current passed through water splits it into hydrogen and oxygen

$$2H_2O \rightarrow 2H_2 + O_2 \qquad (-285\ kJ/mol) \tag{1}$$

As opposite to reaction (1), production of water, heat, light and sometimes sound takes place when normally hydrogen burns reacting with oxygen from air

$$2H_2 + O_2 \rightarrow 2H_2O \qquad (+285\ kJ/mol) \tag{2}$$

Most fuel cells directly generate electricity using the chemical reaction between a fuel (hydrogen) and an oxidant (oxygen). İn the hydrogen fuel cell the chemical reaction is exactly same to reaction (2), but instead of producing light and heat, electrical current is produced.
A fuel cell is an electrochemical energy conversion device which converts the chemical hydrogen and oxygen into water, and in the process it produces electricity. Most fuel cells today use hydrogen and oxygen as the chemicals (Mench, 2008; O'Hayre et al., 2009; Vialstich et al., 2003; Varga, 2007; Ohashi et al.,2010; Basu, 2007; Hoogers, 2003). Fuel cell provides a DC (direct current) voltage that can be used to power motors, lights or number of electrical appliances. The existing fuel cells are usually classified by their operating type and

the type of electrolyte they use. The main types of fuel cells are given in Table 1 (Vielstich et el, 2009).

Fuel Cell types	Fuel	Efficiency (%)	Operating temp. (oC)
PEM	H_2	40-50	80
Direct Methanol (DMFC)	Methanol	35	80
	Ethanol		
Solid Oxide (SOFC)	H_2, CO,	45-55	700
	CH_4		
Molten Carbonate (MCFC)	H_2, CO,	50-60	650
	CH_4		
Phosphoric Acid (PAFC)	H_2	40-50	190
Alkaline (AFC)	H_2	50	50

Table 1. Types of fuel cells

Polymer exchange membrane fuel cell (PEM or PEMFC). The PEM is the most likely candidate for portable and transportation applications. The PEM has a high power density and a relatively low operation temperature (ranging from 80 to 120°C). PEM type fuel cell consists from two electrodes (the anode and cathode) and the electrolyte between them. The electrodes of the fuel cell are thin layers of material, with the platinum catalyst dispersed in carbon, which are applied to each side of the membrane, yielding what is known as a Membrane Electrode Assembly (MEA) (Fig.1). The *anode*, the negative post of the fuel, has several jobs. It conducts the electrons that are freed from the hydrogen molecules so that they can be used in an external circuit. It has channels etched into it that disperse the hydrogen gas equally over the surface of the catalyst. The *cathode*, the positive post of the fuel cell, has channels etched into it that distribute the oxygen to the surface of the catalyst. It also conducts the electrons back from the external circuit to the catalyst, where they can recombine with the hydrogen ions and oxygen to form water. The *electrolyte* is *the proton exchange membrane.* PEM can be made from either pure polymer membranes or from composite membranes where other materials are embedded in a polymer matrix. One of the most common and commercially available PEM material is Nafion, a DuPont product (US PATENT 5266421, granted 2008-11-30). The membrane presents a solid polymer having negative ionic sites attached to polymer chains. When such a membrane is hydrated (having water absorbed into it), hydrogen ions (protons) can move among the sites so that the protons can be transported across the membrane. The membrane only conducts positively charged ions and blocks electrons. For a PEMFC, the membrane must be hydrated in order to function and remain stable.

The *catalyst* (electrode) is a special material that facilitates the reaction of oxygen and hydrogen. It is usually made of platinum nano-particles very thinly coated onto carbon paper or cloth. The catalyst is rough and porous so that the maximum surface area of the platinum can be exposed to the hydrogen or oxygen. When an H_2 molecule comes in contact with the platinum catalyst on the anode side, it splits H_2 into two H^+ ions (protons) and two electrons (e^-). The electrons are conducted through the anode, where they make their way through the external circuit (doing useful work such as turning a motor) and return to the cathode side of the fuel cell. On the cathode side of the fuel cell, oxygen gas (O_2) is being forced through the catalyst, where it forms two oxygen atoms. Each of these atoms has a strong negative charge. This negative charge attracts the two H^+ ions through the membrane, where they combine with an oxygen atom and two of the electrons from external circuit to form a water molecule (H_2O). Chemistry of a fuel cell is following:

$$\text{Anode side:} \quad 2H_2 \rightarrow 4H^+ + 4e^- \tag{3}$$

$$\text{Cathode side:} \quad O_2 + 4H^+ + 4e^- \rightarrow 2H_2O \tag{4}$$

$$\text{Net reaction} \quad 2H_2 + O_2 \rightarrow 2H_2O \tag{5}$$

This reaction in a single fuel cell produces only about 0.7 V. To get this voltage up a reasonable level many separate fuel cells must be combined to form a fuel cell stack. The described principles of fuel cell operation by using a sulphonated polystyrene ion-exchange membrane as the electrolyte are firstly realized by W. Thomas Grubb (General Electric Company, 1955).

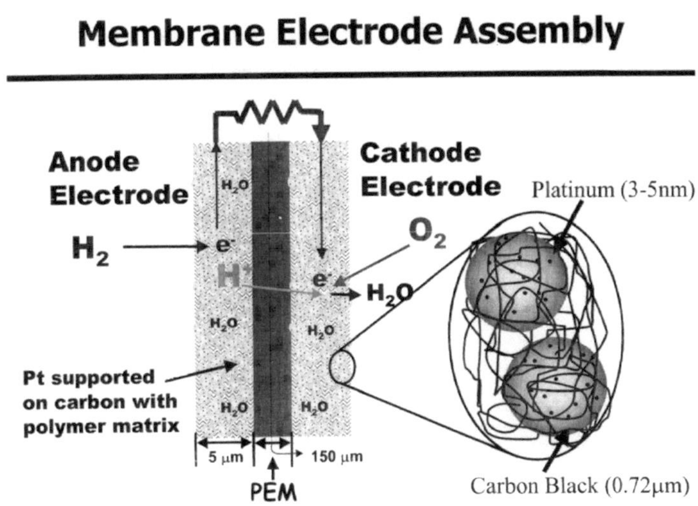

Fig. 1. Proton Exchange Membrane (PEM-type hydrogen fuel cell)

Direct-methanol fuel cell (DMFC) is a subcategory of proton-exchange membrane fuel cell in which methanol is used as the fuel. DMFC operates on dilute aqueous solutions, with reported workable devices allowing no more than 3% methanol in water. Polymer electrolyte membranes are also used to construct Direct Methanol Fuel Cells (DMFC). The DMFC requires a relatively large amount of platinum to act as a catalyst, which makes this fuel cell expensive. The DMFC relies upon the oxidation of methanol on a catalyst layer to form carbon dioxide. Water is consumed at the anode and is produced at the cathode. Positive ions (H^+) are transported across the proton exchange membrane - often made from Nafion - to the cathode where they react with oxygen to produce water. Electrons are transported through an external circuit from anode to cathode, providing power to connected devices. The half-reactions are:

$$\text{Anode (oxidation):} \quad CH_3OH + H_2O \rightarrow 6H^+ + 6e^- + CO_2 \tag{6}$$

$$\text{Cathode (reduction):} \quad 3/2O_2 + 6H^+ + 6e^- \rightarrow 3H_2O \tag{7}$$

$$\text{Overall (redox reaction):} \quad CH_3OH + 3/2O_2 \rightarrow 2H_2O + CO_2 \tag{8}$$

During the methanol oxidation reaction carbon dioxide is formed, which strongly adsorbs onto the platinum catalyst, reducing the surface area and thus the performance of the cell. Platinum is very expensive and contributes to the high cost per kilowatt of these cells. Moreover, disadvantage of DMFC is that the methanol is toxic and flammable. The waste products of this type cells are carbon dioxide and water. The advantage of the DMFC consists in that methanol is a liquid fuel cell and is therefore easer to store and transport than hydrogen for PEM fuel cell. In terms of the amount of electricity generated, a DMFC can currently generate 300-500 mW/cm^2. DMFC uses the same type of electrolyte as PEM fuel cell, but require thicker membranes and greater loading (density) of catalysts. It operates at between 80 and 120°C with an efficiency of up 40%. DMFC based micro fuel cells can be used for portable electronics such as mobile phones, digital camera, laptop computers etc. Military applications of DMFC are an emerging application since they have low noise and thermal signatures. These applications include power for soldier-carried tactical equipment, battery chargers, and autonomous power for test and training instrumentation.

Solid oxide fuel cell (SOFC). In the SOFC design, the anode and cathode are separated by an electrolyte that is conductive to oxygen ions but non-conductive to electrons. The electrolyte is typically made from zirconia doped with yttria. It is best suited for large-scale stationary power generators that could provide electricity for factories or town. This type of fuel cell operates at very high temperatures (700-1000°C) and uses H_2, CO or CH_4 as fuel. Advantages of SOFC are high power density (200-2000 mW/cm^2), fuel flexibility and the possibility of internal reforming. Hydrogen is generated on the anode porous surface by steam reforming of , for example, methane (Bove & Ubertini, 2008)

$$CH_4 + H_2O \rightarrow CO + 3H_2 \tag{9}$$

and subsequent water gas shift

$$CO + H_2O \rightarrow CO_2 + H_2 \tag{10}$$

The transition metals, such as Ni in the anode play role of catalyst. The direct electro-oxidation of the fuel

$$CH_4\,(g) + 4O^{2-} \rightarrow CO_2\,(g) + 2H_2O\,(g) + 8e^- \tag{11}$$

$$CH_4\,(g) + 3O^{2-} \rightarrow CO\,(g) + 2H_2O\,(g) + 6e^- \tag{12}$$

$$CO\,(g) + O^{2-} \rightarrow CO_2\,(g) + 2e^- \tag{13}$$

may take place under dry conditions though processing in elementary steps as does the steam reforming. A single cell operated with hydrogen and oxygen provides at equilibrium the open-circuit voltage of 1.2 V at standard conditions (T=273K, p=1 atm). The cell voltage depends on the used fuel, operation temperature and pressure. The open-circuit voltage of SOFC operating on hydrogen and oxygen is about 0.9 V at 1000°C

Alkaline fuel cell (AFC) with potassium hydroxide (KOH) electrolyte is one of oldest designs for fuel cell. The AFC is very susceptible to contamination, so it requires pure hydrogen and oxygen. It is also very expensive, so this type of fuel cell is unlikely to be commercialized.

Molten-carbonate fuel cell (MCFC) operates in manner of SOFC, expect the electrolyte consists of liquid (molten) alkaline carbonate ($NaHCO_3$), which is an oxidizing agent. *MCFC* is also best suited for large stationary power generators. It uses H_2, CO or CH_4 as fuel and operates at 600°C that makes the design a little less expensive than that of *AFC*.
Phosphoric-acid fuel cell (PAFC) with molten H_3PO_4 electrolyte has potential for use in small stationary power-generation systems. It operates at a higher temperature (200°C) than polymer exchange membrane fuel cell.
At the moment, the further increase in functionality of portable devices is severely hindered by the limits energy sources. Higher energy densities, smaller size, lower cost per watt and faster charging are the wish list. Fuel cells have the potential to increase the energy density of such devices by a factor of 3 to 10 compared to presently used lithium batteries. They are more user friendly since recharging is not required. Most of the fuel cell research activities have been up to now devoted to large power automotive and stationary applications.
During the past years, portable versions of the fuel cells have emerging. The most promising miniature fuel cell is the direct methanol fuel cell (Crawley, G., 2007; Kamardin et el., 2009). DMFC is inexpensive, convenient, does not require pressurized hydrogen gas and provides good electrochemical performance. This is still large in size compared to an electrochemical battery and further reduction will be needed. Charging consists of replacing on the fly. This provides a continued source of energy similar to fueling a car.
A large number of companies continue to work in developing and fabricating of direct methanol fuel cells. Most of such fuel cells are on stage of "in development" or testing phase. Toshiba unveiled a prototype DMFC for a laptop (http://www.batteryuniversity .com/parttwo). The company gave no indication as to when the product would be commercialization available. A direct battery replacement that offers high power, small size and competitive price is still several years away. As good as the fuel cell may look from the outside, 15-years of experiments has not solved number of persistent problems. One is the slow start-up; another is the low electrochemical activity at the anode. This is especially apparent with the DMFC. Deposition of CO_2 , forming during operation of DMFC (see eq.(26)) results in decrease of active the anode surface and thereby decreases the performance and durability of the fuel cell. Each cell produces about one volt and when loaded, the relatively high internal resistance of membrane-electrode-assembly (MEA) sandwich causes the voltage drops quickly. Miniature DMFC of 1 watt for portable communication systems is developed by 'Motorola Labs" (http://www.fuelcells.org/MicroMarket.pdf).
Fuel cells might be the answer to our problems, but first scientists will have to sort out a few major issues: *(a) Cost.* Chief among the problems associated with fuel cells is how expensive they are. Hydrogen is the most common element in the universe. However, hydrogen does not naturally exist on Earth in its elemental form. Engineers and scientists must produce pure hydrogen from hydrogen compounds, including fossil fuels or water. In order to extract hydrogen from these compounds, we have to exert energy. The required energy may come in the form of heat, electricity or even light. This increases cost of energy produced by fuel cells. Additionally, for PEMFC systems, proton exchange membranes, precious metal catalysts (usually platinum), gas diffusion layers and bipolar plates make up 70 percent of a system's cost. In particular, researchers must either decreases the amount of platinum needed to act as a catalyst or find an alternative. *(b) Durability.* Researchers must develop PEMFC membranes that are durable and can operate at temperatures greater than 100°C and still function at sub-zero ambient temperatures. Currently membranes tend to degrade

while fuel cells cycle on and off, particularly as operating temperatures rise. *(c) Hydration.* Because of PEMFC membranes must be hydrated in order to transfer hydrogen protons, researchers must find a way to develop fuel cell systems that can continue to operate in sub-zero temperatures, low humidity environments and high operating temperatures. At around 80°C hydration is lost without a high-pressure hydration system.

In past years, there has been considerable interest in development of micro fuel cells for portable electronic devices due to their advantages over conventional batteries, including rapid recharging and much higher stored energy density. In the next 10 years the international technology roadmap for semiconductors (http://public.its.net) forecasts a decrease of the voltage required to power the working of microcircuits toward 0.6 V. The use of miniature fuel cells appears an attractive way to power portable electronics with a clean and refillable energy source. Among all types of fuel cells only two (PEM and DMFC) are suitable for miniaturization. The limitation principally comes from the working temperature required to be lower than 100°C. One of them is the proton exchange membrane (PEM) fuel cell. A key element of a PEMFC is the membrane that must behave high conductivity for protons and be impermeable to all other present species.

At the present time, the best conductivity (0.08 $S.cm^{-1}$) is reached by Nafion® perfluorosulfonated membranes. However the high cost and the geometric instability during hydration are only some of the severe constraints of such polymers. The nature of the proton conductive process in an ionometric membrane such as Nafion® is not still completely understood. The present consensus (Bunker et el., 1998) is to describe a Nafion® membrane as a skeleton of hydrophobic chains including connected hydrophilic domains containing water molecules. The diameter of the connecting channels is about 3 nm. The low stiffness of this skeleton is responsible for the swelling if the membrane with hydration in response to the molecular interactions. Moreover, this type of polymer membranes cannot be patterned with standard microelectronic technology.

A number of alternative solutions have been proposed to avoid these problems, one of them being the use of organic-inorganic hybrid membranes. The literature reports original research like {hybrid silica/ Nafion®} membranes (Jung, 2002) or a porous silica membrane filled with sulfonated polyelectrolyte (Yamaguchi et el., 2002). Yet these are not suitable to obtain easily integrated miniature cells.

Problems of performance, reliability, durability, a restricted allowable ambient-temperature range for operation and cost of miniature PEM fuel cells, particularly of DMFC, for portable devices were related with:

- materials of polymer membrane and platinum catalyst,
- poisoning the platinum catalyst by carbon dioxide during fuel cell operation,
- the large resistance drop in the membrane-electrode-assembly,
- compatibility of DMFC technology fabrication with standard micro technology.

These problems are not solved and require researches which must be focused on new porous materials and structures. Last time the new porous silicon-based fuel cell has attracted high attention.

3. Preparation and properties of porous silicon

Porous silicon (PS) layer on monocrystalline Si substrate and its manufacture by the technique of electrochemical etching of silicon substrate in HF solution or by chemical etching in HF-HNO_3 mixture are known as early as from 1956 (Canham, 1997; Feng & Tsu,

1994). A great interest to porous silicon arisen in the 90-s of 20-th century is caused by observation of strong visual photoluminescence and electroluminescence at room temperature. Electrochemical etching of silicon is attractive because of the possibility to tune the pore size from a few nanometers to a few tens of micrometers, just by choosing wafer doping level and etching conditions. Moreover, a wide range of porous layer thickness, porosities, surface areas and morphologies can be formed depending on the etching conditions. The bulk silicon was shown modifies during the etching to sponge-like structure with silicon columns and hydrogen covered pores.
The simplest electrochemical cell is shown in Fig. 2. The Si wafer acts as the anode and the platinum is the cathode. The thickness of porous silicon layer on Si substrate is determined by duration of etching. The porosity, i.e. the void fraction in the porous layer is determined by the current density (about 10 - 100 mA/cm^2), composition electrolyte, resistivity and the doping density of Si substrate.

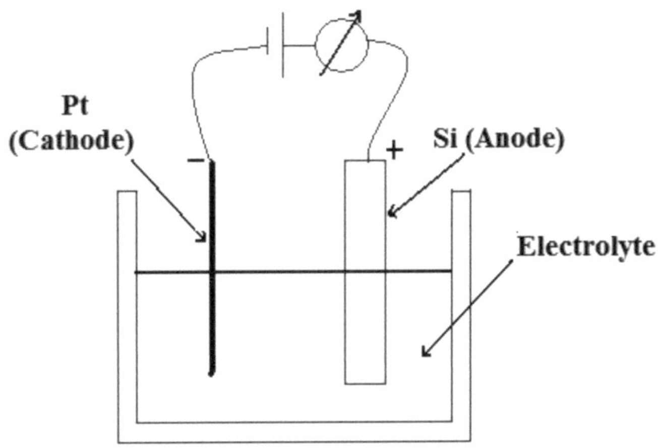

Fig. 2. Cross-sectional view of lateral anodization cell.

The anodic reaction on the Si substrate can be written during pore formation as (Beale et. al., 1985)

$$Si + 6HF \rightarrow H_2SiF_6 + H_2 + 2H^+ + 2e^- \quad (14)$$

Silicon atoms are dissolved as SiF_6^{2-} requires the presence of F^- ions (from HF solution) and positively charges holes (from the silicon wafer) at the silicon interface. Concentration of holes in p-Si is sufficiently high (about 10^{14} - 10^{18} cm^{-3}) and this case the nano-size pores were formed. Concentration of holes in n-Si is very small (about 10^2 - 10^6 cm^{-3}) and therefore generation of holes is possible due to illumination of n-Si substrate.
The structure and size of pores in porous silicon layer formed on n-Si substrate differ from those for layer on p-Si. If electrochemical etching was carried out at relatively low current density (10-80 mA/cm^2), then the local dissolution of silicon surface takes place. Herewith, pore formation begins on surface defects of Si and further growth of pores into silicon substrate proceeds due to the holes diffusion to Si-electrolyte interface. In the case of large current density (0.5 - 0.8 A/cm^2) when the amount of holes moving to Si-electrolyte interface is very high, the etching of top regions of Si substrate is preferred. It ensures the

uniform etching of silicon surface and formation a smooth surface of substrate (the so-called the electropolishing process). The raising the current density above the critical value at the end of anodization process results in a detachment of the porous silicon film from Si substrates. The behavior at high current densities turns out to be useful to produce porous silicon free-standing layers. The anodic reaction during the electropolising can be written as

$$Si + 6HF \rightarrow H_2SiF_6 + 4H^+ + 4e^- \quad (15)$$

Pores, depending on the its diameter, denoted as micropores (R < 2 nm), mesopores (2 nm < R <50 nm) and macropores (R > 50 nm). Under illumination the pore size dependent on doping density and anodization conditions, with diameters in the range 100 nm - 20 μm (macropores). The crystalline structure of porous silicon presents a network of silicon in nano-(micro)-sized regions surrounded by void space with a very large surface-to-volume ration (up to 10^3 m^2cm^{-3}) (Herino, 1998). The structure of porous silicon is like a sponge where quantum effects plays fundamental role (a quantum sponge) (Bisi et el., 2000). The pore surfaces are covered by silicon hydrides and silicon oxides and therefore they are very chemical active. These features of porous silicon (a quantum system, a sponge structure and an extremely large pore surfaces) ensure many possible applications, such as light emitting diode, sensor, hydrogen fuel cell and other applications.

PS layers with a thickness of 10-20 μm and an average porosity of 40 to 80% were prepared on n-type (111) Si substrates ($\rho = 1\times10^{-2}$ Ω cm) by anodic etching in $HF:H_2O$ = 1:3 solution at a dc current of about 15 mA cm^{-2} under white-light illumination (Dzhafatov et al., 2004). For some measurements, the PS films were then detached from the Si substrate by electro polishing in the same solution with a current density of 0.8-1.0 A cm^{-2}. The free-standing PS films were characterized by porosity, thickness and resistivity measurements. The average porosity was measured by a gravimetry technique. Resistivity and charge carrier concentration measurements were carried out on the free-standing PS layers attached to a dielectric substrate (glass) by using the Van der Pauw technique. *In* or *In-Ga* alloy was used as an ohmic contact to the PS layer. Morphological characterizations of the PS surface were performed by scanning electron microscopy.

The average porosity (P), i.e. the avoid fraction in the porous layer, can be obtained by gravimetry using the equation

$$P = (m_1 - m_2)/(m_1 - m_3) \quad (16)$$

Here m_1 is Si sample mass before the anodization etching, m_2 just after etching and m_3 after the removal of the porous layer by electro-polishing or after a rapid dissolution of the whole porous layer in a 3% KOH solution. Guessing the porous silicon mass m_{PS}, the average porosity can be also determined by using the equation

$$P = 1 - m_{PS}/\ \rho S d = (m_1 - m_{PS})/m_1 \quad (17)$$

One can also get the porous silicon layer thickness d using the equation

$$d = (m_1 - m_3)/\rho S \quad (18)$$

where ρ is the Si density (2.33 g/cm^3) and S is the etched surface. Fig. 3. shows the SEM micrographs of porous silicon surface.

The electrical measurements of the free-standing PS layers with 65% porosity (300 K, 45% RH) gave values of $\rho = 1.8\times10^6$ Ω cm for resistivity, $p = 9.6\times10^{12}$ cm^{-3} for hole concentration, and $\mu = 0.36$ cm^2/(V s) for hole mobility (Dzhafarov & Aydin , 2009).

Fig. 3. Scanning electron microscopy of porous silicon layer surface.

Fig. 4. illustrates FTIR spectrum of free-standing PS film of thickness 12 μm measured at room temperature (Dzhafarov & Can, 2000). The peaks related with absorption on vibration of Si-H (2100 cm^{-1}) and Si-O bonds (1100 cm^{-1}) located on pore surfaces were observed from Fig. 4. These bonds play an important role in regulating optical, electrical and gas sensing properties of porous silicon.

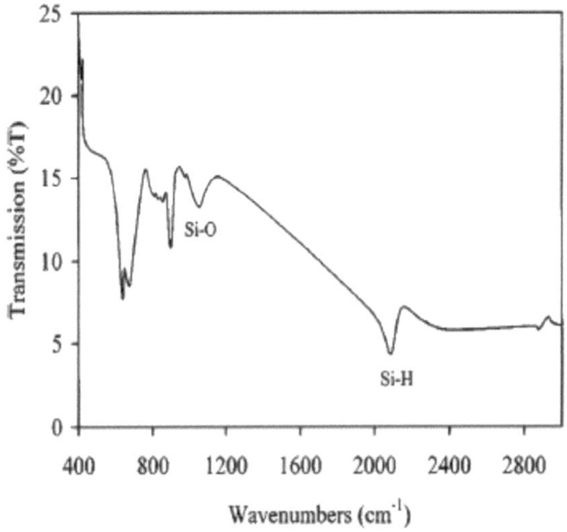

Fig. 4. FTIR spectrum of porous silicon film (300 K).

The effect of isothermal annealing of free-standing PS films on changes of intensity of absorption coefficient of Si-H (2100 cm^{-1}) and Si-O (1100 cm^{-1}) peaks is used for estimation of diffusion coefficient from equation (Abdullaev & Dzhafarov, 1987)

$$Q = 2\,\pi^{-0.5}\,SN_o\,(Dt)^{0.5} \qquad (19)$$

Here Q is the total quantity of hydrogen (or oxygen) penetrating from air into PS film (or out-diffusing from PS film), $N_o = N(0,t)$ is the surface concentration on an external surface of PS film and S is the area of sample. Results of these measurements showed that in the range of 65–185°C the temperature dependence of hydrogen and oxygen diffusion coefficient along the porous surfaces are described as (Dzhafarov & Can, 2000)

$$D(H) = 5x10^{-10} \exp(-0.37 \text{ eV/kT}) \quad (20)$$

$$D(O) = 1.3x10^{-8} \exp(-0.50 \text{ eV/kT}) \quad (21)$$

The activation energy for diffusion of hydrogen along the porous surfaces estimated from response (or recovery) V_{oc} - t curves for Au/PS/Si cells under humid ambient (90%RH) is 0.34 eV (Dzhafarov & Aydin, 2010).

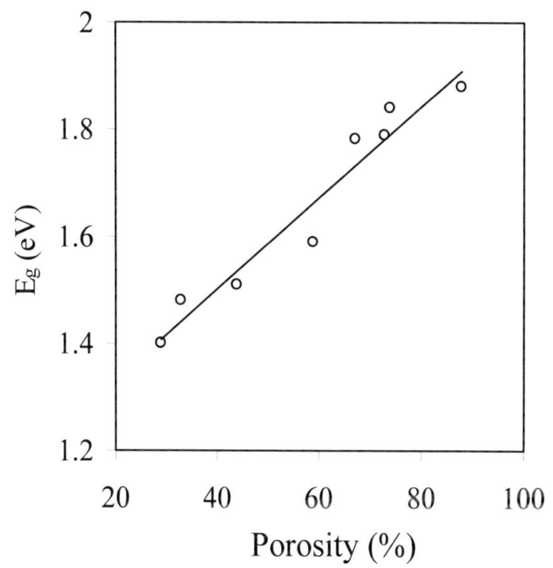

Fig. 5. Energy band gap in depending on porosity of PS film (40% RH, 300 K).

Fig. 5. shows the effective energy gap in dependency on porosity of the free standing PS films, calculated from extrapolation of the high energy part of $\{\alpha^2 (h\nu)^2 - h\nu\}$ spectra (Dzhafarov et el., 2007). Near linear increase of band gap from 1.4 to 1.9 eV with rising of porosity of PS films in the range of 30-90 % is observed. Data on Fig. 5. concerning a increase of the energy gap in dependency on porosity of PS films, can be explained by a model including the quantum confinement of carriers in the PS microcrystallites, causing the widening of the Si band gap.

4. Porous silicon-based hydrogen fuel cells

Most of PEM-type fuel cells have used Nafion® as the proton exchange membrane, yet it does not seem to be right solution. Indeed Nafion® polymer membranes are inadequate to standard micro fabrication techniques. Recent our and other investigations (Dzhafarov et al., 2008, 2010; Dzhafarov & Aydin Yuksel, 2011; Chu et al., 2007; Pichonat et al., 2005; Gyoko et al., 2005; Lee et el., 2007; Moghaddam et el., 2009; Hayase et el., 2004; Starkov et el., 2006) have shown the growing interest to the development of porous silicon-based miniature fuel cells for chip scale power. The main advantage of porous silicon membranes consists in suitable to obtain easily integrated miniature cells.

The potential advantages of porous silicon for fuel cell applications are follows:

a. unique combination of the crystal structure and the huge surface to volume ratio (about of 1000 m^2/cm^3);
b. strong surface chemical reactivity, significant change of porous silicon surface as a result of different kinds of treatment (thermal, chemical, optical etc.);
c. high ionic conductivity over the pores;
d. possibility to make 3D structures and proton-conducting membranes;
e. opportunity to modify the morphology by the control of the porous dimensions from nano- to micro-scale;
f. simplicity and cheapness of the technology that could be compatible with silicon IC technology.

Taking into account this, porous silicon has proved to be a promising material to replace traditional Nafion-based proton exchange membrane, as this material provides a porous matrix that can be functionalized for further proton exchange behavior. Nowadays, investigations of porous silicon based miniature fuel cells is the conceptual stage and the number of reports on fuel cells with porous silicon membrane is limited. Recently some novel applications of porous silicon have been discovered utilizing other interesting characteristics (Stewart & Buriak, 2000). Nanoporous silicon is readily formed by anodic etching in hydrofluoric acid. The nanoscale pore sizes that can be achieved in porous silicon by anodic etching may provide a conduit for proton transport. In an acidic aqueous media, such as a formic acid, the silicon surface is readily oxidized leaving hydroxyl terminated pore surfaces. This hydroxylated surface is known to be acidic (Cook, 1990) and should allow for proton conduction through the nanoporous. Furthermore, the nanoscale diameter of the pores may serve as a diffusion barrier preventing larger molecules from going through. In principle, the viability of porous silicon as a proton conductor as well as a diffusion barrier to fuel molecules can be further promoted by the function of porous silicon (Song & Sailor, 1999). A major advantage of porous silicon is its compatibility with conventional silicon micro-fabrication technology, which presents the possibility of fabricating different components of a fuel cell or even a fuel cell and its supported devices in a monolithic fashion. Porous silicon should be stable at elevated temperatures, unlike many polymeric materials. The use of porous silicon as a gas diffusion layer (Maynard & Meyers, 2000) or a support for catalysts (D'Arrigo et el., 2003) has been demonstrated.

The maximum power density of 47 mA/cm^2 and open-circuit voltage of 0.76V for PEM-type fuel cell with porous silicon membrane and Pt-based catalyst on both side of PS membrane were reported (Moghaddam et al., 2009). This has been achieved due to development of a technique for self-assembly of molecules with functional end group inside high aspect ratio nanoporous (7-10 nm) within a porous silicon membrane.

Micro direct ethanol fuel cell with porous silicon electrodes and Nafion membrane have been described (Aravamudhan et el., 2005). The structure consists of a porous silicon anode (fuel micro-column) and a porous silicon cathode (air micro-column) sandwiching a proton exchange membrane (Nafion 115™). Platinum is deposited on both the micro-columns to act both as a electrocatalyst and a current collector. Ethanol has been used as fuel. The hybrid type {porous silicon/Nafion/porous silicon} fuel cell generated an open-circuit voltage of 260 mV and maximum power density of 8.1 mW/cm^2 at room temperature with a fuel supply of 5 μL of 8.5 M ethanol solution.

The complicated design of fuel cell with porous silicon membrane have been studied in (Chu et al., 2007). The three-layer catalysts {Nafion+Pd/Au+Pd/Au} were painted directly onto both the porous silicon membrane surfaces. A solution of 5 M formic acid with 0.5 M sulfuric acid was used as the fuel at the anode. The cathode was exposed to ambient air for oxygen supply. For this micro fuel cell peak power about of 94 mW/cm^2 have been reached. However, using complicated and high-cost catalysts in this type of fuel cell makes problematic its application.

A new way of making miniature fuel cell, using a porous silicon membrane with a silane grafted on pores walls has been demonstrated in (Pichonat et al., 2005). Tests of fuel cell with the chemical grafting of molecules bearing acid groups on the pore walls of a porous silicon membrane allow to achieve power density of 17 mW/cm^2. Measurements have been carried out under H_2 feeding that is provided by 20% NaOH solution electrolysis and O_2 directly by ambient air.

The porous silicon wafer filled with sulfuric acid solution or Nafion as a proton conduction membrane for micro fuel cells have been developed (Gyuko et al., 2005). For the fuel cell filled with sulfuric acid solution (20%) or Nafion, assembled into the H_2/air ambient the maximum power density of 12.75 mW/cm^2 and 1.27 mW/cm^2 were achieved respectively.

Porous silicon as a gas diffusion layer (GDL) have been used in a micro direct methanol fuel cell (Lee et al., 2007). Porous silicon was prepared by photoelectrochemical etching of silicon wafer. The maximum power density of DMFC of 9.37 mW/cm^2 was obtained under using 2M methanol as fuel.

The hybrid type of {porous silicon layer/ Nafion 112} hydrogen fuel cell has been developed (Hayase et al., 2004). The fuel channels were fabricated by photolithographic pattering and subsequent wet etching of silicon. The porous layer was formed by anodization of silicon from the polymer electrolyte membrane. The maximum power density of 1.5 mW/cm^2 and open-circuit voltage of 840 mV were observed by hydrogen feed.

A layer of carbon nanofibers and nanotubes formed on the surface of macropores in silicon, in which the pore cross section and structure varies with depth, have been offered (Starkov et al., 2006). The obtained structure, in authors opinion, can be serve as a basis for the development of electrodes for polymer cells with a specific of conducting surface area.

Porous silicon also can be used as hydrogen reservoirs. Concentration of atomic hydrogen chemically bound to the specific surface of porous silicon (of 95% porosity with nano-crystallite dimensions of 2 nm) is about 66 mMol/g (Lysenko et al., 2005).

5. Metal/porous silicon/silicon –based direct hydrogen fuel cells

Most previous works on fuel cells using the porous silicon has been focused on direct methanol fuel cell (DMFC) with an expensive Pt, Pt-Ru or Pd as catalyst (Aravamadhan et al., 2005; Gold et al., 2004). New type of fuel cell based on {metal/porous silicon/silicon}-Schottky type structure has been proposed and realized (Dzhafarov et al., 2005; 2008; 2010; 2011). Such fuel cell uses the porous silicon layer as membrane, Au film as electrode-catalyst and operates at room temperature under direct supplying of different hydrogen-containing gaseous and liquid fuels.

Cell fabrication. The Au/PS/Si structures were fabricated by the evaporation of a thin Au film onto the PS surface with a porosity of 65% at room temperature by using the electron-beam technique (Dzhafarov et al., 2004). The thickness of the deposited Au film was 150 nm, as obtained by measurements during the evaporation using a deposition controller (Inficon, Leybold). Current-voltage (I-V) characteristics, open-circuit voltage (V_{oc}), and short-circuit current density (J_{sc}) of the (Au)/PS/Si and PS/Si structures were measured at room temperature in air (300 K, 45% relative humidity (RH)) as well as for different gas atmospheres (humid, CO, and H_2S) in the measuring cell. The concentrations of CO and H_2S were measured using a BW Defender Multi-Gas Detector. The gas-stimulated open circuit voltage (V_{oc}) and short-circuit current density (J_{sc}) for the contacts to the Au film and PS layer (or Si substrate) were measured directly by a Thurlby-1503 digital multimeter. (Fig. 6). RH in the cell was measured with an Extech-44470 hygro thermometer. The photosensitive properties of the Au/PS structures were analysed by measuring the I-V characteristics in the dark, in daylight, and under tungsten-halogen lamp illumination 150 mW cm^{-2}). All the investigated structures exhibited very weak photosensitivity. The value of the open-circuit photo voltage in daylight and under tungsten-halogen lamp illumination was 1-3 mV. Therefore, gas-sensitive measurements were performed under daylight illumination

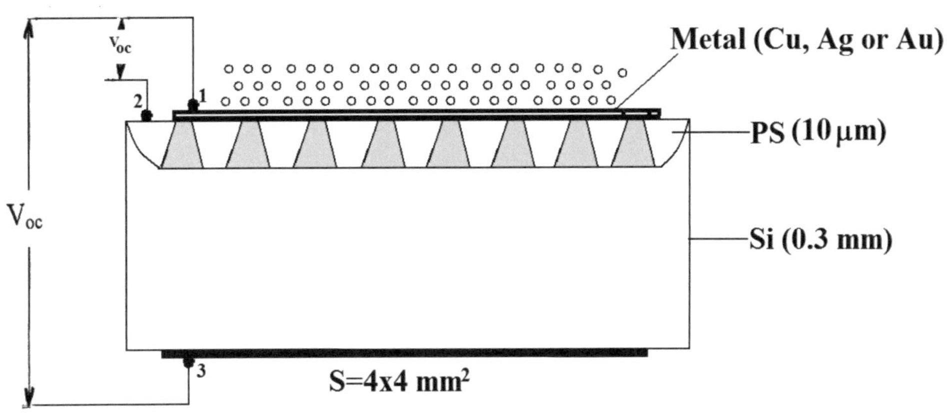

Fig. 6. Schematic sketch of Metal/PS/Si structure.

SEM micrographs of the PS and Au/PS surfaces are shown in Fig.7. Here the bright and relatively dark regions are the tops of silicon walls and intervals containing no of PS, respectively. The larger dimensions of the bright islands on the Au/PS surface than on the PS surface may be caused by 2-3 neighboring islands covered by unbroken gold film. We suppose that gold atoms deposit primarily on the external surface of the porous silicon layer at the initial stage of evaporation. Then, as the gold film thickness on the external surface of the porous silicon increases, the pores are fully closed by gold film.

The current-voltage characteristics of Au/PS/Si structure showed the good rectifying properties. The typical reverse I-V characteristics of Au/PS structure in air ambient at 45 %RH, 70 %RH, 83 %RH, 90 %RH and 99 %RH at room temperature are presented in (Fig. 8) (Dzhafarov et el., 2004).

It is seen that the reverse currents considerably increase with the rising of relative humidity. The value of current at 99 %RH (for 2V) increases in comparison with that at 45 %RH (for

2V) by factor 12. It will be noted that PS/Si structures (without Au film) showed a weak rectifying characteristics and the weak humidity-sensitive properties.

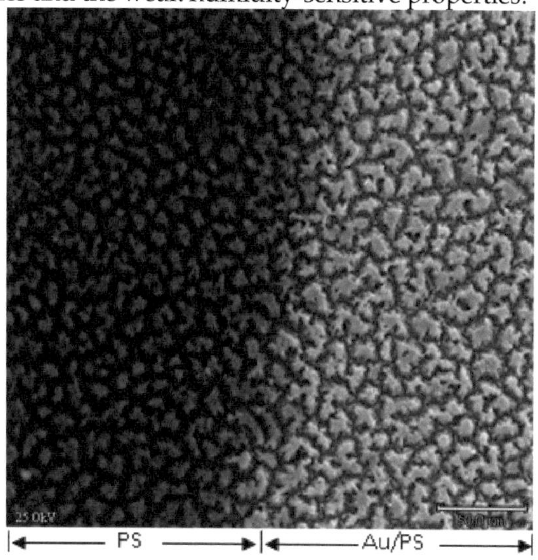

Fig. 7. SEM micrographs of PS and Au/PS surfaces.

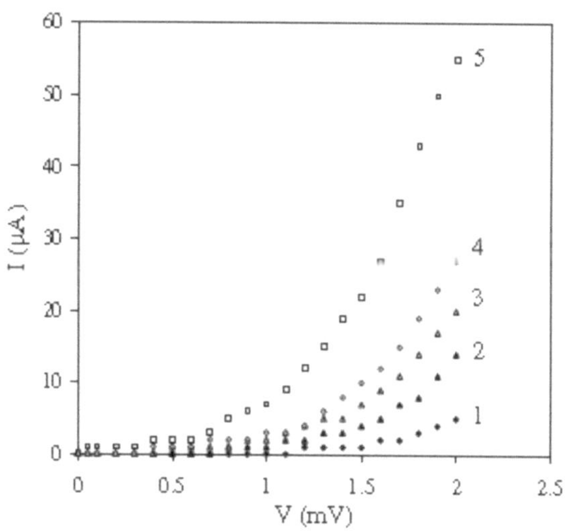

Fig. 8. Reverse I-V characteristics of Au/PS/Si structure in humid ambient: (1) 45, (2) 70, (3) 83, (4) 90 and (99) %RH.

The humidity (hydrogen)-voltaic effect i.e. generation of a the open-circuit voltage between contacts to Au film and Si under humidity (hydrogen-containing gas) exposition is observed for Au/PS/Si Schottky-type structure. The similar effect is also discovered for Ag/PS/Si and Cu/PS/Si structures (Dzhafarov & Can, 2000; Dzhafarov et al., 2002). Fig. 9 illustrates the open-circuit voltage in dependency on the relative humidity for one of such Au/PS/Si structures. It is seen that the V_{oc} approximately linearly increases from 15 mV to 400 mV with rise of the relative humidity from 51 %RH to 95 %RH. The humidity-sensitivity of Au/PS/Si structure is about 10 mV/(%RH). The PS/Si structures without Au film on PS surface displayed the low value of V_{oc} in humid atmosphere (about of 10 mV at 95%RH). In other words, the presence of the Au (Ag or Cu) film on the porous silicon surface is a necessary condition for the generation of electricity in the Au/PS/Si cell in humid atmosphere.

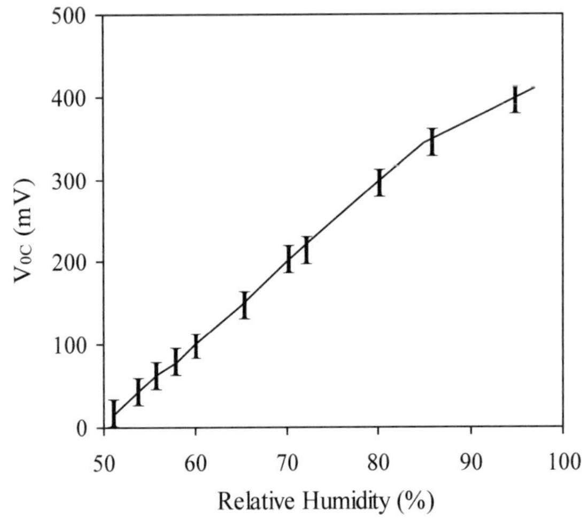

Fig. 9. The open-circuit voltage - RH dependence for Au/PS/Si structure (300 K).

The mechanism of the generation of the electricity in the metal/PS/Si. We suggest that in Au/PS/Si cell, similar to Proton Exchange Membrane (PEM) fuel cell, the Au film and PS layer play the role of the catalytic anode and electrolyte, respectively. The interface region between the porous and crystalline silicon (PS/Si), which is very imperfect and stressed, plays the role of the cathode. Electrons and protons formed in Au catalyst film (anode) after hydrogen splitting (*Anode:* $H_2 \rightarrow 2H^+ + 2e^-$*)*, pass through the external circuit and along the pore surfaces of PS layer (electrolyte), respectively and reach the cathode (PS/Si interface) region. Water molecules and oxygen from air can easily penetrate into the PS/Si interface due to the imperfections in this area. Here the hydrogen is recombined and reacts with oxygen to produce water molecules (*Cathode:* $2H^+ + 2e^- + \frac{1}{2} O_2 \rightarrow H_2O$*)*.

The hydrogen-containing gas - voltaic effect, i.e., the generation of the open-circuit voltage was also observed for the Au/PS/PS structure in H_2S gas atmosphere (Fig. 10) (Dzhafarov et al., 2008). Chemical reactions generation of electricity taking place for the case of the Au/PS/Si cell in the H_2S gas+air atmosphere will be considered in the late. The sensitivity of the Au/PS cell to H_2S gas was about 1.8 mV/ppm.

This phenomenon is reversible, i.e., for Au/PS/Si structures, inserting and removing the structure from the gas is accompanied by the response and recovery of V_{oc} , respectively. Fig. 11 shows the response-recovery behavior of the open-circuit voltage of the Au/PS/Si cell after successive cycles of placing the cell in a H_2S atmosphere (45 ppm). It can be seen that the response time is about 60 s. Sensitivity of Au/PS/Si structure to H_2S gas (45 ppm) defined from the relation

$$S=\{V_{oc}(gas) - V_{oc}(0)\}/V_{oc}(0) \qquad (22)$$

is equal 16.5. Here V_{oc}(gas) and V_{oc}(0) are the open-circuit voltage for concentration of H_2S gas 45 ppm and 0, respectively (Fig. 10).

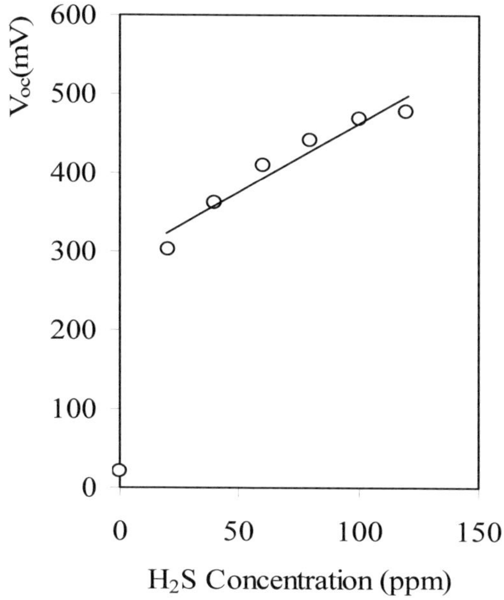

Fig. 10. Voltage generated by Au/PS/Si sensor as a function of the H_2S gas concentration.

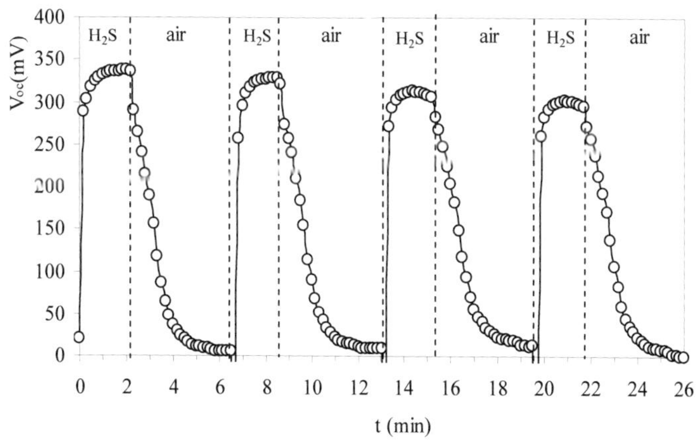

Fig. 11. Voltage response of the Au/PS/Si sensor exposed to H_2S gas (45 ppm).

To overall, the voltage generation mechanism in Au/PS/Si structure at H_2S atmosphere is similar with the above-described mechanism for humid ambient.

Fig. 12 illustrates the dependence of the open-circuit voltage generated in Au/PS structures on the CO concentration at 45 and 58% RH (Dzhafarov et al., 2008). An almost linear increase of V_{oc} is observed with increasing CO concentration. V_{oc} also increases with rising RH. The increase of voltage at higher humidity observed in Fig. 12 can be related to the influence of the increased concentration of water molecules in air. The voltage generated (about 30 mV), at a zero concentration of CO gas is related to the humidity of air.

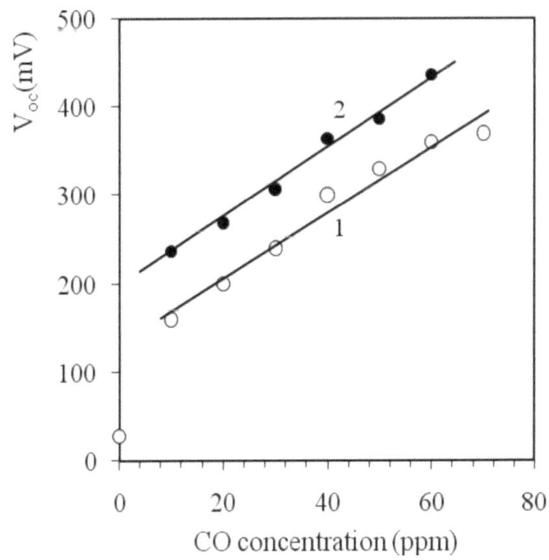

Fig. 12. Voltage generated by Au/PS/Si sensor as a function of the CO gas concentration in (1) 45 % RH and (2) 58 % RH.

Concerning the operating principle of the Au/PS/Si sensor to detect the CO molecules, one can suppose that at the first stage, the water molecules from the humid air in the presence of the Au catalyst interact with the CO, resulting in the formation of hydrogen, according to the electrode reaction (Kohl, 2001)

$$CO + H_2O \rightarrow CO_2 + H_2 \quad (23)$$

$$H_2 \rightarrow 2H^+ + 2e^-. \quad (24)$$

After the reaction described by eq. (23), the formation of electricity proceeds in accordance with eq. (24). Note that in spite of the absence of hydrogen in the CO gas, this gas is, nevertheless, hydrogen-producing, as shown by equations (23) and (24).

Thus, the above results, i.e., the voltage generation in the Au/PS/Si Schottky-type structures in a humid, CO, or H_2S atmosphere indicate both the fuel cell and gas sensor functionalities of these structures.

The effect of the voltage generation was discovered upon dipping the Au(Ag,Cu)/PS/Si structures into different *hydrogen-containing solutions* (KOH (potassium hydroxide), CH_3CH_2OH (ethanol), CH_3OH (methanol), $C_6H_{12}O_6$ (glucose), H_3BO_3 (boric acid), C_5H_{12} – $C_{16}H_{34}$ (benzine), $NaBH_4$ (sodium borohydride), Black sea-water etc.) (Dzhafarov et al., 2004).

Hydrogen as a source of clean energy has a number of attractive advantages, such as high energy density, nontoxic reaction products and abundant natural resources. However, the application of hydrogen for power sources has been greatly restrained due to the lack of safe and convenient generation and storage methods. Hydrogen can be stored in high pressure tanks or liquefied H_2, or by adsorption on activated carbon, carbon nanotubes or in hydrogen-storing alloy (Momirlan & Veziroglu, 2002). Among hydrogen-storing materials, sodium borohydride ($NaBH_4$) due to the high hydrogen content (about of 10% wt or 100 kg/m^3) is suggested as new fuel media supplying hydrogen at normal temperatures. The catalytic hydrolysis of sodium borohydride proceeds as

$$NaBH_4 + 2H_2O \rightarrow NaBO_2 + 4H_2 \quad (25)$$

As result of this reaction, each mole $NaBH_4$ can generate 4 moles of hydrogen (8 protons) gas. This is a very effective system, since we are getting hydrogen out of the water as well. We get eight hydrogen atoms (four molecules) from just four atoms of hydrogen in the $NaBH_4$.. The hydrogen is formed right there on the anode, and is thus immediately used by the fuel cells. $NaBH_4$ is a nonreversible chemical hydride for the one-time hydrolysis generation of H_2. It is shown that $NaBO_2$ byproduct of reaction (20) can be recycle back to $NaBH_4$ using coke or membrane (Kojima & Haga, 2003).

The effect, similar to humidity-voltaic effect (Dzhafarov et al., 2004)], i.e. the generation a voltage between the contacts to Au film and Si under humidity exposition, was also discovered on dipping of Au/PS/Si structure into $NaBH_4$ solution (Dzhafarov et al., 2010). The fuel reactions on anode (Au) surface immersed in $NaBH_4$ fuel and on cathode side (Si) stayed in air is as follows:

$$\text{Anode: } NaBH_4 + 2H_2O \rightarrow NaBO_2 + 8H^+ + 8e^- \quad (26)$$

$$\text{Cathode: } 8H^+ + 8e + 2O_2 \rightarrow 4H_2O \quad (27)$$

The dependence of the $NaBH_4$:H_2O solution conductivity on $NaBH_4$ content measured by using "3200 Conductivity Instrument" (YSI) is presented in Fig. 13. As seen from this figure, the conductivity increases with increasing $NaBH_4$ content. This effect can be caused by hydrogen (proton) production from $NaBH_4$ in water (see eq. (27)).

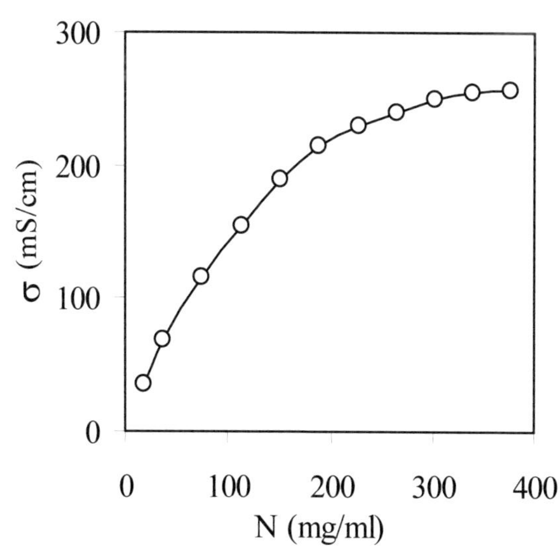

Fig. 13. Conductivity of $NaBH_4$: H_2O solution vs. $NaBH_4$ content.

Fig. 14 illustrates the open-circuit voltage arising in Au/PS/PS structure dependent on the concentration of $NaBH_4$ solution (N) (Dzhafarov et al., 2005; 2010). It is seen that the V_{oc} - N dependence seems to be a curve with a maximum (V_{oc} =550 mV) at N = 30 mg/ml. In the range of concentration of $NaBH_4$ solution up to 30 mg/ml gradient of curve (dV_{oc}/dN) is large (about 13 mV.ml/mg), i.e. voltage generated in Au/PS/Si structure is very sensitive to $NaBH_4$ concentration. The Au/PS/Si structure on dipping into $NaBH_4$ solution also exhibited the short-circuit current (about of 2.5 mA/cm^2). pH measuring of $NaBH_4$ solution showed that with increase of concentration of $NaBH_4$ from 10 to 30 mg/ml, pH of the solution increases from 9.3 to 9.5, i.e. increase of pH of $NaBH_4$ solution results in increase of voltage generated in the cell (Fig. 14). It should be noted that, as opposed to Au/PS/Si structure, in PS/Si structure (without the Au film) the marked generation of electricity on dipping it in $NaBH_4$ solution was not observed.

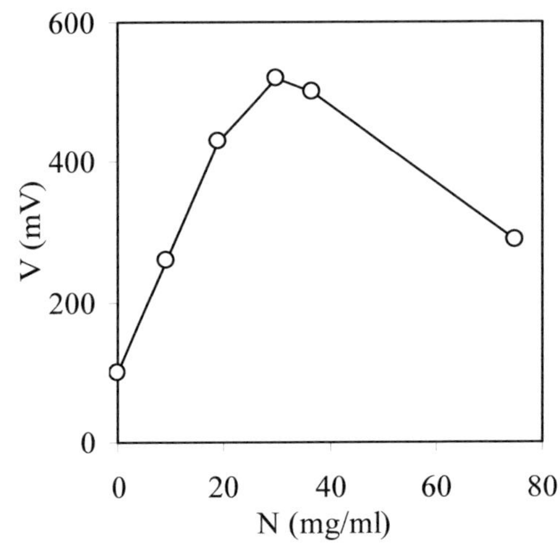

Fig. 14. The voltage generation in Au/PS/Si cell as a function of $NaBH_4$ contains in $NaBH_4$:H_2O.

The V_{oc}-N dependence observed in Fig. 14 can be explained by two concurrent phenomena. The increase of $NaBH_4$ concentration, for the relative low contents of $NaBH_4$ (N<30 mg/ml), is accompanied by increasing of proton concentration (up to 2x10^{11} cm^{-3}) and this results in rising of voltage. We assume that at large content of $NaBH_4$ (N>30 mg/ml), products of reaction (26) precipitate in pores of Au and porous silicon and thereby hindering penetration of protons to interface. Non-observation of the electricity generation in PS/Si structure (without Au film) can be explained by absence of the catalyst Au film stimulating splitting of $NaBH_4$ molecules. On other hand, at large contents of $NaBH_4$ (N>30 mg/ml), products of reaction (26) precipitate in pores of Au and porous silicon and thereby hindering penetration of protons to interface. Thus, Schottky-type Au/PS/Si structures sunk in $NaBH_4$ and other hydrogen-containing solutions generate a voltage up to 550 mV. These data indicate on availability of using Au/PS/Si structures as hydrogen cells.

The use of Au/Porous silicon/ Silicon-Schottky type structure as a direct hydrogen sulphide fuel cell was also demonstrated in (Dzhafarov & Aydin Yuksel, 2011). The porous silicon filled with hydrochloric acid was developed as a proton conduction membrane. The Au/PS/Si cells were fabricated by first creating the porous silicon layer in single-crystalline

Si using the anodic etching under illumination and then deposition Au catalyst layer onto the porous silicon. The performance of the fuel cell was measured at room temperature by supplying hydrogen sulphide solution and the Black sea-water. Current-voltage characteristics of Au/PS/Si cell at normal air ambient (300 K, 40%RH), distilled water and H_2S+H_2O solution were shown in Fig. 15. The hydrogen sulphide-stimulated generation of the open-circuit voltage (up to 400 mV) and the short-circuit current (up to 1 mA) was observed for the I-V characteristics of Au/PS/Si cells sunk in H_2S-containing solution (Fig.15, curve 3), whereas value of V_{oc} generated in the distilled water (without H_2S) is markedly lower (about 120 mV) (Fig.15, curve 2). It should be noted that such effect of electricity formation was not observed in the PS/Si structures lacking the Au film.

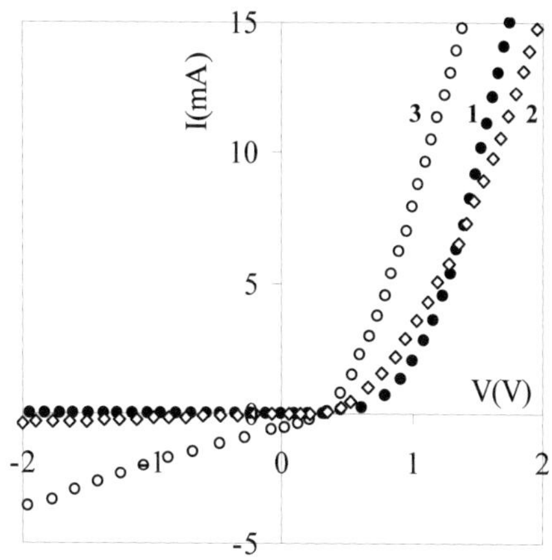

Fig. 15. The current-voltage characteristics of Au/PS/Si structure in (1) air (300K, 40%RH), (2) distilled water and (3) H_2S+H_2O solution (80mM).

The open-circuit voltage generation for the Au/PS/Si cell in H_2S+dH_2O solution at different concentration of H_2S is shown in Fig.16.

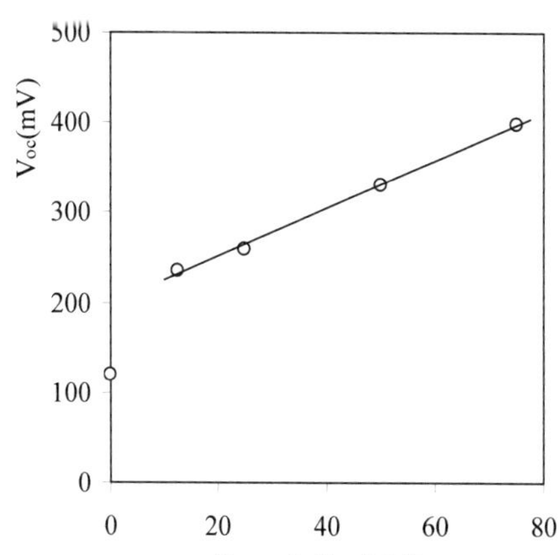

Fig. 16. The voltage generation in Au/PS/Si cell as a function of H_2S concentration in distilled water.

Almost linear increase of voltage (from 230 to 400 mV) is observed with increasing the H_2S concentration from 17 to 75 mM. Value of V_{oc} in pure water (120 mV) is markedly lower than that in hydrogen sulphide-containing solution (Dzhafarov & Aydin Yuksel, 2011).
Fig. 17 shows the power output-current density curves of Au/PS/Si cell in H_2S+H_2O solution. Results were gained at room temperature with 12, 23, 50 and 80 mM hydrogen sulphide. From the polarization curves we found that maximum of the power density of the single direct hydrogen sulphide cell increases with concentration of H_2S and reaches 30 W/m^2 at 80 mM. Such increasing the maximum of the power density with H_2S concentration can be caused by increasing of concentration of protons forming in Au catalyst and diffusing into porous silicon layer.

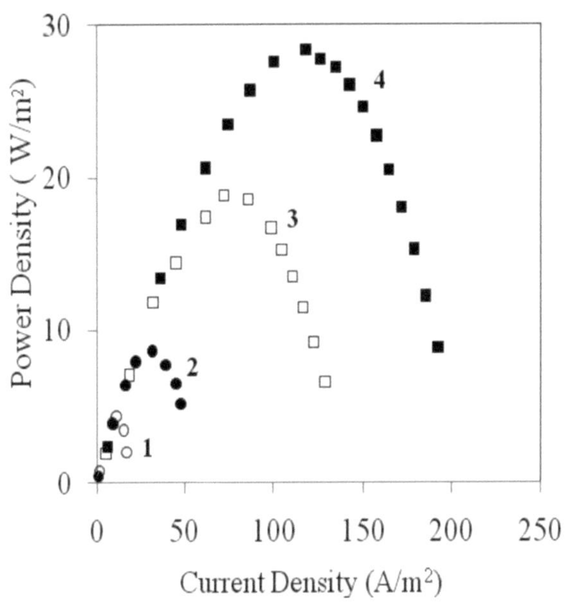

Fig. 17. The power density - current density characteristics of Au/PS/Si cell in H_2S+dH_2O solution of (1) 12 mM, (2) 23 mM, (3) 50 mM and (4) 80 mM (300 K).

It is known that the Black sea-water contains significant concentration of hydrogen sulphide (about 0.05-1.50 mg/ml depending on deep water), therefore we made attempt to use the Black sea-water as fuel for Au/PS/Si cells. Indeed, for the Au/PS/Si cell sunk in the Black sea-water, as in case it dipping into H_2S+H_2O solution generation of electricity up to 0.4 V was observed. It is be noted that successive placing in and removal of the Au/PS/Si cell from H_2S-containing solution is accompanied by the response and recovery of the value of V_{oc} with response time about 40-60 s.
The mechanism of the generation of the electricity in the metal/PS/Si cells under hydrogen-containing ambient has already been proposed (Dzhafarov et al., 2008). We suggest that in Au/PS/Si cell, similar to direct ethanol fuel cell [39], the Au film and PS layer filled with hydrochloric acid play the role of the catalytic anode and electrolyte respectively. In the case of Au/PS/Si direct hydrogen sulphide fuel cell, H_2S+water solution as fuel is supplied at the Au anode and porous silicon layer acts as the electrolyte. The following reactions take place in the presence of a catalyst (Basu, 2007; Hoogers, 2003)

At the anode: $H_2S + 2H_2O \rightarrow SO_2 + 6H^+ + 6e^-$ (28)

At the cathode: $6H^+ + 6e^- + 3/2O_2 \rightarrow 3H_2O$ (29)

The overall reaction can be expressed as

$$2H_2S + 3O_2 \rightarrow 2SO_2 + 2H_2O \quad (30)$$

Electrons and protons formed in the Au catalyst film after hydrogen splitting ($H_2 \rightarrow 2H^+ + 2e^-$), pass through the external circuit and PS layer, respectively and reach the cathode (PS/Si interface) region. The oxygen from air can easily penetrate into the PS/Si interface due to the imperfections in this area. Here the hydrogen is recombined and reacts with oxygen to produce water molecules. Generation of the electricity in pure water without H_2S (Fig.16) can be related with water splitting in part on the Au catalyst layer ($2H_2O \rightarrow 2H_2 + O_2$ and $H_2 \rightarrow 2H^+ + 2e^-$).

Stability of operation for Au/PS/Si cells as a function of storage time in the presence of the hydrogen sulphide solution (80 mM) was observed for about 100 h. More long-time tests of operation stability of the cell are currently under investigation. Moreover, nature of molecules responsible for generation of electricity on placing of the Au/PS/Si cell in the Black sea-water is yet remained non-clear.

These results allow supposing that the Au/PS/Si structures can be used as the hydrogen sulphide miniature fuel cells. In addition to this, the nearly linear dependence the open circuit voltage generated in Au/PS/Si cell in dependence on concentration of H_2S in hydrogen sulphide containing solution (Fig. 16) permits of using such structures as hydrogen sulphide sensors with sensitivity about 2.5 mV/mM.

The similar effect of voltage generation was also observed during dipping of Ag/PS structures in other hydrogen-containing solutions (the distilled water, fresh water, Black sea-water, ethanol and methanol) (Table 2) (Dzhafarov et al., 2008).

Ambient	Conductivity of solution (mS/cm)	V (mV)
Humid (95% RH)		430
Distilled water	0.08	505
Fresh-water	0.47	514
Black sea-water	48.4	559
Ethanol	0.0095	254
Methanol	0.047	380

Table 2. Data on V_{oc} generated in Ag/PS/Si cells in humid and in the hydrogen-containing solutions

Largest value of voltage, generated in Black sea-water for the cell can be caused by contribution of the hydrogen sulphide molecules in sea-water. As can be seen from Table 2, the correlation between changes of the V_{oc} and J_{sc} parameters of Ag/PS/Si cells and the conductivity of hydrogen-containing solution is observed. Such correlation can be tentatively explained by an increase of concentration of the charge particles, deposited on the pore surfaces with increase of conductivity of solutions.

6. Conclusion

Thus, the Metal (Au, Ag or Cu)/Porous silicon/Silicon structures placed in hydrogen-containing gas or liquid ambient generate voltage (400-550 mV) and maximum power density of 30 W/m^2. These results demonstrate the feasibility of development of low-cost small porous silicon-based direct hydrogen fuel cells for portable electronics and medical applications. Moreover, the Au/Porous silicon structures can be used as hydrogen-sensitive sensors.

7. References

Abdullaev, G.B. &. Dzhafarov T.D. (1987). *Atomic Diffusion in Semiconductor Structures*, Harwood Academic Press, New York, ISBN 3-7186-0413-2

Aravamudhan, S., Rahman, A. & Bhansali, Sh. (2005). Porous Silicon Based Orientation Independent Self-priming Micro Direct Ethanol Fuel Cell. *Sensors and Actuators*, Vol.A123-124, (2005), pp. 497-499, ISSN 0924-4247

Basu, S. (2007). *Recent Trends in Fuel Cell Science and Technology*, Springer, New York, ISBN 978-0-387-355537-5

Beale, M.I.J, Benjamin, J.D., Uren, M.J., Chew, N.G. & Cullis, A.G. (1985). An Experimental and Theoretical Study of the Formation and Microstructure of Porous Silicon. *Journal of Crystal Growth*, Vol. 73, (1985), pp. 622- 636, ISSN 0022-0248

Bisi, O., Ossicini, S. & Pavesi, L. (2000). Porous Silicon: a Quantum Sponge Structure for Silicon Based Optoelectronic. *Surface Science Reports*, Vol.38, No.1-126, (April 2000), pp.1-129, ISSN 0167-5729

Bove, R. & Ubertini, S. (eds.). (2008). *Modeling Solid Oxide Fuel Cell*, Springer, Berlin, ISNB-13: 978-1-4020-6994-9, e-ISBN-13: 978-1-4020-6995-6

Bunker, C.E., Simmons, K.J. & Rollins, H.W. (1998). Steady-state and Time-resolved Fluorescence Spectroscopic Probing of Microstructures and Properties of Perfluorinated Polyelectrolyte Membranes. *Journal of Electroanalytical Chemistry*, Vol. 459 (1998) pp.15-28, ISSN 1572-6657

Canham, L.T. (Ed.) (1997). *Properties of Porous Silicon*, IEE-Inspec, London, 0-86341-555-5

Chu, K., Shanon, M.A. & Masel, R.I. (2007). Porous Silicon Fuel Cell for Micro Power Generation. *Journal of Micromechanics* and *Microengineering*, Vol.17, (2007), pp. S243-S247, ISSN 0960-1317

Cook L.M. (1990). Chemical Processes in Glass Polishing. *Journal of Non-crystalline Solids*, Vol.120, No.4, (1990), pp.152-171

Crawley, G. (2007). Direct Methanol Fuel Cells (DMFC). Fuel Cell Today (Informing for Fuel Cell Industry), (August 2007)., pp.1-13

D'Arrigo, G. D., Spinella, C., Arena, G. & Lorenti, S. (2003). Fabrication of Miniaturized Si-based Electrolytic Membranes. *Materials Science and Engineering* C. Vol. 23. No.1-2, (January 2003), pp.13-18, ISSN 0928-4931

Dzhafarov, T.D. & Can, B. (2000). The Diffusion Distribution of Hydrogen and Oxygen in Porous Silicon Films. *Journal* of *Materials Science Letters*, Vol.19, (2000), pp. 287-289, ISSN 0261-8028

Dzhafarov, T.D. & Can, B. (2000). Humidity-voltaic Characteristics of Ag-PS-Si Structures. *Journal* of. *Materials Science Letters*, Vol.19, (2000), pp.2193-2195, ISSN 0261-8028

Dzhafarov, T.D., Can Omur, B., Oruc, C. & Allahverdiev, Z.A. (2002). Hydrogen-sensing Characteristics of Cu-PS-Si Structures. *Journal of Physics D: Applied Physics*, Vol. 35, (2002), pp.3122-3126, ISSN 022-3727

Dzhafarov, T.D., Oruc, C. & Aydin, S. (2004). Humidity-voltaic Characteristics of Au-Porous Silicon interfaces. *Journal of Physics D: Applied Physics*, Vol. 37, (2004), pp. 404-408, ISSN 022-3727

Dzhafarov, T.D., Oruc Lus, C. & Aydin, S. (2005). Metal/Porous Silicon Schottky-type Hydrogen Fuel Cells. *Proc. of Ninth Grove Fuel Cell Symposium*, pp. 17-19, London, (4-6 October 2005)

Dzhafarov, T.D., Aydin, S. & Oruc Lus, C. (2007). Effect of Humidity on Optical and Electrical Properties of Free-standing Porous Silicon Films. *Fizika*, Vol.8, No.5, (2007), pp. 17-20, ISSN 1028-8546

Dzhafarov, T.D., Aydin Yuksel, S. & Oruc Lus, C. (2008). Porous Silicon-based Gas Sensors and Miniature Hydrogen Cells. *Japan Journal of Applied Physics*, Vol. 47, No. 10, (2008), pp. 8204-8207, ISSN 0021-4922

Dzhafarov, T.D., Aydin, S. & Oren, D. (2008). Effect of Diffusion of 1 Group Metal on Characteristics of Metal/Porous Silicon Sensors. *Solid State Phenomena*, Vol.131-133, (2008), pp. 189-194, ISSN 1012-0394

Dzhafarov, T.D. & Aydin, S. (2009). Nano-porous Silicon for Gas Sensors and Fuel Cell Applications. *Journal of Gafqaz University*, No. 25, (2009), pp. 20-35, ISSN 1302-6763

Dzhafarov, T.D. & Aydin, S. (2010). Diffusion of Hydrogen in Porous Silicon-based Sensors. *Journal of Porous Media*, Vol.13, No.2, (2010), pp. 97-102, ISSN 1091-028X

Dzhafarov, T.D., Oruc Lus, C., Aydin Yuksel, S., Caliskan, M. & Yesilkaya, S.S. (2010). Au/Porous Sillicon-based Sodium Borohydride Fuel Cells. *International Journal of Energy Research*, Vol.34, No.15, (2010), pp. 1386-1392, ISSN 0363-907X

Dzhafarov, T.D. & Aydin Yuksel, S. (2011). Silicon-based Direct Hydrogen Sulphide Fuel Cells. *Journal of Nanoscience and Nanotechnology*, Vol.11, No.3, (2011), pp. 843-848, ISSN 1550-7033

Feng Z.C. & Tsu, R. (1994). *Porous Silicon*, World Scientific, ISBN 981-02-1634-3, Singapore

Gold, S., Chu, K., Lu, C., Shanon, M.A. & Masel, R.I. (2004). Acid Loaded Porous Silicon as a Proton Exchange Membrane for Micro Fuel Cells. *Journal of Power Sources*, Vol.135, (2004), pp.198-203, ISSN 0378-7753

Gyoko, N. & Naohiro I. (2005). Porous Silicon as a Proton Exchange Membrane for Micro Fuel Cells. *Electrochemistry*, Vol. 73, No. 11, (2005), pp. 939-943, ISSN 0013-4651

Hahn, R., Wagner, S. & Reicht, H. (2007). Chapter 8. Constraint of PEM Micro Fuel Cells for Portable Electronics, In: *Fuel Cell Electronics Packaging*, Kuang, K. & Easter, K., pp. 145-163, Springer, Berlin, ISBN-13: 978-0-387-47323-9, e-ISBN-13: 978-0-387-47324-6.

Hayase, M., Kawase, T. & Hetsuzawa, T. (2004). Miniature 250 μm Thick Cell with Monolithically Fabricated Silicon Electrodes. *Electrochemical Solid-State Letters*, Vol. 7, (2004), No. 8, pp. A231-234.

Herino, R (1998). Pore Size Distribution in Porous Silicon, In: Canham, L. (Ed.). *Properties of Porous Silicon*, Inspec, London, pp. 89-98, ISNN 0-86341-555-5

Hockaday, R.G. (2000). *Microfuel Cells for Portable Electronics*, Manthattan Scientifics, Power, Santiego

Hoffmann, P. (2002). *Tomorrow Energy: Fuel Cells and the Prospect for a Cleaner Planet.* MIT Press, New York, ISBN-10: 026258221X, ISBN-13: 978-0262582216

Hoogers G. (Ed.). (2003). *Fuel Technology Handbook*, CRC Press, New York, ISSN 1095-4244

Jung D.H. (2002). Performance Evaluation of a Hybrid Nafion®/Silicon Oxide Membrane for Direct Methanol Fuel Cell. *Journal of Power Sources*, Vol. 106, (2002), pp.173-177, ISSN 0378-7753

Kamarudin, S.K., Achmad, F.& Daud, W.R.W. (2009). Overview on the Application of Direct Methanol Fuel Cell (DMFC) for Portable Electronic Devices. *International Journal of Hydrogen Energy,* Vol. 34, No. 16, (August 2009), pp. 6902-6916.

Kojima,Y. & Haga,T. (2003). Recycling Process of Sodium Metaborate to Sodium Borohydride. *Journal of Hydrogen Energy,* Vol.28, (2003), pp. 989-993, ISSN 0360-3199

Kohl, D. (2001). Function and Application of Gas Sensors. *Journal of Physics. D,* Vol.34, (2001) pp. R125-R149, ISSN 022-3727

Lee, C., Lee, S., Huang, R. & Chuang, C. (2007). Application of Porous Silicon on the Gas Diffusion Layer of Micro Fuel Cells. *Key Engineering Materials,* Vol.364-366, (2007), pp. 849-854

Lysenko, V., Bidault, F. & Alekseev, S. (2005). Study of Porous Silicon as Hydrogen Reservoirs, Journal of Physical Chemistry B, Vol.109, No. 42, (2005), pp. 19711-19718, ISSN 1089-5647

Maynard, H.L. & Meyers, J.P. (2000). Small Fuel Cells. *Proc. of 2-nd Annual Small Fuel Cells and Battery Technologies for Portable Power Applications Int. Symp.,* New Orleans, 2000

Mench, M.M. (2008). *Fuel Cell Engines,* Wiley, New York, ISBN 978-0-471-68958-4

Moghaddam, S., Pengwang, E., Masel, R.I. & Shannon, M.A. (2009). Nanostructured Silicon-based Proton Exchange Membrane for Micro Fuel Cells. *Power MEMS,* December 1-4 (2009), pp.116-119

Momirlan, M. & Veziroglu, T.N. (2002). Current Status of Hydrogen Energy. *Renewable and Substainable Energy Reviews,* Vol. 6, No. 1-2, (2002), pp. 141-179, ISSN 1364-0321

Ohashi, M., Guo, Z., Wang, N. & Tseng, K.H. (2010). Small DMFC for Portable Applications, *Fujika Technical Review,* Vol. 1, (2010), pp.44-46

O'Hayre, R., Cha, S.W., Corella,,W. & Prinz, F.B. (2009). *Fuel Cell Fundamentals,* Wiley, New Jersey, ISBN-13: 978-0-471-74148-0, ISBN-10: 0471-74148-5

Pichonat,T. & Gauthier-Manuel,B. (2005). Development of Porous Silicon-based Miniature Fuel Cells., *Journal of Micromechanical and Microengineering,* Vol.15, No. 9, (2005), pp.179-184, ISSN 0960-1317

Song, J.H. & Sailor, M.J. (1999). Chemical Modification of Crystalline Porous Silicon Surfaces. *Comments on Inorganic Chemistry,* Vol.21, No. 1-3, (1999), pp. 69-84, ISSN 0260-3594

Starkov, V.V., Redskin, A.N. Dubonos, S.V. (2006). Carbon Nanofibers in a Gradient-Porous Silicon Structure. *Technical Physics Letters,* vol. 32, (2006), No. 1, pp. 82-83

Stewart, M.P.& Buriak J.M. (2000). Chemical and Biological Application of Porous Silicon Technology. *Advanced Materials,* Vol.12, No. 12, (2000), pp. 859-869, ISSN 0935-9648

US PATENT 5266421-Enhanced membrane-electrode interface, granted 2008-11-30

Varga, A. (2007). Chapter1. Introduction to Fuel Cell Technology, In: *Fuel Cell Electronics Packaging,* Kuang, K. & Easter, K., pp. 145-163, Springer, Berlin, ISBN-13: 978-0-387-47323-9, e-ISBN-13: 978-0-387-47324-6

Vielstich, W., Gasteiger, H.A. & Yokoka, H. (eds.) (2009). *Handbook of Fuel Cells: Advances in Electro-catalysts, Materials, Diagnostics and Durability.* Vol. 5 & 6. Wiley, Hoboken, ISSN 978-0-470-74151-1

Vielstich,W.,Lamm,A. & Gasteiger, H.A. (eds.). (2003). *Handbook of Fuel Cells: Fundamentals, Technology, Applications.* Vol.1-4, Wiley, New Jersey, ISBN: 978-0-71-49926-8

Yamaguchi, A., Ibe, M., Nair, B.N. & Nakao, S. (2002). A Pore-filling Electrolyte Membrane-Electrode Integrated System for Direct Methanol Fuel Cell Application. *Journal of Electrochemical Society,* Vol. 149, No. 11, (2002), pp. A1448-1453, ISSN 0013-4651

http://public.itrs.net/, International Roadmap for Semiconductors, executive summary. (2003). *Technical report,* SIA

14

Hydrogen - The Ecologically Ideal Energy Vector

Mircea Cârdu[1] and Malvina Baica[2]
[1]HERBING S.R.L, Bucharest
[2]Department of Mathematical and Computer Sciences, The University of Wisconsin
[1]Romania
[2]USA

1. Introduction

Strategic analysis of the energy sector of the global economy reveals two very important elements which have a substantial influence on the development process in this domain. These elements are:

The exhaustion of available conventional fuels (natural gas, coal, oil) in the near future. The time limit for this process is estimated to be a few decades for natural gas and oil, and about two centuries for coal. It is evident that, as these fuel reserves are depleted over time, their cost will continuously increase. There has been a tremendous increase in terrestrial atmospheric pollution, from noxious and greenhouse gases, emitted during the combustion process in conventional fuels by the energy industry (thermo power plants), as well as metallurgical, chemical, and construction materials, and from energy for aviation, naval, auto and railway transport. The continuous increase in the average temperature at the earth's surface, due to the greenhouse effect produced by CO_2 accumulation in the atmosphere, has an increasingly intense effect on terrestrial flora and fauna, as well as on climate change.

This apocalyptic scenario of mankind's future should mobilize the international community to take global measures to reduce atmospheric pollution, and to produce useful energy especially electrical energy from non-conventional sources (sun, wind, sea, waves, tide, and geothermal energy).

Although these sources have very little constancy (with a reduced energy density), and to place a value on them is both difficult and unsatisfactory, when viewed against mankind's permanently increasing demand for energy, these sources are well suited for the production of electrical and thermal energy. However, to produce the energy needed to propel all the means of transportation currently available, there is a need for an intermediate, non-polluting energy carrier from unconventional energy sources; one of the most promising of such energy carriers in this situation is Hydrogen.

Hydropower plants (HPPs), and very often nuclear power plants (NPPs), are only suited for production of electrical energy. Hydraulic energy cannot be directly used to power vehicles, and the use of nuclear energy is not feasible on a large scale in the near future, except for ships and submarines. Resources for hydraulic energy are limited, although regions with large energy consumption (Europe, North America) which are equipped to use hydraulic energy resources, the economical advantages are considerable high. As for the majority of

unconventional resources HPPs can produce only electrical energy consequently their contribution to energy for transportation is limited to the production (by electricity) of an intermediate energy carrier, which could be Hydrogen.
Hydraulic energy resources are limited, but as a result of natural water circulation on the planet, they are recyclable. Evidently, producing electrical energy in HPPs does not result in noxious or greenhouse gas emissions into the atmosphere. Thus, from this point of view, HPPs represent a very advantageous electrical energy source. However, HPP's can have a somewhat negative effect due to perturbation of the ecological equilibrium in the local ecosystems where they are located.
Nuclear energy is the future alternative to producing energy from conventional fuels; nuclear fuel reserves are estimated to last for two to three centuries. Moreover, there is also the possibility of recycling these primary nuclear fuels naturally existent by using them to reproduce nuclear reactors. In this way, natural nuclear fuel resources can be amplified. NPPs do not produce noxious or greenhouse gas emissions.
Considering the increasing global energy and fuel demand, and at the same time attempting to maintain a strict limit on atmospheric pollution, the use of primary energy resources such as coal and hydrocarbons should be continued on the condition, that these resources be recycled, in order to limit, as much as possible, the continuing degradation of our environment.
At the present time, many proposals exist for recycling carbon from conventional fuels, using a chemical process for the carbon dioxide (CO_2) resulting from their combustion. This process in turn requires a great deal of energy consumption, but it results in hydrocarbons, alcohol or methanol, high quality fuels.
CO_2 collection for recycling is a practical possibility at stationary energy installations, but for vehicles CO_2 collection and its storage are operations which involve enormous technical problems. For ecological reasons, and because CO_2 is the principal greenhouse gas, studies are being conducted which consider the possibility of storing huge quantities of this gas for later use.
Careful analysis of these two important problems, which have arisen because of the continuing, large-scale use of conventional fuels, will lead to the conclusion that Hydrogen is the ideal energy carrier of the future. Mainly, we refer to molecular Hydrogen, which is stable and can be easily used in practical applications. Further justification for this conclusion will follow.
Hydrogen is not a natural primary energy source, as is, for example, coal. Molecular Hydrogen is found in natural states in the cosmos only where it is free in a proportion of 70%, compared with the total interstellar matter. It occurs in extremely reduced quantities compared with the volume of the stars, including the Sun, where Hydrogen is 70% of their composition. The atmospheres of exterior planets of the solar system (Mars, Jupiter, Saturn Uranus Neptune, Pluto), have 90% hydrogen content; the remaining 10% is helium (He).
On Earth, Hydrogen can come from an inexhaustible source, water, but to obtain Hydrogen from water requires the use of another energy form (solar, electrical etc.) in a quantity of 20-80%, compared with the energy obtained from its use as a fuel. Thus, we consider its denomination as the energy vector, and not as an energy source. Since, by burning Hydrogen we obtain water vapor, it's recycling from water is achieved by an automatic integration of the water cycle in nature. Water vapors are not toxic, and from an ecological point of view, Hydrogen used as fuel represents an ideal energy carrier (Cârdu & Baica, 1999). This offers a real advantage for Hydrogen, if it is obtained from water using a technology where a primary nonpolluting energy source, such as solar, hydraulic or nuclear energy is utilized.

The characteristics of Hydrogen will be analyzed as a stable fuel; methods for production, storage, and transportation, as well as the technical systems employed in using it as fuel, will be examined.

2. Characteristics of hydrogen

In this section, the physical and thermodynamic characteristics of Hydrogen which determine its behavior as fuel, will be discussed. Table 1 (see -page 346) gives for comparison, the principal characteristics of Hydrogen and some hydrocarbons currently used, especially in transportation. In order to proceed systematically, these characteristics will be analyzed and grouped into two categories, advantageous and disadvantageous characteristics respectively, in all aspects of its use as fuel.

2.1 Advantages of hydrogen over other fuels

Water is produced by Hydrogen combustion, and as we know, water is not noxious. Because of the nitrogen content in the atmosphere, burning Hydrogen in the air can produce NO_x which is a noxious gas, but for NO_x content reduction, known and applied methods for any classical fuel combustion process can be used in power plants or thermic engines.
The reduced value of NO_x content in gases resulting from combustion is due to the reduced air quantity for stoichiometric combustion, related to the resultant energy in Hydrogen, compared with other fuels. Thus, knowing the inferior heat power compared to the mass-unit of Hydrogen is 121 MJ/kg, and the air quantity needed for the stoichiometric combustion of a kg of Hydrogen is of 34.2 kg/kg, it follows that the specific air quantity needed in the combustion for each MJ of heat released is, in the case of Hydrogen about 280 g. Some other conventional fuels of superior quality (gasoline, methanol or methane gas), have a higher value of these characteristics, which is about 340 g. Given that Hydrogen has the smallest molecular mass compared with all other elements or chemical combinations, it is the best reactive fluid which can be utilized in a rocket propulsion system; the propulsive force of such a fluid is so much greater due to the fact that the molecular mass of the respective fluid is smaller. Thus, in this case we cannot say that Hydrogen can be used as fuel, but only as jet fluid. This problem will be analyzed in detail in another chapter, in which the problems of using Hydrogen as a fuel will be addressed. The diffusion velocity value in air of Hydrogen is greater than that of any other fuel. This represents an advantage for forming a homogeneous mixture with air, or Oxygen, in a very short time, with a positive effect on the burning quality (a more complete combustion process).
The domain between the two inflammation limits of Hydrogen is very large, especially in the poor mixture zone (very large value $\lambda > 2$ for the air excess). This characteristic represents an advantage, particularly when Hydrogen is used as fuel for thermic engines. Such engines can be more stable functioning in conditions with large air temperature variations; these conditions do not require a highly precise control of the feeding and ignition.

2.2 Disadvantages of hydrogen compared with other fuels

The density of Hydrogen, both in gaseous and liquid form, has very low value compared with all other gaseous or liquid fuels. This characteristic of Hydrogen represents a disadvantage compared with other fuels, especially when Hydrogen is used in transportation, since the volume and the mass of the fuel containers for the vehicles is by necessity very large, and the distance covered between two successive refueling is substantial. The high value of Hydrogen's heat power, relative to the mass unit, reduces this disadvantage to some extent. All these facts lead us to prefer the storage of Hydrogen in liquid form. This storage form in turn represents a disadvantage since it requires the containers to function at very high pressures and very low temperatures. High pressures require heavy containers and low temperatures demand special construction materials, since materials used today, especially steel, become fragile at these temperatures. Moreover, Hydrogen has the ability to diffuse easily through steel at normal temperatures, and so the mechanical characteristics of steel can suffer severe deterioration.

The storage and transportation of Hydrogen in a gaseous form (which can be implemented if it is used for stationary energy installations), also creates important problems because of its very reduced heat power values relative to its volume, as a consequence of its very low density value.

Solving this problem requires large diameter pipes for transportation, and large volume containers for storage. At the same time, to transport Hydrogen using these pipes, and to store it in these containers requires high power compressors.

The characteristics of Hydrogen for ignition energy and burning velocity are advantageous from one point of view, and disadvantageous for other reasons. Thus, fast ignition and burning processes are considered advantages as far as combustion is concerned, especially for engines with compression ignition. On the other hand, fast ignition and high burning velocity can provoke untimely burning initiatives and high pressure gradients inside cylinders, resulting in very high mechanical stress.

The above mentioned advantage can thus be very useful, using Hydrogen as combustion initiator when some conventional fuels (hydrocarbons) mixed with Hydrogen are used.

3. Production, storage and transportation of hydrogen

3.1 Hydrogen production

It has been stated that Hydrogen is not a fuel which can be found naturally on earth, as can the classic fuels (coal, hydrocarbons, and biomass). Hydrogen can be obtained from the chemical combinations in which it exists, just as hydrocarbons or especially, water. It becomes an energy vector, thus it still can be used in its turn as fuel, since its combination with Oxygen is exothermic (combustion), and its final product is water (non-pollutant). Evidently, by extracting Hydrogen from hydrocarbons result in carbon byproducts, typically carbon monoxide (CO) and/or carbon dioxide (CO_2). These byproducts generate pollution; they are noxious and they belong to the category of greenhouse gases. On the other hand, the Hydrogen extraction process involves energy consumption, usually of thermic nature, and its production also requires conventional fuel consumption. This process results in noxious gas emissions in the atmosphere. Thus, Hydrogen produced by this method does not solve either of the problems mentioned in the first section of this paper.

Another way to extract Hydrogen, this time from water, is the gasification of coal or other organic substances. The gasification process involves the consumption of these fuels, in addition to Hydrogen, some pollutant gases, primarily CO_2, are also produced. Consequently, this method of producing Hydrogen, on the basis of carbon (C) or hydrocarbons will not be examined here.

Hydrogen often appears as a secondary product in oil processing, and to use it as fuel also presents an interesting possibility (Peltier, 2007).

In the respective technological processes, Hydrogen appears as a mixture, especially with CO, but also with some other gases, such as CO_2, CH_4, N_2 and water vapors. In these occurrences it is very important to separate Hydrogen from the gas mixture. This separation by membranes is a very frequently used industrial method. Thus, for example, in refineries, Hydrogen results in the production of toluene transformation in benzene at the catalytic isomerism and in the processing of heavy fractions, in order to obtain lighter hydrocarbons. Hydrogen is separated from the gas mixture issuing from these processes with the help of membranes.

The ammonium synthesis process (starting from methane), also results in a very high proportion (about 60%) of Hydrogen in the gas mixture formed. Hydrogen separation can also be realized in this case by using two modules with membranes installed in the cascade.

In the countries with a developed chemical industry, Hydrogen production is very important as a secondary product. Germany is a significant example; the German chemical industry annually produces about 10 billion m^3 of Hydrogen. Hydrogen also results as a secondary product in the heavy water fabrication process, if some special technologies are applied (Peculea, 1995).

Electrolysis is the principal procedure for obtaining Hydrogen from water using electrical energy. If this energy were produced in its turn in power plants using nonpolluting primary energy (hydraulic, nuclear, solar, wind), the problem of obtaining fuel for transportation, could be solved without polluting the atmosphere. The primary energy forms mentioned above (hydraulic, nuclear, etc.) are practically inexhaustible, offering a solution to the exhaustion of global energy resources, and the degradation of the ecological equilibrium, as noted in the first chapter.

The process of producing Hydrogen, using nuclear energy, was also approached in the study *Vision 2020,* produced by the NEI (Nuclear Energy Institute) of the USA. In *Vision 2020* it is shown that sea water desalination and hydrogen production, using the nuclear energy installations, could be as important in the 21st century as electrification was in the 20th century.

The problem of introducing Hydrogen as a fuel on a large scale in contemporary society is not at all simple from the technical and economic point of view.

First, it is evident that using an energy reactor as the primary-source energy circuit continuously, to the final energy utilizer, inevitably results in an energy loss, mainly because of the energy vector production process. Thus, in Hydrogen production by electrolysis, the specific electrical energy consumption is about 4-5 kWh/m^{3}N. Expressed in international units, this specific consumption is about 14,400-18,000 kJ/m^{3}N. The inferior heat power of Hydrogen is 10,880 kJ/ m^{3}N. This means that the electrical energy efficiency to transform Hydrogen in a fuel is $\varepsilon_p = 0.6\text{-}0.75$. In other words, in this process, just for the production process of Hydrogen, about 25-40% of the electrical energy will be lost. This in its turn is produced from a primary energy source with an efficiency $\eta < 1$, depending to its mode of production.

We have to mention that using CNE (NPPs) as energy source to produce Hydrogen by electrolysis presents as disadvantage a risk of radioactive pollution. This thing is more evident now, after what happened at the NPP Fukushima-Japan, as a result of the devastating earth-quake of March 11, 2011.

These considerations of losing up to 40% of the electrical energy when the efficiency is <1 regarding energy source and materials, related to obtaining and preparing water for electrolysis, adapting installations to electrolysis, and determining a price for Hydrogen

(compared to the caloric energy unit produced by combustion) cause the use of Hydrogen to cost more than the price of the hydrocarbons as fuels, respectively. It is agreed that solar energy is a satisfactory choice among the primary non pollutant energy sources for producing Hydrogen by electrolysis. Because of its nature (cyclical variable and variable intensity) solar energy must be accumulated under a special form (Hydrogen storage).
Radiolysis is a physical phenomenon which can be used to decompose water into its component elements. This decomposition is produced in homogeneous nuclear reactors (NR), under the influence of radiation and nuclear particles, which appear as a result of fission in the nuclear fuel. This modality for the production of Hydrogen is of great interest, especially since the process of radiolysis can be organized at relatively reduced values of the NR temperature; thereby increasing security in its operation. However, obtaining Hydrogen by radiolysis is disadvantageous from the point of view of energy efficiency. At a value of about $10^{13}cm^{-2}s^{-1}$ for the neutron flux (normal for a homogeneous NR), the value of this efficiency is only 6%. An increase in efficiency could be gained by increasing the value of the neutron flux in RN.
In thermonuclear fusion it is possible to obtain, for example, neutron flux of (10^{15}-10^{16}) $cm^{-2}s^{-1}$.
There have been some proposals for using photosynthesis (directly or as a model), to produce Hydrogen. It is known that in green plants and algae (especially blue algae), photosynthesis, under the influence of solar radiation, produces a decomposition process of the water molecule into O (emitted in the atmosphere), and Hydrogen, followed by a secondary cycle to recombine Hydrogen and CO_2, thereby obtaining carbon hydroxide, which is needed as plant food.
With this natural process as a model, Professor Jean Marie Lehn, of Strasbourg University, in France invented an industrial system to produce Hydrogen using this method. He proposed the replacement of chloroplasts in the plant cell with a chemical substance based on ruthenium (Ru). This metal (group VIII b in the periodic element table), is capable of absorbing part of the solar radiation. This energy is used to obtain Hydrogen from water, with the help of a solution containing rhodium (Rh), which takes the role of the natural enzyme contained in the plants, and is named Hydrogenise. Research continues in this domain. Also, it is studied the possibility of producing Hydrogen biologically (by fermentation) where is no need for the solar light.

3.2 Storage and transportation of Hydrogen

Hydrogen storage also presents special technical difficulties; these problems result from theoretical considerations, but they are confirmed by NASA's operation staff, which is experienced in storing and handling Hydrogen in liquid form for the Saturn and Centaur rockets.
The simplest way to store Hydrogen is at a high pressure, or at atmospheric pressure, if this is performed when Hydrogen is in a gaseous state. However, given the very low density value of Hydrogen, this method of storage is of no practical value, especially for vehicles, where the volume and weight of fuel transportation reservoirs are especially important. Thus, for an automobile with reduced capacity, about 14 kg of Hydrogen (the equivalent of 40 kg of gasoline), would be necessary and would require 18 gas cylinders with Hydrogen, at a 150 bars pressure. Since such a gas cylinder weighs about 70 kg, the total weight of the fuel storage system would be about 1200 kg greater than the weight of the car.
Up to the present, this problem was solved for rockets, as we have shown, by storing Hydrogen in a liquid state (at -253°C) in very well-insulated thermic reservoirs (cryogenic

reservoirs). Since the density of Hydrogen in a liquid state is much greater than Hydrogen in a gaseous state, the reservoirs have a lower volume and weight, so that Hydrogen storage in a liquid state becomes feasible. This approach is possible in large vehicles, such as trucks and buses. The disadvantage of this storage system is that a cryogenic reservoir is complicated and expensive; it requires a high energy consumption to perform Hydrogen liquefaction; the feeding system is complicated and there is a significant loss during feeding and storage.

The transportation of Hydrogen does not involve more difficulties than for other flammable gaseous liquids however. One example is the actual transportation of gaseous Hydrogen using a pipeline some hundred km long, which existed in Germany for more than 60 years.

One storage method with great future possibilities uses metallic hydrides as the Hydrogen impregnating mass. It is known that some metals dissolve large quantities of Hydrogen at high temperatures. Thus, in inter crystalline interstices of iron (Fe), a volume of Hydrogen can be inserted which is 19 times greater that the volume of Fe, and this characteristic has the value 875 in palladium (Pd). The metals' capacity to react with Hydrogen decreases from left to right in the serial arrangement of the Mendeleev Periodic Table. The elements of the group III b (Sc, Y La, Ac), are most reactive in this situation, and those of group IX b (Co, Rb, It) to X b (Ni, Rd, Pt) are less reactive. A remarkable exception to this rule is Pd of the group X b.

Some economic considerations (the price of some metals), influenced the choice of the metals for Hydrogen storage in the form of hydrides to be made, requiring more than the above rule. Thus, some experiments are presently being performed on two types of alloys, Fe-Ti and

Mg - Ni. These are less expensive alloys, and their Hydrogen absorption capability compared well with the base material mass (the alloy used). Very often a mixture of both alloys is used. Since Hydrogen inclusion in a metal (or an alloy) is exothermic, and Hydrogen liberation is an endothermic reaction, these two processes can be used in air conditioning for homes, industrial spaces, and some vehicles.

Lately, there is more and more research to store Hydrogen using some absorptions of type LiBH4-MgH2 together with Ru under the non particle form as catalyst.

4. Hydrogen as fuel

The possibilities for using Hydrogen as a fuel depend on its characteristics, analyzed in Chapter 2, and their advantages or disadvantages compared with other fuels used in transportation or energy. The most important advantage of Hydrogen is, as we have shown, that Hydrogen combustion results almost exclusively in water vapors which are not noxious; they do not contribute to atmospheric pollution, nor do they have a negative influence on animals and plants. From this point of view, Hydrogen is an ideal fuel. Hydrogen combustion can be used in all types of engines or thermo power plants which presently use known conventional fuels. To a great extent, Hydrogen can be used as fuel for engines and installations in their existing form, without important modifications, and with minimum adjustments.

A practical use of any Hydrogen storage system, described in the previous chapter, includes a modality with characteristics corresponding to the technical domain in which Hydrogen is used as a fuel. In addition to thermo power engines and installations currently used, Hydrogen is very well suited for use in an energy system category which will be employed more and more in the future--namely fuel cells (Apollo Space Program..., 2009). As we know, fuel cells (FC), generate electrical energy directly as a result of the electrochemical reaction of a hydrocarbon (methane gas or methanol) or of Hydrogen. Depending on the FC type, and the temperature at which the electrochemical reaction is produced, efficiencies of 40% to 70% can be obtained for the electrical energy produced. For the case of electrochemical reaction high values for efficiency can be obtained for high power FC (2-3 MW) in the existence of high temperature values (800-1000°C).

4.1 Motor vehicles

Extensive research is presently being undertaken, especially for ecological reasons, to find alternatives to conventional fuels (gasoline, diesel fuel, methanol), which can be used for ground transportation vehicles, whose numbers are continuously increasing. This research is concentrated on two solutions: replacing thermic engines with pistons with electrical motors fed from accumulator batteries, and total or partial replacement of conventional fuels with Hydrogen. Among these types of engines, those with sparking ignitions are best suited to function with Hydrogen. Transforming actual engines for this purpose requires minimal adaptation.

Engines with compression ignitions are less suited to function with Hydrogen. It has been proven that the best storage for Hydrogen, in use with trucks and cars, is in its liquid form, in cryogenic reservoirs. .Injecting liquid Hydrogen into the cylinders is very difficult, since its boiling temperature is very low. Engines with sparking ignitions, operating with Hydrogen have an effective efficiency of about 35%, greater than engines using conventional fuels. Maximum efficiency using Hydrogen can be obtained for high values of excess air ($\lambda > 2$), compared with $\lambda \approx 1{\cdot}1$ in conventional engines. Engines operating with Hydrogen are more economical than gasoline using an inferior load, but by using Hydrogen there is a reduction of the engine's maximum power. Therefore, when the load increases, gasoline participation also increases; thus, for a load more than 85% of maximum, the engine is fed exclusively with gasoline. Using the mixture Hydrogen - gasoline, gasoline savings of 70-75% were obtained. For this reason, the start and function of duel fuel engines, using Hydrogen, is presently being investigated.

These characteristics of the dual fuel engine could be expected to elicit large-scale application to urban transportation. Investigations are also being made into the use of fuel cells (FC) fed with Hydrogen in automotive traction systems. These are able to produce electricity continuously, which is very advantageous for traction systems.

4.2 Aviation

The rationale for using Hydrogen in airplanes equipped with turbo reactor motors is based on the fact that for equal volumes of gasoline, Hydrogen is ten times lighter; but the heating power of Hydrogen, related to the weight unit, is about three times less than that of gasoline. Allowing for the weight of Hydrogen, the volume and thus the dimensions of a plane using Hydrogen would be greater, but the weight would be smaller than planes using conventional fuels. For this reason, the take-off distance for Hydrogen-operated airplanes could be reduced by 50%.

Only minor modifications are required to adapt a turbo-reactor motor to function with Hydrogen. The high energy value in Hydrogen vaporization constitutes an advantage for the cooling structure of high-flight velocity airplanes, where the aerodynamic heating phenomenon is very important. Thus, flying the Concorde at 18,000m, and a velocity characterized by Mach number M = 2, the wing temperature rises to 90°C. For, M = 3, this

temperature reaches 250°C. For this reason, experts consider that high-speed airplanes (M > 5), will require the use of H as fuel. Certainly, as in automobile engines, there is the possibility of using dual fuels for turbo reactor motors.
Research in this direction led to some interesting results in the area of environmental protection. Adding Hydrogen to the gasoline mass flow (5-6) %, and injecting gasoline for the remaining 95% in the combustion chamber, resulted in a four-fold reduction in atmospheric CO and CH emissions. An addition of 12% Hydrogen reduced these noxious gas emissions by a factor of more than 20. All of these elements, which make the use of Hydrogen as fuel for the turbo reactor airplanes very attractive, have resulted in considerable research.

4.3 Space technology

Space technology represents the only domain which presently employs Hydrogen on an industrial scale to propel Saturn and Centaur rockets, used by NASA, by utilizing the reaction force of the motors' rocket, resulting from Hydrogen combustion in its contact with oxygen, both of these components being stored in a liquid state (cryogenic reservoirs).
The opportunity to use Hydrogen as rocket fuel first resulted from good specific traction characteristics (reaction force), realized by using this apparatus. Additionally, it emits almost no atmospheric pollution.
The decision to use Hydrogen as fuel was planned as far back as the ballistic V_2 rocket, used by Germany in the Second World War. In this case, by using methyl alcohol as fuel for the V_2 Rocket, the flue gas evacuation velocity from the jet nozzle was of 2100 m/s, and the operating radius was (theoretically) about 550 km, and practically about 300 km. In order to increase the operation radius by the 1500 km, necessary to reach the United States, it would be absolutely necessary to use Hydrogen as fuel, since, as a result of hydrogen-oxygen reaction, it would obtain a propulsion gas velocity of about 3100 m/s (with O_2 liquid carburant). In space technology it is of great interest to use Hydrogen not only as a fuel but also as a rocket propulsion fluid.
This approach is justified by the fact that the reaction gas exit velocity is inversely proportional to the molecular weight of the corresponding gas; and among all gases, Hydrogen has the smallest molecular weight, as shown in Chapter 1. However, the reaction force in the gas jet propulsion motor is directly proportional to the respective gas evacuation velocity from the jet nozzle.
However, as has been shown, in addition to its dependence on the gas molecular weight, this velocity also depends on the absolute temperature value of the gas at the entrance of the set nozzle. In fact, it is directly proportional to this temperature. Thus, for a value of the corresponding temperature at T_0 = 1773 K (t_0 = 1500°C), the maximum velocity of Hydrogen at the exit from a considered reaction jet, having an exit infinite size section, is 6700 m/s. In order to heat Hydrogen at a high temperature, a heat source contained in a much reduced mass at the rocket board is needed.
Returning to the velocity problem at the exit from the TNR jet nozzle: Using Hydrogen as the reactive fluid, we note that if atomic hydrogen were used, the velocity would be double that of molecular hydrogen. Thus, practically, the limited value of the exit section area of the jet nozzle, would reach values of this velocity for about 10,000 m/s using atomic Hydrogen, compared with about 5000 m/s using molecular Hydrogen. However, it is known that the dissociation of molecular Hydrogen in atomic Hydrogen is very difficult to achieve at very

high temperatures. Thus, the dissociation degree, expressed by the Hydrogen molecule in Hydrogen atoms, compared with the total Hydrogen molecular number, is 1%, for 1900 K, and it increases with the temperature up to 82%, for 4000 K. On the other hand, atomic Hydrogen is very unstable, having a very strong tendency to return to its molecular Hydrogen form.

4.4 Energy

A very large field of Hydrogen utilization in energy is represented by gas turbine installations (GTI); this can be achieved either by Hydrogen with air, or oxygen combustion. Obviously, the principal advantage of using Hydrogen in GTI is of an ecological nature. In Hydrogen combustion with oxygen, water appears as a combustion product and it is not a pollutant. In Hydrogen combustion with air, nitrogen oxides could appear as pollutant products. For this reason, measures must be taken to reduce the content of nitrogen oxides in the flue gases; these measures could include temperature reduction in the combustion chamber (CC), which is already very high. Therefore, during Hydrogen combustion with air (under stichiometric conditions) the maximum combustion temperature is 2503 K (see Table 1). The temperature reduction in CC, up to the desired value, can be achieved by introducing supplementary air, or by water injection. In the case of Hydrogen combustion with oxygen, it is appropriate to consider the Total Water Injection System - TWI (Cârdu&Baica, 2002). In this system the working fluid in the turbine could be a unique gas steam, which after its expansion could be condensed, with complete water recuperation, as in a steam turbine. Of course, such power plants can be technically and economically justified only when Hydrogen is available as a secondary product, resulting from a technological process of the chemical industry, for example. Thus, it is completely illogical to consider producing Hydrogen for energy purposes using a conventional energy primary source (caloric, hydraulic, solar etc.), when this respective primary source can be directly used to produce electrical energy with a greater total efficiency. However, producers of electrical energy are advised to operate with a constant load, as with nuclear power plants (NPPs). It is possible that Hydrogen produced by electrolysis could be profitable when energy demand in the respective power system is reduced. Hydrogen produced in this manner constitutes a not pollutant fuel reservoir, which can be used in a GTI, when electrical demand is greater in that energy system.

Hydrogen can also be an energy accumulator, very important for some electrical energy producers (electrolyze), which, by their nature have widely fluctuating energy demands, as in the case of electricity from the solar or wind energy.

In addition to the ecological advantages of using Hydrogen for GTI in energetics, there are technical elements which justify the assertion for thermo dynamic advantage, however this subject is not the goal of this paper.

In the USA, the Los Alamos National Laboratory is studying an interesting integrated system which produces (not pollutant) Hydrogen from coal in generating thermal and electrical energy (Lackner & Ziock, 2000). The electrical energy is produced using Hydrogen in a high temperature solid oxide fuel cell (HTSOFC). Hydrogen is produced from coal and steam using a conventional gasification process. As a result of this process, ashes are obtained simultaneously with Hydrogen, CO_2. From the mixture of Hydrogen and CO_2, CO_2 is held back by a chemical process, where it combines with CaO to result in a solid form; $CaCO_3$. In its turn, $CaCO_3$ is calcined with the help of the heat which results from the process which takes place in the HTSOFC. As a result of $CaCO_3$ calcination, CO (which is recyclable), and CO_2 are produced. CO_2, in its turn, can be naturally recycled by its combination with magnesium silicate (serpentine), from which results magnesium carbonate, silica (quartz), and water. This is an exothermic reaction and the resultant heat can be also used. This gasification process can be carried out without the contribution of air, and consequently NO_x does not appear. In using coal with sulphur (S) content, a necessary quantity of CaO or $CaCO_3$ is used to retain this sulphur. In this way we eliminate the possibility of sumptuous oxides emission.

5. Conclusions

The previous chapters have demonstrated the following:

5.1. Hydrogen is a gas with physical and chemical properties recommended for use as fuel for thermodynamic engines and plants; it has many advantages over conventional fuels such as coal and hydrocarbons). A very important ecological advantage is that Hydrogen combustion does not produce CO_2 or SO_2, noxious gases or greenhouse gas (CO_2). In its combustion with air, it can produce nitrogen oxides (NO_x) which are noxious gases and greenhouse gases, but the actual technological level ensures minimal NO_x content in the flue gases. Hydrogen combustion produces water vapors almost exclusively, and they are not noxious.

Hydrogen also has disadvantages compared with other fuels, and some of the most important are those concerning technical difficulties regarding its storage.

5.2. Hydrogen is not found in a free state in nature. Thus we cannot speak of Hydrogen as a primary source, as are conventional fuels. To obtain Hydrogen from the substances in which it occurs, another form of energy must first be consumed. For this reason, Hydrogen is in fact an energy vector from its production source (most desirable without producing atmospheric pollution), to the point at which it is used as a fuel.

The progression of Hydrogen from product to tool, is exemplified by its production via electrolysis (using electricity produced by a nuclear power plant), and ending with Hydrogen used to perform rocket propulsion. By using the chemical energy of some fuels to produce Hydrogen (particularly thermal energy), this approach becomes rational if Hydrogen results as a secondary product, based on some technological processes of the chemical industry. Thus, from an ecological point of view, it is justifiable to produce Hydrogen by electrolysis (using electrical energy), by radiolysis (using nuclear particles of radiations energy), or by photosynthesis (using the solar radiation energy), and also by biological way (fermentation).

Moreover, Hydrogen storage can be achieved either in liquid form at a high pressure and extremely low temperature (about 20K), or in the form of metallic hybrids. Storage is one of the most difficult problems in the technology of using Hydrogen for energy.

5.3. Presently, there are still major technical and economical difficulties in using Hydrogen as fuel on large scale, but the increasing demand for protecting the environment will require ever greater use of Hydrogen as fuel. Hydrogen can be used as fuel in all areas presently using conventional fuels (coal and hydrocarbons): transportation, space technology (rockets) and for energy (thermo power plants). The existing experimental installations in all of these fields assure that, should there be an immediate, large-scale demand for Hydrogen as fuel, there will be a technical solution available. Moreover space technology has used Hydrogen for a long time as fuel for Saturn and Centaur rockets in the USA space program. Great importance is also attached to using Hydrogen as a fuel in automotive transportation in the USA.

The characteristics	MU	The fuel					
		Hydrogen	Gasoline	Diesel oil	Kerosene	Natural gas	Methanol
Chemical composition C/H/O	-	- 1.0 -	0.85 0.15 -	0.87 0.13 -	0.853 0.147 -	0.75 0.25 -	0.375 0.125 0.500
Density (gas/liquid)	$\frac{kg}{m^3}$	0.0899 70.8	4.88 700-755	- 820-930	- 800-850	0.717 415	- 791
Diffusion coefficient in air	m/s	$0.66 \cdot 10^{-6}$	-	-	-	-	$0.186 \cdot 10^{-4}$
Feature temperatures							
- boiling start	K	20.2	313	-	423	111·4	337·7
- boiling finish	K	20.2	453	-	523	-	-
- of ignition	K	823-873	773-793	593-653	708	923-973	741
- of freezing	K	11	213	213	213	90.4	175.2
- maximum of combustion in air	K	2503	-	-	2340	2316	2175
Stoichiometric air quantity	kg/kg	34.2	14.9	14.3-14.7	14.8	17.25	6.47
Inflammability limits							
- inferior	% vol.	4	0.59	-	-	5.5	-
- superior	% vol.	75	6	-	-	15	-
Vaporization energy (at 20°C)	kJ/kg	451-476 (at 20.2K)	292-314	18	208	5111	1105
Ignition energy	kJ/kg*)	20	250	-	250	-	-
Inferior heat power (at 15°C; 760 mm Hg)	MJ/kg	121	43.744	42.5	42.85	49.79	19.3
Combustion speed	m/s	2.7	0.4	0.4	0.4	0.36	-

*) 1 kg of fuel in stoichiometric mixture with air

Table 1. The main technical characteristics of the hydrogen in comparision with other fluid fuels (hydrocarbons)

6. Acknowledgement

Partial funding for this work was provided by the University of Wisconsin, Whitewater, WI, USA.

7. References

Cârdu, M., Baica, M., (1999), Regarding a new variant methodology to estimate globally the ecologic impact of thermo power plants, *Energy Conversion and Management (ECM)*, 40, pp. 1569-1575.

Cârdu, M., Baica, M., (2002), Gas turbine installation with total water injection in the combustion chamber, *Energy Conversion and Management (ECM)*, 43, pp. 2395-2404.

Jianfeng, M., et al. (2011), Enhanced hydrogen sorption properties in the LiBH4 - MgH2 system catalysed by Ru nanoparticles supported on multiwalled carbon nanotubes, *Journal of Alloys and Compounds*, Vol. 509, Issue 15, 14 April, pp. 5012-5016.

Lackner, K.S., Ziock, H.J., (2000), From low to no-emissions, *Modern Power System (MPS)*, March.

Peculea, M., (1995), The Romanian Technology for Heavy Water Separation, *Energetica Journal*, 43, No. 1, Serie A (in Romanian).

Peltier, R., (2007), Port Arthur II Integrated Hydrogen/Cogeneration Facility, Port Arthur, Texas, *Power*, Vol. 151, No. 9, September, pp. 2-4.

***Apollo Space Program Hydrogen Fuel Cell, (2009), *Spaceaholic. com.*, Retrieved 2009-09-21.